AF772320

Palgrave Studies in World Environmental History

Series Editors
Vinita Damodaran, Center for World Environmental History, University of Sussex, Brighton, UK
Rohan D'Souza, Graduate School of Asian and African Area Studies, Kyoto University, Kyoto, Japan
Sujit Sivasundaram, University of Cambridge, Cambridge, UK
James Beattie, The Center for Science in Society, Victoria University of Wellington, Wellington, New Zealand

The global climate emergency has stimulated great interest in environmental studies and has encouraged a range of scholars, including historians, to place the environment at the heart of their analytical and conceptual explorations. Environmental history has been of considerable assistance in efforts to comprehend the traumatic environmental difficulties facing us today. An understanding of the history of human interactions with all parts of the world's surface and with living organisms and other physical phenomena, is an essential aspect both of historical scholarship and its adjacent fields, such as the history of science, anthropology, geography and sociology. This is especially so in the light of Covid-19. Environmental History helps us to reconsider the bounds of possibility open to humans over time and space, in their interaction with different environments. This series explores these interactions in research that touch on all parts of the globe and all manner of environments including the built environment through studies that are attentive to the local, regional and planetary scale. Books in this series come from a wide range of fields of scholarship, from the sciences, social sciences and the humanities. The series particularly encourages interdisciplinary projects that emphasize historical engagement with science and other fields of study.

Daniel Rothenburg

Irrigation, Salinity, and Rural Communities in Australia's Murray-Darling Basin, 1945–2020

palgrave
macmillan

Daniel Rothenburg
Scientific Coordinator and Research Associate at the Collaborative Research Center 923 "Threatened Order"
University of Tübingen
Tübingen, Germany

ISSN 2730-9746 ISSN 2730-9754 (electronic)
Palgrave Studies in World Environmental History
ISBN 978-3-031-18450-5 ISBN 978-3-031-18451-2 (eBook)
https://doi.org/10.1007/978-3-031-18451-2

© The Editor(s) (if applicable) and The Author(s), under exclusive license to Springer Nature Switzerland AG 2023
This work is subject to copyright. All rights are solely and exclusively licensed by the Publisher, whether the whole or part of the material is concerned, specifically the rights of translation, reprinting, reuse of illustrations, recitation, broadcasting, reproduction on microfilms or in any other physical way, and transmission or information storage and retrieval, electronic adaptation, computer software, or by similar or dissimilar methodology now known or hereafter developed.
The use of general descriptive names, registered names, trademarks, service marks, etc. in this publication does not imply, even in the absence of a specific statement, that such names are exempt from the relevant protective laws and regulations and therefore free for general use.
The publisher, the authors, and the editors are safe to assume that the advice and information in this book are believed to be true and accurate at the date of publication. Neither the publisher nor the authors or the editors give a warranty, expressed or implied, with respect to the material contained herein or for any errors or omissions that may have been made. The publisher remains neutral with regard to jurisdictional claims in published maps and institutional affiliations.

Cover illustration: Robin Smith/Getty Images

This Palgrave Macmillan imprint is published by the registered company Springer Nature Switzerland AG
The registered company address is: Gewerbestrasse 11, 6330 Cham, Switzerland

Preface

Books are collaborative efforts, which is masked by the single author tradition in the Humanities, even if only one person receives credit and takes responsibility for the final product. Personally, I have no doubt that this book would not have been possible without the moral and intellectual support, ideas, patience, time, and goodwill of many people. It is them who deserve my thanks for having been able to complete a project which, to me, has been a personal and intellectual adventure that has changed the way I look at the world and how I understand my own place in it.

This book is based on my Ph.D. thesis which I submitted in May 2020. Therefore, first and foremost, my thanks and respect go to my supervisors. Ewald Frie entrusted me with this project on a subject that I knew nothing about at the outset. His constructive comments, practical help, and liberal confidence in a young historian and philosopher who often moves in a realm of abstract ideas have been an invaluable assurance throughout the project. Klaus Gestwa's encouraging supervision style and his comments have helped me a great deal to realize the potential of my topic and connect my historiographic interests with the environmental point of view.

My sincerest thanks go to the editors of Palgrave Macmillan's *Studies in World Environmental History* series who were kind enough to include my book. Furthermore, my gratitude goes out to the community of environmental historians in Australia who welcomed me and contributed their knowledge and enthusiasm for my ideas. Tom Griffiths and Libby Robin introduced me to Australian environmental history with their Ph.D. workshop at the Australian National University, in which I participated in 2016. Their generosity was overwhelming. Andrea Gaynor, Rebecca Jones, Heather Goodall, Daniel Connell, Cameron Muir, Don Garden, Graeme Davison, and Peter Russ were kind enough to listen to my ideas and contribute their expertise. Emily O'Gorman also contributed the external review of my thesis. David Harris became a friend for whose moral support and confidence I am most grateful. He also took me on adventures in the bush with his "beast."

Many people and institutions in Kerang, Shepparton, Swan Hill, Pyramid Hill, and Tatura have also contributed to my book with their time, knowledge, and resources. George and Maureen Hardwick kindly invited me to their home in Appin and gave me access to their documents. Elaine Jones, Gwen and Bill Twigg, Bill Maher, Arthur Young, Patricia McMahon, Stuart Simms, Norman Mitchelmore, and John Dainton kindly agreed to interviews. John Girdwood not only agreed to meet me but took along John Poole, who brought his archive. Matt Hawken granted me access to the records of Agriculture Victoria at Kerang and Glen (whom I forgot to ask for his last name!) showed me the Lakes system. The staff at the Swan Hill Regional Library, the Sir John Gorton Library in Kerang, the Shepparton Library, and the Historical Societies in Kerang, Pyramid Hill, Shepparton, and Tatura gave me liberal access to their material.

Back home, I valued my reference groups whose wit and sympathy kept me sane during long months of writing. Several people contributed directly to my thesis. Fernando Esposito, through his sponsorship and friendship, was instrumental in me becoming a historian. He taught me how to connect my theoretical ambitions with empirical work. Cosima Götz, who accompanied my writing process with empathy and support, also put her analytic mind to my emerging manuscript. Timm Schönfelder, with whom I supposedly shared an office, made helpful remarks along the way and was a splendid teaching partner. Irina Rückert patiently organized a constant flow of new books and supported my work with research and her excellent English language skills. Fellow bibliophile Lea Schneider proofread the original manuscript, hunting my grammar and punctuation errors. David Schäfer, my irreplaceable friend who has accompanied me since the early days of my studies, kindly agreed to proofread my thesis. As for Christina Brauner and her "big questions"—how glad I am that you became part of this story! More big things to come.

My colleagues at the Collaborative Research Center 923 "Threatened Order" and its fabulous staff, Cornelia Stoll, Heike Bäder, Andrea Kirstein, and Roman Krawielicki, have made my time in the "Ph.D. incubator" both productive and humane.

Finally, my parents, Jens and Anita Rothenburg, were the real stars of the show. Both in their own way give their unwavering love and support to their well-meaning, but difficult, son. My book is dedicated to them.

Tübingen, Germany
November 2022

Daniel Rothenburg

Acknowledgments

Research for this project was funded by the Deutsche Forschungsgemeinschaft (DFG, German Research Foundation)—Project-ID 170320015 – SFB 923.

Proofreading was partly funded by the Laura Bassi Foundation.

Portions of this book have previously been published as "Too Much Water: How Salinisation Transformed Australia's 'Food Bowl,' 1945–2017." *International Review of Environmental History* 7 (2/2021): 145–158. Used with permission.

Contents

About the Author

Daniel Rothenburg is Scientific Coordinator and Research Associate at the Collaborative Research Center 923 "Threatened Order," University of Tübingen, Germany. He studied History and Philosophy from 2009 to 2015 and received his Ph.D. in Modern History in 2020. His thesis was awarded the Walter Witzenmann Award of the Heidelberg Academy of Sciences and the Doctoral Award of the Faculty of Humanities at the University of Tübingen. From 2020 to 2022 he was Assistant Professor at the Department of Modern History, University of Tübingen. He is the author of various academic publications in the field of Australian history, environmental history, and the history of infrastructures.

Abbreviations

ALP	Australian Labor Party
BP	(Years) Before Present
CoAG	Council of Australian Governments
CSIR(O)	Council for Scientific and Industrial Research; later Commonwealth Scientific and Industrial Research Organisation
GIRDAC	Goulburn Irrigation Region Drainage Action Committee
GMID	Goulburn–Murray Irrigation District
ICM	Integrated Catchment Management
ISAC	Inland Salinity Action Committee
KARF	Kerang Agricultural Research Farm
KESC	Kerang Environmental Study Centre
KIRSAC	Kerang Irrigation Region Salinity Action Committee
KLAWG	Kerang Lakes Area Working Group
MDB	Murray–Darling Basin
MDBA	Murray–Darling Basin Authority
MDBC	Murray–Darling Basin Commission
MDBMC	Murray–Darling Basin Ministerial Council
MRB	Mineral Reserve Basins Scheme
MTF	Ministerial Task Force on Salinity
MV(D)L	Murray Valley Development League; later Murray Valley League
PPWC	Parliamentary Public Works Committee (Victoria)
RMC	River Murray Commission
RWC	Rural Water Commission (Victoria)
SHIRF	Swan Hill Irrigators Research Farm
SPPAC	Salinity Pilot Program Advisory Council
SRWSC	State Rivers and Water Supply Commission (Victoria)

TWE	Transferable Water Entitlements
UDDFPL	Ultima and District Dryland Farmers Protection League
UNEP	United Nations Environment Programme
VFGA	Victorian Field and Game Association
VIRPO	Victorian Irrigation Research and Promotion Organisation
WCS	World Conservation Strategy

List of Figures

CHAPTER 1

Introduction

An Outsider's View

In the winter of 2017, I traveled around rural Victoria for seven weeks. In the small university town of Tübingen in the south-west of Germany, where I came from, it would have been summer. However, contrary to what tourists usually do, I escaped the sunny season to enjoy the rather mild winter in south-eastern Australia. I was an outsider here—someone from the city. That I also happened to be from Central Europe was secondary. I picked up a tiny car on La Trobe Street in downtown Melbourne and drove north via Ballarat and Bendigo all the way to Albury, feeling very vulnerable between all the trucks, pickups, and utility vehicles. It was like I was the first European to come there just to see the Hume Dam. Then my path led me downstream along the Murray, to Yarrawonga and Echuca, and finally to Kerang and Swan Hill—places that up until then had hardly been more than names on a map for me and that came up in the literature I had read in Germany. Usually, they were not named in very favorable contexts; they might well have been post-apocalyptic wastelands created and then ruined by water—and salt. They turned out not to be. In fact, Kerang now holds a special place in my heart with its welcoming and interested people, the stunning Loddon River environment just outside of town, and its friendly little restaurants.

© The Author(s), under exclusive license to Springer Nature Switzerland AG 2023

D. Rothenburg, *Irrigation, Salinity, and Rural Communities in Australia's Murray-Darling Basin, 1945–2020*, Palgrave Studies in World Environmental History,
https://doi.org/10.1007/978-3-031-18451-2_1

It is the outsider's privilege to be amazed at things that locals find commonplace. At home, I would not dream of treating a local newspaper with more attention than skimming through it, but here, I eagerly digested bits of local news, letters to the editor, and notes about cultural events in the area. Yet what caught my eye in July of that year brought me straight back from lunch to my purpose in the area. The feature "Pumped: Who's benefitting from the billions spent on the Murray-Darling?" by ABC TV's program *Four Corners* was causing a stir. In its recent episode, the program alleged that water bought by taxpayers for the environment had been stolen on a large scale by irrigators in the Barwon–Darling area in New South Wales to grow cotton. In addition, *Four Corners* claimed that the New South Wales public service had turned a blind eye to this practice, even giving irrigators inside information they could use to undermine the $13 billion plan to save the MDB, which was introduced in 2012. "Instead of saving the river, it looks like the Murray-Darling Basin Plan has made a fortune for a lucky few," stated Paul Sinclair, Director of the Australian Conservation Foundation. He suggested that the Plan, which was supposed to benefit the environment and the people in the Basin, was being undermined by parochial interests and their political backers. Environment Victoria, the state's peak non-government environment organization, claimed that the real issue was not water theft. Rather, the question was "whether the rules themselves are rigged to favour particular vested interests." The group demanded an inquiry into irrigator influence on decision-making by the New South Wales government and the Murray–Darling Basin Authority (MDBA).[1]

The Federal Minister for Water, Barnaby Joyce, however, saw the feature as an attack on irrigation itself. In a statement clearly aimed at the people of rural towns like Kerang, which receive their irrigation water from the Murray–Darling system, he accused *Four Corners* of taking part in a campaign to take water away from irrigators and "create a calamity

[1] Anonymous, "Pumped: Who's Benefiting from the Billions Spent on the Murray-Darling?," ABC TV, accessed June 10, 2022, http://www.abc.net.au/4corners/pumped/8727826; Gabrielle Chan, "Murray-Darling Basin: Allegations of Water Theft Spark Calls for Judicial Inquiry," *The Guardian*, July 25, 2017; Anonymous, "Explosive Four Corners Investigation into the Barwon-Darling River Systems Must Trigger ICAC Investigation," Australian Conservation Foundation, last modified July 25, 2017, https://www.acf.org.au/barwon_darling_icac_investigation; Juliette Le Feuvre, "Stop Sucking Rivers Dry," Environment Victoria, last modified August 8, 2017, https://environmentvictoria.org.au/2017/08/08/stop-sucking-rivers-dry.

[...] for which the solution is trying to take more water off you, shut more of your towns down." The minister told an audience that the government had purposefully placed the water portfolio in the agriculture ministry "to look after rural Australians and to wrest it back off 'the greenies'." Similarly, the Federal Member for Mallee, Andrew Broad, responded by saying "[w]e're not going to shut down irrigated agriculture just because Four Corners would like us to."[2]

It seemed to me that more was being discussed than the actions of individuals who had compromised the supply of water to the environment and rural towns for their profit. Underneath it all was the question of whether irrigation itself had a legitimate place in the MDB. Environmental advocates used the opportunity to call the irrigation industry as a whole into question, while those dependent on votes from rural people, whose livelihoods are often intimately linked with irrigation, felt they had to defend the industry and blame individuals.

I realized that Kerang was a part of a bitter controversy over irrigated agriculture, the environment, and rural Australia in general. The community of about 3900 people, around 280 km north-west of Melbourne, lies in the Goulburn–Murray Irrigation District (GMID), regarded as Australia's Food Bowl and the country's largest irrigation district. In 2018, this area boasted 1200 dairy farms providing 21% of Australia's milk and 400 km^2 of irrigated crops such as wheat, barley, triticale, corn, and oilseeds. Both industries combined provide $5.9 million to the economy, and they are both highly dependent on a reliable supply of irrigation water, as the annual rainfall is usually insufficient. In fact, the high-value agriculture in the GMID has been developed chiefly through irrigation and was deemed vulnerable to reductions in available water by the MDBA.[3]

[2] Ken Jenkins, "Basin Water Probe," *Gannawarra Times*, August 1, 2017; Angus Verley, "MP Warns Against 'Knee-Jerk Reaction'," *Gannawarra Times*, July 28, 2017; Geoff Adams, "Water Theft Allegations," *Gannawarra Times*, August 1, 2017.

[3] Murray-Darling Basin Authority, *Guide to the Proposed Basin Plan. Appendix C: Goulburn-Murray Community Profile* (Canberra: Murray-Darling Basin Authority, 2010), 779; 806; Anonymous, "2016 Census Quick Stats: Kerang," Australian Bureau of Statistics, 23 October 2017, accessed June 10, 2022, http://www.censusdata.abs.gov.au/census_services/getproduct/census/2016/quickstat/SSC21320; Anonymous, "District Profile. Goulburn-Murray Irrigation District," *Irrigation Leader Magazine*, accessed June 10, 2022, https://irrigationleadermagazine.com/goulburn-murray-irrigation-district.

The GMID is situated in the south of Australia's agricultural heartland, the Murray–Darling Basin (MDB), which is home to more than 2 million people, including more than 46 Aboriginal nations. A further 1.3 million people outside of the Basin depend on its water resources. The Basin's economy produces more than a third of Australia's food. In 2015, its gross value of agricultural production was around $24 billion, which included $9 billion from irrigation. 46% of the gross value of Australia's irrigated products came from the MDB, grown on 1.4 million hectares. Irrigated production accounted for nearly 100% of Australian rice, 93% of its cotton, 76% of its grapes, 29% of its dairy, and nearly 100% of oranges (Figs. 1.1 and 1.2).[4]

By contrast, the relative importance of agriculture in the Australian economy has steadily declined. In the first half of the twentieth century, agriculture accounted for about a quarter of the country's economic output and up to 80% of its exports. In 2012, its share of the Gross Domestic Product had decreased to about a mere 2.4%. The number of agricultural properties has declined by over one-third in the last 30 years. Still, Australian farms produced 93% of the food consumed in Australia, even though 60% of farm produce was exported to other countries.[5]

However, this all comes at a price. In 2017, agriculture claimed 95% of all water diversions in the MDB. At the same time, Australia faces both increasing water use and growing demand by expanding urban centers, such as Melbourne, driven by both population and economic growth. Agriculture, which consumes about 50–65% of all water in Australia, is thus challenged by water demands from others, including the environment, which the Commonwealth represents on the water market. This is paralleled by more competition for land due to expanding urban areas, which, in turn, puts pressure on the nation's food production, which an increasing population demands.[6]

[4] Murray-Darling Basin Authority, *Basin Plan Annual Report 2015–16* (Canberra: Murray-Darling Basin Authority, 2016), 3–4; Murray-Darling Basin Authority, *The 2020 Basin Plan Evaluation* (Canberra: Murray-Darling Basin Authority, 2020), 74.

[5] "Australian Social Trends 2012: Australian Farming and Farmers," ABS, December 11, 2012, last modified December 12, 2012, http://www.abs.gov.au/AUSSTATS/abs@.nsf/Lookup/4102.0Main+Features10Dec+2012.

[6] Australian State of the Environment Committee, *State of the Environment 2016. Inland Water* (Canberra: Australian Government Department of the Environment and Energy, 2016), 14–15; "Water Account, Australia, 2012–13," ABS, last modified November 27, 2014, http://www.abs.gov.au/AUSSTATS/abs@.nsf/Lookup/4610.

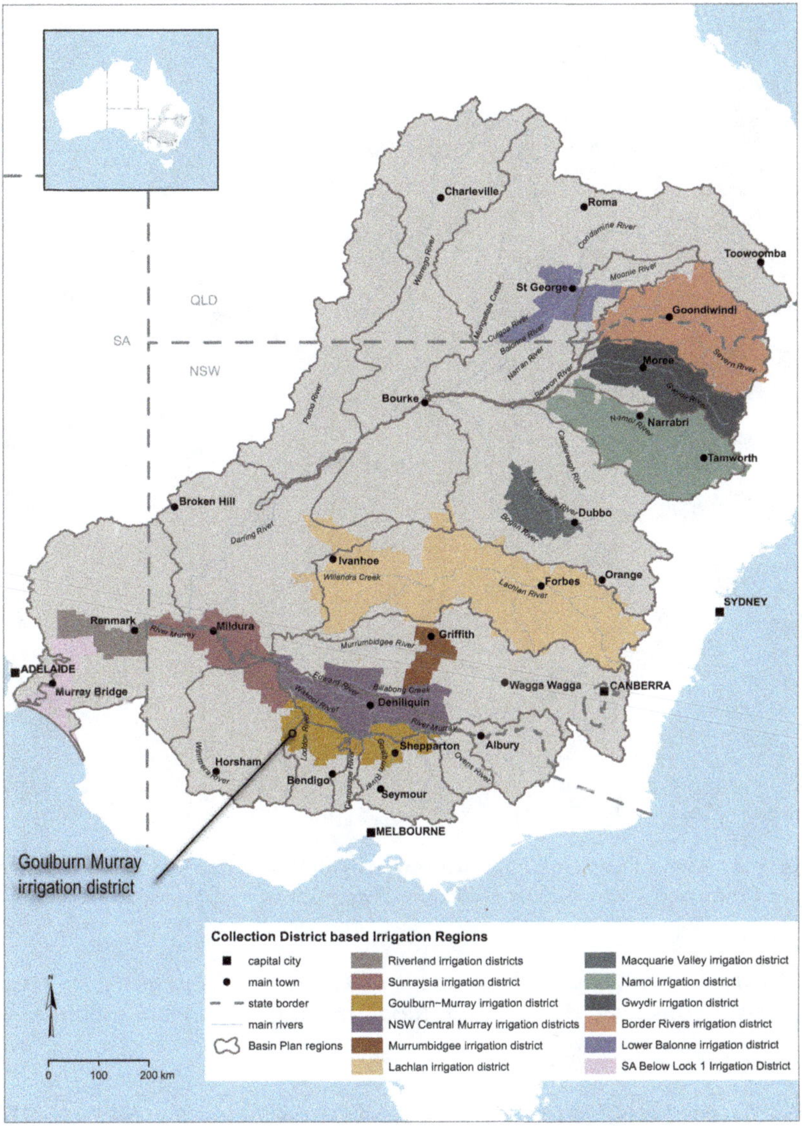

Fig. 1.1 Map of the Murray–Darling Basin with the Goulburn–Murray Irrigation District (*Source* Licensed from the Murray–Darling Basin Authority under a Creative Commons Attribution 4.0 International License)

Fig. 1.2 Map of the Goulburn–Murray Irrigation District in 1967 (*Source* © State Rivers and Water Supply Collection, State Library of Victoria, Melbourne)

Next to increasing competition for water, over a century of effort to satisfy consumer demand for food through intensive agriculture in

0Main+Features12012-13?OpenDocument; "Water Account, Australia, 2015–16," ABS, last modified November 23, 2017, http://www.abs.gov.au/ausstats/abs@.nsf/mf/4610.0; Murray-Darling Basin Authority, *Basin Plan Evaluation 2017* (Canberra: Murray-Darling Basin Authority, 2017), 19; Australian State of the Environment Commitee, *State of the Environment 2016. Overview* (Canberra: Australian Government Department of the Environment and Energy, 2016), 3–4.

inland Australia, in which irrigation has been a key component, have also led to disastrous effects on the environment. Additionally, climate change is expected to exacerbate the existing issues of land and water through more pronounced seasonality, increasing intensity and frequency of extreme weather events such as droughts and floods, less productive rainfall patterns, and productivity changes in some regions. This is particularly evident in the MDB, as a South Australian Royal Commission inquiry found, which was prompted by *Four Corners*' 2017 allegations. In his report, released in January 2019, Commissioner Bret Walker unequivocally declared that "the Basin is in danger of being run down. [...] The threat of its degradation being irreversible is greater than ever. Climate change, alone, has that clear potential. Our exploitation of its water resources magnifies that threat." The Basin Plan ignored the potentially catastrophic risks of climate change, which would likely result in considerably less water run-off into the twenty-first century. A complete overhaul of the scheme was required, including an allocation of more water to the environment and a demand that all states review their water resource plans to expressively recognize and authorize the taking and using of water by Indigenous people in exercise of their native title rights. Highlighting these worries, in January 2019, several major incidents of mass fish deaths at an unprecedented scale occurred in New South Wales on the Lower Darling where hundreds of thousands of animals died. The main cause was deemed to be over extraction of water by the cotton industry from the Barwon–Darling system, which was on the path to collapse, a review by the New South Wales Natural Resources Commission warned.[7] Amidst these "water wars," the future of the MDB and its communities is in doubt.[8]

[7] Australian State of the Environment Commitee, *State of the Environment 2016. Land* (Canberra: Australian Government Department of the Environment and Energy, 2016), 14; Bret Walker, *Murray-Darling Basin Royal Commission Report* (Adelaide: Gouvernment of South Australia, 2019), 35; Anne Davies, "Murray-Darling Basin Royal Commission Report Finds Gross Maladministration," *The Guardian*, January 31, 2019; Anne Davies and Lisa Martin, "Menindee Fish Kill: Another Mass Death on Darling River 'Worse than Last Time'," *The Guardian*, January 27, 2019; Anne Davies, "Barwon-Darling River Ecosystem on Path to Collapse, Review Warns," *The Guardian*, July 24, 2019.

[8] Anne Davies, "Water wars: Will politics destroy the Murray–Darling Basin plan—and the river System itself?" *The Guardian (Australia)*, December 13, 2019.

Against this backdrop, it should be of little surprise that many Australians question irrigation as an enterprise. In 2009, Neil Barr observed that "irrigation has been engaged in a public relations battle over its legitimacy, its own 'licence to operate'," and this still seems valid today.[9] Yet historically, it has been one of the nation's great pursuits. According to Michael Cathcart, nation-building and settlement in inland Australia through irrigation development were the fulfillment of one of white Australians' biggest national ambitions—their Water Dreaming. Massive infrastructural projects, such as the Hume Dam in the 1930s and the Snowy Mountains Scheme during Post-War Reconstruction, were objects of delight and utopian impetus; they were expressions of irrigation's promise to deliver water to a dry land and make it a garden. This was also a promise to realize the settler dream of a closely populated Australia and a modern venture to subjugate the allegedly useless Australian nature and utilize it.[10]

However, since the 1960s, irrigation has been severely criticized as inefficient, uneconomic, and wasteful, which questions its privileged place in the national economy. Most prominently, it has been identified as one of the major causes of an environmental hazard—the salinization of soil and water. Salinization occurs when water, applied to land in large quantities, releases the salt already in the soil or brings new salt onto the land. Salt, in significant concentrations, poisons plants and therefore reduces agricultural production, which has been a disaster for many farming dependent regions in the MDB. At its worst, salinization can render farmland useless for production, leaving a barren land covered with a salt crust. Salinization also entails a loss of habitat for animals in streams, wetlands, and water bodies. Although the problem is partly caused by land-clearing and the replacement of native perennial vegetation with introduced annual crops, it is mostly linked to excessive water use by irrigators. By the 1990s, salinization seriously affected about 10% of the world's irrigated area and was widespread in such diverse countries as

[9] Neil Barr, *The House on the Hill. The Transformation of Australia's Farming Communities* (Canberra: Land & Water Australia in association with Halstead Press, 2009), 110.

[10] Michael Cathcart, *The Water Dreamers. The Remarkable History of Our Dry Continent*, Melbourne 2009; Daniel Rothenburg, "'The Majesty of Concrete' Hume Dam and Australian Modernity," *Australian Studies Journal* 31, no. 2017 (2017).

Russia, the U.S., India, Pakistan, and Egypt.[11] Today, it is estimated that inadequate irrigation causes declining productivity due to waterlogging and salinization in about one-third of the globally irrigated area.[12]

In 1987, salinization was dubbed "Victoria's and arguably Australia's greatest environmental threat."[13] Not only was the legitimacy of irrigation being increasingly called into question by this growing urgency and awareness, but natural resource management in the MDB and communities, such as those in the GMID, underwent a process of rapid transformation. This was why I had come to this area. I wanted to know how rural communities in the GMID that were, and are, highly dependent on irrigated agriculture have been transformed through the environmental and societal change induced by salinization.

I understand salinization as a slow catastrophe, which creeps into a locality insidiously, almost imperceptibly. It usually stems from normal, everyday practices, rather than exceptional circumstances. Its causes and effects are not always clearly linked across time and space, and its impacts can have long and vastly differing lead times. What Rob Nixon, who prefers to speak of "slow violence," observed for a host of "slowly unfolding environmental catastrophes" like "[c]limate change, the thawing cryosphere, toxic drift, biomagnification, deforestation, the radioactive aftermaths of wars, [and] acidifying oceans" is also true for salinization. It presents "formidable representational obstacles that can hinder our efforts to mobilize and act decisively." A slow catastrophe does not produce dramatic events of chaos that can be easily visualized, but rather, wide-ranging consequences that emerge from *longue durée* interactions between human and ecological systems.[14]

[11] William Meyer, *Human Impact on the Earth* (Cambridge: Cambridge University Press, 1996), 77; John R. McNeill, *Something New Under the Sun. An Environmental History of the Twentieth-Century World* (New York, London: Norton, 2000), 157–73; 177–81.

[12] Paul Ekins, Joyeeta Gupta and Pierre Boileau, eds., *Global Environmental Outlook. GEO6. Healthy Planet*, Healthy People (Cambridge: Cambridge University Press, 2019), 215.

[13] Government of Victoria, *Salt Action: Joint Action. Victoria's Strategy for Managing Land and Water Salinity* (Melbourne: Salt Force, 1987), 5.

[14] Neil Barr and John Cary, *Greening a Brown Land. The Australian Search for Sustainable Land Use* (Melbourne: Macmillan, 1992), 66; Government of Victoria, *Salt Action*, 5; Rob Nixon, *Slow Violence and the Environmentalism of the Poor* (Cambridge, MA: Harvard University Press, 2011), 2; Fiona Williamson and Chris Courtney, "Disasters

By tracing the spread of salinization throughout the GMID in the second half of the twentieth century, the problem serves as a lens through which processes of adaption and transformation can be focused. The environmental consequences of unsustainable irrigation farming practices and their repercussions for the citizens of rural communities come into view. Salinization changed attitudes toward the environment and irrigation, which questioned the ideology of Water Dreaming in the face of the consequences of European farming on the continent. The emergence of local self-help and community-based conservation groups, decades before environmental activists discovered the problems of the MDB, can also be attributed to the problem. New approaches in natural resources management, prompted by an increased understanding of the salinity problem, were developed on grass roots and government expert levels. Salinization furthered the implementation of neoliberal policies in water management that were fused with the goal of achieving sustainable water use. Lastly, resource conflicts between rural communities and urban and rural Australians under the conditions of a highly modified natural environment can be traced by looking at salinization.

This book is about communities and their transformations. I understand them as social groups of humans bound together by common interests and worldviews living within similar environmental conditions and sharing an emotional attachment to them, which is not purely instrumental. Of course, communities can be divided by class, race, gender, and lifestyles, and individuals form associations to pursue their interests. Occasionally, communities are pitted against each other because individuals identified as part of one or another. All of these aspects are considered here. However, what chiefly makes a community cohesive for the purposes of this inquiry is the fact that the individual members—in one way or another—were dependent on irrigated agriculture, have lived and worked in the same anthropogenic environment, and endured the environmental, economic, psychological, and social impacts induced by salinization. This, in turn, also made their respective communities the base unit of collective action.[15]

Fast and Slow: The Temporality of Hazards in Environmental History," *International Review of Environmental History* 4, no. 2 (2018), 5.

[15] Gerard Delanty, *Community*, 2nd ed. (London: Routledge, 2010), x–xiii.

The main object of study is Kerang, which lies in the west of the GMID, while other rural towns in the same region, Swan Hill and Pyramid Hill, are also considered. In the late 1980s, the region was renowned as "the home of salinity."[16] Shepparton, located further upstream on the Murray River in the east of the GMID, serves as a useful contrast due to its vastly different environmental conditions and its more favorable economic position. Both towns were rural centers of irrigated agriculture that developed salinization problems. However, the situation in Shepparton never reached the same urgency as in Kerang. What linked them was the Murray, which still is a lifeline to both communities. This common link was both a source of cooperation and conflict, entangling their histories, an example of how the GMID lends itself to a regional perspective on salinization that is sensitive to particularities, commonalities, and differences within the studied region.

This is a regional history of salinization which is intertwined with the history of the twentieth century. Underneath highly aggregated numbers about continuing land degradation, changing rules of discourse about irrigation and the environment, new denominations for statutory institutions, economic upheavals, and the rise of new policy regimes lie the subjectivities and rationalities of ordinary people. In this study, their perceptions of salinization, their environmental awareness, their political agency and ideologies, the changes in their communities and their lives, and the resource conflicts that arose from living in a degrading environment are revealed.

Chapter 2 explores the history, geography, and geology of the MDB and the GMID. It is a highly modified anthropogenic landscape, largely due to its development for irrigated agriculture. The Basin environment is, furthermore, naturally prone to salinization. However, there are important differences between its regions which greatly influenced the scale and effects of their respective salinity problems.

Chapter 3 sketches historical antecedents to the focused period starting with Aboriginal modifications to the hydrology of the Basin and closing with the Great Depression. The main narrative begins with the years of reconstruction after World War II. This time marks a watershed in Australian environmental history. Before 1945, the patterns of expansion, settlement, and environmental impacts were essentially a technologically

[16] Gynlais O. Jones, "We've Done Something About Salt," *The Northern Times*, March 1, 1988.

advanced extension of those from the nineteenth century. However, the post-war years brought record prices for primary products and minerals, the growth of manufacturing, rapid population increase, a sharp rise in the standard of living, the emergence of a mass consumer society, and large engineering projects designed to harness the Australian waters. Much of this was in line with the global trends that characterized the Great Acceleration after 1945.[17]

Chapter 4 addresses the period from the end of World War II to the floods of the early 1970s, which established a new baseline for salinity in the GMID. A post-war boom in infrastructure construction made a rapid expansion of irrigated agriculture possible. Symbolized by the completion of the Big Eildon dam in 1956, the amount of water available to irrigators in the GMID significantly increased. Eildon and other ambitious river regulation works were key factors of an endeavor to achieve security from the unpredictable moods of a highly dynamic natural environment.[18] However, the quest for security had its own dialectic; the escalation of salinization being one aspect. A series of wet years in the early 1970s caused an intensification of salinity problems in the GMID and an upsurge in community agitation and political attention, as well as an intensified search for a solution.

From here, the clear divisions come to an end. The intervals covered in Chapters 5 through 7 intentionally overlap, as they focus on different phenomena that often occurred simultaneously. From the early 1970s, salinization prompted significant upheaval in Northern Victoria's rural communities and occasionally bitterly fought conflicts between upstream and downstream communities, tied together by their common use of the River Murray. This is the focus of Chapter 5. Salinization, furthermore, led to rural activism—both on the cutting edge of the Age of

[17] Don Garden, *Australia, New Zealand and the Pacific. An Environmental History* (Santa Barbara et al.: ABC-CLIO, 2005), 99–100; Bruce R. Davidson, *European Farming in Australia: An Economic History of Australian Farming* (Amsterdam: Elsevier, 1981), 325; John R. McNeill and Peter Engelke, *The Great Acceleration. An Environmental History of the Antropocene since 1945* (Cambridge, MA: The Belknap Press of Harvard University Press, 2016), 4–5.

[18] Joseph M. Powell, *Watering the Garden State. Water, Land and Community in Victoria 1834–1988* (Sydney: Allen & Unwin, 1989), 235.

Ecology and rooted in an older tradition of conservation.[19] The emergence of this activist 'scene'—their core ideas, leaders, and daily work—is the subject of Chapter 6. Lastly, salinization, as analyzed in Chapter 7, became an important factor for reform in natural resource management in the MDB. As the appreciation of the Basin's ecological crisis increased, escalating salinization catalyzed a fusion of sustainable development and neoliberal market expansion and commodification to achieve better environmental and economic outcomes. The period studied in these three chapters concludes with the Victorian government's large-scale intervention in 1988, which heralded the beginning of a new era in salinity management.

Chapter 8 then examines the various legacies of Water Dreaming and salinization in the GMID and traces the developments since 1988 in the broader context of water and environmental issues in the MDB. In this period, in a reversal of the wet years of the early 1970s, the situation once again dramatically changed with the Millennium Drought in southern Australia, which reduced many salinity problems in the short term.

In the early twenty-first century, salinization is no longer an urgent and high-profile environmental issue in the MDB. Consequently, this book turns to a discussion of the possible future impacts of climate change and reduced availability of water for the environment and MDB's communities, such as the GMID.

The consequences of excessive mobilization of water in search for security in a highly dynamic natural environment, which led to salinization, have profoundly influenced the environment and the lives of rural and urban Australians over the past century. This history can be read as part of the pre-history of contemporary conflicts about irrigation, water, and the environment, which will significantly determine how Australians and their natural environment will interact in the future. There can be little doubt that the MDB's future is in grave danger, and a fundamental change to the way Australia values and uses water is urgently necessary.[20]

This is especially true concerning the traditional owners of the Basin, who hold an intimate relationship with the country. According to a statement of their philosophy by the Ngarrindjeri Nation, which is quoted in

[19] Joachim Radkau, *The Age of Ecology. A Global History*, trans. Patrick Camiller (Cambridge et al.: Politiy Press, 2014), 7–8.

[20] Walker, *Royal Commission Report*, 15.

the preface to Commissioner Bret Walker's Report on the MDB, "Our Lands, Our Waters, Our People, All Living Things are connected. We implore people to respect our Ruwe (Country) as it was created in the Kaldowinyeri (the creation). We long for sparkling, clean waters, healthy land and people and all living things."[21] As Walker has phrased it, the "damage and depletion of the water resources, ecosystems and biodiversity of the Murray-Darling Basin since European colonisation, and the trauma and dislocation experienced by Aboriginal people, are part of the same story."[22] The Commissioner continues to write that "we need to fix what we have spoiled, and we are spoiling. The 'we' in that mea culpa, it must be admitted, does not include the traditional owners, the First Nations, the Aboriginal inhabitants of the various Basin countries. The indigenous ancestors of those of us who are Aboriginal citizens of Australia are not, in any decent sense, responsible for the plight that the Basin is in."[23]

This nexus between the colonization of Australia, the dispossession of its inhabitants by Europeans, and the environmental destruction wrought by mismanagement and ignorance, of which salinization is one symptom, is also the reason why there are few mentions of Indigenous people in this book. The spread of salinization in the twentieth century is a story of European Australians, played out among them, without any role by Aboriginal people. While it is their land, it was not their responsibility.

Environmental History for Contemporaries

Histories of past environmental hazards can never pretend to be untouched by the global ecological challenges we face in our own time. Environmental History, inevitably, is for contemporaries. We live in an era of human history in which our potentiality, and therefore our responsibility to the planet, are intertwined like never before. As Tom Griffiths has said, "[i]n the long history of the planet, there has never been a moment as strange or as grave as ours, because it uniquely fuses planetary

[21] Walker, *Royal Commission Report*, 1.

[22] Walker, *Royal Commission Report*, 1.

[23] Walker, *Royal Commission Report*, 14.

destiny and collective human will." What then is the role of Environmental History for the contemporary challenges we face living on planet Earth?[24]

Environmental History has often exhibited a moral purpose and strong political commitments. This was also the case in Australia, where many early writers believed that Europeans had greatly damaged the environment and were interested in knowing how this had come about. They partly hoped that understanding would help ameliorate or avoid future damage. This makes sense, considering that the discipline came into existence at the dawn of the "Age of Ecology," around 1970, which saw the rise of the environmental movement and the popularization of ecological thinking. While the movement undoubtedly lent visibility and a sense of purpose to the field, its intellectual roots can be traced back to the nineteenth century.[25]

Broadly, Environmental History explores the relation of human society to the natural environment. It regards human history as intertwined with environmental conditions, instead of treating environments as "mere backdrops to the dramas of history, but participants in their own right, interacting with all the others."[26] According to Donald Worster's classic manifesto: "Above all, it rejects the conventional assumption that human experience has been exempt from natural constraints, that people are a separate and 'supernatural' species, that the ecological impacts of their past deeds can be ignored."[27]

[24] Tom Griffiths, "Radical Histories for Uncanny Times," ABC, last modified February 27, 2018, http://www.abc.net.au/radionational/programs/bigideas/radical-histories-for-uncanny-times/9478670.

[25] Sarah Brown et al., "Can Environmental History Save The World?," *History Australia* 5, no. 1 (2008) 03.02; Andrew C. Isenberg, "Introduction. A New Environmental History," in *The Oxford Handbook of Environmental History*, ed. Andrew C. Isenberg (Oxford: Oxford University Press, 2014), 4; Radkau, *Age of Ecology*, 7–8.

[26] John R. McNeill and Verena Winiwarter, "Soils, Soil Knowledge and Environmental History: An Introduction," in *Soils and Societies. Perspectives from Environmental History*, eds. John R. McNeill and Verena Winiwarter (Isle of Harris: White Horse Press, 2006), 2–3.

[27] Donald Worster, "Appendix: Doing Environmental History," in *The Ends of the Earth. Perspectives in modern Environmental History*, ed. Donald Worster (Cambridge: Cambridge University Press, 1998), 290.

If environmental histories can be of use today, it is because they tell stories about how humans have brought about, responded to, and understood environmental change in the past. They broaden our understanding of the complex interrelationship between human society and the natural environment. This is especially important since environmental change will, in all likelihood, be one of the central challenges for ensuring and redistributing prosperity on our planet.[28] Thus we will need a change of policies and economic regimes, as well as of attitudes: "As soon as it is no longer possible to abstract from nature, we have to think *with* Gaia."[29]

This is the challenge of the Anthropocene, that the Holocene, in which we had been living in for the past 11,000 years or so, is over and a new epoch has begun—one, in which humanity is the central ecological and geological agent. As Clive Hamilton, Christoph Bonneuil, and Jean-Baptiste Fressoz have recently argued, the new era requires us to rethink Modernity and with it the central pillars of our worldview: the overarching priority of economic growth, the binary between nature and culture, the belief in (linear) progress facilitated by unleashing of the forces of production, and the project of forging societies unbounded by the limits of nature, which have been key signatures of the modern period.[30]

To help achieve this, environmental histories that broaden our horizon of possibilities may be useful. Although, as Michel Foucault has argued, one does not find a solution for present problems in a different epoch.

[28] Ruth A. Morgan, "Histories for an Uncertain Future: Environmental History and Climate Change," *Australian Historical Studies* 44, no. 3 (2013): 354.

[29] Christophe Bonneuil and Jean-Baptiste Fressoz, *The Shock of the Anthropocene. The Earth, History and Us* (London: Verso, 2016), 41. Original Emphasis.

[30] Bonneuil and Fressoz, *The Shock of the Anthropocene*, 19–44; Clive Hamilton, "Human Destiny in the Anthropocene," in *The Anthropocene and the Global Environmental Crisis. Rethinking Modernity in a New Epoch*, ed. Clive Hamilton et al., London (New York: Routledge, Taylor & Francis Group 2015). The term "Anthropocene" was coined by the chemist Paul Crutzen and the marine scientist Eugene F. Stoermer in 2000. See Paul J. Crutzen and Eugene F. Stoermer, "The Anthropocene," *IGBP Newsletter* 41 (2000); The term has since been taken up by many other scholars, although the debate when the Anthropocene began and whether it is an at all appropriate denomination for our era is in full swing. See, for example, Bonneuil and Fressoz, *The Shock of the Anthropocene*; Paul J. Crutzen, "Geology of Mankind," *Nature* 3, no. January 2002 (2002); Clive Hamilton et al., "Thinking the Anthropocene," in *The Anthropocene and the Global Environmental Crisis. Rethinking Modernity in a New Epoch*, ed. Clive Hamilton et al. (London; New York: Routledge, Taylor & Francis Group, 2015); McNeill and Engelke, *The Great Acceleration*.

There are no examples in history. It is *non magistra vitae*. Therefore, as Foucault concludes, the task of thinking is to continuously reconsider the way we think and act in our own period. The responsibility of scholars is not to tell people what to do, but rather to use their analytic work to question the implicit assumptions of the present. Exploring historically different ways of thinking and acting changes the way we behave in the present. This way, environmental histories can connect the past to the present where we need to morally engage. They show that humanity's relationship with its environment is dynamic and not at all natural.[31] That said, Environmental History will not single-handedly save the world, but it should be understood as a part of the timely task of learning to sustainably live on Earth.[32] As such, it "remains at heart a humanistic discipline."[33]

In this sense, this book is anthropocentric, as it is interested in the effects of environmental factors on human society. The concept of a slow catastrophe alone carries these implications. Chris Courtney and Fiona Williamson have argued that related terms, such as "disaster" and "hazard," fail to recognize that disturbances that look catastrophic from the perspective of human beings might not have been as negative when viewed from the ecosystem's perspective. They are, therefore, relative concepts that offer a species-specific rather than a holistic perspective on the environment.[34] One could argue that salinization is an example of environmental degradation that is universally bad for *all* life forms, not just humans.[35] This, however, does not remove the conceptual approach's anthropocentric premises.

The Anthropocene makes nonsense of the boundary between nature and culture. Nonetheless, human societies and natural systems should be understood as entities of different qualities, temporalities, and logics

[31] Michel Foucault, "On the Genealogy of Ethics: An Overview of Work in Progress," in *The Foucault Reader*, ed. Paul Rabinow (New York: Pantheon Books, 1984), 343; Michel Foucault, "Le souci de la verité (1984)," in *Dits et Écrits 1954–1988. IV. 1980–1988*, ed. Daniel Defert and François Ewald (Paris: Gallimard, 1994), 675–677.

[32] Brown, "Can Environmental History Save The World?"

[33] Chris Courtney and Fiona Williamson, "Disasters and the Making of Asian History," *Environment and History* 26 (2020): 1.

[34] Courtney and Williamson, "Disasters," 1–2.

[35] McNeill, *Something New Under the Sun*, xxv.

which are, unavoidably, interdependent. Salinization is, therefore, understood as a phenomenon caused by reciprocal interaction between humans and the natural environment. It would be inadequate to tell this story by only considering the projections of human subjects without considering the reactions by their natural environment or the other way around, without considering the limitations imposed by the dynamics of the natural environment on human affairs.[36]

As Michael Cathcart has laconically put it: "Project whatever you like. If you fail to find water, you will die."[37] This may be something to keep in mind beyond the realm of academia.

[36] Tom Griffiths, "Environmental History, Australian Style," *Australian Historical Studies* 46, no. 2 (2015): 163; Dipesh Chakrabarty, "The Climate of History: Four Theses," *Critical Inquiry* 35, no. Winter (2009): 201–207; Bonneuil and Fressoz, *The Shock of the Anthropocene*, 19–44.

[37] Cathcart, *The Water Dreamers*, 3.

CHAPTER 2

History, Geography, and Geology

The Murray–Darling Basin

The history of the GMID is inextricably linked to the development of irrigated agriculture in the MDB since the late nineteenth century. The Basin is located in south-east Australia and covers approximately one million km^2. Due to this great magnitude—extending over three-quarters of New South Wales, more than half of Victoria, significant portions of Queensland and South Australia, as well as the whole Australian Capital Territory—the MDB encompasses a wide range of climatic conditions and natural environments, from rainforests, to temperate zones, to subtropical areas, to dry and semi-arid ones. Its topography is generally flat to rolling, seldom rising to more than 200 m above sea level, but also includes Australia's highest peaks of more than 2000 m (Fig. 2.1).

The Basin is defined by the catchment areas of the Murray and Darling rivers and their tributaries. The "Mighty Murray" is 2530 km long and runs westward from its source in the Australian Alps in New South Wales along the border to Victoria and then flows into the sea at Lake Alexandrina in South Australia. The Darling, its biggest contributor, is 2740 km long and runs south-westward from Queensland to its confluence with the Murray at Wentworth in Victoria. However, despite the great lengths of these rivers, the surface run-off of the Murray–Darling system is one

© The Author(s), under exclusive license to Springer Nature Switzerland AG 2023

D. Rothenburg, *Irrigation, Salinity, and Rural Communities in Australia's Murray-Darling Basin, 1945–2020*,
Palgrave Studies in World Environmental History,
https://doi.org/10.1007/978-3-031-18451-2_2

Fig. 2.1 Riverine landscape in the Murray–Darling Basin, near Bears Lagoon, Victoria, July 7, 2017 (*Source* Daniel Rothenburg. Licensed under a creative commons attribution-noncommercial 4.0 international license)

of the lowest among the world's major river systems. In much of the Basin, evaporation greatly exceeds rainfall. Many streams only carry water during times of flood and are dry the rest of the time. In times of high flow, most of the streams meander across the floodplains and have much more complex courses than those depicted on maps. This is due to Australia's climate and rainfall. Virtually all of the continent's river systems are subject to considerable flow variability from one year to the next. The Basin's climate, too, is highly variable and characterized by large variations in rainfall between years and extreme events such as floods and droughts. Recent years have nicely illustrated this erratic aspect with near average conditions in 2012–2013, but each subsequent year up to and including 2015–2016 have been increasingly drier and hotter. In May 2016, a distinct return to wetter conditions followed, with above average rainfall and inflows across much of the Basin in 2016–2017. At the end of the decade, many regions were in a state of prolonged drought, while the 2020s have started with a series of wet years. During the time when I made final corrections in my manuscript, the Murray–Darling system experienced major flooding.[1]

In the twentieth century, the MDB became Australia's most important agricultural region and its heartland of irrigated agriculture. The development of infrastructural works for river regulation, chiefly along the Murray, facilitated this. Most of the river's water has been dammed and diverted for a variety of uses, such as irrigation, domestic, and others. However, irrigated agriculture represents the chief aspect of the Basin's developed character. The Murray's seasonal cycle has been reversed. In their natural state, Australian rivers, except for those in the wet corners of the continent, are reduced to low and sluggish flows or chains of billabongs during dry seasons. During wet months they flow wide, flood large areas, and form wetlands.[2] Today, due to regulation and diversion, the Murray runs nearly full in the summer, when the water is needed for irrigation, and low in the winter, when the reservoirs are refilling. The Snowy River, which previously flowed eastward, has been redirected to flow westward into the Murray–Darling system, and many wetlands have been drained for grazing and cultivation.[3]

The Murray has been transformed into a regulated river. Between 1915 and 1974, Lake Victoria Reservoir, Hume Dam, the Snowy Mountains Scheme, five barrages on Lake Alexandrina, 13 locks on the Murray, the Yarrawonga Weir with Lake Mulwala, and other works were completed. Dams were built to "control river flow, ensuring there was water in the river at the right times for irrigated agriculture, [...], and to minimise the effects of drought (Fig. 2.2)."[4]

[1] Peter Crabb, *Murray-Darling Basin Resources* (Canberra: Murray-Darling Basin Commission, 1997), vi, 2–6; Murray-Darling Basin Authority, *Basin Plan Evaluation, 2017*, 17; Joseph M. Powell, *The Emergence of Bioregionalism in the Murray-Darling Basin, 1993*, 7, 9; Davies, *The Guardian*, December 12, 2019; Donna Lu, "'Everything is saturated': what's driving the latest floods in eastern Australia," *The Guardian*, October 17, 2022.

[2] Garden, *Australia, New Zealand, and the Pacific*, 113; Walker, *Royal Commission Report*, 13.

[3] Garden, *Australia, New Zealand, and the Pacific*, 113; Daniel Connell, *Water Politics in the Murray-Darling Basin* (Leichhardt: The Federation Press, 2007), 100.

[4] Emily O'Gorman, *Flood Country. An Environmental History of the Murray-Darling Basin* (Collingwood: CSIRO Publishing, 2012), 137.

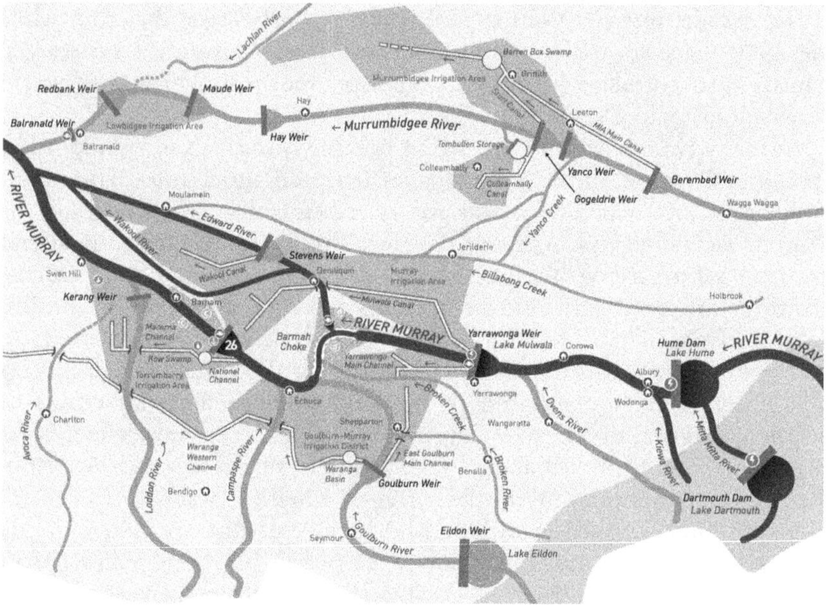

Fig. 2.2 A "system" view of the River Murray (*Source* Licensed from the Murray–Darling basin authority under a creative commons attribution 4.0 international license)

The MDB is, therefore, in a "highly engineered state of affairs" in which its rivers have "ineradicable artificial qualities" which make it virtually impossible to return to a "pre-development" state.[5] This, however, does not mean that nature's "independent dynamic" is completely in control, as salinity in the Basin vividly shows.[6]

[5] Walker, *Royal Commission Report*, 13.

[6] Powell, *Bioregionalism*, 7.

A Deep History of Salinity

Salt and saline waters are natural features of the dry, ancient Australian landscape. The MDB, too, is a saline environment in terms of its soils, geology, surface water, and groundwater.[7] The latter is generally saline, except near the recharge zones and in the limestone aquifers of the south-western section, partly because of leaching from marine deposits, but mainly because of evaporation. The major source of salt in the soil is from rainfall and, to a lesser extent, from marine sediments and the weathering of rocks.[8]

For humans, salinity is a "sleeping giant of the underworld" that is awakened sporadically during one of nature's cyclical moods of climate and geology.[9] Aboriginal peoples experienced major episodes of salinization caused by cycles of flooding and drought. When they arrived in Australia, then part of the landmass of Sahul, some 50,000 to 60,000 years ago, they found much of the Murray Basin to be a mostly verdant landscape of trees, shrubs, and grasses with a wet and cold climate. The south Australian Alps were then covered with an ice sheet. About 25,000 years later, approaching the peak of the last glacial cycle, accelerated winds, hot summers, and low rainfall, combined with episodes of drought, turned the landscape highly saline, diminishing their food resources. The climate reached its present balanced state about 15,000 years ago. Trees returned to the landscape and salinity receded. For the next 10,000 years, a general vegetation-groundwater equilibrium was sustained, and no evidence exists of major events of salinization. Changes brought about by the Europeans, however, disturbed this state.[10]

[7] Crabb, *Murray-Darling Basin Resources*, 150; Peter Russ, *The Salt Traders. A History of Salinity in Victoria* (East Melbourne: Department of the Premier and Cabinet, State of Victoria, 1995), 14.

[8] Ann R.M. Young, *Environmental Change in Australia since 1788*, 2nd ed., (Melbourne et al.: Oxford University Press, 2004), 47–48, 51–52.

[9] Russ, *The Salt Traders*, 20.

[10] Russ, *The Salt Traders*, 20–22; Peter Hiscock, *Archaeology of Ancient Australia* (London: Routledge, 2008), 44; Peter Veth and Sue O'Connor, "The Past 50,000 Years: An Archaeological View," in *The Cambridge History of Australia*, eds. Cambridge: Cambridge University Press, 2013), 19; 27–29; J.M. Bowler, "Environmental and Salinity History of the Murray Basin in the Last 500,000 Years" in *Abstracts Murray Basin 88 Conference, Canberra, 23–26 May 1988* (Canberra: Department of Primary Industries & Energy, Bureau of Mineral Resources, Geology & Geophysics, 1988), 19.

Fundamentally, the MDB "sits on an almost closed groundwater basin with only one outlet at the Murray mouth in South Australia." The catchment is very flat, and rainfall is low and variable, except in the east. Most parts of the area do not supply any run-off to major streams. As a result, the system has little capacity to purge itself of salts and sediments. Much of the salt mobilized into streams is not flushed out of the Murray mouth, but redistributed elsewhere: "By contrast, much of Europe and North America is well drained by powerful fast-flowing rivers."[11]

Barr and Cary have, accordingly, likened the MDB to a bathtub filled with soil to explain how salinization occurs. There are three ways of disposal for incoming rainfall: overflow, evaporation from the top layers of the soil, or evapotranspiration, i.e., the absorption of water within the root profile by grasses or trees. Rainfall that seeps beyond the water table settles at the bottom of the bath and seeps out of the "plug hole," the deep lead aquifers. These are valleys created by ancient rivers which had been gradually filled with coarse sand and gravel, covered by fine sediments. In this state, the water cycle is in rough equilibrium, where rainfall coming into the tub is matched by the water leaving. The water table fluctuates around its long-term average depth, depending on rainfall variations.

Irrigation upset this balance because it dramatically increased the amount of water pouring into the bath. More water overflowed and seeped into the ground. As the deep-rooted native grasses and the trees growing in the "bathtub" had been replaced with shallow-rooted irrigation pastures, more water that entered the ground seeped through the shorter root zone of these plants down to the water table. Inevitably, the water table has risen, sometimes spectacularly in wet years. Although it does recede in dry periods, with irrigation, the long-term trend is a steady increase.

When the water table is within one to two meters of the ground surface, capillarity draws water upward, where it evaporates. This brings the bathtub system back into balance, but also carries salts to the surface, resulting in a concentration of salt in the surface layer of the soil—the root zone for grasses. Since salt is toxic to them, they stop drawing water from the soil and die from water stress. The high concentration of salt also burns plant tissue. Another effect of high water tables is waterlogging,

[11] Connell, *Water Politics*, 18.

which discourages biological activity and soil fauna, reducing soil fertility and degrading the soil's structure.[12] The effects are greatest where soils and water are naturally saline and water tables are close to the ground surface. Most of the soils used for MDB irrigation are red-brown earths and grey, red, and brown clays with high levels of salinity in their subsoils. The salt is mostly common salt-shaker salt, or sodium chloride.[13]

Even though they are closely linked, it is useful to distinguish between two types of salinization: irrigation and dryland. Irrigation salinity occurs when water is applied to the soil in greater quantities and frequency than under natural rainfall conditions. Not all of this extra water from irrigation is used by plants, and some of it seeps down into the soil and can cause the water table to rise, as the bathtub model illustrates. Dryland salinity is due to land-clearing and the replacement of deep-rooted native perennial vegetation—Mallee eucalypts and grasses which dominated Northern Victoria—with shallow-rooted annual pastures and crops. These introduced species use less water than the native ones and thus lower the evapotranspiration rates, causing more water than before to seep through the soil, mobilize stored salt, and drain down to the groundwater. Dryland salinity adds to the problems of irrigation salinity. Moreover, it has been a widespread issue on non-irrigated land, contributing to erosion and poor plant growth and lowering the water's quality. Both salinity types have been a formidable challenge to agriculture and human livelihoods in the Basin, particularly in parts of the GMID.[14]

[12] Neil Barr and John Cary, *Greening a Brown Land. The Australian Search for Sustainable Land Use* (Melbourne: Macmillan, 1992), 221–22.

[13] Young, *Environmental Change*, 46–50.

[14] Young, *Environmental Change*, 46–54; Barr and Cary, *Greening a Brown Land*, 61. Dryland salinity has also been an issue in the Western Australian wheatbelt. See Quentin Beresford et al., *The Salinity Crisis. Landscapes, Communities and Politics* (Crawley: University of Western Australia Press, 2004); Quentin Beresford, "Developmentalism and Its Environmental Legacy: The Western Australian Wheatbelt 1900–1990s," *Australian Journal of Politics and History* 47, no. 3 (2001); Andrea Gaynor et al., "Looking Forward, Looking Back: Toward an Environmental History of Salinity and Erosion in the Eastern Wheatbelt of Western Australia" in *Country. Visions of Land and People in Western Australia*, ed. Andrea Gaynor et al. (Perth: Western Australia Museum, 2002).

The Goulburn–Murray Irrigation District, Kerang, and Shepparton

The GMID sits in the south-west of the MDB. Its major towns are Shepparton, Swan Hill, and Echuca. There are other important ones – Kerang, Cohuna, Pyramid Hill, and Boort—which are somewhat smaller. Today, the area extends from Swan Hill in the west, to Yarrawonga in the east, all the way to Murchison, Bridgewater, and almost to Wycheproof on its southern boundaries. In the north it is delineated by the River Murray. Other major rivers include the Goulburn, Campaspe, and Loddon.In 2010, the GMID had a population of about 134,000 people and around 12,600 farm businesses.[15] It encompasses the Kerang Lakes Area in the west which has a large number of wetlands, swamps, lakes, and waterways with high environmental values.[16]

The development of an irrigation system with a vast network of channels to carry water from reservoirs to irrigation farms has created, in the words of Peter Davies and Susan Lawrence, "a new Cartesian geometry," overlaying and partially replacing waterways' natural curves.[17] This network, which supplies the six irrigation areas, was constructed between 1900 and 1950. It comprises 6300 km of channels, 800 km of natural waterways, and about 23,000 water outlets for irrigation, stock, and domestic water, serving an area of approximately 9900 km^2. Today, much of the irrigation system is still manually operated, using a time-consuming and heavy labor process essentially unchanged over its 100-year existence, which has just recently received substantial upgrades to automate water delivery.[18]

The GMID was formed in 1959 through the amalgamation of 20 irrigation districts, some of which had been established as far back as

[15] Murray-Darling Basin Authority, *Guide to the Proposed Basin Plan*, 779; 802; 812.

[16] Kerang Lakes Area Working Group, *Communities in Partnership. Kerang Lakes Area Draft Salinity Management Plan* (Kerang: Kerang Lakes Area Working Group, 1992), vi.

[17] Peter Davies and Susan Lawrence, "Engineered Landscapes of the Southern Murray-Darling Basin: Anthropocene Archaeology in Australia," *The Anthropocene Review* 6, no. 3 (2019): 190.

[18] Murray-Darling Basin Authority, *Guide to the Proposed Basin Plan*, 781; 784; Irrigation Leader Magazine, "District Profile.".

1886.[19] With an area of about 3340 km^2 irrigated in 1962–1963, it represented about 80% of Victoria's total irrigated area. Water from the rivers Murray, Goulburn, Campaspe, and Loddon was distributed throughout 7690 km^2, enabling intensive irrigation for dairying, sheep, fat lambs, orchards, and market gardens. Its dimensions at that time are not identical to those of today, though the general shape is largely similar. Most notably, irrigation has been extended well beyond Boort in the south-west of the district. The 40-kilometer gap of non-irrigated land between Pyramid Hill and the area around Lockington has been closed, eliminating the former separation of the eastern and western parts.[20]

In 1963, the GMID had a total reservoir capacity of about 7.75 billion m^3, though in real terms, only 3.39 billion m^3 were available. This was due to the water being shared with New South Wales and South Australia, other Victorian irrigation districts, private diverters, and town water supplies. As the Victorian Year Book for 1965 matter-of-factly reports, only about half of this volume of water was recorded as delivered onto the land. The rest was lost in distribution by unmeasured deliveries, evaporation, seepage, and "unavoidable waste."[21] All of the rural towns explored in this research were growing at that time, with the population of Swan Hill rising from 5197 in 1957 to 7376 in 1966. Kerang grew from a population of 3227 to 4165 and Shepparton from 10,848 to 17,523 over the same period.[22]

In today's political context, the GMID is a part of the MDB, which is both a physiographical area, the catchment of Australia's greatest river basin, as well as a political and administrative entity. It is managed by the MDBA, a cooperation of the five "Basin States"—Victoria, New South Wales, Queensland, South Australia, and the Australian Capital Territory—and the Australian Commonwealth.[23] Before the MDBA was

[19] A.G. Robertson, *Report on Goulburn Murray Irrigation District Inquiry* (Melbourne: State Rivers and Water Supply Commission, 1977), 6.

[20] V.H. Arnold, *Victorian Year Book 1965* (Melbourne: Commonwealth Bureau of Census and Statistics, Victorian Office, 1965), 498–99.

[21] Arnold, *Victorian Year Book 1965*, 500.

[22] Arnold, *Victorian Year Book 1965*, 117–18; V.H. Arnold, *Victorian Year Book 1968* (Melbourne: Commonwealth Bureau of Census and Statistics, Victorian Office, 1968), 128–29.

[23] "Governance," Murray-Darling Basin Authority, accessed August 16, 2022, https://www.mdba.gov.au/about-us/governance.

founded in 2008, the development of irrigated agriculture in the district had been under the jurisdiction of the River Murray Commission (RMC), later the Murray–Darling Basin Commission (MDBC), the MDBA's precursors, and the Victorian State Rivers and Water Supply Commission (SRWSC).[24] As will be shown in greater detail, an environmental history of this region cannot disregard these political circumstances. State power and resources have played a crucial role in the transformation of the environment, the patterns of settlement, and the living conditions of the people who inhabit the region. In the case of the MDB and the GMID, state-induced and financed infrastructure and administration were, and are, vital to the development of irrigated agriculture, the economic structure of the region, and the management of salinity.

During the period studied in this book, the district encompassed an area that was roughly identical to the Kerang and Shepparton regions on the Victorian side of the Riverine Plains, as they were defined in the first comprehensive study of the nexus of salinity, irrigation, and drainage in the Murray Valley—the Gutteridge Report of 1970.[25] It is convenient to use these definitions here because the report subsequently became an authoritative document on the salinity issue. Successive examinations of the problem, and the actors involved, continued to use its regional divisions. Here, they are helpful to detail the regions studied throughout the book. The many differences in geology, environmental conditions, land use, and economic structure between the two regions are of particular interest. Moreover, the boundary between Kerang and Shepparton became a political divide. Their different positions regarding provision of government finance, technical assistance, and political attention was the cause of considerable contestations between activists, engineers, farmers, and citizens.

In the Gutteridge Report, the southernmost part of the Murray catchment is split into zones differentiated by their geological history, namely, the manner in which their underlying sediments were deposited. Kerang and Shepparton lie on the Victorian side of the Riverine Plains zone, where rivers that flowed down the surrounding highlands have deposited

[24] "A Path to Water Reform: Timeline of Water Resource Management in the Murray-Darling Basin," Murray-Darling Basin Authority, accessed August 16, 2022, https://www.mdba.gov.au/publications/products/river-management-timeline-poster.

[25] Gutteridge, Haskins & Davey, *Murray Valley Salinity Investigation. Volume 1. The Report* (Canberra: River Murray Commission, 1970).

sediments for millions of years. The present drainage pattern is largely due to two eras of deposition by ancient rivers, the prior streams (about 25,000 to 14,000 before present [BP]) and ancestral rivers (14,000 to 7000 BP), which crisscross much of the Basin. Due to these deposits, the area is prone to salinization. Salts are liberated through the decomposition of bedrock in the headwaters of the streams. Partly, cyclic salt is carried over the sand dunes from the sea. Groundwater salinities are generally high, especially near the rivers. By contrast, in the Mallee zone further west, salinity in the groundwater is derived mainly from sea water trapped in marine sediments, and to a lesser extent from soluble salts that are dissolved as groundwater passes through aquifers, as well as cyclic salt.[26]

The Riverine Plains zone is further divided into five regions; two of them are Kerang and Shepparton. The Kerang region comprises 11 irrigation areas on nearly 3240 km^2: Swan Hill, Mystic Park, Fish Point, Kerang, Cohuna, Koondrook, Third Lake, Tragowel Plains, Calvil, Dingee, and Boort. It is an alluvial plain formed by the Loddon prior stream system whose sediments are predominantly cover floodplain clays that are highly saline. These soils are, in general, too heavy for horticulture, so the region has always been devoted almost entirely to pastures. In the mid-1960s, the area was dominated by dairying and sheep grazing for fat lambs and coarse wool. The only horticultural areas were Tresco and Woorinen, near Swan Hill. There are large tracts in the center of the region which contain only a few, if any, aquifers.[27]

The Shepparton region, covering about 5260 km^2, comprises six irrigation areas: Murray Valley, Shepparton, Rodney, Tongala-Stanhope, Deakin, Rochester, and Campaspe. Like the Kerang region, Shepparton is also a level clay pan, but with a slightly steeper surface gradient and a higher proportion of lighter-textured soils and sediments. Most of the agricultural difficulties in the region were flooding and waterlogging due to the surface sediment conditions. Salinity was not a major problem because, in the areas where the water table was near the surface, the groundwater was not saline. Its main livestock enterprise was dairying.[28]

[26] Gutteridge, Haskins & Davey, *Murray Valley Salinity Investigation: A Summary* (Canberra: River Murray Commission, 1970), 4–8; Colin Pardoe, "Riverine, Biological and Cultural Evolution in Southeastern Australia," *Antiquity* 69, no. 1995 (1995): 698–699; Powell, *Bioregionalism*, 8–9.

[27] Gutteridge, Haskins & Davey, *Murray Valley Salinity Investigation—Report*, 226–28.

[28] Gutteridge, Haskins & Davey, *Murray Valley Salinity Investigation—Report*, 273–75.

Shepparton was the only Riverine Plains region to support a significant amount of horticulture, as the home to 95% of Norther Victoria's irrigated orchards.[29]

Generally, the Kerang region had a much greater intensity of irrigation and smaller farms compared to much of Shepparton, with its numerous small dairy farms that relied largely on intensively irrigated perennial pastures in certain areas, notably on the Swan Hill Flats and in Cohuna. A look back into the environmental history of the region will provide further insight into this situation.[30]

[29] Gutteridge, Haskins & Davey, *Murray Valley Salinity Investigation—Report*, 2; 47.

[30] Gutteridge, Haskins & Davey, *Murray Valley Salinity Investigation—Report*, 2; 227–28.

CHAPTER 3

Antecedents

Aboriginal Water Engineering

Irrigation—before it was associated with a multitude of environmental issues—was depicted as a response of human ingenuity to the continent's climatic variability—a "panacea for the irregularity of Australian seasons."[1] According to its proponents, Australia did not have a shortage of water, but rather maldistribution. If water could be stored and distributed when needed, it would even out the seasons and convert arid land into productive farms and orchards. In the twentieth century, engineers and water managers went about implementing this promise on a grand scale.[2]

Still, the colossal transformations brought about by water engineering in the past century should not obscure the view of the long human history of water management in Australia. J. M. Powell insists on the "continuity of human endeavour" evident in Aboriginal engineering: "[M]ajor social and economic characteristics of Aboriginal settlement were partly based on complex manipulations of the physical environment; those manipulations included some significant interpretations and adaptions of the

[1] Garden, *Australia, New Zealand, and the Pacific*, 110.

[2] Garden, *Australia, New Zealand, and the Pacific*, 110.

© The Author(s), under exclusive license to Springer Nature Switzerland AG 2023

D. Rothenburg, *Irrigation, Salinity, and Rural Communities in Australia's Murray-Darling Basin, 1945–2020*, Palgrave Studies in World Environmental History,
https://doi.org/10.1007/978-3-031-18451-2_3

hydrological system which should be filed in the same category of tenacious interventionism that has been ascribed for the European invaders. The surprising Aboriginal achievements in water engineering probably pre-date the first European efforts by more than a century."[3]

Despite the European settlers' ignorance or arrogant disregard of Aboriginal modifications, Powell insists that Aboriginal people were "managers," "incipient farmers," "quasi-engineers," and "ecological agents living off, and not simply with, or in spite of, the physical environment."[4]

Robyn Ballinger has vividly described how the first peoples on the Northern Plains of what today is called Victoria—the Barapa Barapa, the Wemba Wemba, the Dja Dja Wurrung, the Ngurai-illamwurrung, and the Yorta Yorta—adapted to the "seesawing shifts in climatic conditions" over thousands of years. Before the peak of the last glacial period, about 25,000 BP, they hunted megafauna and camped on the shores of lakes, where they caught fish and harvested mussels. With the onset of the most intense phase of the ice-age, these lush environments cooled, dried, and became more saline. The climate became increasingly colder, windier, and more arid, diminishing streams' food sources. At the end of this glacial period, about 14,000 years ago, climatic conditions became stabler and ancestral rivers more productive. Over the last 2000 years, a wetter, warmer climate has provided abundant resources.[5]

Responding to the unpredictability of the climate, Aboriginal peoples moved seasonally between swamps, creeks, rivers, hills, and grasslands.

[3] Joseph M. Powell, *Watering the Garden State. Water, Land and Community in Victoria 1834–1988*, Sydney et al. 1989, 22.

[4] Powell, *Garden State*, 24. In a similar fashion, scholars have debated the impact of Aboriginal fire regimes to manage the land by controlled burning. See Barr and Cary, *Greening a Brown Land*, 7–9; Bill Gammage, *The Biggest Estate on Earth. How Aborigines Made Australia* (Sydney: Allen & Unwin, 2011); Eric Rolls, *A Million Wild Acres. 200 Hundred Years of Man and an Australian Forest* (Melbourne: Thomas Nelson, 1981); Karl W. Butzer and David M. Helgren, "Livestock, Land Cover, and Environmental History: The Tablelands of New South Wales, Australia, 1820–1920," *Annals of the Association of American Geographers* 95, no. 1 (2005); Ian Abbott, "Aboriginal Fire Regimes in South-West Western Australia: Evidence from Historical Documents," in *Fire in Ecosystems of South-West Western Australia. Impacts and Management*, ed. Ian Abbott and Neil Burrows (Leiden: Backhuys Publishers, 2003).

[5] Pardoe, "Riverine, Biological and Cultural Evolution," 696–701; Robyn Ballinger, *An Inch of Rain. A Water History of Northern Victoria* (North Melbourne: Australian Scholarly, 2012), 4–5.

The latter were burnt to create new plant growth and to drive out kangaroos and emus to be caught and speared. They collected plant food, made bread, and hunted birds and other animals for their meat. The location and sequence of water sources were memorized and passed on to the next generation in stories and songlines—ritual chants or songs—and mapped onto tools for hunting[6]: "In seasons of abundant rainfall, Aboriginal clans moved through [C]ountry and established semi-permanent base camps near inland water sources such as ephemeral creeks and lake complexes."[7] In times of flood they relocated to higher ground.[8]

Aboriginal people made use of seasonal drainage patterns by engineering channels to redirect floodwaters through weirs made of stakes that slowed down the flow, allowing fish to be caught more easily as the water ebbed. Nets, or fences of boughs, were suspended across waterways between trees to catch ducks and other waterfowl.[9] Bruce Pascoe has argued that aquaculture "was well-established in Australia long before the first colonists arrived." A series of clay dikes was placed across the River Murray's floodplain "to prevent it from receding too quickly during summer and thus ensure retention of fish stocks." The warm, shallow water provided good conditions for breeding fish.[10]

In times of scarce rainfall, people congregated at more permanent rivers and waterholes fed by underground springs. Water could be found in hollow tree trunks and tree roots. They enlarged natural soaks by digging into the soft sediments above a slowly permeable clay sub-soil or at the foot of a sloping rock that disappeared into the plain. Aboriginal people "deepened natural stone depressions and chipped channels into the surrounding rock to divert water into the hole. To reduce evaporation and keep out animals, a flat slab of rock was then placed over the well." Local peoples established pathways through country according to the location of these wells. Water was carried in animal skins. In times

[6] Ballinger, *An Inch of Rain*, 5–6.

[7] Ballinger, *An Inch of Rain*, 6.

[8] Ballinger, *An Inch of Rain*, 6.

[9] Ballinger, *An Inch of Rain*, 6.

[10] Bruce Pascoe, *Dark Emu. Black Seeds. Agriculture or Accident?* (Broome: Magabala Books, 2014), 53–54.

of very low rainfall, people migrated to join relatives, or moved along a chain of water sources to a different region.[11]

When the European invaders arrived, and subsequently displaced the Aboriginal peoples, they too struggled with "Nature's firmest promise: a guaranteed uncertainty," but differently.[12] Northern Victoria's newly arrived pastoralists, and later farmers, would soon try to alleviate shifting climatic conditions through their own engineering endeavors and irrigation.

European Invasion, Squatting, and Agriculture

The Surveyor-General of New South Wales, Major Thomas Mitchell, and his party reached the Northern Plains in June 1836. They did not come unannounced. English words had traveled along the trading routes and catastrophic smallpox epidemics had decimated Aboriginal populations even before Europeans had set foot into the area.[13] Mitchell had been ordered by Governor Sir Richard Bourke to finish tracing the course of the Darling to the Murray River to find agricultural land suitable for permanent settlement. However, Mitchell strayed from his route to explore what he found to be an inviting country, a veritable Australia Felix.[14] He praised the "park-like appearance" of the country with its creek ponds, streams, extensive grasslands, and abundance of trees, highlighting its potential for settlement by farmers and graziers, to a welcoming readership. The allure of the landscape he described, "almost without any aboriginal inhabitants" and therefore "in the best state for

[11] Ballinger, *An Inch of Rain*, 6–7.

[12] Powell, *Garden State*, 30.

[13] Ballinger, *An Inch of Rain*, 7. See Judy Campbell, *Invisible Invaders. Smallpox and other Diseases in Aboriginal Australia 1780–1880* (Melbourne: Melbourne University Press, 2002); Noel Butlin, *Our Original Aggression. Aboriginal Populations of Southeastern Australia 1788–1850* (Sydney: Allen & Unwin, 1983).

[14] Ballinger, *An Inch of Rain*, 7–12. Mitchell's accounts were published in 1838 and in a revised edition in the following year. See Thomas L. Mitchell, *Three Expeditions into the Interior of Eastern Australia, with Descriptions of the Recently Explored Region of Australia Felix, and of the Present Colony of New South Wales, Second Edition, Carefully Revised*, Vols. 1 and 2 (London: T. & W Boone, 1839).

the reception of British emigrants," was well received by his audience.[15] This beautiful land, ostensibly ready for taking, was a managed landscape, shaped by millennia of Aboriginal burning to foster feed for kangaroo and other game. Mitchell's prophecy of a settled agricultural landscape, thanks to its suitability for channels, its permanent watercourses, and plentiful rainfall, was widely promoted in Australia and Britain. Although his assessment was strongly influenced by his arrival in a year of heavy rainfall, and rejected thousands of years of experience by Aboriginal peoples based on a journey of two weeks, the Major's vision of the Northern Plains as a promising landscape awaiting improvement directed future white settler interaction with the country.[16]

The first phase of European settlement in Victoria was dominated by livestock grazing, under a system of illegal occupancy known as "squatting." Mitchell's news of a splendid country was followed by a "grass rush" into the south-east of Australia. In little more than 15 years, fortune-hunters overran most of the favored areas of what was then the Port Phillip District of the colony of New South Wales. Dispossessing the native inhabitants, less than 1000 commercial graziers divided up the country between them until 1848. As most of the better-watered land had already been claimed by 1839, and land-seekers were forced to move further afield, the first wave of squatters moved into the Northern Plains from 1841 to 1853. A second wave followed from 1853 to 1869. Often owned by absentees, these new runs were used to fatten sheep and cattle en route to the Melbourne and Bendigo markets, or to shear sheep and transport the wool to the ports of Echuca and Swan Hill.[17]

The squatters led a semi-nomadic life of sheep and cattle raising, which involved moving several times between areas, wherever the next most favorable location would be. Their occupation entailed severe landscape transformations, including the redirection of rivers and probably the first successful experiments with irrigation. Their trial-and-error approach to

15 Thomas L. Mitchell, "Account of the Recent Exploring Expedition to the Interior of Australia," *Journal of the Royal Geographical Society of London* 7, no. 1837 (1837): 277.

16 Richard Broome, *Aboriginal Victorians. A History since 1800* (Sydney: Allen & Unwin, 2005), 69; Ballinger, *An Inch of Rain*, 7–16; Barr and Cary, *Greening a Brown Land*, 49–53; 238–39.

17 Ballinger, *An Inch of Rain*, 17; 21–25; Powell, *Garden State*, 40–42.

land use proved very damaging to the environment.[18] High stocking rates led to the destruction of native grasses, saltbush, and heathlands: "Victoria's soils and native vegetation had been subjected, for the first time, to the impact of millions of cloven hooves and the selective feeding habits of domesticated animals." Along with the more insidious ecological changes, "the gullies, land-slips, disrupted watercourses and saline 'springs' were alarming harbingers of the coming crises."[19] The shrinking of their food sources and habitat by the millions of sheep and cattle led to a significant decline of native animals—kangaroos, emus, quail, and turkeys.[20]

Within the Plains' variable climate, European and Aboriginal economic systems relied on the same areas of land. Consequently, as the squatters moved in, the Barapa Barapa, Wemba Wemba, Dja Dja Wurrung, Ngurai-illamwurrung, and Yorta Yorta not only lost their homelands, but their (water) resources were diminished. In the dry years of 1838–1840 and 1842, violence increased between Europeans and Aboriginals, and also among Aboriginal peoples. Subsequently, they were driven from their land into protectorates. In 1836, protectorate stations were established in Victoria, including one at Franklinford on the Loddon River in 1841, and a Moravian mission station at Lake Boga in 1851. Here, Aboriginal people were supposed to be induced to assume more civilized, settled lives, encouraged to take up cultivation, attend school, and be instructed in Christianity. Some Aboriginal peoples were removed to Coranderrk, near Healesville, in 1863. Others came to depend on supply depots operating in a number of places across the Northern Plains.[21]

Multiple population shocks hit Aboriginal people from 1800. Their population was dramatically reduced by colonization. Although the number of pre-contact inhabitants in Victoria is unknown, modeling suggests a pre-1788 population of 60,000. Smallpox alone possibly halved and then halved the population again in two shocks to about 10,000–15,000 people before European settlers even entered Aboriginal land. It is estimated that this contact population declined to around 1907, which would mean a decline of 80% in just two decades. This collective trauma was brought about by violence, diseases, the disruption of food supplies,

[18] Powell, *Garden State*, 44–46; Barr and Cary, *Greening a Brown Land*, 10–18.

[19] Powell, *Garden State*, 46.

[20] Ballinger, *An Inch of Rain*, 27.

[21] Ballinger, *An Inch of Rain*, 26–29.

and cultural customs. Low birth rates, due to poor nutrition; venereal disease; loss of land and faith in the future prevented the population from recovering. Between 1839 and 1859, deaths outnumbered births by almost five to one.[22]

Nonetheless, Aboriginal peoples adapted to the new conditions. They lived and worked on sheep and cattle runs established by the Europeans, and still maintained ties to their homelands. Especially after the gold rushes of the early 1850s, squatters and Aboriginal people had strong working relationships: men were employed as shearers, woodcutters, shepherds, water carters, and fencers; women as needle workers, midwives, and bark cutters. There were personal connections and cultural exchanges between indigenous people and white settlers. The latter were taught canoe-making, cooking, the local language, and adopted fire farming practices, which had been used by Aboriginal peoples to manage the grasslands.[23]

Meanwhile, an increasing number of European migrants hungered for more land. Instigated by the urban bourgeoisie and the miners, and fueled by an influx of British Chartists, Irish rebels, and other radicals into Victoria, there was a considerable push for land reform. This led to a series of Land Acts for selection from the 1860s to the mid-1880s. Their aim was to unlock public land that was still held on temporary licenses by big pastoralists, for small-scale farming[24]: "Squatting was a rich man's game. It required significant capital and it involved risks that the undercapitalised could not afford."[25] But the very prospect of available land was what had lured many poor immigrants to Australia. Now various colonies passed legislation to promote the settlement of yeomen—family farmers living off small freehold properties, largely self-sufficient, with the land passed down to the next generation. The land reformers followed a philosophical ideal of agrarianism in their project to create a peaceful, productive rural society based on the small freeholder.[26]

[22] Broome, *Aboriginal Victorians*, 90–93.

[23] Ballinger, *An Inch of Rain*, 29–31.

[24] Powell, *Garden State*, 40; 62–63.

[25] Barr and Cary, *Greening a Brown Land*, 207.

[26] Barr and Cary, *Greening a Brown Land*, 206–11; Marylin Lake, *The Limits of Hope. Soldier Settlement in Victoria 1915–1938* (Melbourne: Oxford University Press, 1987), 11–24.

A new Land Act in 1869 was followed by rapid selection. By the end of 1878, nearly 4415 km^2 were being farmed, mostly in Gippsland and on the Northern Plains. Almost all land open for selection in the Kerang district had been allotted by 1875. This swift uptake was also due to the failing of gold yields, causing many miners to look for other occupations. Established farms were struggling because the soils had been exhausted, yield limits had been reached with the available wheat varieties, and labor costs were mounting. Additionally, heavy rainfalls on the Northern Plains between 1870 and 1875 helped entice settlers to "take a shot" at farming.[27]

With the breaking up of runs through the selection and subsequent fencing of the land, Aboriginal people became more dependent on the white economy. At least two Aborigines took up land themselves under the 1869 Act. The Aboriginal Protection Act of the same year provided for the setting up of six reserves and allowed the Governor to prescribe where and how Aboriginal people lived and to take charge of orphaned or neglected children. It was then estimated that a mere 60 Aboriginal people lived in the Boort–Loddon district and a further 80 in the Gunbower and Terrick Terrick districts.[28]

The Northern Plains were dramatically transformed by the selection. Land was cleared, townships established, roads built, and cultivation extended on a large scale. But the uncertainty of weather patterns prevailed with long dry periods from 1867, including numerous El Niño episodes between 1877 and 1885, and floods at Kerang, Rochester, and Sandhurst (Bendigo) between 1876 and 1884. Debt was widespread among farmers. The populations of the shires of Echuca, East Loddon, and Swan Hill decreased from 22,400 in 1877 to 19,200 in 1882. This, in turn, led to increased farm sizes. As farmers were dependent on volatile international markets to sell their produce, fluctuations in wheat prices made their situation more difficult. By removing native vegetation, altering the hydrology of the country, and over-cultivation, the European

[27] Ballinger, *An Inch of Rain*, 35–39; Gynlais O. Jones et al., *Fifty Years of Achievement. Department of Agriculture and Rural Affairs Kerang 1989* (Kerang: Department of Agriculture and Rural Affairs, 1989), 6.

[28] Ballinger, *An Inch of Rain*, 46–49.

settlers increased the country's fragility and their vulnerability to drought conditions. Victorians questioned whether the Northern Plains were at all suited to settlement farming.[29]

Irrigated Landscapes

Irrigation commenced in several Australian colonies in the 1860s and 1870s, in New South Wales, Victoria, and South Australia. Victoria, however, emerged as the forerunner.[30] The earliest examples were Chinese market gardeners at Echuca and Rochester who used horse works to pump water from the Campaspe River to grow vegetables. In 1864, a selector at Kerang irrigated about 52,600 m^2 of land from the Loddon River. Wheat, peas, maize, sorghum, lucerne, and fruit were grown on irrigated land in the area by 1886. In 1877, an aqueduct from the Coliban River to Sandhurst (Bendigo) and Castlemaine was completed, proving that water could be redirected to drier regions. An unassumingly named Grand Victorian North-Western Canal Company presented an audacious scheme to the Victorian government in 1871 to irrigate six million acres (about 24,280 km^2) west of the Goulburn River by conserving flood waters in reservoirs. The project never proceeded, even though parts of it were built later, yet it remained a darling of the advocates of irrigation and was revisited multiple times.[31]

Meanwhile, the drought from 1877 to 1881 triggered new government action regarding the Northern Plains. With thousands of small-farming families in distress, a commission was appointed, and in 1881, the Urban and Rural Waterworks Trusts were established. Their chief objective was to supply water for stock and domestic use by a modest

[29] Ballinger, *An Inch of Rain*, 53–58; see also Don Garden, *Droughts, Floods and Cyclones: El Niños That Shaped Our Colonial Past* (North Melbourne: Australian Scholarly Publishing, 2009). El Niño, or El Niño Southern Oscillation (ENSO), is an erratic phenomenon with profound climatic impacts across the globe that results from changes in ocean currents in the Pacific Ocean, which are interrelated with changes in air pressure, airflow, and wind patterns. In Australia, it causes failing rainfall and drought conditions in large areas of the country. Sometimes an El Niño episode is followed by its opposite, known as La Niña, resulting in periods of higher-than-average rainfall. Garden, *Australia, New Zealand and the Pacific*, 325–27.

[30] Garden, *Australia, New Zealand and the Pacific*, 110.

[31] Ballinger, *An Inch of Rain*, 39–42; 60; Russ, *The Salt Traders*, 52–53; Jones et al., *Fifty Years of Achievement*, 7.

construction program of dams and canals. However, in 1883, following intensive lobbying, new legislation provided for Irrigation Trusts to be established.[32] In the northern districts, the desire to irrigate was particularly strong, and the boosters' promise found receptive ears. At the first conference of irrigators held in Melbourne in 1890, an aptly named John Garden, who had initiated irrigation at Cohuna in 1882, claimed that irrigation would make the districts "blossom as the rose, and [fill] with the teeming population of prosperous farmers surrounded by every social comfort and enjoyment."[33]

After the Victorian Irrigation Act passed in 1886 under the leadership of Alfred Deakin, community-based, state-supported Irrigation Trusts were established. In Deakin's view, irrigation would revitalize the yeoman ideal and finally break the squatters' power by forcing the subdivision of the land as it would require more intensive cultivation and establish a superior class of small farmers on independent family smallholdings. Economic pressures from other primary producers, such as India and South America, combined with price fluctuations in wool and wheat, convinced Deakin that Australian farmers' safety lay in the variety of their products. To grow an assortment of more water-intensive crops, he concluded, they needed irrigation. With the construction of the Laanecoorie Weir on the Loddon, the Goulburn Weir, and the Waranga Channel, the core of the late twentieth century's irrigation system took shape. By 1896, 25 Irrigation Trusts managed an area of 3150 km^2 on the Northern Plains. About 15% of this area was irrigated, mainly in the Goulburn and Torrumbarry systems.[34]

At the same time, two Canadian–American entrepreneurs, George and William Chaffey, negotiated deals with the Victorian and South Australian governments to establish irrigation colonies at Mildura and Renmark. These managed to supply urban markets with citrus, grapes, and dried fruits by the early twentieth century. However, their schemes ultimately

[32] Powell, *Garden State*, 98–101; Ballinger, *An Inch of Rain*, 61–68.

[33] Quoted in Ballinger, *An Inch of Rain*, 75; Russ, *The Salt Traders*, 51–52.

[34] Powell, *Garden State*, 117; Hilary Susan Howes, "The Spectre at the Feast. The Emergence of Salt in Northern Victoria's Irrigated Districts," *Environment and History* 14, no. 2 (2008): 221–22; Edwyna Harris, "Development and Damage: Water and Landscape Evolution in Victoria, Australia," *Landscape Research* 31, no. 2 (2006): 174–76.

failed due to bad planning, unsuitable soils, and unreliable river flows and rainfall.[35]

Still, the development of irrigation schemes that used water from the Murray and its tributaries and infrastructure works, now gathered momentum. On the Northern Plains, major work was undertaken at Reedy Lake, Lake Charm, Cullen's Lake, Kangaroo Lake, and Lake Tutchewop—seasonal lakes that had occasionally been filled with irrigation water under the Trust system. After high evaporation rates in the dry year of 1885, and the destruction of works in 1893 and 1894 by floods, regulating weirs were constructed and supply channels to the lakes built. The Kerang Lakes National Works opened in 1904 to provide supplies for irrigation.[36]

The early irrigation schemes were troubled by a lack of capital and expertise. They struggled with substantial water losses through evaporation and channel leakage.[37] In 1899, there were about 90 Irrigation and Waterworks Trusts in Victoria—most of them in grave difficulties. Unlike the irrigation schemes of the twentieth century, the Trusts' systems did not radically transform the landscape and were dependent on native hydrology. Consequently, they were plagued by the very problems of seasonal uncertainty they had wanted to overcome. When the demand for water peaked during droughts, there was not enough water; yet in wet years, irrigators did not need the water provided by the Trusts. Between 1886 and 1892, except for the season that followed the drought of 1888, was a time of high prices and yields. Most wheat farmers on the Northern Plains had no need for irrigation, nor were they persuaded to investments into infrastructure and employing labor. Most significantly, farmers had developed a culture of cultivating wheat on large holdings that was adapted to the variable rainfall on the Plains. In short, the Irrigation Trusts were a commercial disaster, and a Relief Act was passed in 1899 to write off 75% of their liabilities. The future of irrigation was in doubt.[38]

Enter the Federation Drought, one of the worst since colonization, which held the eastern states in grip from 1895 to 1902. It added fresh

[35] Garden, *Australia, New Zealand and the Pacific*, 109–11; Connell, *Water Politics*, 70–77.

[36] Ballinger, *An Inch of Rain*, 97–98.

[37] Powell, *Garden State*, 101–4.

[38] Powell, *Garden State*, 118–20; Ballinger, *An Inch of Rain*, 82–83.

incentive to irrigation development and prompted pressure from farmers who campaigned for a steady supply of water. Engineering the rivers, they hoped once more, would mitigate droughts in the future.[39] Furthermore, the ambition to harness the Murray for navigation, irrigation, and water conservation became a considerable impetus to Federation, as cooperation between the colonies that shared the river was necessary to achieve it.[40]

In Victoria, rural development became reason of state. A powerful new public agency, the SRWSC, was founded in 1906. Tasked with developing the Murray for irrigation, it was driven by a fresh impetus and commanded unprecedented authority. Rural water supply was centralized to establish the state in "untrammeled control of 'all natural resources of water supply'."[41] The RMC, consisting of three state representatives and a federal government appointed chair, was founded in 1917, "empowered to supervise the construction and operation of all regulatory facilities."[42] As the states were required to undertake the intended engineering program and manage the operational tasks, New South Wales and South Australia, like Victoria, founded their own water authorities.[43] Together, these public agencies set about a concerted effort to canalize the river, "to make it into a channel to support water supply from dams to irrigation settlements and, initially, to improve river navigation."[44]

James C. Scott has stressed how super agencies, such as the SRWSC or the Tennessee Valley Authority (TVA) in the U.S., became key drivers of environmental transformation in the twentieth century, merging states' power with the scientific endeavor to tame and control nature in order to put the project of modernity into practice. The agencies were empowered

[39] O'Gorman, *Flood Country*, 119; Clement J. Lloyd, *Either Drought or Plenty. Water Development and Management in New South Wales* (Parramatta: Department of Water Resources, New South Wales, 1988), 181.

[40] Don Wright, "The River Murray. Microcosm of Australian Federal History," in *Federalism in Canada and Australia. The Early Years*, ed. Bruce W. Hodgins et al. (Waterloo, Ontario: Wilfrid Laurier University Press, 1978).

[41] Powell, *Garden State*, 147.

[42] Powell, *Garden State*, 142.

[43] Powell, *Garden State*, 142–43.

[44] Emily O'Gorman, "Unnatural River, Unnatural Floods? Regulation and Responsibility on the Murray River in the 1950s," *Australian Humanities Review* 48 (2010). On the debate on navigation and irrigation, see Connell, *Water Politics*, 56–61.

to invest huge sums of capital, grant loans, expropriate private property holders, and resettle thousands of people.[45]

Preceding the TVA, the SRWSC was developed into a textbook example of a super agency, responsible for irrigation and water supply, and controlled by an expert commission. Running water was declared public property, which could not be owned by a private person, being in the class of things common to the people of the state. Ownership of the beds and banks of water frontages and watercourses were vested in the state, and all private rights acquired after 1886 were abolished. Compulsory water rates were introduced and allocated to each holding, regardless of whether the owner used the facility or not.[46]

In its first decades of existence, the Commission successfully established itself as a central resource management agency and obtained myriad responsibilities. It was empowered to establish irrigation districts without the affected landowners' approval. Its portfolio included hydrological works, the monitoring of supplies, research and extension services, the laying-out of subdivision patterns over new frontiers, preliminary mapping, soil surveys and conservation, siting and layout of new townships, and the abandonment of old ones. In 1912, it gained primary management functions for irrigated estates, which it retained until 1932.[47]

Even though irrigation was not at first associated with "closer settlement," but rather seen as a way to bolster an existing development pattern, in the early twentieth century, it became one of the key means for further settlement.[48] This was a new theme of reformist land legislation first mentioned in the 1870s. It meant "the intensification of land use, population distribution and landholding [...] on Crown land, [...] by the purchase, subdivision and re-sale of large freehold estate."[49] During the

[45] James C. Scott, *Seeing Like a State. How Certain Schemes to Improve the Human Condition Have Failed* (New Haven; London: Yale University Press, 1998), 94–95; James C. Scott, "High Modernist Social Engineering. The Case of The Tennessee Valley Authority," in *Experiencing the State*, eds. Lloyd I. Rudolph and John Kurt Jacobsen (Oxford; New York: Oxford University Press, 2006).

[46] Powell, *Garden State*, 144–48.

[47] Powell, *Garden State*, 147; 155; 166; 177–78.

[48] Powell, *Garden State*, 152–53.

[49] Powell, *Garden State*, 98.

depression of the 1890s, it led to the creation of the first government-sponsored colonies and villages for the unemployed in Melbourne.[50] Under the leadership of the Water Commission's chairperson, Elwood Mead, irrigation became "the backbone of closer settlement." The policies' success should halt rural depopulation. This was needed to justify the massive investments in infrastructure. If unsuccessful, the state's investment would have been wasted and the promise of irrigation unrealized. After a poor response from within Australia, the state turned to marketing its irrigated estates overseas, which were dependent on a large supply of agricultural laborers, in Europe and the U.S. to lure immigrants to Victoria.[51]

After World War I, optimism prevailed under boom conditions, due to high demand and prices for Australian agricultural produce and new trading partners such as the U.S. and Japan. This also promoted an expansion of irrigated settlements. The Northern Plains became a focus of government-induced closer settlement initiatives: The Returned Servicemen Settlement Scheme and the British Empire Migration Program. By 1919, irrigated districts were farmed by soldier settlers at Barham, Cohuna, Echuca, Koondrook, Dingee, and Kerang.[52] These post-World War I settlements schemes are, however, generally considered to have been failures due to the small size of the allotments, the settlers' lack of training, their unsuitability due to war trauma, and the fall in prices for primary produce.[53]

Irrigation development, nonetheless, continued. In the 1920s, the state embarked on a substantial building program to drought-proof Northern Victoria. In 1923, the Torrumbarry Weir and Lock supplied water to about 2430 km^2 of land. Murray water was channeled from the weir to permanently fill Kow Swamp and supply summer irrigation water to the districts of Koondrook, Gannawarra, Cohuna, and Kerang. The enlargement of the Waranga Basin and the construction of the Sugarloaf

[50] Powell, *Garden State*, 117.

[51] Powell, *Garden State*, 150–64.

[52] Ballinger, *An Inch of Rain*, 114–15.

[53] Powell, *Garden State*, 174–77; 225; Lake, *The Limits of Hope*; Richard Waterhouse, *The Vision Splendid. A Social and Cultural History of Rural Australia* (Fremantle: Curtin University Books, 2005), 198–208; Joseph M. Powell, *An Historical Geography of Modern Australia: The Restive Fringe* (Cambridge et al.: Cambridge University Press, 1988), 90–120.

(Eildon) Reservoir were also underway. The SRWSC completely remodeled the existing irrigation districts. The Rochester Irrigation District was expanded, and two new ones, at Leitchville and Third Lake, were created. In 1926, the Waranga Basin was strengthened and extended via a concrete embankment. The Waranga Western Channel was enlarged and lengthened to reach the Loddon in 1923, and further westward to reach the Mallee in 1928–29. Timber structures were replaced with concrete head checks, syphons, subways, stops, and culverts. Outdated outlets were replaced by the new Dethridge Wheels. New channels were excavated and existing ones cleaned. The beds of the Loddon, Pyramid Creek, and Reedy Creek were cut and graded in 1928–1929. Drainage systems to lower the water table for Cohuna and Leitchville were completed in 1928. By that year, 30 of the 37 irrigation districts in Victoria were located in the north. Up to 1921, their emphasis had been on dairying and fruit-growing. However, in the period until 1928, in the face of declining fruit prices, there was an increase of dairying and mixed farming, including pig raising.[54]

Between 1914 and 1940 the irrigated area in Victoria increased from 405 km^2 to nearly 2430 km^2, and from 240 km^2 to 1540 km^2 between 1920 and 1940 in New South Wales, while Queensland and South Australia had smaller areas under irrigation.[55] With the increased dependence of a growing population on irrigated agriculture, fundamental changes to the hydrology of the Plains were made. How water flowed across the country was dramatically altered through the construction of roads, railways, levee banks, and irrigation channels. The SRWSC was forging nothing less than a *new* landscape that should be "shielded from economic, agricultural, and climatic unpredictability."[56]

In spite of the Depression of the 1930s, the area under cultivation in Victoria more than doubled in this period, from an average of 14,000 km^2 in 1895–1905 to 32,000 km^2 in 1931–1935. The completion of the Hume Dam in 1936, which had been under construction since 1919, and several smaller works laid the foundations for the post-war boom

[54] Ballinger, *An Inch of Rain*, 119; 122–23.

[55] Garden, *Australia, New Zealand and the Pacific*, 111.

[56] Ballinger, *An Inch of Rain*, 110.

in construction.[57] However, an environmental side-effect of this offensive on state-directed expansion of irrigation development had already been underway for decades. An underground flood, that would put these efforts in doubt, was rising.

The Significance of Salinity

In the late nineteenth and early twentieth centuries, salinization was less extensive worldwide compared to today, and scientific understanding was relatively limited. While irrigation salinity was better understood and documented, dryland salinity was not described until the 1920s and not widely accepted for several decades. Crucially, irrigation salinity had been recognized even before the introduction of irrigated agriculture in Victoria, as it had been observed in North America, India, and Europe. Alfred Deakin also noted the development of waterlogging and salinization due to insufficient drainage in Egypt and Italy during a study tour and recommended preventive management and education. But even though drainage was generally recognized as necessary, it was rarely implemented, even after salt problems had become apparent.[58]

Human-induced salinity and erosion was recorded in Victoria as early as 1853, 19 years after the official settlement of the colony had begun. A Western District squatter, John G. Robertson, wrote with dismay to Governor La Trobe about extensive soil erosion, the appearance of saline springs, loss of native grasses, and landslips on his property.[59]

The early appearance of irrigation salinity did not go unnoticed, either. Only ten years after the Victorian Irrigation Act of 1886, the Mildura Royal Commission found that numerous farmers were experiencing salinity problems. In 1907, A. S. Kenyon, an agricultural engineer, drew attention to over-irrigation and insufficient drainage in Victoria. He noted seepage and deaths of fruit trees in Mildura and Bendigo, a loss of soil structure, and damage to cereal crops. Kenyon drew on cases from India and the U.S., which reported on the nexus of irrigation and

[57] Powell, *Garden State*, 137; 143–44.

[58] Howes, "The Spectre at the Feast," 220; 223; see W.E. Wood, "Increase of Salt in Soil and Streams Following the Destruction of the Native Vegetation," *Journal of the Royal Society of Western Australia* 10, no. 7 (1924).

[59] Russ, *The Salt Traders*, 31–32; Robertson's letter is partly quoted in Powell, *Garden State*, 47.

salinity.[60] Around Kerang, the first evidence of soil salting appeared along the Macorna Channel soon after its opening in 1890. The water tables were six to ten meters below the surface when irrigation commenced in the 1880s. By 1965, they had risen to two meters.[61]

Further inquiries confirmed the increasing extent of salinization. In 1916, the luxuriously titled "Final Report from the Royal Commission on Closer Settlement as to the Working of the Closer Settlement Acts in the Irrigable Districts and a General Review of the Finances of Closer Settlement together with Appendices" discussed the "alkaline trouble" at the Mead settlement in Cohuna. It reported that "due to an immensely greater volume of water being put upon the land than it was accustomed to [...] the water table rose, and the alkali being a deadly poison to plant life, desolated the areas affected."[62]

Salt had accumulated just two years after the commencement of irrigation in the area. The problem was attributed to channel seepage, irrigation, and a subsoil that hindered subterranean drainage. The Commission recommended greater care of water use, greater attention to surface tillage, better preparation, grading of the land before irrigation, and a systematic inquiry into the best means of drainage.[63]

The SRWSC had been well aware of patches of salt-affected land in the district since the days of the old Irrigation Trusts, but had felt no rush to implement remedial measures until the end of 1910, pondering "whether the cost of [.] a [drainage] system would be warranted against the abandonment of affected portions."[64] Reluctantly, it did offer the settlers in the areas most affected by salt to excise the affected parts of their allotments, to be given additional areas for work, free water to flush out the salt, and financial advantages for land reclamation. By 1914, saltwater at Cohuna was drained into Leitchs Creek and Barr Creek.[65]

[60] A.S. Kenyon, "Drainage and Irrigation," *Journal of Agriculture Victoria* 5 (1907): 206. Quoted in Howes, "The Spectre at the Feast," 229.

[61] Gutteridge, Haskins & Davey, *Murray Valley Salinity Investigation—Report*, 229.

[62] John Glass Johnstone et al., *Final Report from the Royal Commission on Closer Settlement as to the Working of the Closer Settlement Acts in the Irrigable Districts and a General Review of the Finances of Closer Settlement together with Appendices* (Melbourne: Government Printer, 1916), 28.

[63] Johnstone et al., *Closer Settlement*, 28–30.

[64] Johnstone et al., *Closer Settlement*, 28.

[65] Howes, "The Spectre at the Feast," 229–31; Ballinger, *An Inch of Rain*, 110.

Salinization continued to spread on the Northern Plains in the 1920s and 1930s. Farmers on the Tragowel Plains experienced a deterioration of their pastures and the failure of their crops in 1920, some 30 years after the completion of the Laanecoorie Weir on the Upper Loddon River.[66] After the Torrumbarry Weir became operational, cheap water became available in unmeasured and unrestricted supplies. As a result, shallow water tables spread.[67] On the Swan Hill Flats, wet periods in 1939 seem to have brought the water table close to the surface, but salinization never became as widespread as in the Kerang–Cohuna area, due to less saline sediments and water tables, and virtually all of the land being under perennial pastures.[68]

The 1925 Royal Commission on Soldier Settlement noted that "on irrigation areas, blocks, which when taken up appeared suitable, have proved to be unsuitable owing to salt trouble after watering."[69] Once again, drainage works were recommended. A minority report even claimed that some of the areas utilized for soldier settlement had *never* been suitable for irrigation, including the districts of Cohuna, Tresco, Woorinen, Merbein, and Mildura.[70] In the same year, the Council for Scientific and Industrial Research (CSIR) and the SRWSC commenced an extensive investigation into the problems of the Murray Valley, including seven detailed soil surveys in irrigation districts to investigate waterlogging and salinity over the next 14 years. As a result, farmers were instructed to use water less excessively; adopt border irrigation systems of larger, open bays; and grade land in order to improve water run-off.[71]

By the 1930s, a permanent rise in the water table had occurred in the Northern Plains. Many unsuccessful irrigators abandoned their land in the Cohuna, Kerang, Swan Hill, and Mildura areas and left the soil in a desolate condition. Fields with little vegetative cover were encrusted with a white salt crust. Rabbits plagued the inhabitants. On forsaken

[66] Russ, *The Salt Traders*, 71.

[67] Jones et al., *Fifty Years of Achievement*, 8;Gutteridge, Haskins & Davey, *Murray Valley Salinity Investigation—Report*, 229.

[68] Gutteridge, Haskins & Davey, *Murray Valley Salinity Investigation—Report*, 231.

[69] James Turnbull et al., *Report of the Royal Commission on Soldier Settlement Together with Appendices* (Melbourne: Government Printer, 1925), 17.

[70] Turnbull et al., *Soldier Settlement*, 49–50; Howes, "The Spectre at the Feast," 231–32; Barr and Cary, *Greening a Brown Land*, 261.

[71] Ballinger, *An Inch of Rain*, 130.

farmland in the Kerang area, native salt-tolerant halophytes such as samphire (*Salicornia australia*), seablite (*Suaeda maritima*), and Dillon bush (*Nitraria schoberi*) appeared.[72]

In stark contrast to the offensive in the development of infrastructural works, only essential drainage measures were undertaken in the 1930s, such as digging salt drains, cutting burrs, clearing timber, and cutting wood for steam-driven river dredges. Sustenance payments were given to unemployed men to extend drainage lines in the Shepparton Valley, but World War II ended these efforts, and large areas remained undrained. In the Kerang area, a network of surface drains was constructed between 1933 and 1936. However, in many parts, the SRWSC had not installed drainage because it was considered too expensive.[73]

At first glance, salinization was the flipside, the unintended consequence, of irrigation development that was done without adequate drainage measures. However, it was more than a technical problem. J. M. Powell has argued with exquisite hyperbole, yet perceptibly, that [t]he grim spectacle of salinization could be interpreted as a warning of "primeval" natural purifications, a violent purging of presumptuous intruders. [...] "Salinisation" [...] represented an inquisitorial rebuke of our very tenure of Australia: figuratively, of course, it seemed to demand that we "turn back"—to first principles—and that we begin reform with a heart-felt *mea culpa*.[74]

Put more prosaically, salinization questioned a model of national development based on irrigated agriculture. It challenged one of the leading national ideologies, the closer settlement vision, and its dream to populate the interior by making it productive through modern technology. It posed a tangible material threat to many rural communities that relied on irrigated agriculture. Finally, and perhaps most gravely, it could be understood as a symptom of an almost criminal disregard of Australia's ecological realities; that the Europeans' very mandate to the continent had been revoked.

[72] Russ, *The Salt Traders*, 74; Ballinger, *An Inch of Rain*, 132; Jones et al., *Fifty Years of Achievement*, 3.

[73] Ballinger, *An Inch of Rain*, 132; Barr and Cary, *Greening a Brown Land*, 223; Jones et al., *Fifty Years of Achievement*, 8; Russ, *The Salt Traders*, 96.

[74] Powell, *Bioregionalism*, 75. Original emphasis.

CHAPTER 4

The Dialectic of Environmental Security (1945–1976)

Dreaming Security

In post-World War II Australia, water was promoted as liquid gold. This was also the title of a glossy book published by the Australian Publicity Council in 1960, exhibiting the jewels of the Goulburn Valley, north-eastern Victoria, and the Snowy Mountains Hydro-Electric Scheme—"the biggest and most exciting thing in Australia."[1] Begun in 1949 and under construction for 25 years, the scheme's major aim was to store the waters of the Snowy and the Eucumbene Rivers and to divert them inland to the Murray and Murrumbidgee. It involved no less than 17 giant dams plus several smaller ones, nine power stations, about 130 km of aqueducts, almost 160 km of tunnels, and a new road network.[2]

Officially endorsed by the Premier of Victoria, Henry Bolte, the story went that Australians were "transforming potentially fertile areas of low

[1] Australian Publicity Council, *Liquid Gold Australia* (Melbourne: Australian Publicity Council, 1960), 112.

[2] Powell, *Garden State*, 233–35. See also Lionel Wigmore, *Struggle for the Snowy. The Background of the Snowy Mountains Scheme* (Oxford: Oxford University Press, 1968); George Seddon, *Searching for the Snowy. An Environmental History* (St. Leonards: Allen & Unwin, 1994); Grahame Griffin, "Selling the Snowy: The Snowy Mountains Scheme and National Mythmaking," *Journal of Australian Studies* 79 (2003).

© The Author(s), under exclusive license to Springer Nature Switzerland AG 2023

D. Rothenburg, *Irrigation, Salinity, and Rural Communities in Australia's Murray-Darling Basin, 1945–2020*,
Palgrave Studies in World Environmental History,
https://doi.org/10.1007/978-3-031-18451-2_4

rainfall to rich areas of closer settlement, important to food producers not only for Australia, but the world."[3] An SRWSC pamphlet from 1964 declared that "The Development of Victoria depends on Water." Water conservation allowed for greater production, increased population, larger national income, more public revenue, reduction of drought losses, closer settlement, and better living conditions. Each new irrigation farm meant two extra families on the land, two in nearby towns, and four extra families in the city.[4]

River regulation was regarded as a benevolent force and environmental deterioration as a necessary loss in the pursuit of the common good.[5] The special aura of irrigation influenced national goal-setting and defense interests. The importance of community preparedness, national self-sufficiency, and the need for more secure strategic deployment of the population were all repeatedly cited in its favor.[6]

It was the heyday of "High Modernity" in Australia, both as a material process of transformation and an appreciation of the modernity of the present, which embraced the possibilities of an open future.[7] The ideology of "High Modernism" to James C. Scott was "a strong, one might even say, muscle-bound, version of the self-confidence about scientific and technical progress, the expansion of production, the growing satisfaction of human needs, the mastery of nature (including human nature), and, above all, the rational design of the social order commensurate with the scientific understanding of natural laws."[8]

[3] Australian Publicity Council, *Liquid Gold Australia*, 2.

[4] State Rivers and Water Supply Commission, *The Development of Victoria Depends on Water* (Melbourne: State Rivers and Water Supply Commission, 1964).

[5] Paul Sinclair, *The Murray. A River and its People* (Melbourne: Melbourne University Press, 2001), 79.

[6] Powell, *Garden State*, 225.

[7] Ulrich Herbert, "Europe in High Modernity. Reflections on a Theory of the 20th Century," *Journal of Modern European History* 5, no. 2007 (2007): 9–11; Lutz Raphael, "Ordnungsmuster der 'Hochmoderne'? Die Theorie der Moderne und die Geschichte der europäischen Gesellschaften im 20. Jahrhundert," in *Dimensionen der Moderne. Festschrift für Christof Dipper*, ed. Lutz Raphael and Ute Schneider (Frankfurt am Main: Lang, 2008), 76–77; 85–86. See also Christof Dipper, "Docupedia-Zeitgeschichte: Moderne (English Version)," Version: 2.0, accessed November 28, 2018, http://docupedia.de/zg/Dipper_moderne_v2_en_2018.

[8] Scott, *Seeing Like a State*, 4.

Faith in progress reached its climax during this period. Tradition and the past were considered impediments on the road to prosperity, including the nearly limitless ambition to transform nature to suit humanity's purposes. Overcoming them was the way into a utopian future.[9]

Doing their part in this project, the state construction authorities intensified their effort to canalize the River Murray,[10] and in just one generation, it was transformed into a regulated river. This was facilitated by numerous river regulation works: "The Lake Victoria Reservoir, Hume Dam, Snowy Mountains Hydro-Electric Scheme, five barrages on Lake Alexandrina blocking the Murray mouth, thirteen locks spanning the Murray, Lake Mulwala and other works were completed between 1915 and 1974."[11]

Through this, the Murray was not only transformed materially but conceptually. A new kind of relationship between people and the river developed. Previously, it had been regarded as a natural, and therefore erratic, entity with its own dynamic. However, as early as 1939, the river's flow had been substantially modified by the operation of the Hume Dam. The RMC could usually determine when water would flow in the river and how much. Subsequently, the Murray's seasonal flows were reversed for irrigation, and eastward flowing rivers were turned westward by the Snowy Mountains Scheme. During the 1950s, the period of the most dramatic environmental interventions, it came to be perceived as an engineered system, under control and monitoring by centralized government authorities and experts[12]: "With dams and other forms of river regulation, governments became intertwined with river flow: they were, in a very real sense, embedded in the river. The distinction between natural and unnatural river flow became increasingly blurred."[13] Super agencies and experts had taken control of the river. To use Emily O'Gorman's term, they, in turn, assumed responsibility to deliver "environmental security" to its users.[14]

[9] Scott, *Seeing Like a State*, 93–97.

[10] O'Gorman, *Unnatural River*, 3.

[11] O'Gorman, *Flood Country*, 137.

[12] O'Gorman, *Flood Country*, 137–38; 143; Sinclair, *The Murray*, 68–69; O'Gorman, *Unnatural River*, 8.

[13] O'Gorman, *Unnatural River*, 8.

[14] O'Gorman, *Flood Country*, 137.

Tradition, in modern Australia, never enjoyed many rights in the head-long rush to transform the land into a productive estate. European settlement proceeded without concessions to traditional culture, settlement patterns, or land use.[15] In the words of Tom Griffiths, their arrival "exploded a capsule of accelerating change." It initiated a process much less peaceful and much more radical than settlement, by now re-envisaged as invasion, seems to suggest. Australia experienced a "compressed double-revolution": the near coincidence of colonization and industrialization: "A settler society, whether or not numerically dominant, was an invading, investing, transforming society with an internal frontier, both natural and cultural."[16]

In the twentieth century, a popular ideology legitimated the quest to radically transform the Australian continent, which Michael Cathcart called "Water Dreaming." The water dreamers hailed hydro-engineering as a means to facilitate the settler dream of a closely populated Australia. At times, this endeavor was fueled by fears that the hands of Asian invaders would reach out for the wide open and sparsely populated spaces of the continent.[17] "Energised by the titanic achievements of civil engineering, the water dreamers championed the capacity of hydro-engineering to redeem and animate the great silences of inland Australia. [...] [H]ydroengineering [w]as an exercise in nation-building driven by an assurance that the emptiness could be filled—that human ingenuity would transform the sullen bush into a wonderland of orchards, farms, towns and cities."[18]

Their faith rested on the legal doctrine that the continent was *terra nullius* in the full sense of these words—belonging to no one. This was "one of the shared—almost unconscious—myths of white Australia."[19] All Aboriginal land was considered property of the Crown. According to this view, Aboriginal peoples could not be owners of this land, because

[15] John Gascoigne, *The Enlightenment and the Origins of European Australia* (Cambridge: Cambridge University Press, 2002), 71; Henry Reynolds, *Frontier: Aborigines, Settlers and Land* (Sydney: Allen & Unwin, 1996), 194.

[16] Tom Griffiths, "Ecology and Empire: Towards an Australian History of the World," in *Ecology & Empire. Environmental History of Settler Societies*, ed. Tom Griffiths and Libby Robin (Edinburgh: Edinburgh University Press, 1997), 4–6; 10.

[17] Cathcart, *The Water* Dreamers, 210–11; 231.

[18] Cathcart, *The Water Dreamers*, 247.

[19] Cathcart, *The Water Dreamers*, 54.

they were neither Christians nor ruled by Christians. However, more importantly, they did not use the land's productive potential: "Tilling of land, breaking the soil, turning it over, became the point of difference."[20] Unimproved land was *terra nullius* until someone removed it from its state of nature, added labor, and made property of it. Improvement, the European colonizers felt, would mend the strangeness and harshness of the continent they had taken to their needs. This view underpinned the new settlements and linked their endeavors with a faith in progress[21]: "Before *terra nullius* was a legal concept in Australia, it was a state of mind."[22]

Australia was not just considered to be a country without owners, but one without history: "The country we describe is as yet without a history, without traditions, and indeed without associations. Its past is a veritable blank," as the Tasmanian explorer James E. Calder put it in the 1840s.[23] With the arrival of the Europeans, history came to the ostensibly timeless land. In the colonizers' view, the Australian continent and its inhabitants, caught in a cycle of unchanging nature, would finally be conquered by civilization and "historyless 'wilderness' [...] put to the cool, productive blade of agriculture."[24]

Not only was the continent considered to have no human history, or at least none to speak of, but its nature was also wasteful and inefficient. The intermittent flows of Australian rivers were understood as improvident. Waste, like water running into the sea without being harnessed for human purposes, was considered an inherent quality of nature. Instead, the future possibilities of the continent were valued. In a new land with "nothing to destroy [...] landed on a naked shore to form, to found, and to create," an invading, transforming European society claimed the freedom to invent the future. Water control became a weapon of conquest to transform a landscape perceived to be of no value and remake it into an image that

[20] Cameron Muir, *The Broken Promise of Agricultural Progress. An Environmental History* (London: Routledge, 2014), 92.

[21] Gascoigne, *European Australia*, 8–13.

[22] Cathcart, *The Water Dreamers*, 54. Original emphasis.

[23] Quoted in Richard Flanagan, *A Terrible Beauty. History of the Gordon River Country* (Richmond: Greenhouse, 1985), 67.

[24] Sinclair, *The Murray*, 36.

conformed to European sensibilities.[25] Hydro-engineering was to develop the crude and unformed "raw elements" of Australia and refashion them into a "civilised, natural landscape."[26]

However, *terra nullius* could be turned against the white settlers themselves unless they occupied and used the land. According to Cathcart, the doctrine was undoubtedly self-serving, but not necessarily cynical. The European settlers considered it a fundamental rule of human affairs, and hydro-engineering would solve this problem. Dams and irrigation channels would finally banish the specter of drought, mitigate floods, and make Australia fit for white settlement. It would allow white Australians to lay claim to Country, facilitate radical social reform, break the squatters' land monopoly, and enable ordinary citizens to work their own lands and run their own farms.[27]

"Irrigationists," such as Alfred Deakin, saw it as an agent of moral and intellectual improvement, which would forge responsible citizens "capable of controlling democratic institutions and undertaking local self-government."[28] A general belief in the superiority of rural life and the importance of maintaining a strong rural population added to this. With appeal to biblical and historical precedent, farming was romanticized as the original and fundamental human occupation, while urban life was considered a deterioration from this natural state.[29]

In the early twentieth century, "conserving the flood waters by locking up the rivers with dams" became the "crescendo of a new song of nationalism."[30] It received popular expression in the book *Australia Unlimited* (1918), a survey of the country's primary industries, by journalist and writer Edwin J. Brady. In his view, Australians were engaged in a battle against the "last walls of nature."[31] They would storm them with an army of immigrants and settlers. He argued that Australia would need an increase in population from 100 to 500 million in just a few generations

[25] Sinclair, *The Murray*, 58–63.

[26] Cathcart, *The Water Dreamers*, 199.

[27] Cathcart, *The Water Dreamers*, 201; 210; 216.

[28] Howes, "The Spectre at the Feast," 222.

[29] Howes, "The Spectre at the Feast," 222; Lake, *The Limits of Hope*, 11–24.

[30] Cathcart, *The Water Dreamers*, 200.

[31] Edwin J. Brady, *Australia Unlimited* (Melbourne: G. Robertson, 1918), 446.

to defend the nation against the supposed ambitions of Asia. Hydro-engineering, he claimed, had turned the "useless" deserts of Australia into what they actually were—gardens.[32]

In the interwar years, the journalist and author Ernestine Hill championed river regulation and irrigation with a similar appeal. Her popular book, *Water into Gold* (1937), reprinted 11 times until 1958, is a romanticized history of irrigation along the Murray. Hill claimed that with the "annihilation of deserts," nothing of value was destroyed. On the contrary, the best was brought out of the land.[33]

The Great Depression and the war effort had halted the construction of new waterworks. However, after World War II, under improved economic conditions, the dream was pursued with fresh vigor. The most prominent symbol of the revived "development imperative with its emphasis on both material wealth and the taming of wayward turbulent nature" was the Snowy Mountains Scheme.[34] It was part of the plan of the Post-War Reconstruction's architects for "security, prosperity and an end to the division of society into winners and losers." To Cathcart, it was "white Australia's triumph over the willfulness of nature," the realization of more than a century of water dreaming.[35] However, for the GMID, the Big Eildon Dam, completed in 1956, then Australia's largest storage reservoir, proved to be even more significant.[36]

More Water: The Post-War Boom

"Eildon makes History" *The Age* cheered on July 19, 1956, the occasion being the opening of the upgraded Eildon Dam on the Goulburn River. Not only was history made, but civilization itself, understood as man's ability to store water in abundance, was invoked as the paper celebrated the reservoir's filling as the "climax" of "another grand design in the long history of water conservation and irrigation projects" which would

[32] Cathcart, *The Water Dreamers*, 219–21.

[33] Ernestine Hill, *Water into Gold* (Melbourne: Robertson & Mullens, 1937), vii; Sinclair, *The Murray*, 77–78.

[34] Powell, *Garden State*, 222; Connell, *Water Politics*, 99.

[35] Cathcart, *The Water Dreamers*, 239–40.

[36] Murray-Darling Basin Authority, *Guide to the Proposed Basin Plan*, 783; Garden, *Australia, New Zealand and the Pacific*, 117.

"cushion years of drought." Those who gazed down its 14 million cubic yards of earth and rock embankment were asked to "spare a thought for the future."[37]

This enthusiastic celebration of the anticipated benefits of a new dam was to some extent a continuation of the interwar fascination for the "Majesty of Concrete"—a laudatory title for the Hume Dam, which was completed in 1936.[38] Dams were hailed as pyramids, cathedrals, and signifiers "of a thoroughly Australian confidence in the future."[39] In this regard, Australia conformed to the global trend of state-induced, large-scale environmental transformation in the twentieth century—the era of dam building.[40] About 800,000 small and 45,000 large (over 15 m) dams were built worldwide throughout the century to extend irrigation, facilitate flood control, and generate hydro-electricity.[41] Eildon was part of a worldwide dam-building spree which began after 1945. During the 1960s, more than one large dam was completed every day, and by the 1990s, about two-thirds of the Earth's streamflow passed over or through dams of one sort or another.[42] Governments valued the image they suggested: "an energetic, determined state capable of taming rivers for the social good."[43] Dams became a cultural obsession that was rooted in the modern significance of technology. Scientists and engineers became the incarnation of a modern Prometheus, "the One man who would stand alone" and complete the project "to tame, control and discipline nature."[44] As fascinating "objects of delight in themselves, signs and wish

[37] "Eildon makes History," *The Age*, July 19, 1956.

[38] Rothenburg, "The Majesty of Concrete," 101.

[39] Joseph M. Powell, "Snakes and Cannons: Water Management and the Geographical Imagination in Australia," in *Environmental History and Policy: Still Settling Australia*, ed. Stephen Dovers (Melbourne: Oxford University Press, 2000), 61.

[40] Klaus Gestwa, *Die Stalinschen Großbauten des Kommunismus. Sowjetische Technik- und Umweltgeschichte, 1948–1967* (München: Oldenbourg, 2010), 19. My translation.

[41] McNeill, *Something New Under the Sun*, 190.

[42] McNeill, *Something New Under the Sun*, 157–59.

[43] McNeill, *Something New Under the Sun*, 157.

[44] Maria Kaika, "Dams as Symbols of Modernization: The Urbanization of Nature Between Geographical Imagination and Materiality," *Annals of the Association of American Geographers* 96, no. 2 (2006): 276.

images of a better society that was yet to arrive," they were understood as symbols of technology's emancipatory promise, to rid society of want.[45]

In this period of dramatic environmental interventions, governments, through agencies like the SRWSC, set out to deliver environmental security to the users of their engineered landscape, the Murray Valley.[46] The GMID, formed in 1959 by amalgamation of 20 irrigation districts, would benefit greatly from this (Fig. 4.1).[47]

Eildon was a key part of the Victorian post-war expansion in construction activity, which was driven by favorable economic conditions and political imperatives. With the onset of the Long Boom, lasting from the 1940s to the early 1970s, development accelerated like never before. This extraordinary period was based on the recovery and expansion of the industrial economies, the restoration and gradual liberalization of the mechanism of the global economy, and the abundant availability of cheap energy.[48]

Worldwide, agriculture grew rapidly in the post-war years. For Australian farmers, the three decades after the war were the most prosperous ones. The expansion of production was partly due to favorable prices, but also to important technical changes. Subterranean clover pastures, synthetic fertilizers, pesticides, and trace elements were used widely. The use of large machinery proliferated, leading to a decline in the agricultural workforce and the ability to work in giant areas. Additionally, advances in the control of rabbits and soil erosion were made.[49]

Between 1950 and 1953, Australian farmers received record prices for wheat and butter, as well as high prices for wool, sheep, and cattle. This resulted in high farm incomes. The outbreak of the Korean War, prompting the U.S. government to upgrade its military resources,

[45] Kaika, *Dams as Symbols of Modernization*, 296.

[46] O'Gorman, *Flood Country*, 137–38; 143; Sinclair, *The Murray*, 68–69; O'Gorman, *Unnatural River*, 8.

[47] Robertson, *Report on Goulburn*, 6.

[48] Powell, *Garden State*, 222; 225; Barrie Dyster and David Meredith, *Australia in the Global Economy. Continuity and Change*, 2nd ed. (Cambridge: Cambridge University Press, 2012), 160; Davidson, *European Farming in Australia*, 325; McNeill and Engelke, *The Great Acceleration*, 38–40.

[49] Davidson, *European Farming in Australia*, 337–43; McNeill and Engelke, *The Great Acceleration*, 38–39; 52; 88–89; 144.

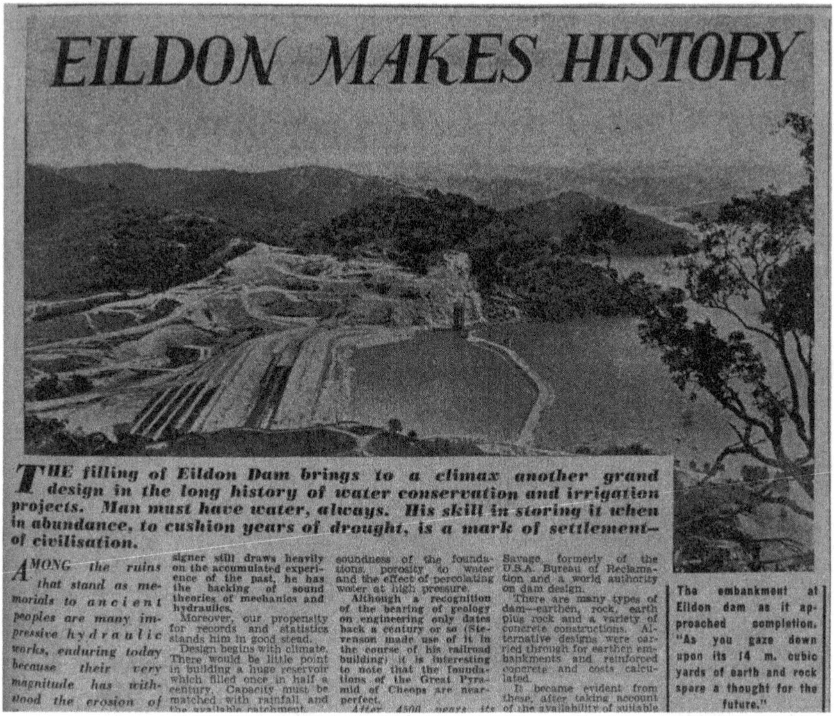

EILDON MAKES HISTORY

THE filling of Eildon Dam brings to a climax another grand design in the long history of water conservation and irrigation projects. Man must have water, always. His skill in storing it when in abundance, to cushion years of drought, is a mark of settlement—of civilisation.

AMONG the ruins that stand as memorials to ancient peoples are many impressive hydraulic works, enduring today because their very magnitude has withstood the erosion of …

… signer still draws heavily on the accumulated experience of the past, he has the backing of sound theories of mechanics and hydraulics.

Moreover, our propensity for records and statistics stands him in good stead.

Design begins with climate. There would be little point in building a huge reservoir which filled once in half a century. Capacity must be matched with rainfall and the available catchment …

… soundness of the foundations, porosity to water and the effect of percolating water at high pressure.

Although a recognition of the bearing of geology on engineering only dates back a century or so (Stevenson made use of it in the course of his railroad building) it is interesting to note that the foundations of the Great Pyramid of Cheops are near-perfect.

After 4500 years its …

… Savage, formerly of the U.S.A. Bureau of Reclamation and a world authority on dam design.

There are many types of dam—earthen, rock, earth plus rock and a variety of concrete constructions. Alternative designs were carried through for earthen embankments and reinforced concrete and costs calculated.

It became evident from these, after taking account of the availability of suitable …

The embankment at Eildon dam as it approached completion. "As you gaze down upon its 14 m. cubic yards of earth and rock spare a thought for the future."

Fig. 4.1 "Eildon makes History." The Age, July 19, 1956 (*Source* © The Age, Melbourne. The use of this work has been licensed by Copyright Agency except as permitted by the Copyright Act, you must not re-use this work without the permission of the copyright owner or Copyright Agency)

demanded vast amounts of wool, causing an upsurge in prices. This Golden Age of the 1950s ended when world prices for agricultural products declined in 1958, while those for requisites rose.[50] The decline in prices was, however, offset by increases in productivity. From 1943 to 1965, the total value of agricultural output increased by 93%. This allowed farmers to maintain their incomes. Australia had achieved an increase in production at a far more rapid rate than the growth of its

[50] Davidson, *European Farming in Australia*, 344–51; Rob Linn, *Battling the Land: 200 Years of Rural Australia* (Sydney: Allen & Unwin, 1999), 154.

population and was a net exporter of all commodities most farmers produced.[51]

During this period, farmers enjoyed increasing subsidies. By the mid-1950s, the dairy, sugar, and egg industries received home support prices. After 1958, these were joined by wheat, which also received a direct subsidy for some years. However, there was little or no financial support for wool and sheep meats and beef producers.[52] Dairy in particular received top priority from governments as it was seen as an effective means of closer settlement and had been subsidized from 1926. In a dual-price scheme, domestic consumers paid a higher price for butter and cheese than was earned from exports. From 1942, the Federal Government paid a bounty to producers, and, at times, even guaranteed a minimum price. The liquidmilk market was regulated by the states.[53]

In these conditions, the development of river regulation works enjoyed a high priority. New soldier settlements with over 1000 farms were founded in the Goulburn Valley at Cobram and Numurkah, in Central Gippsland, and at Robinvale.[54] The SRWSC embarked on an ambitious building program which drastically increased the amount of water available to irrigators. It planned the doubling of irrigation development in ten years as an urgent post-war task. This goal was not achieved; however, the Commission did double the storage capacity of its reservoirs. Big Eildon was upgraded from 444 million m^3 to 3.39 billion m^3 storage capacity. The project was tendered internationally, partially financed with the help of credit from the U.S., and built by the Utah Construction Company. When it was completed in 1955, it was Australia's largest storage reservoir.[55]

This major upgrade to Eildon greatly increased the water availability in the GMID's irrigation system. The system was designed to deliver, on average, 130% of water rights. As the resource was now significantly greater than the water used, irrigators had an effectively unlimited supply.

[51] Davidson, *European Farming in Australia*, 369; 388.

[52] Davidson, *European Farming in Australia*, 351.

[53] Ted Henzell, *Australian Agriculture. Its History and Challenges* (Collingwood: CSIRO Publishing, 2007), 139.

[54] Russ, *The Salt Traders*, 70.

[55] Powell, *Garden State*, 233–35; State Rivers and Water Supply Commission, *Fifty-Second Annual Report, 1956–57* (Melbourne: Government Printer, 1957), 7.

However, it did eventually show some limitations. During dry conditions in the 1960s and 70 s, allocations only reached 160% at one time.[56] Additionally, the successive completion of the Snowy Scheme between 1949 and 1974, while principally designed to generate electricity, directed huge volumes of irrigation water into the River Murray system.[57]

With the dam operating at its full capacity, the Commission increased its water deliveries by a generous 40% the following year. The enlargement was complemented by an extensive program of channel enlargement to deliver the increasing quantities of water, which was completed in 1963. In 1957, the Cattanach Canal was opened, which conveyed water from the Goulburn Weir to the Waranga Reservoir, duplicating the existing Goulburn–Waranga Channel and thereby doubling the quantity of water delivered from Eildon to the irrigation districts of the Goulburn Valley. The use of irrigation water in this area increased by 58%. Shepparton received a plus of about 2.46 million m^3, Kerang about 3.7 million m^3, while Pyramid Hill and Swan Hill even received 40% and 25% more water, respectively.[58] The GMID's total reservoir capacity increased nearly fivefold with the enlargement of Eildon and the Eppalock reservoirs and the construction of the Cairn Curran and Tullaroop reservoirs.[59] The area irrigated and water supplies increased steadily up until the drought restrictions of 1967–1968. In 1945, about 2124 km^2 was irrigated. In 1962–1963, with a rise to about 3340 km^2, the GMID represented about

[56] Murray–Darling Basin Authority, *Guide to the Proposed Basin Plan*, 783; Powell, *Garden State*, 225.

[57] Garden, *Australia, New Zealand and the Pacific*, 117.

[58] State Rivers and Water Supply Commission, *Fifty-Second Annual Report, 1956–57*, 7–8; 18; State Rivers and Water Supply Commission, *Fifty-Third Annual Report, 1957–58* (Melbourne: Government Printer, 1958), 7; 15–17; State Rivers and Water Supply Commission, *Fifty-Eight Annual Report, 1962–63* (Melbourne: Government Printer, 1963), 30.

[59] State Rivers and Water Supply Commission, *Fifty-Sixth Annual Report, 1960–61* (Melbourne: Government Printer, 1961), 17.

80% of Victoria's total irrigated acreage.[60] In 1964, the water rights allocated to irrigators increased substantially, from 1.26 billion m^3 to 1.52 billion m^3.[61]

More was underway: The year before, the state had committed to a ten-year building program for ten additional water storages worth $76 million—on top of the planned expenditure for town water supplies and further expansion of irrigation—to "ensure adequate supplies for irrigation and other purposes in Northern Victoria, and [.] greatly boost the economy" as the Premier, Henry Bolte, promised.[62] Overall, by the mid-1960s, the area under irrigation had risen to nearly 5260 km^2 in both Victoria and New South Wales, to 1210 km^2 in Queensland, and to 400 km^2 kilometers in South Australia.[63]

Between 1945 and 1959, the storage capacity for irrigation in Victoria had increased almost threefold. The area irrigated and production expanded at unprecedented rates, from 2690 km^2 in 1945 to 4260 km^2 in 1959–1960. There was a 32% increase in the area actually irrigated in the 1950s. However, at this time only about 25% of water was reported to be used for growing plants. The remaining 75% were lost by seepage from reservoirs and channels, evaporation, and the excessive application of water by farmers.[64] The SRWSC had commenced a program of improvements to reduce these excessive water losses after the war. Earthen channels were being replaced by concrete ones to reduce leakage. Existing structures came under maintenance. In the 1960s, efficiency modifications to the Torrumbarry irrigation system were carried out and completed in 1973. These measures did cause a slow, steady improvement. In 1963, in the GMID, about 50% of water was actually delivered on land.[65] Still, massive amounts of water were lost every year:

[60] Arnold, *Victorian Year Book 1965*, 498–499; Gutteridge, Haskins and Davey, *Murray Valley Salinity Investigation—Report*, 46.

[61] State Rivers and Water Supply Commission, *Sixty First Annual Report, 1965–66* (Melbourne: Government Printer, 1965), 14; State Rivers and Water Supply Commission, *Fifty-Ninth Annual Report, 1963–64* (Melbourne: Government Printer, 1964), 7.

[62] "£37 1/2 Million Water Supply Plans. State Programme for the Next 12 Years," *The Age*, July 12, 1963; Russ, *The Salt Traders*; 70; 79.

[63] Garden, *Australia, New Zealand and the Pacific*, 111.

[64] State Rivers and Water Supply Commission, *Fifty-Fourth Annual Report, 1958–59* (Melbourne: Government Printer, 1959), 9; Powell, *Garden State*, 225.

[65] Arnold, *Victorian Year Book 1965*, 500; Russ, *The Salt Traders*, 76–77.

In the Kerang region, the main losses that contributed to the rise of water tables were due to seepage from carriers, irrigation and farm channels, lakes, and overwatering by farmers. Adding the losses from evaporation and evapotranspiration, a total of about 511.9 million m^3 were lost every year in the mid-1960s—substantially more than the entire capacity of the old Eildon reservoir.[66] For the Shepparton region, similarly, it was estimated that the losses from open channels by leakage, evaporation, and evapotranspiration, plus those attributed to farms, amounted to about 573.6 million m^3 per year.[67]

The creation of environmental security primarily meant increasing the amount of water in the system. With each drought, the next storage was built with more capacity to hold water in order to achieve a greater system capacity which would allow for an extension of the water supply to more Victorian farmers.[68] Abnormal seasonal conditions were compensated with more water releases.[69]

Meanwhile, in Swan Hill and Kerang, local citizens and farmers organized to tackle their salinity problems, building on previous attempts. Before the war, Alan Morgan, agricultural researcher at the State Agricultural Research Farm in Werribee, had been sent to Kerang in the drought year of 1939 to head the newly established Kerang Agricultural Centre. "Morgy" had worked with farmers and set up experimental plots on salt-affected land in Swan Hill, Kerang, and Cohuna to investigate treatments and improvement of soils.[70]

[66] Gutteridge, Haskins and Davey, *Murray Valley Salinity Investigation—Report*, 241–44.

[67] Gutteridge, Haskins and Davey, *Murray Valley Salinity Investigation—Report*, 278–82.

[68] Russ, *The Salt Traders*, 78.

[69] For example, during to the drought of 1967–1968, the SRWSC's deliveries of irrigation water reached a new record during the 1966–1967 season, the GMID receiving 11% more than in the previous one. Despite restrictions, this high level was maintained in the following drought year. State Rivers and Water Supply Commission, *SixtyThird Annual Report, 1967–68* (Melbourne: Government Printer, 1968), 11; State Rivers and Water Supply Commission, *Sixty-Second Annual Report, 1966–67* (Melbourne: Government Printer, 1967), 11.

[70] Jones et al., *Fifty Years of Achievement*, 4; 9; State Rivers and Water Supply Commission, "Experimental Plots," October 17, 1939, VPRS 6008/P0006, File No. 57/20441/Pt.0001, Public Record Office Victoria, Melbourne; Russ, *The Salt Traders*, 97.

These efforts had been interrupted by the war effort, although Eric Bond, on whose closer settlement property one of the plots had been established, had continued the work in his private capacity.[71] In late November 1945, a mere three months after the cessation of hostilities in the Pacific, the search for local solutions to salinization had gained new impetus. In Cohuna, a Salt Reclamation and Agricultural Improvement Committee was formed to "undertake any necessary steps to assist Soil reclamation and Agricultural improvement, and [.] recommend investigations into problems associated with pasture improvements and irrigation." It included representatives from the shire council, the Agricultural Society, the Irrigators' Advisory Board, the Young Farmers' Club Council, the Department of Agriculture, and the SRWSC. The committee requested the resumption and extension of experimental work and soil reclamation on Mr. Bond's property.[72] It cooperated with Alan Morgan and the Agricultural Centre, pushed for a survey of the district's salt-affected, and encouraged inquiries into drainage feasibility.[73]

Following these revived attempts, the most important breakthrough at self-help of people in the Goulburn–Murray region was the founding of two community-owned research farms. After a meeting at the Swan Hill town hall in August 1954, a proposal to establish an "irrigators' research farm" on the Tyntynder Flats was launched, which led to the establishment of the Swan Hill Irrigators Research Farm (SHIRF).[74] Seventy Kerang irrigators followed suit two years later and adopted a chamber of commerce suggestion to establish their own research farm,

[71] State Rivers and Water Supply Commission, "Cohuna District," March 8, 1940, 6008, VPRS 6008/P0006, File No. 57/20441/Pt.0001, Public Record Office Victoria, Melbourne; State Rivers and Water Supply Commission Cohuna to State Rivers and Water Supply Commission Melbourne, "Eric T. Bond—Experimental Plot," July 5, 1944, VPRS 6008/P0006, File No. 57/20441/Pt.0001, Public Record Office Victoria, Melbourne.

[72] "Cohuna Soil Reclamation and Agricultural Improvement Committee to State Rivers and Water Supply Commission," November 23, 1945, VPRS 6008/P0006, File No. 57/20441/Pt.0001, Public Record Office Victoria, Melbourne; Cohuna Salt Reclamation and Agricultural Improvement Committee to State Rivers and Water Supply Commission, "Constitution, July 1, 1946," VPRS 6008/P0006, File No. 57/20441/Pt.0001, Public Record Office Victoria, Melbourne, 1.

[73] Anonymous, "Resumption of Experimental Work," *The Kerang New Times*, December 21, 1945; Anonymous, "Soil Investigation at Cohuna," *The Kerang New Times*, December 10, 1956.

[74] Anonymous, "Enthusiastic Support for Research Farm," *The Guardian (Swan Hill)*, August 16, 1954.

"to be established with local money, and government departments to work in conjunction with farmers on the district problems, biggest of which is the salting of land because of irrigation."[75] This was the birth of the Kerang Agricultural Research Farm (KARF), located on the Murray Valley Highway between Kerang and Cohuna on a partly salted block with characteristic soil types.[76]

Salinity problems were then well-known in Kerang, Cohuna, Swan Hill, and on the Tragowel Plains. Upon his arrival in 1949, Alan Morgan's successor, Ken Garland, had found Kerang to be "the end of the earth." Productivity losses had reached 50% by 1950, large tracts of farmland were salinized, and many farmers on the Tragowel Plains were living like paupers. Just as in the Shepparton area, the floods in the years between 1951 and 1956 put good land out of production within a short time in the Kerang and Swan Hill areas.[77]

During this time, there was a profound uncertainty about the extent of salinization. In 1961, the Victorian Parliament's Public Works Committee conducted an inquiry into "What methods would be necessary to overcome the serious damage to productive land by a high water table and salting of the land in the GMID, particularly in the Kerang Districts."[78] The Soil Conservation Authority had estimated in 1958 that there were over 40 km^2 of land affected by dryland salting, mostly confined to the southern and western parts of the GMID. However, the areas in which salting could readily occur were approximated to be a further 4000 km^2.[79]

Concerning the irrigation areas at this time, there were no detailed long-term records to illustrate the rise of groundwater levels in the main salt-affected areas of Kerang, Cohuna, Tragowel, and Calivil. More to

[75] Anonymous, "Research Farm at Kerang," *The Age*, April 20, 1956.

[76] Anonymous, "Research Farm Site to be Bought," *The Kerang New Times*, June 8, 1956; Anonymous, "Proposed Site for Research Farm," *The Kerang New Times*, May 4, 1956.

[77] Russ, *The Salt Traders*, 96–100; Jones et al., *Fifty Years of Achievement*, 3–4; 8–9; Gutteridge, Haskins and Davey, *Murray Valley Salinity Investigation—Report*, 231.

[78] J.W. Moore and A.M.I.E. Aust, *Report on Goulburn-Murray Irrigation Water Tables and Salting. Part II, Drainage Requirements due to Irrigation Development* (Melbourne: State Rivers and Water Supply Commission, 1961), 1.

[79] Soil Conservation Authority, "Soil Salting of Non-Irrigated Soils," *S.C.A. Farmers' Bulletin* 10 (1958): 1.

the point, the information available was markedly sparse. For Kerang, the permanent groundwater table was said to have been at about six to nine meters beneath the surface prior to irrigation, and it was known that in 1931 the water table levels in 23 wells dug by the SRWSC ranged from about 0.5 to 1 m from the surface. For Shepparton, even less was known about groundwater levels.[80] However, since 1961, systematic drilling was carried out in many GMID irrigated areas, revealing that in the Tragowel Plains, Calivil, and Dingee irrigation areas, water table depths varied from about 0.6 to about 3 m. Moreover, a good correlation between salt-affected areas and those underlain by a high water table was recognized. On the Tragowel Plains, about 18 km^2, or 8%, were visibly salt affected, while water table levels in North Shepparton were found to range widely.[81]

For the Swan Hill area, the Department of Agriculture conducted a survey from 1962 to 1964, which concluded that severe salting had developed in conjunction with high water tables. In the Tresco and Nyah irrigation areas, considerable tracts had gone out of horticultural production due to salinization, and the decline in vines was mainly attributed to salt and alkalinity. The Tyntynder Flats, predominantly dairying areas, also suffered declining productivity. About 5% were found to be salt affected to some degree. The highly saline soils of Mystic Park, too, had developed extensive salting problems.[82]

Not only was the available knowledge about salinization in the GMID patchy at best, the installation of drainage facilities had not kept up with the rapidly accelerating pace of water supply development. A surface drainage system had been installed between 1933 and 1936, but not extended to the whole of Kerang and Tragowel. While huge funds were mobilized to extend the irrigated area, the problem remained that the SRWSC preferred to drain areas with high agricultural value, like the intensive horticultural areas around Swan Hill and Shepparton, rather

[80] W.P. Dunk, *Report on Goulburn-Murray Irrigation District Water Tables and Salting. Part III, Causes and Prevention of Salting in the Goulburn-Murray Irrigation District* (Melbourne: State Rivers and Water Supply Commission, 1961), 3; 5.

[81] A. Webster, *Report on Goulburn-Murray Irrigation District Water Tables and Salting. Part V, Current Investigations on the Control of Underground Water and Land Reclamation* (Melbourne: State Rivers and Water Supply Commission, 1963), 1–3.

[82] Department of Agriculture, *Victoria, Soils and Land Use near Swan Hill, Victoria* (Melbourne: Government Printer, 1966), 13–14.

than those with poor soils and low yields, like Kerang and the Tragowel Plains.[83]

The Commission had started to install a drainage system on the Swan Hill Flats and in the Kerang-Cohuna area in 1959, which caused steady beneficial effects. However, less than half of the farmers on the Flats had connected with the network in the mid-1960s. Outside the area, very little was being done to slow the approach of drainage and salinity problems. On the Tragowel Plains, subsequent additions had been made to the surface drains installed in the 1930s, but overall, there were few efforts to counter the rapidly developing salinization.[84] Localized attempts had been made at dewatering by pumping from groundwater aquifers in Kerang, Swan Hill, and Calivil, and the efficiency improvements made by the SRWSC in the 1960s had important remedial effects.[85] By contrast, Shepparton already had an extensive drainage system for most of the region by the mid-1960s, which in some areas successfully prevented waterlogging. A large number of groundwater bores and pumps existed throughout the region, reducing the groundwater problem.[86]

Thus, in the midst of the boom in water supply development that characterized the 1950s and 1960s, the Goulburn–Murray region, on the one hand, benefited greatly from the huge amount of extra water, and the resulting extension of irrigated agriculture. On the other hand, drainage was not keeping pace with water delivery, and excessive amounts of water were lost in the irrigation system. Adding to this were the facts that water was sold to irrigation farmers at compulsory charges, which were generously subsidized, covering only the operating costs of the newly built schemes, while the capital costs were borne by the community at large. Irrigators had no incentive to increase efficiency because they had to pay for the water whether they used it or not, and it was cheap. Governments had to encourage water use to justify their investments in infrastructure. Except in times of drought, controlling or even monitoring diversions

[83] Gutteridge, Haskins and Davey, *Murray Valley Salinity Investigation—Report*, 251; Russ, *The Salt Traders*, 89.

[84] Gutteridge, Haskins and Davey, *Murray Valley Salinity Investigation—Report*, 251; 231.

[85] Gutteridge, Haskins and Davey, *Murray Valley Salinity Investigation—Report*, 251.

[86] Gutteridge, Haskins and Davey, *Murray Valley Salinity Investigation—Report*, 286.

was not a high priority. Salinization, by now a long-standing problem, was promoted further.[87]

With the absence of sufficient activity by the Water Commission to alleviate their problems, associations of farmers, local government, and business in the Kerang and Swan Hill areas revived their pre-war efforts to help themselves. Salinization was, then, not the object of substantial public, or political, attention in Victoria. Detailed long-term studies about the social, environmental, and economic repercussions were not available, save for a few exploratory inquiries. One result is that the consequences of six decades of salinity problems for the rural communities in the GMID are very hard to delineate, even in broad strokes. This would not change until the drought of 1967–1968, when salinity appeared on the agenda of national politics.

Enter Salinity: The Drought of 1967–1968

On March 31, 1968, at 4 p.m., the thermometers in Kerang hit 46 degrees Celsius. As town and district slowed down to a crawl, electrical stores reported excellent sales of fans, water coolers, and air conditioners. At 10 a.m. and 39 degrees, some keen women set out to play tennis. When the temperature reached 47 in the clubrooms, they called it a day. It was the second hottest day on record.[88] During the drought, which lasted for 15 months, the levels of major irrigation storages supplying the GMID were falling rapidly. Lake Hume had been at only 50% capacity at the beginning of the 1967–1968 irrigation season and received only 9% of its average inflow in 1967. At the end of the season, it was emptied almost completely.[89] The first farmers ran out of water in late December 1967. More followed by the end of January the next year.[90] The Forest Commission reported that about 100,000 eucalypts would perish, and young ibis were testified to be dying in a Mooroopna swamp as the stream

[87] Davidson, *European Farming in Australia*, 370; Harris, "Development and Damage," 176–77; Connell, *Water Politics*, 124.

[88] Anonymous, "A Scorching 115 Degrees," *The Kerang New Times*, February 2, 1968.

[89] Anonymous, "Save Water Warning to Towns," *The Kerang New Times*, January 9, 1968; Jenny Keating, *The Drought Walked Through: A History of Water Shortage in Victoria* (Melbourne: Department of Water Resources Victoria, 1992), 194.

[90] Anonymous, "Drought Co-ordinator Reviews the Season," *The Kerang New Times*, July 19, 1968.

that fed it stopped flowing. In Shepparton, fish were removed from a lake with an all-time low water level to a reservoir filled from the Goulburn River.[91] It was deemed the worst fire danger period since the Black Friday fires of 1939 with long periods of total fire bans.[92] In September 1967, the Victorian government declared 35 shires drought relief areas, offering eligible farmers concessionary rates on fodder and stock transport, and special loans if they could not get credits from banks. Those who could no longer work their land could apply for unemployment relief work. In the Kerang shire, 40 men cleared Mallee shrubs from roads, and others were employed for tree planting, road forming, and building structures on the Avoca River.[93]

Despite all of this, irrigated production in the GMID for some products still increased, facilitated by excessive water use. The output of milk and butterfat increased by 8%, even though production per cow was down. The production of canning fruits was also rising, with peaches up by one-fifth. By comparison, wheat production, where only 1% of the area was irrigated, dropped by 40%. The Commission's response to the drought was to increase its irrigation water deliveries. In 1966–1967, they reached a new record for the third season in succession. The GMID received 1.82 billion m^3, 11% more than in the previous season. This level was almost retained in the following one, aided by supplementary releases from the Snowy scheme and Lake Eildon, even though the restrictions imposed in 1965–1966 were upheld.[94] In this year of a statewide drought, the SRWSC proudly declared that "it was mostly only

[91] Keating, *The Drought Walked Through*, 201.

[92] Anonymous, "A Scorching 115 Degrees," *The Kerang New Times*, February 2, 1968; "Heatwave Worst for Many Years," *The Kerang New Times*, February 27, 1968. On the 'Black Friday' fires, see Tom Griffiths, *Forests of Ash. An Environmental History* (Cambridge: Cambridge University Press, 2001), 129–49.

[93] Keating, *The Drought Walked Through*, 187–188; Anonymous, "Kerang Shire to List Drought Work," *The Kerang New Times*, March 15, 1968; Anonymous, "$16,450 in Drought Grants," *The Kerang New Times*, March 15, 1968.

[94] Keating, *The Drought Walked Through*, 194–198; State Rivers and Water Supply Commission, *Sixty First Annual Report, 1965–66* (Melbourne: Government Printer, 1965), 13; State Rivers and Water Supply Commission, *Sixty-Third Annual Report, 1967–68*, 11; 16–17; Anonymous, "Production is Up Despite Drought," *The Kerang New Times*, July 23, 1968; State Rivers and Water Supply Commission, *Sixty-Second Annual Report, 1966–67* (Melbourne: Government Printer, 1967), 11.

those streams supported by the Commission's reservoirs which carried flows of any useful significance."[95]

Shortly after good rains in April 1968 had heralded the end of the drought, Gynlais Oughton ("Gyn") Jones, irrigation research officer at Kerang, wrote in the local newspaper, *The New Times*, that "[w]ater, as has been shown this year must become the limiting factor in irrigated farm production. [...] In the future the demands of world hunger, insatiably thirsty metropolis and the demands of other irrigators must hang over the irrigator who applies water uneconomically."[96]

His conclusion: Irrigation farmers had to learn to save water. It was not a popular statement in his community during a time when L. R. Vincent, President of the Goulburn and Waranga Water Users' United League, denounced a "definitive threat" to irrigation water supplies by "certain metropolitan interests." Vincent warned his fellow irrigators that, due to Melbourne's water shortage, they might fall victim to "pirating" of their irrigation supplies.[97] In the metropolis, water restrictions had been imposed for almost a whole year, between September 1967 and 1968, with subsequent bans on fixed sprinklers, hosing on private gardens, and filling of swimming pools, as well as restrictions on bucket watering and car washing. Industry water consumption had decreased by almost 30% in December and January 1967. Voices demanding to tap the Big River in East Gippsland for Melbourne's water supply made themselves heard, causing a fierce debate regarding the conflicting interests of country versus city over water supplies.[98] An *Age* editor criticized that "a small irrigation area" was allowed to "dictate" the water supply of a city of two million and argued that tapping the Big River was the only way of ensuring that Melbourne would not again be caught short on water.[99] Conversely, Mr. T.A. Kendall of Kerang declared that the "Northern districts of Victoria must not be deprived of water which could be diverted to Melbourne," emphasizing that "towns such as Mildura and

[95] State Rivers and Water Supply Commission, *Sixty Third Annual Report, 1967–68*, 11.

[96] Anonymous, "Making Economical Use of Water," *The Kerang New Times*, May 10, 1968.

[97] Anonymous, "Sees Threat to Irrigation Water," *The Kerang New Times*, January 16, 1968.

[98] Keating, *The Drought Walked Through*, 167–85.

[99] "A Dry Argument," *The Age*, December 6, 1967.

Shepparton would 'be off the map' but for the irrigation schemes." He denounced Melbourne as a "selfish city" and moved that the delegates to the annual conference of the Victorian Wheat and Woolgrowers Association, whom he was addressing, object to any diversion of water from north of the Great Dividing Range to augment her supplies.[100]

The background of these debates was a fundamental critique of the water dreamers project: "In every state the sacred cow of irrigation was under attack [...] The development imperative was publicly reviled and caricatured as a whoring, simple-minded, antediluvian, anti-nationalist and morally bankrupt approach."[101] The case was prominently put forward from an economic rationalist point of view by protagonists like Bruce R. Davidson and K. O. Campbell, two agricultural economists at the University of Sydney. The SRWSC stood firmly by its established *credo* that water conservation "on an imaginative scale" was of vital importance for Australia to keep her present standard of living: "Irrigation development is not simply an isolated business undertaking, and in advanced countries it is not so regarded. It is nation-building in the real sense."[102] In the Commission's view, irrigation had supported decentralization, increased production, and returned various taxes back to the state. Campbell's opposition case went as follows: Irrigation's role of promoting regional growth and affecting income distribution was being exaggerated. It was in fact not as socially desirable as other forms of investment. Evaluations of public irrigation development should examine its potential contribution to national economic growth and the broader structure of the economy. Irrigation should be viewed not as the most important but one of many facets of the national agricultural policy, rather than the

[100] Anonymous, "Local Man Fights to Retain Water," *The Kerang New Times*, March 22, 1968.

[101] Powell, *Garden State*, 242.

[102] A.L. Tisdall, "Irrigation in Victoria," in *Water Resource Use and Management. Proceedings of a Symposium held at Canberra by the Australian Academy of Science, 9–13 September 1963*, ed. Australian Academy of Science (Melbourne: Melbourne University Press, 1964), 458.

sine qua non of crop and livestock production.[103] He urged a "continual reappraisal of the place of irrigation in a rapidly growing economy."[104]

At a National Water Symposium in Canberra held in September 1963, the economist took on Ronald East, the Victorian Water Commission's long-standing chairperson. He argued that there was now a reduced need for new irrigation projects as the economy had become more urbanized and diversified. Shortage of water was now less of a constraint on national development than it had been in the past. The political institutions of irrigation development were legacies of days when the resources picture in rural industries had been profoundly different. Campbell expressed doubt as to whether opening up new irrigation areas was the most effective way of utilizing manpower and capital to achieve increased rural output needed to meet growing domestic food requirements and increase foreign exchange earnings.[105]

Davidson, like his colleague, was critical of the fact that taxpayers were tacitly assumed to bear the capital costs for infrastructure development to expand irrigation. He made this a central point of his hard-nosed, no-nonsense analysis of the nation's irrigated agricultural economy arguing that "none of Australia's irrigation schemes operates profitably and [.] the Australian people would have a higher standard of living if the area of irrigated land had been much smaller."[106] According to his analysis, irrigation was largely inefficient and for the most part only profitable because the state subsidized its agricultural products. Far higher returns could be obtained by opening up new regions for dryland farming given the advances in farming practices, fertilizers, and insecticides. This way, Australia could produce all food and fiber it needed without irrigation and operate a profitable system of agriculture. Davidson added that only 10%

[103] "Water Symposium Clash Over Irrigation Cost and Benefits," *The Age*, September 14, 1963; Powell, *Garden State*, 247–50.

[104] K.O. Campbell, "An Assessment of the Case for Irrigation Development in Australia," in *Water Resource Use and Management. Proceedings of a Symposium held at Canberra by the Australian Academy of Science, 9–13 September 1963* (Melbourne: Melbourne University Press, 1964), 450–51.

[105] "Water Symposium Clash Over Irrigation Cost and Benefits," *The Age*, September 14, 1963.

[106] Bruce R. Davidson, *Australia Wet or Dry? The Physical and Economic Limits to the Expansion of Irrigation* (Melbourne: Melbourne University Press, 1969), v.

of agricultural commodities were produced on irrigated land while irrigation used over 90% of the water, and cities like Melbourne and Adelaide were suffering from a lack of water supplies.[107] In an upfront attack on irrigation's *raison d'être*, the economist concluded: "It is the conservation of water for uneconomic irrigation schemes, rather than allowing water to flow harmlessly into the sea, that has wasted the nation's resources."[108]

Ronald East replied to such criticism that irrigation was not a normal business undertaking that could be subject to cost–benefit analysis. Could such analyses be applied, he asked, to defense, education, health, parks, public libraries, and the like?[109] He challenged anyone to point out any major Victorian water conservation project they thought should not have been built. Critics who wanted all projects to be judged on the basis of cost–benefit ratios ignored the indirect benefits of irrigation.[110] The Premier, Henry Bolte, in any case, uncompromisingly backed the SRWSC. With an election on the horizon in 1964, he courted rural voters by categorically announcing that "he would not allow 'one drop' of water to be taken from north of the Divide to augment Melbourne's supplies." This served to perpetuate the status quo, and for the next 20 years "Bolte's Divide" stood strong.[111]

The drought of 1967–1968, therefore, intensified the existing conflicts about water distribution in Victoria, which had been fueled by the offensive on further irrigation supplies for the countryside, while Melbourne's water supply had shown dramatic inadequacies. It provided further ammunition for those who questioned the promise of irrigation development with increasing vigor. However, it also highlighted another conflict over water that had been brewing throughout the 1960s. The drought saw the entrance of salinity onto the national stage. It was a prelude to the bitter struggles of the 1970s and 1980s over the responsibilities of polluters upstream of the river against the users further downstream.

[107] Davidson, *Australia Wet or Dry?* v; 1–4; 98–107; 114–117.

[108] Davidson, *Australia Wet or Dry?* 245.

[109] Powell, *Garden State*, 251–255; Campbell, *An Assessment*, 467–469.

[110] "Water Symposium Clash Over Irrigation Cost and Benefits," *The Age*, September 14, 1963.

[111] "Premier Will Not Let City Have Northern Waters," *The Age*, April 22, 1964; Powell, *Garden State*, 243.

Just north-west of where the Murray reaches the sea at Lake Alexandrina at the end of its journey through the south-east of the continent, lies Adelaide. In the late 1940s and 1950s, Murray water began to be distributed in the state, and a pipeline reached the city in 1954. This, in turn, made South Australia and its capital highly dependent on Murray water and the political goodwill of the upstream states. In the 1960s, confronted with the continued expansion of irrigation in New South Wales and Victoria, which was perceived as a threat to its own water security, South Australia had begun to plan a major storage on the Murray at Chowilla to capture the anticipated extra water from the Snowy Scheme, and alleviate the dependence on the upstream states. Concerns about escalating costs, and also potential salinity problems due to the planned reservoir's shallow construction and the resulting high evaporation rates, as well as the saline nature of the area, were expressed.[112] Growers in Mildura demanded that Chowilla should not be considered until the salinity problem in the Murray had been overcome, worrying that the dam "may become a gigantic salt pan."[113] After a heated debate, the RMC decided to discard Chowilla and have a different reservoir built, the Dartmouth Dam in Victoria.[114]

Awareness of South Australia's vulnerability was reinforced and made visible by the drought. High river salinities caused serious impacts along the lower reaches of the Murray and Adelaide.[115] The demand for water, drainage flows, and groundwater levels throughout the Murray Valley had been increasing in the recent years. The SRWSC had been forced to increase the amount of water deliveries not only for irrigation purposes, but also to provide dilution flows to reduce the salinity of the water in the river and in irrigation channels. In late 1967, up to a quarter of the capacity of the Hume Dam was used to flush out saline flows in order to reduce the salinity levels in the Murray to acceptable levels. During the drought, despite the Commission's record water releases, with low flow in the Murray, salinity levels in the river increased. Particularly damaging to

[112] Connell, *Water Politics*, 99–104; Powell, *Garden State*, 237–240; 262–266.

[113] "Huge Dam Plan Challenged," *The Age*, June 12, 1967; Sunraysia Salinity Committee to Minister for National Development, "Recent Meeting," December 18, 1967, AA1976/642, 198, National Archives of Australia, Canberra.

[114] Connell, *Water Politics*, 99–104; Powell, *Garden State*, 237–40; 262–66.

[115] Connell, *Water Politics*, 106.

agriculture were saline "slugs"—sudden flushes of highly saline water—that went down the river at times.[116] Citrus growers on the lower Murray claimed that "the whole economy of irrigated areas and its industries that depend upon them are threatened by increasing salinity," emphasizing that their products had the lowest salt tolerance of all, and they were, thus, most vulnerable.[117]

This leads back to Kerang and, the soon to be infamous, Barr Creek. This stream rises around Leitchville, enters the Loddon River near Swan Hill and its waters, ultimately, reach the Murray. At the beginning of the twentieth century, its catchment was a natural, shallow depression on the Murray floodplain with swamps and lagoons. Fresh water at the top end supplied irrigation water. Many species of native fish thrived, and its banks were well endowed with red gums, black box trees, and herbage. In the 1920s, however, the creek became a drain for irrigation farmers from the Wee Rup, Cohuna, and Koondrook areas. In the early 1930s, following farmers' demands for drainage, the SRWSC built deep surface drains in the Cohuna irrigation district, and dredged the creek to increase its capacity as a drain. So deep, in fact, that it intercepted saline groundwater, which began entering the river. The salt pushed the fish out and soon killed the trees at the banks.[118]

Barr Creek became the main collector drain for much of the Kerang region and the largest individual source of salt in the entire Murray Valley. In 1964, following complaints from fruit growers downstream at Woorinen, a weir was brought into operation to divert water from the Loddon–Barr system backward into the Murray to improve the quality of their water. Salinities in the creek varied widely, reaching a maximum of 20,000 parts per million in winter and a minimum of 500 parts per

[116] Connell, *Water Politics*, 106; Department of National Development, "Notes on the River Murray Salinity Reduction Projects in Northern Victoria," A571, 1968/2238, National Archives of Australia, Canberra, 1; Minister for National Development to Murray Valley Development League, "Chowilla Dam Article," November 11, 1967, AA1976/642, 192, National Archives of Australia, Canberra.

[117] Australian Citrus Growers' Federation to Minister for National Development, "River Murray Salinity—Finance," December 5, 1967, AA1976/642, 198, National Archives of Australia, Canberra; J.P. Cresp to Minister for National Development, "Emergency Meeting," November 6, 1967, AA1976/642, 198, National Archives of Australia, Canberra.

[118] Russ, *The Salt Traders*, 90–91; George E. Hardwick, "A Historical Report of the North Central Catchment Region," George and Maureen Hardwick, Appin, 11.

million in summer. The former is considered to be a severe case of extremely saline water which is suitable for emergency use for salt-tolerant crops only, while the latter is in the range of medium salinity water suitable for plants with moderate salt tolerance. Barr Creek contained a number of pools in which saline water accumulated in large quantities, which, if suddenly released by rain or a storm, could increase its salt load from the average of about 400 to as much as 3000 tons of salt per day, as it did in 1965. In 1966–1967, the stream discharged about 190,000 tons into the Murray.[119]

This dubious role of Barr Creek was what directed the eyes of ministers and representatives in Melbourne and Canberra on the Kerang region. Faced with the aggravated situation during the drought, the Victorian government decided to submit three salinity reduction projects for funding by the Commonwealth's National Water Resources Development Programme. It provided grants totaling $50 million over five years to states for water conservation projects.[120] The Commonwealth approved and made available funds for two of them: Firstly, for pumping of saline water from Lake Hawthorn, which was being used as a drainage basin for the Mildura area, and delivering it through a pipeline to a natural depression nearby to evaporate, worth $1.6 million. Secondly, another $2 million went to construct two small weirs near Kerang to divert water to a pumping station on Barr Creek at the junction to the Loddon which would deliver water to Lakes William, Kelly, and ultimately Lake Tutchewop to evaporate. Tutchewop was a natural, shallow lake used for swimming and recreation, which had become saline due to the increased level of saline underground water in the region.[121] Now, it was supposed to be converted into an evaporation basin for the greater good of the Murray Valley.

[119] Gutteridge, Haskins and Davey, *Murray Valley Salinity Investigation—Report*, 234; Department of National Development, "Salinity Reduction Projects," 2–3; State Rivers and Water Supply Commission, "River Murray. Salinity Reduction—Barr Creek," A1960, 1973/774, National Archive of Australia, Canberra, 1. On the classification of saline water, see: Agriculture Victoria, "Water Quality for Farm Water Supplies," last modified June 9, 2020, https://agriculture.vic.gov.au/farm-management/water/managing-dams/water-quality-for-farm-water-supplies. See "Measuring Salinity" for details.

[120] "Developing Our Resources," *The Age*, March 3, 1968.

[121] David Fairbairn, "Victoria Grant (River Murray Salinity) Bill 1968. Second Reading. Speech. Wednesday, 1 May 1968, 993," in *Parliamentary Debates. House of Representatives*, ed. Commonwealth of Australia; Hardwick, "A Historical Report," 11–12.

But it was contentious who would benefit from sacrificing the lake and who would bear the consequences. The Minister for National Development, David Fairbairn, pointed out the difficulties by irrigators and other water users whose supply situation had been aggravated in recent years by increasing salinity in the middle and lower reaches of the river. Furthermore, he told the House of Representatives, that substantial releases by RMC structures had been required to control water quality. With the "abnormal conditions" during the drought the problem had assumed great urgency. Therefore, it was not possible to make a normal assessment of the economic merits of the proposals. Barr Creek was known as one of the major sources of saline water, including the famous slugs, which caused trouble in the irrigation areas downstream. Hawthorn was presently leaking and was not able to hold all drainage water it received. Fairbairn's bottom line was that both projects helped to save water.[122] To his department, the benefits of the Barr Creek scheme were obvious. The volume of water required for dilution in the Murray would be reduced, by an estimated 22.2 million m^3 per year, thus saving a "considerable amount of water which would be used for production purposes." The scheme would achieve an overall improvement of water quality which would be of economic advantage to irrigators in the lower reaches of the Murray. Faced with the possibility of a higher degree of salinity downstream in South Australia and an anticipated defoliation and death of citrus trees, the interdepartmental committee investigating the matter also saw strong grounds for immediate action.[123]

Some people in the Kerang region were not convinced. Take it from Mr. Frank Silvester of Mystic Park, who claimed that the "whole scheme 'stinks' in its present concept."[124] These quotation marks were added by a careful *New Times* editor, but the point was clear. Mystic Park and Fish Point farmers called a meeting which unanimously opposed the scheme and organized a deputation to the Minister of Water Supply, Bill Borthwick, to protest. They considered the evaporative area too small to be of

[122] Fairbairn, "Victoria Grant, Second Reading, 1 May 1968," 1–4.

[123] Department of National Development, "Salinity Reduction Projects," 3; Interdepartmental Committee on National Water Resources Development Programme, "Salinity Reduction Projects in the River Murray," A571, 1968/2238, National Archives of Australia, Canberra, 1–3; 6.

[124] Anonymous, "Protest over Salt Pumping," *The Kerang New Times*, October 30, 1968.

any real value considering the magnitude of salinity problems, and also expressed concern about flooding from the Avoca River. Excess water would normally go into Tutchewop, but where would it go when the lake was full of saline water? The farmers worried that their lands might be affected by saline inflows brought about by rising groundwater levels in the region. On the other hand, they feared the loss of their local salt industry.[125] In their view, the Tutchewop project was a government fabrication to gain influence in irrigated areas affected by salt. The deputation claimed that most local residents considered the scheme a "blunder which will only increase the standing of some Victorian members of Parliament by its propaganda value downstream to the detriment of this area."[126]

The deputation asked for the scheme to be deferred until further investigations had been made. In response, the government was quick to reassure the farmers that groundwaters in the area were unlikely to be affected but were, of course, closely monitored, and that there was no intention to interfere with existing salt harvesting. Though the government insisted that the scheme should go ahead, the deputation did win some concessions. The maximum level in Lake Tutchewop would be kept lower to avoid spillover and seepage, and three other smaller salt lakes would not be used in the scheme to circumvent disruption of the salt industry. Regarding the possible loss of pondage for flood waters, the minister asserted that the project included an outfall channel into the Little Murray.[127] Still, the SRWSC felt compelled to hold a community meeting at the Lake Charm Hall to explain the scheme to local residents. Mystic Park and Fish Point farmers were told that it would not damage their properties. Furthermore, a series of bores would be sunk

[125] Anonymous, "Minister is Told of Threat to 30 Farms," *The Kerang New Times*, February 2, 1968; Mystic Park Progress Association to Minister for National Development, "Barr Creek/Tutchewop Project," January 20, 1968, AA1976/642, 198, National Archives of Australia, Canberra.

[126] Anonymous, "Minister is Told of Threat to 30 Farms," *The Kerang New Times*, February 2, 1968; Mystic Park Progress Association to Minister for National Development, "Barr Creek/Tutchewop Project."

[127] Anonymous, "Minister is Told of Threat to 30 Farms," *The Kerang New Times*, February 2, 1968; M. for National Development to Mystic Park Progress Association, "Re Barr Creek/Tutchewop Project," February 23, 1968, AA1976/642, 198, National Archives of Australia, Canberra; Anonymous, "Concessions on Barr C. Plan," *The Kerang New Times*, February 9, 1968.

in lake Tutchewop. If they indicated that the use of the lake for salt-water storage was affecting surrounding areas, protective drains would be constructed. Further hydrological and geological investigations over the whole area would be carried out. When in doubt, the Commission said it was prepared to isolate Tutchewop and Fish Point country.[128] "We are not doing this for South Australia's benefit—this money is being spent to benefit irrigation in Victoria," the engineers claimed. Mr. J. Sporle of Fish Point remained sour: "If there is doubt why not put in the drainage scheme now? Who are you to put us out of business?".[129]

For the people in the Kerang region, the situation was a paradox: The drought, for all its hardships, caused the water tables to lower considerably due to the decrease in rainfall and the restriction of irrigation supplies, providing some relief from their long-standing salinization problems. Yet, somewhat ironically, it served as a catalyst to make salinity an issue, which received the attention of the state and Commonwealth government and justified the allocation of substantial funds for mitigation works.[130] The states, through the water commissions as its agents, had assumed responsibility for the reliability of the Murray as an engineered system, which, in turn, had become the object of conflicting interests. For now, this problem was supposed to be solved by ever increasing water supplies with added efficiency improvements. "The future of Australia is going to hinge on how we are going to trap and use water that is running to waste," as Professor T. D. J. Leech, a former engineer-in-charge of the scientific services division of the Snowy Mountains Hydro-Electric Authority, put it, when he came to Kerang in August 1968.[131]

However, the government's intention was not to find solutions for the problems of an irrigation region that had been experiencing salinity problems for close to 70 years, but to maintain, or respectively restore, the environmental security of citrus growers in Mildura and Renmark, and urban dwellers in Adelaide. The suspicions by Swan Hill farmers, that

[128] Anonymous, "Barr Creek Salinity: Assurance on Scheme," *The Kerang New Times*, February 23, 1968; Anonymous, "Claims Salt Scheme is Political Manoeuvre," *The Kerang New Times*, February 23, 1968.

[129] Anonymous, "Claims Salt Scheme is Political Manoeuvre," *The Kerang New Times*, February 23, 1968.

[130] Gutteridge, Haskins and Davey, *Murray Valley Salinity Investigation—Report*, 232.

[131] Anonymous, "Our Future Rests on Water Storage," *The Kerang New Times*, August 20, 1968.

they had to bear the consequences of satisfying the interests of other users downstream who demanded to receive good quality water in sufficient quantities, were not entirely unjustified. At this time, the Sunraysia region, which included Mildura as the center of a vast horticultural region between Red Cliffs and Wentworth, had not suffered substantial yield reductions from salinization, while Kerang had.[132]

The drought, therefore, did not cause salinity to enter the stage because of the fates of irrigation communities in the GMID. Daniel Connell has argued that much of the rethinking of the aims and methods of water management during the 1970s and 1980s was the product of a growing awareness of salinity problems during the considerations about Chowilla. He has also pointed out that, viewed from the early twenty-first century, the Murray was still in a comparatively healthy condition in the 1960s. Irrigation diversions in 1970 were about half compared to the early 2000s. As a result, there was a lot more uncaptured and undiverted water moving. The seasonal flow pattern would have been a lot closer to pre-development conditions. Salinization resulting from land-clearing would have been a lot less than today. According to Connell, it is not surprising that salinity did not become a major issue for the RMC until the late 1960s when the drought set in and caused short-term spikes in salinity levels.[133] Viewed from the Goulburn–Murray region, though, this was a markedly late reaction. It did, however, cause fundamental improvements for the Kerang region itself, and not just for water users downstream, because the Barr Creek-Tutchewop scheme removed four to five times the annual salt increment of 50,000 tons then brought into the region by irrigation water.[134] Still, the amount of water which was lying on the ground after the drought broke and made way for wet weather highlighted the continued lack of drainage.[135] On the Tragowel Plains, the benefits received from the SRWSC's high water deliveries turned sour

[132] Gutteridge, Haskins and Davey, *Murray Valley Salinity Investigation—Report*, 342–43.

[133] Connell, *Water Politics*, 105–06.

[134] Gutteridge, Haskins and Davey, *Murray Valley Salinity Investigation—Report*, 251.

[135] Anonymous, "Poor Drainage in the District," *The Kerang New Times*, June 16, 1968.

when people found they had nowhere to dispose the surplus water which drained onto the roads, spoiling both earth and metal formations.[136]

Another important result of the debate about Chowilla and the drought was the commissioning of the "Gutteridge Report," formally the "Murray Valley Salinity Investigation," by the engineers of Gutteridge, Haskins and Davey. This was the first serious attempt to examine the issues of salinity, irrigation, and drainage in a comprehensive manner.[137] They were tasked with recommending "tolerable levels of salinity in the River Murray [...] appropriate to present and future use of the water for irrigation and other purposes," and to investigate the factors contributing to salinity, including land use, irrigation and drainage practices, and river regulation. Furthermore, they were to assess the impact of Chowilla and other future storages on River Murray salinity to determine the conditions under which high salinity levels occurred and to recommend actions to mitigate salinity in the future.[138]

The Consultants presented an elegant, comprehensive 550-page report, which was awarded the annual Engineering Excellence Award of the Australian Association of Consulting Engineers. Their investigations took two and a half years of work, considering the geological and human history of the Murray Valley and the vastly differing salinity positions, agricultural industries, and economic conditions of its regions.[139] However, they also produced an example of the kind of systemic reductionism which James C. Scott has called "state simplification," conceptualizing the Murray Valley and its inhabitants as interlinked operators exchanging cubic feet per second and parts per million while generating gross values of production.[140]

They found that two-thirds of the Kerang region had shallow water tables within two meters of the surface, mostly in the northern and northwestern parts. The rate of their spread had been reduced, but in the southern part it was continuing slowly, and was predicted to increase with

[136] Anonymous, "Tragowel Plains in New Irrigation Area," *The Kerang New Times*, August 9, 1968.

[137] Connell, *Water Politics*, 106.

[138] Gutteridge, Haskins and Davey, *Murray Valley Salinity Investigation—Report*, vii.

[139] Anonymous, "Salinity Study Wins High Award," *The Riverlander* 26 (March–April 1972): 17.

[140] See Scott, *Seeing Like a State*, 11–83.

the upcoming further augmentation of water supplies. The problem of shallow water tables was made more difficult by the very saline nature of the groundwater, resulting from the high salt content of most of the sediments through which the water tables rose. Additionally, a good deal of extensive irrigation took place in areas underlain by saline cover floodplain material. A trend toward smaller farms increased the intensity of irrigation in the region. The engineers estimated that the area with shallow water tables in the region would rise from about 2025 km^2 to some 2832 km^2 by 2020 if the problems were not checked. These would cause a loss of production within the region of about $3 million to $7 million per year, rising to over $6 million per year by 2020. If a suggested surface drainage system was installed, the losses were estimated to be lower, about $2 million to $3 million per year at the time of the study, rising to about $3 million to $6 million per year in 2020. Converted into current values, in the worst case, the Kerang region would lose about $70 million per year to salinization by the year 2020.[141]

In the Shepparton region, significantly, salinity problems were then still virtually nonexistent, thanks to low-saline soils and relatively non-saline groundwater. Rather, the region was experiencing difficulties due to flooding and waterlogging, although many parts had already received surface drainage installations. Prior to irrigation, water tables seemed to have been about 15 to 18 m below the surface in the Murray Valley Irrigation Area. Since irrigation had begun on a significant scale, the water tables had risen rapidly, about 0.6 m per annum on average, leading to an area with shallow water tables of more than 100 km^2, and a further 200 km^2 in smaller areas within the region. Groundwater salinity was expected to remain low in the future, with small effects on pastures, but more notable ones on horticulture.[142]

These significantly different positions of Kerang and Shepparton were further highlighted by the evaluation of the solutions that the Consultants put forward. Ultimately, the report's recommendations relied on economic cost–benefit analyses. There was very little consideration of

[141] Gutteridge, Haskins and Davey, *Murray Valley Salinity Investigation—Report*, 228–29; 271–72. Calculated using Diane Hutchinson and Florian Ploeckl, "MeasuringWorth: Five Ways to Compute the Relative Value of Australian Amounts, 1828 to the Present," accessed August 16 2022, http://www.measuringworth.com/australiacompare.

[142] Gutteridge, Haskins and Davey, *Murray Valley Salinity Investigation—Report*, 275–76; 295.

the social repercussions of continuing salinization. Nor was there an assessment of the impacts of the suggested solutions on aquatic ecosystems, vegetation, and wildlife. Two options for the salinity problem were discussed: Firstly, a combination of improved farm management and surface drainage via the extension of the Lake Tutchewop scheme to include Lake Tyrrell, a natural salt lake about 120 km from Kerang. Secondly, improved farm management combined with surface drainage and dewatering by groundwater pumping. For Kerang, the Consultants concluded that all possible options were economically unviable to the economy and the state, and even promised only marginal returns to the region itself. By contrast, for Shepparton's valuable horticultural areas with shallow water tables, precluding salinity problems was an appealing investment. More cost-intensive groundwater pumping, which could be combined with water re-use due to the low-saline nature of the groundwater, would yield even greater returns. For its livestock and grain areas, too, groundwater pumping and re-use was considered beneficial from all economic viewpoints.[143]

However, Gutteridge, Haskins and Davey also pointed out that farmers' incomes were unlikely to be able to absorb the anticipated reductions in returns of over 50%, which they predicted for the salt-affected areas of Kerang if nothing was done about salinization. From the point of view of a strict economic cost–benefit analysis, helping the region was not an attractive proposition, but failure to do so would entail considerable human costs: Farmers would migrate from the salinized areas, affecting all persons whose employment was associated with agriculture, causing many people who were only used to living on the land to shift to urban centers where they might be virtually unemployable. Furthermore, the consultants stressed the serious economic consequences arising from the farmers' resettlement in urban areas and the resulting redundant rural infrastructure—schools, roads, electricity delivery, etc.—and not to forget

[143] Gutteridge, Haskins and Davey, *Murray Valley Salinity Investigation—Summary*, 47–52. For a rough comparison of the gross value of production of dairy versus horticulture, consider that the 1345 km^2 of irrigated area used for dairying in the Riverine Plains Zone generated, on average, \$47.3 million, while the mere 130 km^2 used for citrus and deciduous fruits, almost exclusively in the Shepparton area, generated \$28.1 million. Gutteridge, Haskins and Davey, *Murray Valley Salinity Investigation—Summary*, 45; 55.

the costly irrigation works itself. Helping Shepparton was an attractive investment. Helping Kerang was a matter of policy.[144]

Considerable pressure was put on the RMC by farmers' and irrigators' interest groups to release the report, showing that salinity had assumed a degree of urgency. Five hundred copies of the two-volume report were printed, and a further 1000 copies of the summary. The full report was distributed publicly to government bodies, libraries, institutions, and technical bodies, while the condensed version was made available to Members of Parliaments of the Commonwealth, New South Wales, Victoria, South Australia, and the press.[145] The RMC itself considered it a "pre-requisite for the planning of any remedial measures that might prove to be justified," although it was doubtful how far it was legally responsible for maintaining water quality in the Murray.[146]

The "Gutteridge Report," one of the most important legacies of the drought, proved influential in the long run, being the reference point of all further major studies about salinity problems in Victoria in the following two decades. Triggering a further substantial increase in the GMID's salinization problems was, however, something entirely different: flood.

Too Much Water: The Wet Years of the Early 1970s

The 90th annual Kerang and District Agricultural Society Show was opened by the Victorian Attorney-General, Vernon Wilcox, on October 9, 1974. It was an excellent late afternoon. The distinguished guest was accompanied by Alan Wood, Legislative Assembly member for Swan

[144] Gutteridge, Haskins and Davey, *Murray Valley Salinity Investigation—Summary*, 48.

[145] Sunraysia Salinity Committee to Minister for National Development, "Salinity Investigation," July 25, 1970, A1690, A1969/742, National Archives of Australia, Canberra; Sunraysia Districts Co-operative Society Limited to Minister for National Development, "Re: Murray Valley Salinity Report," May 15, 1970, A1690, A1969/742, National Archives of Australia, Canberra; Victorian Irrigation Research and Promotion Organisation to Minister for National Development, "Re: River Murray Salinity Report," August 10, 1970, A1690, A1969/742, National Archives of Australia, Canberra; River Murray Commission to John Gorton, "River Murray Salinity Investigation," August 18, 1970, A1690, A1969/742, National Archives of Australia, Canberra, 2–3.

[146] River Murray Commission to John Gorton, "River Murray Salinity Investigation," 3.

Hill, and welcomed by community leaders such as the society's president Keith Chamberlain and the Shire President, Councilor Brian Drummond. Amidst a seemingly ubiquitous talk about the rural crisis, cuts to subsidies for primary industries, angry accusations of the Federal Labor government being hostile toward rural Australia, and worries about inflation, Mr. Wilcox brought optimism and encouraging words with him as he addressed the audience: "I encourage you, the rural community, not to be downhearted with all the troubles facing us. The common sense of the Australian people will lift the rural community into sound conditions again." And concluding: "We have to look at the problem from a long term point of view. We have got to be optimistic and confident."[147]

Confidence was in high demand during this time. In 1974, the industrial economies entered a severe recession, the first since the end of the war. Inflation soared, economic growth declined, and unemployment increased. The first oil crisis triggered measures to reduce economic growth to negative levels. "Stagflation," a novel combination of stagnant economies and rampant price inflation, appeared on the scene. In 1975, the volume of world trade declined for the first time since 1945.[148] In Australia, the Long Boom ended at the same time as it did in other developed economies. The increase in protectionism in the 1970s and 1980s, *inter alia* in international trade in foodstuffs, made it harder for Australian exporters to market their goods. By 1974–1975 the economy was in stagflation with over 16% inflation and unemployment at 4.6%. Between this time and 1982–1983, real economic growth averaged at 2% per annum with an inflation rate of 12%.[149]

Furthermore, the terms of trade for Australian products had changed drastically following the UK's decision to join the European Economic Community (EEC). By 1972, Japan had replaced Britain and Europe as the principal market for Australia's wool and wheat, and also a substantial proportion of its sugar exports. The U.S. had replaced the UK as the principal importer of beef and sugar. Thus, when Britain finally joined the

[147] Anonymous, "Optimism in Rural Outlook," *The Northern Times*, October 10, 1974.

[148] Dyster and Meredith, *Australia in the Global Economy*, 232–235.

[149] Dyster and Meredith, *Australia in the Global Economy*, 239; 247–53.

EEC in 1973, the effect was at first viewed as not great because Australia had found other markets.[150]

However, it was much more difficult to find alternative markets for fruit, butter, and rice, which were a significant proportion of the products of irrigated agriculture. In 1972–1973, the UK imported over 50% of Australia's butter. In 1973–1974 it imported none. The market proved difficult to replace, as the U.S. supplied its own butter, Japan consumed very little, and the EEC was a net exporter. Domestic and export markets for the product were further reduced as it was partially replaced by margarine. Attempts were made to overcome this by switching to cheese and dried milk. As a result, smaller farmers left the industry in the late 1960s and early 1970s. Although the average farm size remained constant, the numbers of cows milked increased substantially, as did the cows' productivity.[151]

Many horticultural industries also declined. Between 1973–1974 and 1979–1980, the production of canning fruits decreased by over 20%. Apple production decreased by 16%, due to a large extent of farmers leaving the industry. Special subsidies were offered to change from canning and dried fruits to vines and other products, while commodities less dependent on export, such as citrus fruits, bananas, and pineapples, continued to increase in production.[152]

In the same period, the intricate system of protection for Australian agriculture was dismantled. Previously, farmers and graziers had enjoyed a privileged position compared with manufacturers, as they could directly approach the Minister for Primary Industry for financial aid. In 1973, the new Federal Labor government introduced an Industry Assistance Commission to which all industries had to apply, and assistance was only granted after a public inquiry. Additionally, most concessions to farmers and graziers were recalled and not restored after Gough Whitlam's dismissal in 1975.[153] In this time of uncertainty for Australian farming communities, Vernon Wilcox's message is likely to have found

[150] Davidson, *European Farming in Australia*, 388.

[151] Davidson, *European Farming in Australia*, 389; 394–95; 404–05; Linn, *Battling the Land*, 159.

[152] Davidson, *European Farming in Australia*, 371; 412–13.

[153] Davidson, *European Farming in Australia*, 396; Linda Botterill, "From Black Jack McEwen to the Cairns Group. Reform in Australian Agricultural Policy" (National Europe Centre Paper No. 86, Australian National University, 2003), 11–14.

a thankful reception. Another of his statements may have been slightly tongue-in-cheek, although it was perceptive: "Looking at some of your farm produce here today I see that it is of a high quality, but you have one problem here. When I flew in saw you had an abundance of one commodity—water."[154]

It was not just the water supplied by the SRWSC, either. In fact, during the season of 1973–1974 irrigation water usage had been the lowest since the post-war explosion in development, safe for two systems. It was rain water that was still lying around throughout the region—on farmland, in swamps and depressions—waiting to be pumped out.[155] The MDB is "Flood Country," but the early 1970s were exceptionally wet years for Northern Victoria.[156] While Victoria went through a drought in winter, spring, and early summer of 1972–1973, leading to low stream flows, high salinities, and unprecedented demands for irrigation water. February of 1973 brought heavy rains which continued through autumn.[157] In August, the Avoca, Loddon, Campaspe, Goulburn, and Broken Rivers went into prolonged flooding. Pyramid Creek, the Little Murray, the Lower Loddon, and the Kerang Lakes were in flood, and the urban areas of Kerang were threatened by the rising waters after record rainfalls.[158] In Swan Hill, too, 1973's total rainfall was the highest since the beginning of records in 1869.[159] Cadell Street, named after "Captain Cadell of river-boat fame," turned into a waterway after a weekend of heavy July rain. Several low-lying areas were flooded for two days in the city. On August 4, the Little Murray narrowly topped its peak level on Pental Island which had been established during the 1956 floods. Volunteers and unemployed relief workers were at work strengthening and building up levees near Fish

[154] Anonymous, "Optimism in Rural Outlook."

[155] Gynlais O. Jones, "Gun Shots—Some Warm Days," *The Northern Times*, November 19, 1974; State Rivers and Water Supply Commission, *68th Annual Report, 1973–74* (Melbourne: Government Printer, 1974), 15.

[156] O'Gorman, *Flood Country*.

[157] State Rivers and Water Supply Commission, *67th Annual Report 1972–73* (Melbourne: Government Printer, 1974), 10–11.

[158] Anonymous, "Record Wet," *The Northern Times*, January 11, 1974; State Rivers and Water Supply Commission, *68th Annual Report, 1973–74*, 14.

[159] Anonymous, "Rainfall Figures Lag on Last Year," *The Guardian (Swan Hill)*, August 30, 1974.

Point using thousands of sandbags. In the end, the fortifications held back the tide.[160]

Another period of floods followed in May 1974, which was particularly severe in the Ovens and Goulburn Rivers. Urban residential areas in Shepparton, Mooroopna, Wangaratta, Seymour, and Nathalia suffered major flooding. The Goulburn levees were breached, leaving vast areas of flood plains vulnerable.[161] As in Mooroopna, streets were flooded in the Kerang region as early as January 1974 when rain fell almost non-stop for 24 h in most centers. The Murray rose about 1.5 m over a single weekend. Children in Cohuna saw the lighter side and switched to canoes, but wheat farmers were badly affected because they were held up during the harvest, and some lost crops because rain and wind flattened and bleached the wheat.[162] In May, Kerang became a town under siege, with the Loddon River and Pyramid Creek both rising to critical levels. Hundreds of heads of stock were moved from properties adjacent to these streams, and farmers were warned to prepare for evacuation. Sandbag levees held back the swollen rivers from the town, but many farmers experienced widespread flooding of their properties.[163]

The floods of 1973–1974 contributed to the existing salinity problems of the GMID, but in a different way than the drought of 1967–1968 had done. The heavy rains promoted excessive groundwater recharge, in turn raising water tables throughout the district. In areas without drainage or proper land layout, the water remained for a long time even after the rains had stopped, which reduced pasture production through waterlogging. To make things worse, it seeped downward, further adding to groundwater recharge.[164] Hence, a poem from the Tragowel Plains

[160] Anonymous, "Northern Waterway," *The Guardian (Swan Hill)*, July 20, 1973; Anonymous, "Floods Top 1956 Peak," *The Guardian (Swan Hill)*, August 13, 1973; Anonymous, "Rising Rivers," *The Guardian (Swan Hill)*, August 3, 1973.

[161] State Rivers and Water Supply Commission, *68th Annual Report, 1973–74*, 15.

[162] Anonymous, "Rain, Rain…," *The Northern Times*, January 15, 1974.

[163] Martin Summons, *Water: The Vital Element. 150 Years of Shepparton's Growth* (Shepparton: Shepparton Heritage Centre, 2010), 124–25; Anonymous, "Town Besieged by Floods," *The Northern Times*, May 24, 1975; Anonymous, "Town Winning Flood Battle," *The Northern Times*, May 28, 1974.

[164] Anonymous, "Flood Danger for 6 Months," *The Northern Times*, May 2, 1974; Anonymous, "Waterlogging Drops Pasture Production," *The Guardian (Swan Hill)*, July 25, 1973; State Rivers and Water Supply Commission, *68th Annual Report, 1973–74*, 6.

goes: "The rains on the plains/Without any drains/Will lay and stay/And not get away."[165]

Gyn Jones, asking his *Northern Times* readers whether they still had water lying around on their properties, did not miss the opportunity to point out that the rains illustrated the merits of land forming that promoted run-off.[166] In Swan Hill, irrigation research officer Ken Sampson noted that paddocks that were well laid out were draining between each rain, while flat, poorly drained land was still covered by sheets of water after the rains had passed. With the onset of warmer weather, he predicted, evaporation would remove the water but leave behind salt on the surface of much low-lying farmland close to the water table.[167] When the Minister of Water Supply, "Jock" Granter, visited the district in late August 1974, he saw levee banks that were still breached and many affected farms still flooded. Representatives from the Swan Hill Irrigators Research Farm used the occasion to take the minister to an inspection of the uncompleted drainage sections of the Flats and make their case for a full drainage scheme.[168] In 1975, it showed that their problems had advanced further. All parts of the Kerang region for which records were available had shallow water tables and it was assumed that this was the case for almost 75% of the whole region.[169]

Drainage, or lack thereof, was an increasingly pressing issue in Kerang. Surface drains then served one-third of the region. Drainage experiments by the Commonwealth Scientific and Industrial Research Organisation (CSIRO) and the Kerang Agricultural Research Farm had been made in the early 1960s. However, as the water inevitably drained into the Murray, a ban had been imposed on additional groundwater pumps in 1968. Except for a few experimental bores, no further progress had been made.

[165] Anonymous, "They Looked at Progress," *The Northern Times*, July 23, 1974.

[166] Gynlais O. Jones, "Gun Shots—Big Rains," *The Northern Times*, January 1, 1974.

[167] Anonymous, "Waterlogging Drops Pasture Production," *The Guardian (Swan Hill)*, July 25, 1973; "Salt Damage Risk High," *The Guardian (Swan Hill)*, August 29, 1973.

[168] Anonymous, "Minister Surveys Flood Problems," *The Guardian (Swan Hill)*, August 26, 1973.

[169] State Rivers and Water Supply Commission, *Salinity Control and Drainage. A Strategy for Northern Victorian Irrigation and River Murray Quality* (Melbourne: State Rivers and Water Supply Commission, 1975), 47–48; 177.

Tile drainage in the region was virtually confined to the horticultural areas of Woorinen and Tresco near Swan Hill.[170]

Drainage, however, could also become a reason for aggravation. As in 1968, the SRWSC was severely criticized for the modifications it had made to its engineered river system. Again, the question was who would benefit from salinity control works and who would bear its negative consequences. When a Parliamentary Public Works Committee was assembled to inquire into the northern floods of 1973–1974, representatives of the Victorian Field and Game Association (VFGA) argued that the Commission's drainage and water conservation works severely affected the ecology of the Kerang region. The flooding of 1973, they claimed, was greatly aggravated by public and private drainage works, which had resulted in water running off more quickly and causing more widespread flooding downstream. In some cases, they argued, drainage works had actually worked in reverse and caused water to flow up the drainage channels and flood adjacent areas. The effects of irrigation supply and salinity control work on changing flood characteristics of streams and the ability of an area to cope with the floods had not been given proper consideration. Their conclusion was that salinity control works should not be allowed to further encroach on wetlands and reduce the area of natural pondage available. However, this was exactly what the "Gutteridge Report" recommended be done.[171]

Like in an echo to the concerns voiced by the Mystic Park and Fish Point farmers six years prior, the VFGA claimed that the Barr Creek-Tutchewop scheme had caused the "greatest flooding ever known" around the Lakes area because it had cut off the course of the Avoca River's outfall into the Murray, leaving nowhere for the Kerang Lakes' water to go when the Avoca's water was diverted onto the floodplain.[172] Another modified stream was also attracting attention: Pyramid Creek, which was used as an irrigation water carrier, running for 60 km between Kow Swamp and the Kerang Weir to serve the Mystic Park Irrigation District. From 1967 to 1970 the creek underwent major restructuring

[170] State Rivers and Water Supply Commission, *Salinity Control and Drainage*, 4; 37–38; 41–43.

[171] Anonymous, "Drainage Works Affect Our Ecology," *The Northern Times*, May 10, 1974.

[172] Anonymous, "Drainage Works Affect Our Ecology."

to speed up the flow of water. Along with Box Creek, it was deepened considerably. Reminiscent of the fate of Barr Creek, the depth of the dredging went into sand layers underneath the creek, allowing saline water from the underground water table to enter its waters, making it a significant contributor of salt. The problem was not officially identified until the late 1980s, although local opinion had been suspicious much earlier. Drainage water that was being discharged into the creek added to this.[173] The VFGA, in the same inquiry, claimed that during lower flows in off irrigation periods, the salinity level of the creek was adversely affected by seepage of groundwater. A group of Pyramid Creek farmers demanded that the creek's salinity readings be published weekly because they felt that its salinity levels had increased considerably in the years prior. Both parties contended that under wet conditions the risk of flooding had been located from inhabitants upstream to those downstream. As a result of the dredging, water was now running down the creek faster than before and its levels rose so high that salt drains in the lower reaches flowed in reverse and flooded adjacent land. Farmers said they were unable to irrigate because their pump sites were underwater.[174] Again, the question of responsibility was on the table: "the creek is no longer a natural stream, but an irrigation channel, and a drainage channel reconstructed by and controlled by the SR&WSC."[175]

From a political point of view, it was crucial that much of the flooding of 1973–1974 happened in the Shepparton region. Record rains caused severe waterlogging. About 300,000 peach trees were destroyed. Over 120 orchards were affected and exhibited symptoms of salinization on a scale and intensity never experienced before. There was also an increasing number of reports about salinization on pastural properties. There had been troubles with orchards and pastures in and around natural depressions in the past. Earlier horticultural developments around Rochester and Tongala had been almost completely abandoned due to salinization. However, with the floods the issue assumed a new quality. It was estimated that 46% of the irrigated area now suffered shallow water tables.

[173] Hardwick, "A Historical Report," 13–15; Murray-Darling Basin Authority, *Pyramid Creek Salt Interception Scheme* (Canberra: Murray-Darling Basin Authority, 2011), 1; Anonymous, "Pyramid Creek Flood Angers," *The Northern Times*, February 2, 1974.

[174] Anonymous, "Drainage Works Affect Our Ecology;" "Pyramid Creek Flood Angers," *The Northern Times*, February 2, 1974.

[175] Anonymous, "Pyramid Creek Flood Angers."

In intensively irrigated areas, they were approaching the surface, heralding that the problems would intensify in the future. Shepparton, for the first time, had serious salinity and water table problems.[176]

The fact that the optimistic outlook for the Shepparton region, which the Consultants of Gutteridge, Haskins and Davey had presented in 1970, had been radically questioned due to the floods became relevant in two ways. Firstly, it caused the SRWSC and the Victorian government to focus their attention on salinity in the Shepparton region. It was a highly productive area, boasting large fruit and dairy industries. The former accounted for 20%, the latter for 11% of Australian production. It also produced significant amounts of lamb, pork, and vegetables. For comparison, the Kerang region produced a wide range of commodities, such as beef, lamb, mutton, pork, wool, and cereals, but its relevance was markedly smaller, claiming only 2.5% of Australian dairy and 3.9% in wine production.[177]

In response to the state's valuable horticultural industries under threat of waterlogging and salinization problems, the government granted a special allocation of $630,000 to the SRWSC to accelerate the extension of surface drainage works, initiate the installation of a grid of groundwater control pumps to serve the affected areas. Another $150,000 was granted to finance the operation of private groundwater bores to supplement this program.[178] Furthermore, in May 1975 in the aftermath of the floods, when the Commission tabled a tenyear strategy in entitled "Salinity Control and Drainage," its focus was to solve Shepparton's problems.[179]

Salinity—or, more broadly, the water quality of the River Murray—continued to be a major political issue, especially in South Australia. This led to the establishment of the interjurisdictional River Murray Working Party. It was the first in a series of attempts to revise the River Murray Waters Agreement of 1914 to take account of water quality issues. This process eventually culminated in the major institutional changes of the mid-1980s. Initiated by Don Dunstan, Labor Premier of South

[176] State Rivers and Water Supply Commission, *Salinity Control and Drainage*, 45–47; 51; 81–82; State Rivers and Water Supply Commission, *68th Annual Report*, 15.

[177] State Rivers and Water Supply Commission, *Salinity Control and Drainage*, 23–24.

[178] State Rivers and Water Supply Commission, *70th Annual Report, 1974–75* (Melbourne: Government Printer, 1975), 6.

[179] Russ, *The Salt Traders*, 151.

Australia, and Gough Whitlam, the new Labor Federal Prime Minister, the Working Party had been commissioned in March 1973. It was to report on short and long-term solutions to salinity problems in the Murray, taking into consideration the effects of river management on flora and fauna along the river. Its final report was untimely delivered in October 1975 amidst the political crisis that brought down the Whitlam government. Although influential in the long term, it had no immediate consequences beyond the agreement to commission further research. The Working Party had recommended that the maintenance of water quality become an overall objective for the RMC. Salinization was identified as the main form of pollution and high nutrient concentration as a potential problem. Furthermore, the report had called to specify standards as soon as practicable for each reach of the Murray that would reflect an acceptable balance of economic, social, and environmental costs and benefits. Victoria's submission to the Working Party was the SRWSC's ten-year strategy—its "40 Million Dollar Plan."[180]

This was the sum deemed necessary for salinity control and drainage in the Northern Victorian irrigation districts. In current values, this would have amounted to more than $1 billion! It was a recognition that action on a larger scale was necessary to protect industry and water quality, and of Victoria's interstate obligations as a user of the River Murray.[181] The SRWSC proposed an enlargement of the Woorinnen Lakes evaporation basins and an upgrade of the Barr Creek–Tutchewop scheme by minor works in order to intercept highly saline groundwater seepage to the Murray in the Mildura–Merbein area, and more importantly, by extending the scheme to include Lake Tyrrell, as Gutteridge, Haskins and Davey had recommended. Their findings, which were based on environmental factors and cost–benefit analyses that favored Shepparton, had foreshadowed that the region would have priority in the proposed program. Another crucial fact for the SRWSC was that the Shepparton region had significantly lower groundwater salinity levels than the Kerang region—an average of 1900 versus 30,000 parts per million, which permitted the re-use of

[180] Connell, *Water Politics*, 107–108; River Murray Working Party, *Report to Steering Committee of Ministers. October 1975* (Canberra: Australian Government Publication Service, 1977), 3; 5.

[181] State Rivers and Water Supply Commission, *Salinity Control and Drainage*, 1; Anonymous, "Spend $40 m. or Suffer," *The Northern Times*, August 1, 1975. Current value calculated using Hutchinson and Ploeckl, "MeasuringWorth."

drainage water. And indeed, for Shepparton, the Commission's strategy did plan for an extension of surface drainage and the installation of a comprehensive grid of groundwater bores. Disposal of saline water would be facilitated through re-use schemes. Surpluses would be channeled into evaporation basins and the Murray. Water quality would be held within acceptable limits up to Swan Hill, and below that point the Lake Tyrrell Scheme and salt interception works would take effect. These works were deemed economically justifiable.[182]

On the contrary, given the realization "that we will not be able to eliminate entirely the saline environment," as commissioner David Constable admitted, the plans for the Kerang region were largely agricultural.[183] The Strategy did not include engineering solutions like the completion of the existing surface drainage scheme, which only served one-third of the region, because it was not considered economically justifiable with respect to the agricultural benefits to be gained. Kerang would, however, receive the bulk of the envisaged aid to augment extension services and financial assistance for better farm management. The Lake Tyrrell scheme would have surplus evaporation capacity and natural drainage lines would be gradually reinstated. Meanwhile, all areas of the region for which records were available had shallow water tables. Salinization was "obviously well advanced [...], and only complete control of water table levels together with leaching would reverse it on a regional scale."[184]

The authors of the Strategy were aware that continuing salinization was a potential cause for grave social disturbance. If the proposed program was not realized, the SRWSC argued, "irrigation development will survive but at a level of agricultural production in the Kerang and Shepparton regions well below present levels." However, the consequences would go far beyond a mere decline in agricultural production. They would include a disruption of economic and social structures and the need to retrain

[182] State Rivers and Water Supply Commission, *Salinity Control and Drainage*, 1–6; Anonymous, "Salt Answers on the Farm?" *The Northern Times*, September 21, 1976.

[183] David Constable, "We Need Understanding and Goodwill," *The Riverlander* 31 (November 1977): 10.

[184] State Rivers and Water Supply Commission, *Salinity Control and Drainage*, 4; 47–48.

and resettle displaced farmers. Further salinization would mean abandoning a significant portion of a limited natural resource—arable land—to a virtually irreversible process of degradation.[185]

Reactions to the Commission's proposals quickly followed. After the protests, the Parliamentary Public Works Committee's inquiry into the strategy widened its Terms of Reference to also include concerns of Kerang and dryland farmers, while before only solutions for Shepparton were to be investigated. The inquiry was determined to be contentious.[186]

The second reason why the Shepparton region's newly acquired water table troubles became politically relevant, was that the wet years precipitated the formation of the first Salinity Action Group, a new mode of community organization which soon spread through the GMID—the Goulburn Irrigation Region Drainage Action Committee (GIRDAC). The committee was founded at a meeting of representatives from shires, fruit grower and dairy farmer organizations, and the fruit and dairying industry in Mooroopna on June 13, 1974 after "months of water-logging, salinity build-up and widespread local storm and rain" with consequent "flooding […] of the cities, towns, orchards, crops and pasture areas of Northern Victoria."[187]

GIRDAC's first aim was to alleviate the high water tables under the Goulburn region's orchards by groundwater pumping and drainage installations. The group wanted a comprehensive drainage system and a salinity control program for the Shepparton region. Further, it sought to promote the implementation of the recommendations in the "Gutteridge Report," obtain financial support for this task, and to hasten the then running inquiry of the River Murray Working Party in order to enable the RMC to interstate coordination in order to maintain water quality.[188]

Similar self-organization efforts were underway in the Kerang region but had not yet come to fruition. At a meeting in November 1974 called

[185] Anonymous, "Murray Control—Action by MVDL," *The Northern Times*, August 12, 1975; State Rivers and Water Supply Commission, *Salinity Control and Drainage*, 9–10.

[186] Russ, *The Salt Traders*, 150.

[187] Ernest M. Jackson to Minister of Water Supply, June 3, 1974, VPRS 6008/P0006, File No. 74/20650/Pt.0001, Public Record Office Victoria, Melbourne.

[188] D.J. Constable, "Note for File," June 19, 1974, VPRS 6008/P0006, File No. 74/20650/Pt.0001, Public Record Office Victoria, Melbourne.

by the local branch of the Victorian Farmers Union, which followed earlier attempts at gathering community support to formulate a policy toward the control of water tables and salinity, only 13 people attended. A disappointed Gyn Jones speculated that the average farmer may not have a big problem, however, unless the community in general realized that there was one, an integrated overall policy could not be formulated.[189] Three subsequent meetings on the subject were also virtually flops due to the lack of support from the community, even though it was managed to assemble a five-person pilot group to discuss a name for an action committee, possible aims and membership, and methods of publicizing efforts and research.[190] Were the majority of Kerang farmers in apathy, preferring to farm on hoping that no one would notice the problem? Or was the lack of public attention a result of the gradual nature of salinization itself?[191] Kerang was struggling to accept a long-standing slow catastrophe, while Shepparton had received a sudden cold shower. These attempts to launch a similar committee in the Kerang region were not successful until late February of 1976, almost two years after GIRDAC's inception the Kerang Irrigation Region Salinity Action Committee (KIRSAC) was founded.[192] The idea spread. By the late 1970s, there were about 30 similar organizations in Victoria.[193]

In the early 1970s, the quest for environmental security, which had intensified with the post-war boom in irrigation development, had produced its own dialectic. More water was being used by irrigators than ever before. Although efficiency was slowly improving, there were still massive amounts being wasted. With the absence of drains, the floods had escalated the situation. Salinization, a result of "Too Much Water," had become a major issue which received widespread attention. For the

[189] Anonymous, "More Drainage Moves by VFU," *The Northern Times*, August 20, 1974; Anonymous, "Who Cares about Drainage Here?" *The Northern Times*, November 11, 1974.

[190] Anonymous, "Our Survival Depends on Drainage Scheme," *The Northern Times*, February 17, 1976; Anonymous, "Salinity Apathy," *The Northern Times*, December 12, 1974.

[191] Anonymous, "Salinity Problem," *The Northern Times*, September 3, 1974.

[192] Anonymous, "Community Call to Fight the Salinity," *The Northern Times*, March 2, 1976.

[193] ACIL Australia et al., *Causes, Extent and Effects of Salinity in Victoria* (Melbourne: Salinity Committee of the Victorian Parliament, 1983), 153.

first time, it was not only the Kerang region that was affected but the economically valuable Shepparton region. In the irrigation communities, agitation reached new heights and led to the formation of organizations to alleviate the situation. The following years would see an intensification of the conflicts entailing salinization. The future of GMID would be marked by environmental *in*security.

CHAPTER 5

A Time of Crisis (1976–1982)

The Complexity of Salinity

Kerang Memorial Hall, July 13, 1976. The witness picked drastic words to make his point: "If salinity continues its rampage through the irrigation area in this district, there will be no future for industrial and commercial development in Kerang, and the Shire's main centres because there will be no future full stop!"[1] The "salinity menace," which combined with the prevailing rural recession, had to be contained, because otherwise it would destroy a significant area of irrigation country in the Kerang region. Solutions were available, and large-scale control work had to be carried out for the benefit and survival of the area's rural economy.[2] The SRWSC's proposed $40 million strategy was nothing less than "life-saving help."[3]

David Wallace, a local farmer, painted a dire picture to make his case on behalf of the Kerang Development Committee before the Parliamentary Public Works Committee, as the touring Salinity Inquiry came to town for the first time in the winter of 1976. The inquiry not only brought

[1] Parliamentary Public Works Committee at Kerang, "David Wallace," July 13, 1976, VPRS 241/P0000, pp. 64–131, Public Record Office Victoria, Melbourne, 76.

[2] Parliamentary Public Works Committee at Kerang, "Wallace," 76–78.

[3] Parliamentary Public Works Committee at Kerang, "Wallace," 78.

© The Author(s), under exclusive license to Springer Nature Switzerland AG 2023

D. Rothenburg, *Irrigation, Salinity, and Rural Communities in Australia's Murray-Darling Basin, 1945–2020*,
Palgrave Studies in World Environmental History,
https://doi.org/10.1007/978-3-031-18451-2_5

together various government experts from different disciplines to share their knowledge, it was also "supposed to defuse a growing political issue for the government and to calm community unrest."[4]

A large number of community views were obtained as many interested parties took the chance to voice their concerns during the inquiry's multiple rounds through Northern Victoria, with further stops at Mildura, Irymple, Swan Hill, and Shepparton. These testimonies offer insights into the consequences of salinization during a peak period of the slow catastrophe that had been unfolding for decades. They focus the worries and problems of local people in Northern Victoria, highlight their interpretations of salinization, and vividly illustrate the conflicts between individuals and communities that resulted from it. They provide a glimpse of what was underneath the numbers in the highly aggregated diagnoses by technical consultants and agricultural engineers. With this in mind, Ian Meacham, divisional engineer at the SRWSC, called on irrigators in Kerang to voice what their problems were and what would happen if they were not solved: "You can point out community and social problems—the type of thing we could not present in terms of facts and figures."[5]

The Northern Times, too, had done its best to highlight the importance of the inquiry for the people of Kerang with the unambiguous warning that "[d]isaster looms from the Goulburn to Mildura." The "enemy" was "SALT." Readers were urged that "[t]his concerns you—the man on the land, the people in the town and the country, the family, the schools, the employer, the worker. All are threatened with the salt problem." The article implored them to make the inquiry "a mammoth meeting" and show support for their representatives.[6] Jones, on the same day, published an appeal to "hurry up and get down to the Memorial Hall [...] and show this Committee just how concerned you are."[7]

Wallace certainly was concerned. He was eager to illustrate the social and economic effects of salinization in his region to the members of the committee. According to him, it was adding to the hardships brought about by the rural crisis, even though he considered salinization to be the bigger problem in the long run. He argued that the whole area's economy

[4] Russ, *The Salt Traders*, 150.

[5] Anonymous, "Salt Answers on the Farm?," *The Northern Times*, September 21, 1976.

[6] Anonymous, "Your Future is at Stake!," *The Northern Times*, July 9, 1976.

[7] Anonymous, "Gun Shots—To Sow or Not," *The Northern Times*, July 13, 1976.

was in decline, with unemployment having risen from 1.44% in 1971 to 10.13% in 1976—double the national unemployment figure of 4.8% at the time. Wallace added that the population in the borough and shire of Kerang had dropped from 9428 to 8750, by 7.1% or 678 persons, from 1966 to 1974. Worse, a survey of the Kerang Technical High School indicated that, of 208 graduates, 52% left the district in 1975, while of those who remained, 32% were unemployed. Salinization, therefore, was a major obstacle impeding the development of Kerang, at present a service center for the rural community around it, to also become a center of secondary industry.[8] If it was "not controlled and therefore allowed to spread, [...], the whole future economic development in Kerang and district will be bleak."[9]

The newly founded community organization, KIRSAC, whose first objective had been to assemble a submission to the Salinity Inquiry, also proved to be a vocal advocate of Kerang's case.[10] However, Gyn Jones and Des Thomas were less focused on the potential development of the region that was being impeded by salinization. Instead, their strategy was to convince the government and the Water Commission of the importance of protecting the Kerang region. They argued that an "asset" of public and private investment that had been created principally by the development of irrigation systems, along with the social structure of 120,000 people in the rural community, was at stake.[11] With the projected loss of 50% of irrigated farmland, the result would be a "calamity."[12] Thousands of people, they predicted, would be displaced

[8] Parliamentary Public Works Committee at Kerang, "Wallace," 75–77. Figures from the Australian Bureau of Statistics vary slightly but do indicate the same trend: In 1966 the Borough and Shire of Kerang had a population of 9426 which had dropped to 8720. Arnold, *Victorian Year Book 1968*, 122; N. Bowden, *Victorian Year Book 1978* (Melbourne: Commonwealth Bureau of Census and Statistics, Victorian Office, 1978), 198. According to the Australian Year Book for 1975–1976, in May of 1976, 4.1% of the total labor force were unemployed. R.J. Cameron, *Official Year Book of Australia. No. 61, 1975 and 1976* (Canberra: Australian Bureau of Statistics, 1976), 701.

[9] Parliamentary Public Works Committee at Kerang, "Wallace," 77.

[10] Anonymous, "Be Counted for the KIRSAC," *The Northern Times*, June 8, 1976.

[11] Parliamentary Public Works Committee at Kerang, "Gynlais Oughton Jones," July 13, 1976, VPRS 241/P0000, pp. 64–131, Public Record Office Victoria, Melbourne, 82–83.

[12] Curiously, even though the witnesses claimed that this figure was a quote from the SRWSC's Strategy, it is nowhere to be found in the document.

through the loss of farming and employment in commerce and industry, resulting in a "social crisis of tremendous magnitude."[13] They cited communities such as Cohuna and Leitchville, which were highly dependent on the dairying industry with Murray Goulburn and Kraft as major employers. Further loss of population would endanger local government employment opportunities in Cohuna, East Loddon, Gordon, and the Kerang Borough. The homeowners and primary producers who would remain in the area would be stranded in a ghost environment, unable to re-establish themselves elsewhere. The investment in existing recreational and urban facilities, such as sporting complexes and tourist resorts, which would become useless, would be "astronomical." Lastly, the committee was asked to consider the cost of relocating thousands of displaced people and re-establishing public facilities like electricity, telephone, and hospital beds, which would have to be borne by the public.[14] Like David Wallace, Jones and Thomas contended that "the flow-on effects" of Kerang's decline would "hit Victoria's and Australia's rural economy."[15]

Some witnesses from Pyramid Hill offered graphic insights into the consequences of salinization in their district. Donald Andrew Leed's was particularly vivid: "[O]ur pastures as we knew them 15 years ago are very isolated. Whole clumps of trees have died, or are badly affected. [...] Our roads suffer to such a state that our vehicles depreciate at a higher rate than normal. Retailers do not like to mention our district from which they come. Our heritage is being corroded away, the very foundations of our homes are being eaten away by the salt, not to mention our bank balances and the problems connected with survival. [...] [N]eighboring townspeople refer to Pyramid Hill District as the 'desert.'".[16]

Sharman Stone, a sociologist born in Pyramid Hill, also drew attention to the social hardships brought about by continuing salinization in her hometown and the Shire of Gordon.[17] She contended that "[m]ost of the rural population will have to adapt their life expectations to city living"

[13] Parliamentary Public Works Committee at Kerang, "Jones," 84.

[14] Parliamentary Public Works Committee at Kerang, "Jones," 82–85.

[15] Parliamentary Public Works Committee at Kerang, "Wallace," 77.

[16] Parliamentary Public Works Committee at Kerang, "Donald Andrew Leed (1)," July 13, 1976, VPRS 241/P0000, pp. 64–131, Public Record Office Victoria, Melbourne, 114.

[17] "Stone, Sharman Nancy, (the Hon.), (Dr) (1951–)," National Library of Australia, accessed August 16, 2022, https://nla.gov.au/nla.party-727075.

whereas presently, only school leavers were forced to do so. Not only displaced farmers would require relocating, unless reclamation works were soon financed. Businessmen, professionals, and workers in rural settlements would also have to find other employment.[18] The communities in the salt-affected two-thirds of the shire were "at a critical stage in their struggle for continued existence" given the combination of falling product prices and the land's declining productivity, forcing the "farmers' sons to seek employment off the family property. A young man with a family cannot afford to work harder each year for ever diminishing returns,"[19] while, older farmers might stay in the district on reduced income. Hence, these communities experienced a process of depopulation, causing local churches, schools, sporting associations, and businesses to fail or reduce their activities. Yet, young and inexperienced farmers bought salt-affected properties, attracted by the land price and the promise of irrigation. Many farms became absentee-owned. The shire's social fabric was dissolving.[20]

The witnesses' contributions to the case for the Shepparton region reflected that their positions were significantly different from those in Kerang, as salinization had only "reared its head all of a sudden" during

[18] Parliamentary Public Works Committee at Kerang at Melbourne, "Sharman Nancy Stone," July 4, 1978, VPRS 11,559/P0001, 24.07.27, PT4, Public Record Office Victoria, Melbourne, 814.

[19] Parliamentary Public Works Committee at Kerang at Melbourne, "Stone," 815.

[20] Parliamentary Public Works Committee at Kerang at Melbourne, "Stone," 815–16. Sharman Stone graduated from La Trobe University (Melbourne) in 1978 with an M.A. in Rural Sociology. In her thesis, she wrote about rural communities in the Gordon Shire. She later worked for the Rural Water Corporation (the remodeled SRWSC), the Victorian Farmers Federation, and the Victorian Department of Agriculture. She became a Member of Parliament for Murray and retired from politics after 20 years in 2016. "Graduate Profiles. Dr Sharman Stone MP, Master of Arts 1978," La Trobe University Melbourne, October 19, 2004, accessed August 16, 2022, https://web.archive.org/web/20060920215533/www.latrobe.edu.au/alumni/profiles/stone.html (archived website); "Sharman Stone: Liberal MP for Murray Retires from Politics After 20 Years," *ABC News*, March 26, 2016, last modified March 26, 2016, https://www.abc.net.au/news/2016-03-26/sharman-stone-murray-victoria-standing-down-from-politics/7277610. See Sharman Nancy Stone, "Rural Communities. Facility Development, Attitudes and Self-Determination in the Gordon Shire" (Unpublished M.A. Thesis, School of Social Sciences, La Trobe University Melbourne, 1977).

to the wet years of 1973–1974.[21] Eric Merrigan of the region's Salinity Action Group, GIRDAC, made this clear when he told the committee that, to the dairy industry in the region at this point, only marginal losses due to salinization had occurred.[22] Its immediate problem was "inadequate surface drainage which can cause serious loss of production and even death of summer pastures."[23] There was no indication that the fruit industry or graziers holding beef cattle had a salinity problem, either. The witness did claim that large areas had become unsuitable for lucerne hay production, because of high water tables and salinization, which experienced serious losses in 1973–1974, as the crop required good drainage. Although, overall, high water tables were to be prevented, not lowered.[24]

This, however, did not result in a rhetoric that was less drastic when it came to the anticipated repercussions of salinization: "[C]ities and towns of the district are dependent for their survival on the agricultural production of the region."[25] The livelihood of the approximately 90,000 people in the region had to be maintained, and their social structure preserved.[26] In case of a do-nothing scenario (which was never an option for the SRWSC) Merrigan claimed that "at least 80 per cent of the Shepparton Irrigation Region will be irreversibly damaged [...] to the extent that the growth of commercial plants or trees will no longer be possible."[27]

An orchardist from the Murray Valley Irrigation Area, Ivan Roland Routley in his statement urged the committee that "[t]ime is running out, [...] the problem will certainly become worse—a disaster in the making." The threat from the sudden rise of the water tables and "subsequent salting is serious and insidious for the whole area." Despite this, during the cross-examination, he conceded that "[s]alinity as such is only just

[21] Parliamentary Public Works Committee at Kerang at Shepparton, "Eric Merrigan," July 14, 1976, VPRS 11,559/P0001, 24.07.27, PT1, Public Record Office Victoria, Melbourne, 149A.

[22] Parliamentary Public Works Committee at Kerang at Shepparton, "Merrigan," 141–42.

[23] Parliamentary Public Works Committee at Kerang at Shepparton, "Merrigan," 142.

[24] Parliamentary Public Works Committee at Kerang at Shepparton, "Merrigan," 142–46.

[25] Parliamentary Public Works Committee at Kerang at Shepparton, "Merrigan," 146.

[26] Parliamentary Public Works Committee at Kerang at Shepparton, "Merrigan," 149.

[27] Parliamentary Public Works Committee at Kerang at Shepparton, "Merrigan," 147.

appearing as a real problem."[28] A Cobram shire councilor claimed not to have been greatly perplexed or worried by salinization to date. Substantial losses of peaches in 1973–1974 had been due to high water tables when surplus water had been lying around for months rather than salinity. The irrigation area's main problem was waterlogging. Drainage was a means to preempt salinization "before it is too late to save the most productive horticultural area in the State."[29]

Meanwhile, in other areas of the Shepparton region, salinization was, apparently, well underway. Graham Walter Hacon, a grazier from Tongala, claimed that several hundred hectares of the Deakin Irrigation Area were severely affected by salt, all of which had occurred within the previous ten years, "[s]ome country being completely devoid of any vegetation whilst other parts are only sparsely covered with grasses." Much of the land had no aquifers from which to pump in order to provide dilution flows. These tracts he considered "lost to production forever as the damage caused is irreversible."[30] "On my own farm, the country which I once regarded as my best grazing land, now contains many bare and sparsely grassed areas, and with a water table a little over a metre below the surface the affected area is spreading rapidly with devastating results. Other pastures, although not showing any visible signs of salt, do not produce the abundance of growth they [formerly] did, and it is only due to frequent irrigation leaching the soil of soil that these areas continue to produce."[31] Hacon considered salting to be the "most serious problem that has ever confronted the land owner" and worried that, if action was not taken, "the next few years will see many of our farms and eventually most of Northern Victoria become a desert."[32]

[28] Parliamentary Public Works Committee at Kerang at Shepparton, "Ivan Roland Routley (2)," February 8, 1977, VPRS 11,559/P0001, 24.07.27, PT2, 279-280B, Public Record Office Victoria, Melbourne.

[29] Parliamentary Public Works Committee at Kerang at Melbourne, "Philip Andrew Pullar," September 29, 1976, VPRS 11,559/P0001, 24.07.27, PT1, Public Record Office Victoria, Melbourne, 161; Parliamentary Public Works Committee at Kerang at Melbourne, "Ivan Roland Routley (1)," September 29, 1976, VPRS 11,559/P0001, 24.07.27, PT1, Public Record Office Victoria, Melbourne, 198.

[30] Parliamentary Public Works Committee at Kerang at Shepparton, "Graham Walter Hacon," June 28, 1978, VPRS 11,559/P0001, 24.07.27, PT4, Public Record Office Victoria, Melbourne, 790.

[31] PPWC at Shepparton, "Hacon," 790.

[32] Parliamentary Public Works Committee at Kerang at Shepparton, "Hacon," 790.

Clearly, the Kerang and Shepparton regions were complex within themselves. Fundamentally, salinization was very unevenly spread throughout the regions, which was difficult for the ambitious, but abstract, diagnoses of Gutteridge, Haskins & Davey, and the Water Commission to represent. The severity of salting depended on the topography, soil type, and intensity of irrigation, accounting for many of the differences between the two regions. But other factors further complicated the issue. Many farmers watered their properties selectively, prioritizing their best land, which they watered well and controlled salinity through dilution, while they virtually abandoned other areas. As such, even profitable farms might have had symptoms of serious land salinization. Other farmers used a different approach and spread irrigation water thinly over a large area because they knew that unirrigated land rapidly became salt affected.[33] For example, much of the south of the Kerang region, including the Tragowel Plains, with low irrigation intensities were very much affected by salting. Yet, even within such areas there was highly productive, non-saline land, especially where the terrain was slightly elevated and more intensively irrigated. By contrast, the north and east sections were less affected by salt given higher irrigation intensities and more elevated land. Here, salt-affected areas were more isolated and usually existed due to channel seepage and evaporation from adjacent dryland areas.[34] In areas which had access to a drainage network, such as Cohuna, salinization could be under control despite patches of salt-affected land.[35]

Consequently, production losses—and the social consequences of salinization—were not uniform throughout the Kerang and Shepparton regions. Gyn Jones, speaking from his experience as an irrigation officer in Kerang, told the Inquiry that "[y]ou can drive along and find one place which is virtually an oasis, with a terrific production and carrying capacity. Quite often that particular farmer has got rid of the salt and is controlling it. Therefore, he obtains a great production." However, "I can take you to areas at the moment where farmers cannot pay their [water] rates. [...] [Y]ou can go through miles and miles of country where it is virtually

33 ACIL Australia et al., *Causes, Extent and Effects of Salinity*, 129; Parliamentary Public Works Committee at Kerang at Kerang, "Jones," 92.

34 ACIL Australia et al., *Causes, Extent and Effects of Salinity*, 74–75.

35 Keith Thompson, "Cohuna Salt Under Control," *The Riverlander* 32 (March/April 1978): 16.

impossible at the moment for farmers to survive, what with the burden of salinity, plus other economic problems."[36]

Then, there was the issue of silence. When, in 1978, the Minister for National Development, Kevin Newman, said that there was "now a very keen awareness throughout the Murray Valley and the population at large of the threat of salinity and drainage problems pose to the economic and social well-being of the Murray Valley," this was probably correct.[37] However, it did not mean that farmers were prepared to speak about it. Salting, the consultants of ACIL Australia and others wrote, was a much more frightening process than drought or economic depression because it was not cyclic: "It appears as an unstoppable and irreversible threat to the productive base."[38] Farmers might not be prepared to admit they had a salt problem, "for immediately they do, the value of their farm drops dramatically and the possibility of selling becomes extremely difficult."[39] Alternatively, farmers might have problems, but would not know it was salt. Their "threshold of awareness" was usually only passed when salt was already obvious on the property. When undeniable signs of salting were noticed, however, they were likely to describe salinization as the most serious concern for the future.[40] Adding to this was the farmers' lack of knowledge. Leon Heath, a representative of GIRDAC, suggested that the majority of farmers lack a fundamental understanding of basic problems associated with soils, irrigation, and salinity, and that 95% of farmers in the Shepparton region did not use water judiciously, had no contact with government departments, nor the resources to improve their irrigation layouts, install groundwater pumps, and re-use schemes.[41] Instead,

36 Parliamentary Public Works Committee at Kerang at Kerang, "Jones," 92.

37 Anonymous, "Salting, Drainage Study Is on," *The Riverlander* 32 (March/April 1978): 6.

38 ACIL Australia et al., *Causes, Extent and Effects of Salinity*, 151.

39 ACIL Australia et al., *Causes, Extent and Effects of Salinity*, 151; Parliamentary Public Works Committee at Kerang at Shepparton, "Hacon," 791.

40 Parliamentary Public Works Committee at Kerang at Shepparton, "(R.) Henry Holschier," June 28, 1978, VPRS 11,559/P0001, 24.07.27, PT4, Public Record Office Victoria, Melbourne, 781/782; ACIL Australia et al., *Causes, Extent and Effects of Salinity*, 152.

41 Parliamentary Public Works Committee at Kerang at Melbourne, "Leon G. Heath (2)," December 13, 1978, VPRS 11,559/P0001, 24.07.27, PT5, Public Record Office Victoria, Melbourne, 1059; 1061.

some farmers saw tree planting as the absolute answer to salinization, or even resorted to "doubtful cures" like "soil regeneration using bacteria, stopping the use of fertilizers, or deep ripping of the subsoil."[42]

Those who made themselves heard were organized. Whereas, as Sharman Stone argued, "the present population is largely alienated from the political processes that determine their future."[43] Of the large number of community views that were put before the Public Works Committee in Northern Victoria's rural centers, only a fraction came from private individuals. Despite the best efforts of *The Northern Times* and Gyn Jones to rally the people of the Kerang region to the inquiry, it is unlikely that its meetings were large public events that attracted broad attention. When not even 20 people attended during the third hearing in 1978, Jones expressed his frustrations about the lack of attention from his community complaining that "most people will scan the headlines with a shrug and say they weren't involved."[44]

The majority of evidence came from shire representatives, delegates from irrigators' and farmers' organizations, and members of the newly founded Salinity Action Groups. These acted as spokespersons, aggregating community interests and assembling detailed, and often complex, submissions—indeed "a formidable and time-consuming task for ordinary citizens to undertake."[45] But were they not just "squealing to get the government to do something for them" or "giving the area a bad name"?[46] Such was the complexity of salinity.

The Politics of Salinity

Salinization was excessively political. As such, it had been causing friction between communities in the past. The major line of conflict had been the disagreement between water users upstream and downstream of the Murray over who would receive the benefit and who would bear

[42] ACIL Australia et al., Causes, *Extent and Effects of Salinity*, 152.

[43] Parliamentary Public Works Committee at Kerang at Melbourne, "Stone," 816.

[44] Gynlais O. Jones, "Gun Shots—All in on Salinity," *The Northern Times*, July 4, 1978.

[45] Parliamentary Public Works Committee at Kerang at Kerang, "George Ernest Hardwick," July 27, 1978, VPRS 11,559/P0001, 24.07.27, PT4, Public Record Office Victoria, Melbourne, 474.

[46] ACIL Australia et al., *Causes, Extent and Effects of Salinity*, 153.

the consequences of salinity mitigation measures. In the late 1960s, farmers from Mystic Park and Fish Point had resisted the Barr Creek-Tutchewop scheme because they were concerned about losing flood pondage and rising groundwater levels in their region, claiming that the scheme was only worth its propaganda value downstream. Now, the conflict between the Kerang and Shepparton regions, into which the consultants of Gutteridge, Haskins & Davey had divided the Victorian side of the Riverine Plains in 1970, had become the line of demarcation. The consultants had defined the two regions based on their geological history, soil and water conditions, and economic profile. Both were, basically, level clay pans. However, Kerang, having highly saline soils and groundwater, was devoted almost entirely to pastures with only Woorinen and Tresco boasting a significant amount of fruit-growing. Shepparton, by contrast, had a higher proportion of lightly textured soils and relatively non-saline groundwater, supporting a significant amount of valuable horticulture next to its extensive dairy industry.[47]

Salinization divided the two regions politically. This was aided by the fact that both the "Gutteridge Report" and the "Salinity Control and Drainage" strategy had prioritized the Shepparton region based on its favorable environmental conditions and economic significance. Notably, the two Salinity Action Groups, GIRDAC and KIRSAC—in-principle sister organizations—identified themselves as representatives of their own respective region, as defined by the consultants, and claimed to articulate its respective interests.[48]

Those speaking on behalf of the Kerang region made their statements from a position of vulnerability, having "lived with much greater problems of salinity and waterlogging for some 50–55 years."[49] They felt disadvantaged compared to the Shepparton region in terms of economic priority and political muscle. The Commission's strategy did not include any extra drainage for the Kerang region. But what about the two-thirds of the Kerang region which were not drained? Why was Shepparton prioritized

[47] Gutteridge, Haskins & Davey, *Murray Valley Salinity Investigation—Report*, 226–38; 273–75; State Rivers and Water Supply Commission, *Salinity Control and Drainage*, 45; 47.

[48] Parliamentary Public Works Committee at Kerang at Shepparton, "Merrigan," 139; Parliamentary Public Works Committee at Kerang at Kerang, "Jones," 80.

[49] Parliamentary Public Works Committee at Kerang at Kerang, "Jones," 91.

when "both areas could benefit in the long term by improving drainage of the Kerang Region"?[50]

"[C]oncerned by the lack of concern by the Commission," speakers from the region were eager to stress the existential threat that salinization posed to their livelihoods and that the window of opportunity to turn the situation around was closing fast.[51] "[T]he sands of time are running out on this district" as John Regan, a dairy farmer from Donald, said, "and unless immediate remedial action is taken a lot of farms will not recover from increasing salinity." He added with a sigh—which even translated into a written transcript of evidence—that he probably would not live to see better times, though his son might.[52] Mick Clarke, an Albury grazier, made his case less elegiac but with moral fervor when he denounced that the strategy dismissed the Kerang region "as not worth saving in pure economic terms." He challenged the committee: "Unless each of you is brave enough to recommend writing off Kerang and be prepared to accept the economic and social chaos of such a decision, then you must reject the $40 million plan [...] as inadequate."[53] Or, as Gyn Jones and Des Thomas put it: "Our crisis is now." They could not afford to wait.[54]

Upstream, Shepparton was accused of being a polluter, which hurt Kerang downstream. While KIRSAC gave support in principle for the provisions made in the Commission's strategy, at the same time, the group condemned it because, in their view, it did nothing for the Kerang region. Rather, there were aspects where the region was likely to suffer to the

[50] Parliamentary Public Works Committee at Kerang at Kerang, "Jones," 88.

[51] Parliamentary Public Works Committee at Kerang at Kerang, "Donald Andrew Leed (2)," July 27, 1978, VPRS 11,559/P0001, 24.07.27, PT4, Public Record Office Victoria, Melbourne, 632.

[52] Parliamentary Public Works Committee at Kerang at Shepparton, "John Arthur Regan," June 28, 1978, VPRS 11,559/P0001, 24.07.27, PT4, Public Record Office Victoria, Melbourne, 785.

[53] Parliamentary Public Works Committee at Kerang at Melbourne, "Ernest Edward Clarke," July 04, 1978, VPRS 11,559/P0001, 24.07.27, PT4, 834, Public Record Office Victoria, Melbourne. Clarke would later write a book about the deterioration of the Murray called "River in Danger" that stressed "the need for central, adequately funded control of the river system," outlining "mistakes, mismanagement and sheer bloody-mindedness which have adversely affected the Murray." Anonymous, "Another River Murray Book Written in Anger," *The Riverlander* 35 (February 1981). See Ernest E. (Mick) Clarke, *River in Danger* (Albury-Wodonga: Border Morning Mail [Vic.], 1981).

[54] Parliamentary Public Works Committee at Kerang at Kerang, "Jones," 88.

benefit of the Shepparton region.[55] KIRSAC projected that, due to the proposed measures, more saline water would be diverted into the already saline Torrumbarry Irrigation System—at least 100,000 tons extra per year. Although the Kerang region was at present losing 190,000 tons of salt per year, this was only a small proportion, and they could not afford more salt: "Our already poorly drained area cannot tolerate the long-term effects of being a major sink for salt from other areas."[56]

Mallee dryland farmers from the Kerang region made themselves heard, too. At a meeting of 120 people, called by the Ultima and District Dryland Farmers League at Ultima, near Kerang, $3600 were raised to engage professional help in putting together a case to oppose the Tutchewop-Tyrrell scheme. All local parliamentary representatives were present, as well as the Shadow Minister for Water Supply, Jack Ginifer. Speakers at the meeting pointed out that the scheme was not intended to solve the drainage problems of the Kerang region but merely to divert existing drainage water to Lake Tyrrell. Furthermore, the league opposed an open channel to transport saline water on account of this being a major threat to farming areas along its route. The plan would increase the risk of salting dryland areas and would not reduce the salinity problem of Kerang's irrigation areas.[57] According to Andrew Rank, a farm consultant, the proposal was "intended to get Shepparton out of a sticky problem and would do nothing for the Kerang region." The favorable economic conditions for Shepparton outlined in 1970 had since changed greatly with the rural crisis and changing market prospects, and thus the salt drain would only "postpone the day of reckoning" for the region.[58]

GIRDAC appreciated these problems connected with removing saline groundwater and the effects of the disposal of saline water on farmland and Murray water quality. However, it dismissed much of the opposition to the Water Commission's strategy as based on "emotive, subjective and spurious grounds and without viable alternatives."[59] The group

[55] Parliamentary Public Works Committee at Kerang at Kerang, "Jones," 80; 87.

[56] Parliamentary Public Works Committee at Kerang at Kerang, "Jones," 87.

[57] Anonymous, "'Dry ' Farmers' $3600 for Anti-Salt Case," *The Northern Times*, August 15, 1975; Anonymous, "Ultima Is Opposing Channel," *The Northern Times*, July 16, 1976.

[58] Anonymous, "Dry 'Farmers' $3600 for Anti-Salt Case."

[59] Parliamentary Public Works Committee at Kerang at Melbourne, "Heath," 1057.

insisted that, due to its economic value, the Shepparton region should be protected first. Leon Heath candidly declared that, given financial limitations, there had to be some casualties. He hesitated to name it, but ultimately suggested that, in the long term, there should be a consideration of whether the Kerang region should be given up as a means of preserving the Murray. Farmers should be given the option of leaving and be compensated.[60] As everyone agreed that salt loads had to be restricted, he concluded, "I do not think that irrigation water should be supplied to people who abuse it and cause problems to other people, and the community, and perhaps the nation."[61] He referred to the fact that the salt loads in the River Murray increased substantially downstream, as the river collected more and more saline effluent from the irrigation areas around Kerang and Swan Hill. Comparatively, the Shepparton region's drainage outfall was a small addition to the Murray's salinity. The actual problem was that an area with a highly saline environment was allowed to irrigate. In other words, the real polluter was the Kerang region. According to this logic, it was an area of low productivity, which was threatening the viability of the whole of irrigated agriculture along the Murray, and therefore could be sacrificed for the greater good.[62]

There was some merit to this argument. The SRWSC estimated that the salt loads from its proposed drainage works would add about 3000 tons per year to the river. This was small compared to the contributions from Barr Creek alone, which carried an average amount of 150,000–200,000 tons of salt into the Murray; albeit with huge variations and dilution flows by the Loddon. The crucial difference was that much of the salt loads from the Kerang region were due to groundwater seepage caused by the naturally saline conditions and the damage done by dredging Barr Creek and Pyramid Creek. Shepparton, by contrast, added much of its salt via artificial drainage lines used to prevent waterlogging problems.[63]

According to Peter Hallows, who was associated with Gutteridge, Haskins & Davey, it had been "clearly established that irrigation should

[60] Parliamentary Public Works Committee at Kerang at Melbourne, "Heath," 1057–58; 1066/1067–69.

[61] Parliamentary Public Works Committee at Kerang at Melbourne, "Heath," 1069.

[62] Parliamentary Public Works Committee at Kerang at Melbourne, "Heath," 1057–59.

[63] State Rivers and Water Suppy Commission, *Salinity Control and Drainage*, 92; Gutteridge, Haskins & Davey, *Murray Valley Salinity Investigation—Report*, 233–37; 278.

have never been introduced into the Kerang Region."[64] But was the solution really to convert it into a forest area and relocate the farmers elsewhere? Who was allowed to pollute the river? And who would be given assistance to decrease their share? As in 1968, these were questions of policy.

The Promise of Irrigation

Irrigation in Australia had been in dire straits before salinization reached the worrying level of the mid-1970s. The economic rationalist critique of irrigation as national development had severely dampened the optimism invested in its further expansion, though not arrested dam building and the extension of irrigated land. Still, it had become apparent that the introduction of further irrigation schemes and the application of scientific discoveries did not necessarily guarantee success in the development of new agricultural projects and created serious environmental consequences. Academic experts began to emphasize the need for conservation, the prevention of degradation, and the sustainability of existing cropland, rather than the expansion of agricultural output. However, it was perhaps the failure of the Ord River project in northern Western Australia that "did most to dent faith in the capacity of science, capital and enterprise to guarantee the success of large-scale rural development."[65] This river, which was mostly dry for much of the year, but carried vast amounts of water during the tropical wet season, was seen as a classic case of water running to waste, and dams were built. Irrigation crops like rice, cotton, peanuts, sunflower seeds, sugar, and others were planted. However, there were problems with pests, soil, quality, and economics. Once again, Bruce Davidson was a fervent critic, claiming the returns would not justify the capital outlay.[66]

Against such criticism, the SRWSC reasserted its self-conception that water conservation was vital to Victoria's development. Its work

[64] Anonymous, "Murray Control—Action by MVDL," *The Northern Times*, August 12, 1975.

[65] Waterhouse, *The Vision Splendid*, 216–17.

[66] Garden, *Australia, New Zealand and the Pacific*, 118–19; "The North. Is It Worth Developing?," *The Age*, July 11, 1966. See Bruce R. Davidson, *The Northern Myth. A Study of the Physical and Economic Limits to Agricultural and Pastoral Development in Tropical Australia* (Melbourne: Melbourne University Press, 1972).

warranted that wealth was generated from an "endless" variety of irrigation products to the benefit of every community member. Irrigation had enabled the continent to carry a greater population, made possible closer settlement, encouraged decentralization, increased public revenue and national income, and reduced drought losses.[67] Irrigation systems had transformed "potentially fertile areas of low rainfall to rich areas of closer settlement, important food producers not only for Australia but for the world."[68] The deficiencies of Australian nature had been overcome by the application of science. The promise of irrigation, indeed the water dreamers' vision of an Australia Unlimited, had been fulfilled for the common good.[69] Victoria's expenditure on irrigation development, then £75 million (about $14.4 billion in current values), had given the state "a very great asset with which we would not part."[70]

Salinization threatened to destroy precisely this asset. Therefore, according to the Commission's official rationale for "Salinity Control and Drainage," $40 million were considered a "deferred instalment in the overall cost of large-scale irrigation development, established by massive public and private investment over the past ninety years, which has determined the pattern of prosperity of rural and urban settlements."[71] Financing for the program would be in accordance with Victoria's established principles of irrigation development, meaning that the capital costs for drainage works would be borne by the public, whereas operation and management costs would be met by revenue from drainage rates. To encourage improvement in farm management, concessional loans would be offered.[72]

Thereby, the Water Commission's achievement would be upheld. Permanent white settlement in rural Australia could continue and expand. Irrigated agriculture would earn export cash and ensure national self-sufficiency. The public would continue to be asked to accept financial losses as the price for rural settlement was understood as national interest.

[67] State Rivers and Water Supply Commission, *Development of Victoria*, 1.

[68] Australian Publicity Council, *Liquid Gold Australia*, 2.

[69] Australian Publicity Council, *Liquid Gold Australia*, 3.

[70] "Water Symposium Clash Over Irrigation Cost and Benefits," *The Age*, September 14, 1963. Current value calculated using Hutchinson and Ploeckl, "MeasuringWorth."

[71] State Rivers and Water Supply Commission, *Salinity Control and Drainage*, 1.

[72] State Rivers and Water Supply Commission, *Salinity Control and Drainage*, 8.

On the contrary, in the view of Davidson's colleague K. O. Campbell, the irrigation communities that should now be saved from salinization had been built up on the basis of "discriminatory benefits" from public investment. In other words, their interests were not national but, rather, parochial.[73]

Therefore, whatever their other divisions might have been, representatives from the GMID valiantly defended the value of their communities and irrigation for the nation, claiming that the "responsibility of Governments [...] is clear—in the area of preserving the productive capacity of the land—and in the area of maintaining a necessary sense of security for the people."[74] Governments, then, had a twofold responsibility. It was the nation's, and hence the state and federal governments, responsibility to safeguard the livelihoods of the rural communities dependent on irrigation. Likewise, it was the State Rivers and Water Supply Commission's responsibility to maintain or restore their environmental security, having assumed control over the Murray by turning it into a regulated system. With the state in "untrammeled control of 'all natural resources of water supply'," and tasked with using this power for the common good, it was now asked to meet its obligations.[75]

To substantiate these claims, Jones and Thomas took refuge in the lore of Thomas Mitchell's arrival on the Northern Plains. They invoked the Major's Vision of June 29, 1836 that "this indeed seemed to me to be a country where canals would answer well, not so much perhaps for inland navigation, as for the better distribution of water over a fertile country, enclosed as this by copious rivers."[76] Mitchell predicted that people would irrigate, and they did. Then, "by universal request from the community," governments stepped in to increase reservoirs and water allocations. Irrigation, thus, became "the major significant factor in the economic and social development of the Kerang region," and was without a doubt "the lifeblood of the Northern district economy."[77] Their other

[73] "Water Symposium Clash Over Irrigation Cost and Benefits," *The Age*, September 14, 1963.

[74] Parliamentary Public Works Committee at Kerang at Shepparton, "Routley (2)," 280A.

[75] Powell, *Garden State*, 147.

[76] Parliamentary Public Works Committee at Kerang at Kerang, "Jones," 81; Mitchell, *Three Expeditions, Vol. 2*, 158.

[77] Parliamentary Public Works Committee at Kerang at Kerang, "Jones," 81.

key witness, Ronald East's successor, Alfred L. Tisdall, declared that the "pattern and prosperity of the rural and urban settlement [...] in the Northern Irrigation districts have been determined by, and remain virtually dependent on, the practice of irrigation."[78] Their survival depended upon its continued viability.[79]

What was the asset's worth? KIRSAC put forward an argument which, in the twenty-first century, would become popular as "Too Big to Fail." In an attempt to assign a monetary value to the Kerang region, which had been depreciated so much from a purely economic point of view, they contended that the 1820 km^2 developed for irrigation in the Cohuna, Kerang, Boort, Tragowel Plains, Fish Point, and Tresco districts, which presently needed drainage relief, had a replacement value of no less than $200 million (about $4.4 billion in current values). The cost to re-establish a similar irrigation area elsewhere was, therefore, 18 times the volume of the recommended scheme to maintain the viability of the Kerang region, securing the livelihoods of 120,000 people in rural communities and an annual farm gate production of $170 million.[80] A deferred installment of "a mere $40 million," according to this logic, was a comparatively small investment.[81] This would preserve the "massive private and public investment" created by the development of irrigation systems.[82] GIRDAC, among others, used similar arguments to demonstrate the Shepparton area's economic relevance to the state and the nation, boasting approximately 90,000 people living in the area, not to mention its huge dairy and fruit industry.[83]

A frequent catchword was "decentralization"—that "desire to redress the concentration of population and power in Australia's coastal cities, [...] an oft-expressed, but largely unfulfilled, aspiration of reformers and

[78] Parliamentary Public Works Committee at Kerang at Kerang, "Jones," 83.

[79] Parliamentary Public Works Committee at Kerang at Shepparton, "Merrigan," 146.

[80] Parliamentary Public Works Committee at Kerang at Kerang, "Jones," 81–82; 85–86; Parliamentary Public Works Committee at Kerang at Shepparton, "Merrigan," 148–49. Current value calculated using Hutchinson and Ploeckl, "MeasuringWorth."

[81] Parliamentary Public Works Committee at Kerang at Kerang, "Jones," 83.

[82] Parliamentary Public Works Committee at Kerang at Kerang, "Jones," 86.

[83] Parliamentary Public Works Committee at Kerang at Shepparton, "Merrigan," 141–42; 146.

politicians" since the 1880s.[84] According to Eric Merrigan, the Goulburn region was "a successful decentralisation venture" with people "living in well-ordered and functional communities." Their livelihood had to be maintained and the social structure within the town and the farm preserved.[85]

Mick Clarke added, there "should not be too much emphasis on cost. The benefits will continue when the costs are forgotten."[86] The asset was also of immaterial value. It was "Australia's most productive irrigated region," but also "our most successful closer settlement venture on a national scale" that was at stake.[87] Small country towns and farms "with [their] own social structure and community inter-relationships" were of "national importance—both to the Australian economy and the Australian way of life."[88] Despite the many claims to the high productivity of irrigated agriculture, to some extent salinity control was "not a profit making exercise." It was about arresting the "loss of Victorian cultural diversity, and a highly satisfying life style."[89]

To the members of the Public Works Committee, these statements had a familiar ring. The advocates of the irrigation communities invoked the family farm, which was passed on to the next generation, and the joy of rural life itself. This cultural ethos of farming as a way of life, once expressed in the ideal of yeoman farmer, had strongly influenced selection and closer settlement policies. Richard Waterhouse has argued that, in the decades after World War II, this became more or less irrelevant. Farming came to be understood "as just another business, nothing more, nothing less."[90] Yet, the witnesses appealed to a popular ideology that

[84] Graeme Davison, "Decentralisation," in *The Oxford Companion to Australian History*, ed. Graeme Davison et al. (Melbourne: Oxford University Press, 1999), 176; Parliamentary Public Works Committee at Kerang at Melbourne, "Clarke," 835; Parliamentary Public Works Committee at Kerang at Swan Hill, "Peter Stanley Fisher," October 3, 1978, VPRS 11,559, 24.07.27, PT5, Public Record Office Victoria, Melbourne, 977.

[85] Parliamentary Public Works Committee at Shepparton, "Merrigan," 148–49.

[86] Parliamentary Public Works Committee at Melbourne, "Clarke," 835.

[87] Parliamentary Public Works Committee at Kerang at Swan Hill, "Fisher," 976–76A.

[88] Parliamentary Public Works Committee at Kerang at Shepparton, "Merrigan," 147; Parliamentary Public Works Committee at Kerang at Swan Hill, "Fisher," 976–76A.

[89] Parliamentary Public Works Committee at Kerang at Melbourne, "Stone," 814.

[90] Waterhouse, *The Vision Splendid*, 209.

was deeply rooted in Australian culture—a strong belief in the virtues of county life. In colonial Australia, the concentration of population in the coastal cities had been considered an unnatural state to be replaced by a society of small farms and towns spread evenly across a prosperous rural landscape. To achieve this, the bush had to be remade into "countryside." The brown land had to be turned into a green one, giving a powerful cultural impetus to the Water Dreamers' vision of a closely populated Australia—and thus irrigation.[91] Implicitly, it meant an appeal to the importance of white settlement in rural Australia, a late echo from the notion that the continent was *terra nullius* unless occupied and used.

In the twentieth century, most Australians still regarded city life and country life as quite distinct. Country people considered theirs "a more wholesome, independent, genuine, sociable and satisfying existence than the rushed, artificial, unhealthy and nervous life of the town-dweller."[92] And long into the 1960s and 1970s, many city people agreed, lending support to a set of popular beliefs that Don Aitkin has termed "countrymindedness."[93] According to this ideology, Aitkin writes, Australia depended on its primary producers for its high standard of living. Therefore, all Australians should support policies aimed at improving the position of its primary industries in order to encourage people to settle in the country, not the city. Rural life brought out the best in people, while city life was competitive, nasty, and parasitical. The characteristic Australian was the countryman because the core elements of the national character came from the struggle of country people to tame their environment and make it productive.[94]

"Countrymindedness" drew on the moral credit of the pioneers to support the task of building a modern rural country.[95] Now their ghosts

[91] Jay Mary Arthur, *The Default Country* (Sydney: UNSW Press, 2003), 26; Graeme Davison, "Country Life. The Rise and Decline of an Australian Ideal," in *Struggle Country. The Rural Ideal in Twentieth-Century Australia*, ed. Graeme Davison and Marc Brodie (Melbourne: Monash University Press, 2005); Howes, "The Spectre at the Feast," 222; Cathcart, *The Water Dreamers*, 247.

[92] Davison, "Country Life."

[93] Davison, "Country Life;" Don Aitkin, 'Return to 'Countrymindedness'," in *Struggle Country. The Rural Ideal in Twentieth-Century Australia*, ed. Graeme Davison and Marc Brodie (Melbourne: Monash University Press, 2005).

[94] Don Aitkin, "'Countrymindedness': The Spread of an Idea," *Australian Cultural History* 4 (1985): 35; Davison, "Country Life."

[95] Davison, "Country Life."

were conjured to make a case for the intrinsic value of rural communities under the threat of salinization. There can be little doubt that this was sincere, but it was still shaky ground. Australia's post-war expansion had brought, and was still bringing, immigrants into the country who had no Arcadian view of the continent. The family farm, the economic unit that had underpinned the ideology, was in decline, and the affluence and technological changes that were introduced reduced the real and psychological distance between country and city.[96] As Graeme Davison has put it, "the belief in the moral and social superiority of rural life died hard, but it had now to be asserted, as an article of faith, in the face of much evidence to the contrary."[97]

The appeal for help in the present crisis was a vision for a prosperous future. With the aid of one final injection of public investment to control the water tables, the habitat seen by Major Mitchell could be restored, and the Northern Plains "be made more productive than ever before."[98] Irrigators would be able to make full and adaptive use of irrigation for the first time.[99] The investment would, finally, be completed, and the irrigation regions would "continue to function and prosper indefinitely."[100]

People Versus Birds

To most, salinization was about people. The Water Commission affirmed its responsibility for the living conditions of the inhabitants of the Murray Valley, who were dependent on the viability of irrigated agriculture within its jurisdiction, and the effects that resulted from its long-term modifications of the natural landscape. Above all, the productivity and efficiency of the Commission's working landscape were to be preserved and augmented, and environmental security retained or restored for its users. Water use could then further increase, and irrigation development could continue. It was an endeavor to complete the irrigation system and promote further development, and the majority of people

[96] Aitkin, "Countrymindedness," 40.

[97] Davison, "Country Life."

[98] Parliamentary Public Works Committee at Kerang at Kerang, "Jones," 91.

[99] Parliamentary Public Works Committee at Kerang at Shepparton, "Merrigan," 147.

[100] Parliamentary Public Works Committee at Kerang at Shepparton, "Merrigan," 149.

in the salt-affected regions agreed. However, environmental considerations could no longer be disregarded, and they found forceful advocates. GIRDAC's representative, Leon Heath complained that it was "unfortunate that human beings did not warrant as much attention as the birds and the plants."[101] While the preservation of wildlife and wetlands against drainage requirements was the object of contention during these conflicts, a more fundamental issue was at stake: Should salinization be addressed by a technological fix to retain and augment the present use of the Murray Valley or should that use be limited in the interest of the environment? Much like the fierce contentions over public funds for engineering works, these were struggles by different parties over the increasingly limited resources of the Valley.

The backdrop to these controversies was an altered role for conservation in Australia, which emerged in the late 1960s and early 1970s, and increased attention to the environmental consequences of rural development. This entailed an appreciation of the degraded condition of the Murray.[102] Libby Robin has argued that the five-year period from 1968 to 1973 embraced a significant change in self-perception; a move from the height of a national minerals boom to the depths of an international petroleum crisis. During this time, the perception of the role of conservation was dramatically altered. Not unlike Richard Nixon, who famously declared that the 1970s would be the "decade of the environment" and initiated the creation of the Environment Protection Authority, Victoria's Premier Henry Bolte paid lip service to environmental concerns in his 1970 election speech. Following the publication of *Silent Spring* by Rachel Carson, in which the author denounced the indiscriminate use of pesticides, specifically DDT, ever-expanding chemically dependent agricultural development became a concern during the 1960s. Notions about the bush were shifting. To many suburban Australians, the Frontier became fragile, rare, and in need of defense, not conquest.[103] Landmark titles of the eco-age, expressing neo-Malthusian concerns about finite resources and population growth, such as Paul Ehrlich's *Population Bomb* (1968) and the Club of Rome report *Limits to Growth* (1972),

[101] Parliamentary Public Works Committee at Kerang at Melbourne, "Heath," 1057.

[102] Waterhouse, *The Vision Splendid*, 216; Sinclair, *The Murray*, 223; 228.

[103] Libby Robin, *Defending the Little Desert. The Rise of Ecological Consciousness in Australia* (Melbourne: Melbourne University Press, 1998), 42; 134; 141–42; Rachel Carson, *Silent Spring* (New York: Ballantine, 1962).

had profound impact and promoted a stronger sense of urgency about environmental issues.[104] Ecology was popularized as a general interdisciplinary science of the interaction of all biological species in an ecosystem. Gradually, the term was adopted in different contexts, drifting from the sciences to become a new worldview.[105]

This "rise of ecological consciousness," which Robin understands as "a growing awareness of the political dimensions of concerns about the natural world and the place of people in nature," was marked by a key event: The Little Desert campaign.[106] The Little Desert Settlement Scheme of 1968 was one of the last in the long line of proposals for the settlement of Australia's semi-arid lands. In the 1950s and 1960s, development pressures had not only threatened new areas, but also undermined previous gains of the conservation movement. The dispute caused a remobilization in Victoria, which proved successful, and the bushland was retained instead of being developed. However, the dispute was not simply about saving nature, but about introducing a moral sense of limits to development. A local antecedent of the "think globally, act locally" environmentalism, it was nonetheless run by an older conservation movement. It was also the last great campaign for many of its protagonists, and the high point of their campaigning careers. Soon, they felt marginalized by the new environmentalists of the 1970s, who used different language, had different worldviews, and failed to acknowledge their predecessors' achievements.[107]

From a global point of view, Joachim Radkau has argued that the period around 1970 was marked by an environmental revolution. The

[104] Drew Hutton and Libby Connors, *A History of the Australian Environment Movement* (Melbourne: Melbourne University Press, 1999), 126; Paul Warde, Libby Robin and Sörlin Sverker, *The Environment. A History of the Idea* (Baltimore: Johns Hopkins University Press, 2018), 48–72.

[105] Robin, *Little Desert*, 141–42.

[106] Robin, *Little Desert*, 4.

[107] Robin, *Little Desert*, 2; 11; 41–42; 138–39; Hutton and Connors, *A History of the Australian Environment Movement*, 108–12. On the Little Desert dispute, see also Libby Robin, "Of Desert and Watershed. The Rise of Ecological Consciousness in Victoria, Australia," in *Science and Nature: Essays in the History of the Environmental Sciences*, ed. Michael Shortland (Oxford: British Society for the History of Science, 1993); Joseph M. Powell, "'Action Analysis' of Resource Conflicts: The Little Desert Dispute, Victoria, 1963–72," in *The Making of Rural Australia. Environment, Society and Economy: Geographical Readings*, ed. Joseph M. Powell (Melbourne: 1974).

"networking, wide impact and global horizons that developed from 1970 were more or less new. Especially if one remembers some of the grotesque phenomena associated with the technocratic planning fever of the 1960s, the subsequent period appears as a real watershed."[108] After the exhaustion of the great ideologies, popular ecology emerged as the only large-scale answer to mounting environmental issues, such as the continuous growth of motorized road and air travel and the pollution of the soil, water, and atmosphere. Radkau insists that in this regard it was justified to speak of the advent of an ecological age, in which environmental awareness became part of the mainstream.[109]

Accordingly, Earth Day, first celebrated on April 22, 1970, was not a sub-cultural happening but one that united sub- and high-culture, even though the environmental movement retained its protest potential and counter-cultural elements.[110] Environmental talk of "one world" became essential: "[T]hose who campaigned against the pollution of a little stream did so in the proud belief that they were doing their bit to save the world."[111] Environmentalism relied on both movements on a broad scale and institutions. International conferences played a crucial role, especially the Stockholm conference in 1972, which was the first occasion on which the political, social, and economic problems of the global environment were discussed at an intergovernmental forum. The United Nations Environment Programme was created, and a debate between rich and poor countries was brought out into the open.[112]

However, Radkau stresses that this ecological revolution cannot not be explained by one particular causality, but grew out of a conceptual association of various risks. In 1970, at least half a dozen concerns were addressed: nuclear risks, rubbish, water and air pollution, overpopulation,

[108] Radkau, *Age of Ecology*, 7.

[109] Radkau, *Age of Ecology*, 8. Donald Worster, Ramachandra Guha and John McCormick have developed similar arguments regarding a transformation of environmentalism and the spread of ecological thinking around 1970. See Donald Worster, *Nature's Economy. A History of Ecological Ideas* (Cambridge: Cambridge University Press, 1994); Ramachandra Guha, *Environmentalism. A Global History* (Oxford; New York: Oxford University Press, 2000); John McCormick, *The Global Environmental Movement*, 2nd ed. (Chichester: John Wiley and Sons, 1995).

[110] Radkau, *Age of Ecology*, 98; 106.

[111] Radkau, *Age of Ecology*, 93.

[112] Radkau, *Age of Ecology*, 91–99; McCormick, *The Global Environmental Movement*, 107–8.

and species loss. Furthermore, at stake was not only seemingly selfless love of wilderness, which had been a leitmotif of the old conservationism, but also human well-being or even survival. Concern over quality of life issues was characteristic of the eco-era.[113] The conclusion of world wars and the crumbling of Cold War fronts allowed for common problems facing humanity after decades of unprecedented post-war growth—demographically, economically, and regarding atmospheric emissions and fossil fuel burning—to come to the fore. Environmental protection became a task *sui generis* at a time when anthropogenic environmental change had accelerated rapidly.[114]

The revolution immediately leaped to Japan. The People's Republic of China sent representatives to the Stockholm conference, and the experience made a lasting impact on its Chinese participants. By contrast, the USSR and its satellites boycotted the conference, but the Soviet Union participated, *inter alia*, in UNESCO's "Man and the Biosphere" program, which launched in 1971. An awareness of global environmental interrelations developed among Soviet intellectuals and scientists. Environmental protection was on the institutional agenda. However, there was no popular ecological turn in the USSR. It was overwhelmingly Western.[115]

The first Australian environmentalists "emerged out of the radical end of the trade union movement."[116] They abhorred the old boys' networks and gentleman agreements of the old conservationists. Issues such as global destruction (nuclear power and the greenhouse effect), urban living (multiculturalism, heritage, and urban space preservation), and wilderness preservation were merged and blended together under the umbrella-term "green." However, at the outset, they were more

113 Radkau, *Age of Ecology*, 99–110.

114 Radkau, *Age of Ecology*, 111; McNeill and Engelke, *The Great Acceleration*, 184–85; Guha, *Environmentalism*, 65–68.

115 Radkau, *Age of Ecology*, 91–92. Timm Schönfelder, *Roter Fluss auf schwarzer Erde. Der Kuban und der agromeliorative Komplex. Eine sowjetische Umwelt- und Technikgeschichte, 1929–1991* (Paderborn: Brill Schöningh, 2022), 194–200; Klaus Gestwa, "Ökologischer Notstand und sozialer Protest," *Archiv für Sozialgeschichte* 43 (2003): 354–56.

116 Robin, *Little Desert*, 145.

concerned with urban environmentalism and quality of life issues than nature conservation.[117]

In the 1970s, the Green Bans movement in Sydney campaigned against unrestrained urban development, and the environmental movement forged close relationships with anti-nuclear trade unions and Aboriginal groups, although these would be strained by the coming wilderness campaigns. Of these, the most spectacular ones took place in Tasmania. The first was an unsuccessful attempt to stop the inundation of Lake Pedder in 1973. Still, it marked a coming of age of the Australian movement and led to the foundation of the Australian Green Party. The movement managed to mobilize large numbers of activists and supporters and developed the skills and resources to force its way into the mainstream of Australian political life. The second campaign would see a successful bid against the Franklin Dam in 1983, a resounding victory for the movement.[118]

While these struggles were about saving pristine places from people and not about arresting the degradation of working landscapes, such as the MDB, they did shape changing popular conceptions of dams and challenged state and federal governments' environmental and natural resource management frameworks. Still, it is striking that environmental activists would rarely connect with the problems in the MDB until the 1990s.[119] Paul Sinclair has argued that this was because Murray was not wild enough to be a focus of popular environmental activism. By 1970, it "was widely recognised that the Murray was governed more by human-made mechanisms than its natural processes, even though the river remained, at least on the surface, a very beautiful and often isolated place."[120] It was difficult to advocate for the preservation of its flora and fauna based on its value as wilderness because the river was physically and conceptually bound to irrigation. It was neither completely artificial nor natural.[121]

[117] Robin, *Little Desert*, 134–35; 144–45; Hutton and Connors, *A History of the Australian Environment Movement*, 125.

[118] Hutton and Connors, *A History of the Australian Environment Movement*, 118–63; Martin Mulligan and Stuart Hill, *Ecological Pioneers. A Social History of Australian Ecological Thought and Action* (Cambridge: Cambridge University Press, 2001), 217.

[119] O'Gorman, *Flood Country*, 193–95.

[120] Sinclair, *The Murray*, 179.

[121] Sinclair, *The Murray*, 179–82.

Nonetheless, ecological consciousness was not confined to scientists and (urban) environmental activists, and by the 1970s, it had permeated public debate in regional areas like the Murray Valley.[122] The magazine *Riverlander*, which was published by the Murray Valley Development League (MVDL), provides a useful indication. The MVDL, which was founded in 1944 at Yarrawonga on the Murray and is now known as the Murray Darling Association, was an organization dedicated to promoting regional development in the Murray Valley. It battled to advance decentralization programs in favor of regional Australia, and strongly advocated an interstate authority to manage the Valley's resources.[123] Although a fierce advocate of irrigation, the league also retained a keen interest in conservation issues and promoted them to the magazine's readers. It regarded the success of the Little Desert campaign as a "victory."[124] In 1970, it published an obituary for Francis Ratcliffe, a CSIRO entomologist who was among the founders of the Australian Conservation Foundation, without whose "authority and prestige [...] the battle to keep Australia interesting and humane will be harder."[125] *The Riverlander* commented on various issues, such as a looming energy crunch, the depletion of ozone layer, and soil degradation.[126] The magazine also directed attention to some of the measured warnings of an imminent environmental crisis issued by scientists. It published a manifesto by the Australian and New Zealand Association for the Advancement of Science which predicted that "within the lifetime of our children civilisation will face a crisis of survival," and a summary of the "Blueprint for Survival" issued in the magazine *The Ecologist* in January 1972, which was written

[122] Sinclair, *The Murray*, 223; 228.

[123] Adrian Wells, *Up and Doing. A Brief History of the Murray Valley Development League and the Murray Darling Association from 1944 to 2014* (Albury: Murray Darling Association, 2014), 22; 59–66.

[124] Anonymous, "Victory in the Desert," *The Riverlander* 24 (January 1970): 22.

[125] A.B., "Obituary. Francis Ratcliffe," *The Riverlander* 24 (December 1970): 21; Hutton and Connors, *A History of the Australian Environment Movement*, 107.

[126] Radkau, *Age of Ecology*, 91–92; Anonymous, "Caring for Our Environment," *The Riverlander* 31 (May 1977): 9; Peter Simple, "The Passing Show," *The Riverlander* 31 (July 1977): 9; Anonymous, "An Energy 'Crunch' is Coming," *The Riverlander* 33 (January/February 1978): 3.

by Edward Goldsmith and Robert Allen and signed by over 30 scientists.[127] Earth Day on April 22 and World Environment Day on June 5 were honored because "our generation is challenged as never before to give steady and searching regard to [Earth's] well-being."[128]

With no apologies to Edward Gibbon, the "Decline and Fouling of the Murray" received regular attention.[129] In mid-1978, *The Riverlander* reported that salinization was "fast becoming Victoria's major agricultural problem," having cost farmers some $125 million in reduced land values, threatened investment worth more than $1 billion, and causing "untold economic and social costs" to the community.[130] As salinization became an important problem, "water quality" became a key term linked to the rising concern about pollution. Human-induced salinization was now categorized as one form of water pollution, like the one induced by DDT and other pesticides, the water hyacinth, and the European carp. Clearly, environmental issues had become an object of public attention in urban and rural Australia.[131]

Experts reacted to the changing times, too. Unlike the "Gutteridge Report," the SRWSC's strategy included environmental considerations.

[127] Anonymous, "You Have Been Warned: It Can't Go On," *The Riverlander* 25 (June 1971): 6; Anonymous, "You Have Been Warned—Once More," *The Riverlander* 26 (June 1972): 6. See Edward Goldsmith and Robert Allen, *A Blueprint for Survival* (London: Penguin, 1972).

[128] Anonymous, "A Moment for Mother Earth," *The Riverlander* 32 (January/Feburary 1978): 4; Anonymous, "Earth Day; Holy Day," *The Riverlander* 33 (September/October 1979): 4.

[129] Anonymous, "Decline and Fouling of the Murray," *The Riverlander* 28 (March 1974).

[130] Nannette Oates, "Salt has Reduced Victorian Land Values by $125 Million," *The Riverlander* 32 (May/June 1978): 16.

[131] Anonymous, "Pollution—Alarm Bells," *The Riverlander* 25 (July 1971): 13; Anonymous, "Keeping Our River System Clean," *The Riverlander* 26 (January/February 1972); Anonymous, "The Environment: Concern for All," *The Riverlander* 26 (May 1972); Anonymous, "The Quality of Our Water," *The Riverlander* 29 (April 1975): 9. "Pollution," according to the Oxford Dictionary of Ecology, is "the defilement of the natural environment by a pollutant" which is a "by product of human activities which enters or becomes concentrated in the environment, where it may cause injury to humans or desirable species." This embraces chemical substances, noise, vibration, and alterations to the ambient temperature. Anonymous, "Pollutant, Pollution," in *A Dictionary of Ecology*, ed. Michael Allaby, 4th ed. (Oxford: Oxford University Press, 2010), 300.

In 1972–1973, the Commission adopted the principle that the environmental impact of new projects would be considered regarding "both the effects its work will have on the environment and the contribution they will make to the environment and to the quality of life of those people whom they serve." It would hence delineate the likely environmental effects of proposed projects, compare alternative courses, and enable the public to express a viewpoint.[132]

To the Commission, in regard to the irrigation areas of Northern Victoria the "main environmental concern" was "the conservation of a high-quality *artificial* environment."[133] The premise was that the natural environment had already been transformed by long-term European settlement. It was this artificial, modified, environment that had to be retained. Non-intervention was not an option—only further modifications could stop its deterioration. Production was not the only aspect: "Artificially modified landscapes are as essential to our civilization as is a concern for the conservation of 'nature'." An irrigation district, in good order, had a positive scenic value as irrigation had created an ordered lushness and diversity that contrasted favorably with the "apparent austerity and monotony of the adjoining dryland landscapes."[134] The asset was based on a landscape that had been improved upon and made productive—even aesthetically pleasing.

However, the Commission assured, "protection and enhancement of nature-conservation values are given full weight in planning and design, and specific proposals submitted for approval will be subject to environmental assessment."[135] When the strategy was published in 1975, a preliminary survey of wetlands and water bird resources had already been conducted. However, further environmental studies were required for additional drainage works with particular reference to the importance of changes in water level and salinity in water bird habitats. Generally, in the late 1970s and early 1980s, there was still little information about the environmental impacts of increasing salinity: "Changes in rivers and

[132] State Rivers and Water Supply Commission, *67th Annual Report 1972–73*, 4.

[133] State Rivers and Water Supply Commission, *Salinity Control and Drainage*, 9. My emphasis.

[134] State Rivers and Water Supply Commission, *Salinity Control and Drainage*, 72.

[135] State Rivers and Water Supply Commission, *Salinity Control and Drainage*, 9.

wetlands have been due to the compound effects of river regulation, diversions for irrigation and other uses, flood control and mitigation, drainage of wetlands, barrages to prevent salt water inflows, etc." To separate out the effects of increasing salinity levels was nearly impossible. There was virtually no useful information available on the impacts of salinization on terrestrial ecosystems.[136] In short: "A combination of fact and opinion will provide the basis for action in combating salinity for a long time to come."[137]

To some, salinization was about birds. Their spearhead, or rather bayonet, was the Victorian Field and Game Association (VFGA), which put forward the most radical counterpoint to the SRWSC's recommendations. Most community representatives, in principle, agreed to the combination of engineering and agricultural solutions for salinization that the Commission proposed. The VFGA, however, took up the case of the natural environment against the requirements of drainage. While it was "aware of the economic and social problems caused by salinity" it was "also acutely aware of the even greater environmental damage which can be caused by works designed to reduce the economic and social problems unless these works are planned with proper long term regard for the environment."[138]

This hunters' interest group, founded in 1958 at Sale, had been formed to help counter losses of swamps and breeding grounds for game. One of its first moves had been to press for the introduction of game licenses in order to raise money to ensure water supply to key duck-shooting swamps, growing of duck food in barren waters, and the management of breeding swamps by controlling water levels.[139] The loss of habitat and game population should be arrested so that the "sport can be protected

[136] ACIL Australia et al., *Causes, Extent and Effects of Salinity*, xxii.

[137] ACIL Australia et al., *Causes, Extent and Effects of Salinity*, xxv.

[138] Parliamentary Public Works Committee at Kerang at Kerang, "Kenneth John Hooper (1)," July 13, 1976, VPRS 241/P0000, pp. 64–131, Public Record Office Victoria, Melbourne, 118.

[139] Anonymous, "New Game Shooters' Association," *Australian Clay Target Shooting News* 12, no. 1 (1958): 9–11; L.R. Cambrey, "Victorian Field and Game Association, Evidence to Parliamentary Public Works Committee Inquiry on Salinity in Northern Victoria," July 13 1976, 11,559, 24.07.27 PT1, Public Record Office Victoria, Melbourne, 1.

and enlarged still further."[140] The VFGA boasted the patronage of a long-time Premier, Sir Henry Bolte, and his successor, Dick Hamer (later Sir Rupert). On its letterhead stood a wisdom by King George VI exhibiting a sense of stewardship: "The wild life of today is not ours to dispose of as we please. We have it in trust. We must account for it to those who come after."[141]

During the disputes about salinization, the association took the Commission's past and proposed drainage practices head-on. It denounced that engineers, "trained in engineering techniques," had shown no consideration for the needs of flora and fauna in their measures for salinity control since the 1920s: "The criterion was to use the cheapest method to construct feeder channels from rural salt affected lands to our natural wetlands." Barr Creek, the infamously largest individual source of salt in the entire Murray Valley, and Pyramid Creek were cited as examples, having been dredged deeply, the former to increase its capacity as a drain, the latter as an irrigation water carrier. The reason why these creeks were now given very low ratings for birdlife was that they had been "ecologically destroyed by previous State Rivers and Water Supply Commission actions." The VFGA strongly objected to "writing these streams off as having low ratings but giving no indication of the destruction which has already taken place." The result would be that further streams would be compromised, leading to "the total environmental destruction of the area."[142]

It accused the Commission of having done other extensive environmental damage to streams and wetlands in the Kerang area with surface drainage works, as it considered that "publicly owned wetlands cost nothing."[143] Proof was a list of a minimum of eleven areas that had been "destroyed" by drains, claiming that this had been so intense that if this type of environment was to be retained, plans for salinity control had to be based on the premise that no further destruction of streams or wetlands would be tolerated. Considerable values for tourism, recreation,

140 Anonymous, "New Game Shooters' Association," 11.

141 L.R. Cambrey, "Victorian Field and Game Association, Evidence to PPWC," 1.

142 Parliamentary Public Works Committee at Kerang at Kerang, "Hooper (1)," 2.

143 Parliamentary Public Works Committee at Kerang, "Kenneth John Hooper (2)," February 2, 1977, VPRS 11,559/P0001, 24.07.27, PT2, Public Record Office Victoria, Melbourne, 290.

pest control, and general ecological significance were at stake.[144] Therefore, the Commission's plan to divert additional Barr Creek flows away from the Murray to be able to add more salt from the Shepparton region was "environmentally wrong." Instead, further study should be made into alternative disposal methods.[145]

In the VFGA's view, salinization was a problem whose causes dated back to the European seizure of Australia, claiming that the issue "results from land use changes introduced by the white man." There was abundant evidence of long Aboriginal settlement in the affected areas: "It appears that the aborigine used the wetlands as his main source of food and left the dryland undisturbed. There were no salinity problems reported by the first white settlers so the aborigine had lived in harmony with his environment." This Arcadian world was said to have been turned on its head with the arrival of the European settlers in the late 1850s: "The white man started to farm and graze dry land and looked upon the wetlands as waste areas to be drained and turned into dry land. This was the opposite to the aboriginal land use."[146] When he quickly turned to irrigate the dry land, in a matter of "a few years the water table was raised to less than six feet from the surface and the first damage began to show."[147]

According to this interpretation, salinization was the epitome of the fall of European man by the sin of his land use. In this sense, the positions put forward by the VFGA were in line with the ideas of the old preservationist movement that had emerged in Australia in the late nineteenth and early twentieth century. This style of conservationism, identified with protagonists such as Romeo Lahey and Miles Dunphy, successfully fought for the establishment of a system of national parks in areas that otherwise would have been substantially changed. Just like John Muir in the U.S., Dunphy campaigned for a shift in public perception toward wilderness.[148] The VFGA also shared some ground with the green activists who campaigned to save scenic places from development,

[144] Parliamentary Public Works Committee at Kerang at Kerang, "Hooper (1)," 3–4.

[145] Parliamentary Public Works Committee at Kerang at Kerang, "Hooper (1)," 5.

[146] Parliamentary Public Works Committee at Kerang at Kerang, "Hooper (1)," 118.

[147] Parliamentary Public Works Committee at Kerang at Kerang, "Hooper (1)," 119.

[148] Mulligan and Hill, *Ecological Pioneers*, 137–43.

though not their militant campaigning style. However, like them, it effectively discarded Aboriginal landscape modifications.[149] They understood wilderness as primeval, non-human landscapes, "as Europeans supposedly found them—and as Aborigines made them—and it [called] them untouched, pristine."[150]

The concerns that the organization addressed, however, were not unfounded. Again, there were uncertainties involved. ACIL et al. observed a general paucity of data that could be used to predict the impact of man-induced salinity on lake ecosystems. Provided that water birds had alternative wetlands, the increase of salinity in some of them was believed to have little effect, as most water birds in Australia were adapted to visiting several habitats. For example, Lake Tutchewop continued to support large populations despite high salinity levels. However, as more wetlands were being used for evaporative disposal or saline drainage water, the Kerang area was considered likely to provide reduced wildlife value in terms of food, refuge, and breeding sites. Then again, the increased area for irrigation had probably provided greater food opportunities to some bird species as channels and storages associated with irrigation also provided water bird resting sites.[151]

The critique was not new. After the recommendations made in the "Gutteridge Report," which had included the Tutchewop-Tyrrell scheme, had been made public, the Nature Conservation Society of South Australia was alarmed because it foresaw gross damage to the river scene and wildlife. The Society urgently demanded a survey of natural values, and how they were affected by present management, as well as any proposed works, before any measures were implemented. It denounced that river regulation had greatly changed the river environment, drowned Red Gum trees, and reduced breeding areas for wildlife. Evaporation basins had already done much of the damage. Now, with several proposals

[149] Mulligan and Hill, *Ecological Pioneers*, 217; Hutton and Connors, *A History of the Australian Environment Movement*, 159–64.

[150] Tom Griffiths, "History and Natural History. Conservation Movements in Conflict?," in *The Humanities and the Australian Environment*, ed. D. John Mulvaney (Canberra: Australian Academy of the Humanities. Symposium, 1991), 93.

[151] ACIL Australia et al., *Causes, Extent and Effects of Salinity*, 119–24.

that were designed to reduce river salinity, "complete destruction of the natural environment" was at hand.[152]

Furthermore, the VFGA was not the only party to advocate a cautious use of engineering measures for salinity control. The Fisheries and Wildlife Division of the Department of Conservation stressed the urgent need to effectively manage and conserve the remaining public wetland habitats, as their area had dramatically declined due to land-use practices.[153] In principle, the City of Swan Hill supported the need to dispose of saline water, if and only if this did not interfere with the ecological and economic balance of any area. Until it could be proven that the Tutchewop-Tyrell scheme did not disturb the balance of any location, the city demanded that it should not commence.[154] This conformed with the Ultima farmers, who had protested the scheme early on. However, their criticism was based on economic, not environmental, grounds.

Either way, the majority of parties did not subscribe to the preservationist view that the Victorian Field and Game Association put forward, which was criticized as "static" since it discarded the numerous changes in the Kerang area's environment prior to the advent of Europeans.[155] Instead, most sided with the SRWSC. From the point of view of many irrigators, the economic benefits of salinity control were deemed more important than its effects on the environment. Importantly, irrigation water was at stake. Parallel to the Salinity Inquiry, the Public Works Committee had been tasked to inquire into water allocations in Northern Victoria to determine the surplus water that would be available with the completion of the Dartmouth dam to irrigators, towns, flora, and fauna.[156] The Torrumbarry System Irrigators' Association told the inquiry, aiming at the VFGA, that white settlement

[152] Anonymous, "Salinity—And the River's Life and Looks," *The Riverlander* 25 (June 1971): 8–9.

[153] Anonymous, "Wetlands Habitat Need Water," *The Northern Times*, July 14, 1978.

[154] Parliamentary Public Works Committee at Kerang at Swan Hill, "Gary John Mennie," October 3, 1978, VPRS 11,559/P0001, 24.07.27, PT5, Public Record Office Victoria, Melbourne, 931–32.

[155] Parliamentary Public Works Committee at Kerang at Melbourne, "Harvey Hayward Bawden," September 29, 1976, VPRS 11,559/P0001, 24.07.27, PT1, Public Record Office Victoria, Melbourne, 191.

[156] Salinity Committee, *Fourth Report to Parliament. Final Report on: Water Allocations in Northern Victoria* (Melbourne: Salinity Committee of the Victorian Parliament, 1984), 6; 93–94.

and irrigation development had not had an overall reductive effect on wildlife and wetlands. Yes, irrigation had changed the environment, but it had *improved* wetlands and provided feeding grounds such as irrigated paddocks to wildlife, whereas, before European settlement, the natural environment of Northern Victoria had been "harsh and dry," and waterways then quickly shrunk back to water holes, while lakes nearly dried up.[157] This was, rather, an echo of the water dreamers' idea of improving the crude and unformed "raw elements" of Australia and refashioning them into a "civilised, natural landscape."[158] The irrigators claimed that environmental considerations should be in accordance with "the reality of present-day demands of closer settlement and greatly increased and reliable food supply along with a reasonable area for flora and fauna." The environment, in their view, was already catered for in terms of water supply.[159]

Other irrigators' interest groups conformed. The Rodney Irrigators League, in a rigorous statement in defense of the irrigation dream, claimed that "[t]remendous pressures are being exerted by Conservation and Environmental bodies which have a great political influence. They seek more water for wild life and river flow, giving little consideration to irrigators' requirements." Meanwhile, the price of water was increasing, and Melbourne users had their eyes on Northern water. Every irrigator was called to help "combat these other bodies who are, in effect, attacking our livelihood."[160] Likewise, the Goulburn–Waranga Irrigators League, while generally in agreement about conserving flora and fauna, considered it a priority to increase the amount of water available to family farm units in order for them to be viable. Water should only be allocated to the natural environment when surpluses were available and declined specific allocations in preference to the allocation of water for food production.[161]

Other positions were more differentiated. The grazier Graham Hacon was adamant that "[e]verybody within the community is in some degree

[157] Anonymous, "'Farms Before Fauna'—TSIA," *The Northern Times*, July 7, 1978.

[158] Cathcart, *The Water Dreamers*, 199.

[159] Anonymous, "'Farms Before Fauna'–TSIA."

[160] P.J. Fleming,"Rodney Irrigators League," VPRS 11,559/P0001, 24.07.27, PT7, Public Record Office Victoria, Melbourne.

[161] Parliamentary Public Works Committee at Kerang at Shepparton, "Francis John Farrell," February 8, 1977, VPRS 11,559/P0001, 24.07.27, PT2, Public Record Office Victoria, Melbourne, 385–87.

responsible for creating the situation." Irrigators had been "guilty of a sin" because "what we have done is twist the arm of the Commission and say: give us more water, and more water, with the result that we are killing ourselves off with underground water." It was now imperative for the water table to be significantly lowered before any further water allocations were granted. Irrigators should learn to live with the present rights and a little more.[162]

Among almost all parties invested in the salinity issue there was a general consensus that the Murray Valley was an anthropogenic working landscape that required a further modification in order to retain its present use. A small, but vocal, minority maintained that the modifications made since European occupation had degraded a once-pristine environment and advocated the preservation of what was left of the area's wild spaces. However, this was not a generally shared position in Northern Victoria's irrigation communities. On the contrary, the redesign of the Murray Valley's environment by the SRWSC was deemed beneficial. A technological fix was demanded to ensure its future viability. Salinization should be solved by engineering and agriculture so that irrigation could continue and expand. In this regard, most community representatives agreed with the view put forward by the Commission.

However, notwithstanding some of the angry interventions by irrigators' organizations in defense of their water against all other demands, the environmental impacts of irrigation through its great water use could no longer be disregarded. It was recognized that irrigation was contributing to soil salinization and the sacrificing of habitat for drainage disposals.

Paul Sinclair has argued that in the 1940s and 1950s, the Murray was seen as a resource with unlimited capacity to sustain agricultural production, facilitated by sophisticated technological change, whereas environmental deterioration was regarded as a necessary loss in the pursuit of the common good.[163] By the mid- to late 1970s, the deterioration was obvious and acknowledged to be causing grave disturbances. It had to be addressed—sometimes emphatically, sometimes reluctantly, or by way of lip service, as in the case of the Torrumbarry irrigators who recognized that "our community needs a healthy environment" while at the

[162] Parliamentary Public Works Committee at Kerang at Shepparton, "Hacon," 790–93.
[163] Sinclair, *The Murray*, 76–80.

same time declining any concession toward the water needs of flora and fauna.[164]

Perhaps the balancing act between holding up the promise of irrigation and recognizing the importance of environmental considerations in the face of widespread degradation is best illustrated by a statement given by Peter Fisher, the Federal Member for Mallee, before the Salinity Inquiry at Swan Hill. Fisher pointed out that the Murray was "being used beyond its capacity in a situation where its natural flow is often reversed," resulting in major conflicts between maximum irrigation benefits and flood mitigation needs. Demands for more irrigation had the potential to pose irreversible damage to the environment. Land salinization, waterlogging, riverbank erosion, and the impairment of water quality along with the associated social and economic implications were just some of these conflicts in water and land use. Water quality, he added with an appeal to the issues and mood of the eco-age, was "possibly more critical to Australia's future quality of life than even crude oil and while using it to the maximum, we must appreciate it as a limited commodity." It was imperative to "ensure that the Murray system, which encompasses a diverse and unique habitat and a sensitive ecology, is preserved for future generations." This, Fisher maintained, meant saving "Australia's most productive irrigated region and our most successful closer settlement venture on a national scale." It could be achieved by a major technical and physical response similar to the Snowy Mountains Hydro-Electric Scheme which would ensure that the Murray, which supported "an intensively successful decentralisation effort remains stable and also has the ability to absorb future growth."[165] In the same manner, *The Riverlander* could celebrate George Chaffey as the "visionary who gave us irrigation" while simultaneously condemning the deterioration caused by salinization and alerting its readers to the pressing global environmental issues of the time.[166]

Ecological consciousness and water dreaming were not mutually exclusive; they permeated each other. While it was clear that the Murray could no longer be regarded as an unlimited resource—this very notion had

164 Anonymous, "'Farms Before Fauna'—TSIA."

165 Parliamentary Public Works Committee at Kerang at Swan Hill, "Fisher," 967–77.

166 Anonymous, "The Visionary Who Gave Us Irrigation," *The Riverlander* 31 (March–April 1977): 6.

contributed to the whole resource now being at risk—the communities in the GMID were still very dependent on irrigation's economic viability. Furthermore, their identities were closely tied to the promise of irrigation, which had arguably been delivered. They were not quick to part with the conviction that, despite all damage caused, irrigation was a benevolent endeavor that could be improved to safeguard their economic and social future, as well as their environment.

The possessive pronoun is important. Paul Warde, Libby Robin, and Sverker Sörlin have argued that the crucial conceptual change to our contemporary use of "environment" is that it came to be understood as a global object with a fundamental unity. It connects the local and the global, it scales up and down, inside and out, whereas before, conservation had been a predominantly localized issue to a planetary environment.[167] In the case of the GMID's irrigation communities, despite a certain public sensitivity for global environmental issues and the infiltration of popular language with the vocabulary of ecology, it was always a *specific* environment—the Murray Valley—that was at stake. However, it had become regarded as endangered and limited, an environment of quantitative restraints, "understood as a numerically specified dwindling supply of raw materials and an ever-growing mass of pollutants."[168]

In retrospect, R. J. Shannon, the Director of Engineering Services at South Australia's Engineering and Water Supply Department, was too quick to announce in 1977 that "the days of development for development's sake are in the past."[169] But the question of developmental limits was the unresolved environmental issue that salinization introduced into the Murray Valley.

[167] Warde et al., *The Environment*, 11–12.

[168] Warde et al., *The Environment*, 51–52.

[169] R.J. Shannon, "Things are Moving on the Murray," *Riverlander* 31 (December 1977): 30.

CHAPTER 6

Community Activism (1966–1988)

On Mastering a Slow Catastrophe

Slow catastrophes pose a daunting challenge to those working to arrest them. In 1984, the rural sociologist Sharman Stone summarized the problem of salinization as follows: "We have soil and water salinisation which is to an unknown extent affecting an unspecified proportion of the best farm land and major rivers of Australia. We do not know how quickly the problem is spreading, and we are not quite sure yet how to best deal with it. Estimated losses in production through salting range from two to 70 per cent (sic) and no one is quite sure how many millions of dollars that costs, or how many people or what things suffer."[1]

This vagueness was painfully evident to people like Gyn Jones, an irrigation officer at Kerang, who reflected in 1976 that it was "hard to set the community alight with enthusiasm when problems are chronic. It's disaster which brings action and the disaster is going along quite

[1] Anonymous, "Salt 'Disaster' Warning for City People," *The Northern Times*, February 24, 1984.

© The Author(s), under exclusive license to Springer Nature Switzerland AG 2023

D. Rothenburg, *Irrigation, Salinity, and Rural Communities in Australia's Murray-Darling Basin, 1945–2020*, Palgrave Studies in World Environmental History,
https://doi.org/10.1007/978-3-031-18451-2_6

slowly here at present."[2] Salinization, that "slow, insidious, creeping disaster" had been insistently, but inconsistently, spreading through Northern Victoria since the 1890s, rendering soils useless, water hazardous, reducing farm productivity, and causing gradually rising unrest in its communities.[3] For decades, the official responses were phlegmatic and inconsequential. Some communities resorted to self-help, setting up research farms to investigate the reclamation of salted land, largely by their own means. It had taken the visible drought of 1967–1968 to garner salinization greater attention. However, events do not speak for themselves. They must be interpreted in a discursive and political space to gain meaning and have material consequences.

The drought, however, did not raise attention over salinization due to the fate of the irrigation communities affected by it. Rather, it was pressure from South Australia, which was worried about Adelaide's water supply, and the RMC's impetus to save water amidst a storm of criticism of irrigation's alleged inefficiency that made salinity an issue for governments and the public. In the mid-1970s, activists in the GMID looked for ways to make a stronger case for the problems in their own communities. Then, they found an opportunity. For the first time, the floods of the early decade caused a veritable salinity problem for the Shepparton region. The Water Commission tabled its Salinity Control and Drainage strategy, and Victorian parliament provided a stage to voice concerns with its Salinity Inquiry. A group of dedicated people seized the chance to set up GIRDAC. Their counterparts in Kerang, lacking a comparable highlight, were unable to launch a similar community group until almost two years later.

While the action committee in Shepparton was born out of the "salinity panic," Kerang's situation was more difficult. GIRDAC member Leon Heath reflected that "you have had a salinity problem for a long time and it lends itself to a certain amount of apathy."[4] Still, activists in Shepparton, in the long run, had to address similar problems. Heath claimed that "[t]he majority of farmers lack fundamental understanding of basic

[2] Gynlais O. Jones, "Gun Shots—Moves are afoot for KERDAC," *The Northern Times*, February 10, 1976.

[3] Anonymous, "'Survival' Key Word for Local Area," *The Northern Times*, July 2, 1976.

[4] Anonymous, "Community Call," *The Northern Times*, March 2, 1976.

problems associated with soils, irrigation and salinity."[5] Despite the best efforts of those who cared, the Public Works Committee's hearings were not mass meetings, and few people voiced their concerns. Against this backdrop, both Salinity Action Groups shared the same basic challenges. Firstly, to educate the community on the effects of salinization due to a perceived lack of realization that there was a problem.[6] Secondly, they devoted much of their energies to lobby politicians and governments for attention and money toward solving their salinization problems. This was especially relevant for the Kerang region, which was considered economically negligible and environmentally difficult. In the words of Gyn Jones, "our region is being forgotten and it is up to us to remind those who have push and finance and resources that we have an important place in the Victorian, indeed Australian, community."[7]

The character of these community groups is complex. However, at a basic level, they can be usefully understood as conservation groups, although it would be a stretch to call members environmentalists. Despite their emergence during the advent of the green period around 1970, their political methods and modes of organization belonged to the old conservationism rather than the new environmentalism. These activists largely relied on representations and quiet persuasion rather than open protest. They often bitterly condemned the environmental movement's predominantly urban protagonists for their advocacy of saving pristine wilderness from people. Nevertheless, they adopted the language of popular ecology and were receptive to the emerging discourses about the global environment that were both signatures of the Age of Ecology, while still retaining the position that conservation was a predominantly localized issue.

Salinity Action Groups, lastly, are not properly understood as proponents of the power of civil society in conservation issues. They were closely intertwined with both government and industry representatives and interests. Joachim Radkau has reflected that, in many countries, the phenomenon of environmental movements could not be understood without considering government support. This was due to the practical consideration that many environmental problems could not be solved

[5] Parliamentary Public Works Committee at Melbourne, "Heath," 1059; 1061.

[6] Anonymous, The Northern Times, 27.08.1974, "MV Community Should Fight Salinity," *The Northern Times*, August 27, 1974; Anonymous, "Community Call."

[7] Gynlais O. Jones, "Gun Shots—Moves are afoot for KERDAC," *The Northern Times*, February 10, 1976.

without state regulation. This is certainly true for the anthropogenic working landscape that was the Murray Valley, which had been engineered and continued to be administered by powerful government agencies. Radkau further argues that, from this point of view, a focus on the state is virtually essential for environmentalism, which often had to be understood as a hybrid of state and civil society. This final point becomes clear when tracing the roots of the Salinity Action Groups, which, once again, leads to the conflicts about irrigation emerging in the 1960s.[8]

The Birth of Salinity Action Groups from the Spirit of Ernest Jackson

Moves of "history-making importance" are announced too often to not justify a certain skepticism. However, when the Goulburn and Waranga Water Users' United League decided to form a new body to promote irrigator interests on January 4, 1966, it is hard to deny the retrospective importance of this event. The new organization was to: "develop public relations and promotional activities to ensure that Parliament and the public was kept aware of the importance of irrigation, the importance of developing adequate water supplies for the irrigation industry, and to answer attacks on irrigation, and support research and extension."[9]

The initiators postulated that all sections of the industry had to find a way to work together to improve and promote irrigation. This task called for the creation of a body to represent growers, secondary industries, and government departments associated with irrigation farming. The move was considered an "answer to people who make attacks on irrigation, or who have designs on the essential water supplies which make possible the huge productivity of the irrigation industry." Its proponents should "fight back," and "prove to the public everywhere that irrigation farming [is] an indispensable ingredient in Victoria's prosperity."[10]

[8] Joachim Radkau, *Die Ära der Ökologie. Eine Weltgeschichte* (Bonn: Bundeszentrale für Politische Bildung, 2011), 19–20. The English translation is missing the paragraph referred to here.

[9] "New Organisation Promote Irrigation," *Pyramid Hill Advertiser*, January 12, 1966.

[10] "New Organisation Promote Irrigation," *Pyramid Hill Advertiser*, January 12, 1966.

The resulting body was named the Victorian Irrigation Research and Promotion Organisation, or VIRPO.[11]

Given the upfront attacks on irrigation's *raison d'être* by economists such as Bruce Davidson, who provocatively stated that "the establishment and continued expansion of irrigation in Australia can only be declared as a mistake," that caused such a stir during the 1960s, it was not hard to guess in which direction these first bullets had been fired.[12] Furthermore, those who founded VIRPO jealously guarded their irrigation water supplies against Melbourne's interests. However, they clearly understood that the industry had problems that needed addressing by the cooperation of irrigators, secondary industries, and government departments. They found that communication between irrigators was weak, and there was a need for more extensive research on irrigation problems in order to make the industry operate with greater efficiency and profitability.[13]

At a meeting of about 200 attendees representing irrigator organizations, primary and secondary industries associated with irrigation, and government departments at the Civic Centre in Shepparton on March 18, 1966, VIRPO was founded. This broad attendance reflected the impetus to bring together all parties interested in the defense and further expansion of irrigated agriculture. The SRWSC was represented by its commissioner K. D. Green and three other officers. The Fisheries and Wildlife Department, CSIRO, the Forests Commission, the Department of Crown Lands, the Soil Conservation Authority, the Department of Agriculture, and the MVDL also sent representatives. So did 15 irrigator leagues, all from the Shepparton region. Commercial interests also responded. Inter alia, the Shepparton Preserving Company (SPC), Murray Goulburn, BASF, Campbells, Heinz, and Mid Murray Dairy Co. were represented. The Honorable P. V. Feltham, Legislative Assembly member on behalf of SPC, pledged support, including financial contributions for an initial trial period. The meeting unanimously decided to form

[11] "It's Called VIRPO," *Kyabram Free Press*, March 23, 1966.

[12] Davidson, *Australia Wet or Dry?*, 14.

[13] "New Organisation Promote Irrigation," *Pyramid Hill Advertiser*, January 12, 1966; Goulburn and Waranga Water Users' United League, "The Objects and Organisation of the Proposed Organisation for Promotion and Research of Irrigation," VPRS 6008/P0001, 66/0245, Public Record Office Victoria, Melbourne, 1; Victorian Irrigation Research and Promotion Organisation, "Constitution, 1966," VPRS 6008/P0001, 66/0245, Public Record Office Victoria, Melbourne, 1–2.

VIRPO and elect an Executive. A chairman was complemented with 18 members. Six were to come from irrigator organizations; another six from commercial interests, with SPC, Tatura Milk Products, and Stanhope Milk Co-Op already on board; and a further six were to come from government departments. Members were sought from the Water Commission, CSIRO, the Department of Agriculture, the Forests Commission, and the Rural Finance and Settlement Commission.[14]

While Commissioner Green assured the meeting of the SRWSC's support for the new organization, the Commission's internal assessment of VIRPO was rather cautious. A memo authorized by Green reads that the SRWSC was prepared to support planned promotional works to foster awareness of irrigation's importance by providing basic information, venues, and conducting tours of works and projects to build the irrigation story. While it also supported the new organization's objectives in principle, it believed that VIRPO "would do the irrigation image great harm if it sought to develop this awareness by injudicious attacks on the water needs of the metropolis." The basic need was for all water users "to recognise and acknowledge the requirements of other water users." Irrigators had to accept the fact that the exploitation of water resources "must provide for all consumers in a manner which will allow for balanced development and growth of all sectors of the nation."[15]

W. B. Gayfer, Rural Finance and Settlement Commission officer at Shepparton, was more openly critical, reporting after he had attended the first executive meeting that he "could not help but feel that greed is foremost in the minds of the irrigators concerned [...], in that, any suggestions of spreading available water further was quickly stamped out." Furthermore, the officer was concerned about a third of the voting power in the executive being held by government departments, "as while it is

[14] State Rivers and Water Supply Commission, "Notes on Inaugural Meeting to Form Irrigators' Promotion and Research Organisation," March 18, 1966, VPRS 6008/P0001, 66/0245, Public Record Office Victoria, Melbourne, 1–2; F. J. Leete and A. O. Brown, "Formation Meeting of the Victorian Irrigation Research and Promotion Organisation," March 18, 1966, VPRS 6008/P0001, 66/0245, Public Record Office Victoria, Melbourne, 5–9.

[15] K. D. Green, "Goulburn and Waranga Water Users United League. Proposed Promotional, Educational and Research Organization, March 17, 1966," VPRS 6008/P0001, 66/0245, Public Record Office Victoria, Melbourne, 1–2; 5.

claimed that the organization will never be used as a pressure group, I fail to believe that this ideal will be maintained."[16]

VIRPO's first moves were to seek close alignment with commercial interests and government departments for promotional activities. In April 1966, it appointed R. A. Gardner and Associates to promote VIRPO to secondary industries and to approach potential donors to fundraise for the organization's operating costs. The firm was further appointed to create a suitable slogan to advertise irrigation on product labels, to produce a booklet on irrigation for general distribution, to strategize on how to communicate with irrigators, and to discuss with bus companies tours of irrigation areas. Plans to extensively cooperate with the Water Commission were made to place signs and display boards at important irrigation locations. VIRPO also considered commissioning a study on irrigation's economics by an acknowledged expert "who could wear the mantle of independence" to defend irrigation against its critics.[17]

Its early close liaison with commercial interests went so far that, for a time, VIRPO collaborated with the Coca-Cola Corporation and produced three booklets about *The Story of Water in Australia* to defend irrigation in Victoria's classrooms. Although no mention of the organization is made in the booklets, the original concepts and most materials came from VIRPO. The first brochure was researched and written by Frances Leete in his capacity as an employee of Coca-Cola's Sales Promotion Department, who was then also the chairperson of VIRPO. Aided by the Water Commission, these booklets, of which 60,000 copies were printed, were considered valuable irrigation promotion ventures and were distributed to high schools throughout Victoria as school material with the hope that it would be adopted in classes.[18] While some of the booklets' contents are indeed educational—covering basic aspects of the nature

[16] W. B. Gayfer, "Re-Victorian Irrigation Research and Promotion Organisation," May 2, 1966, VPRS 6008/P0001, 66/0245, Public Record Office Victoria, Melbourne.

[17] Victorian Irrigation Research and Promotion Organisation, "Minutes of Executive Meeting," April 29, 1966, VPRS 6008/P0001, 66/0245, Public Record Office Victoria, Melbourne, 1–2; Victorian Irrigation Research and Promotion Organisation, "Untitled," 1966, VPRS 6008/P0001, 66/0245, Public Record Office Victoria, Melbourne, 1–5.

[18] Ernest M. Jackson, "Victorian Irrigation Research and Promotion Organisation, Executive Officer's Report—November/December 1968," December 24, 1968, VPRS 6008/P0001, 68/0629, Public Record Office Victoria, Melbourne, 1; Ernest M. Jackson, "Victorian Irrigation Research and Promotion Organisation, Report of Executive Officer. May–June 1969," June 27, 1969, VPRS 6008/P0001, 68/0629, Public Record Office

of water, the water cycle, its distribution in Australia, and the history of Australian irrigation—they also convey the embattled irrigation dream that its proponents were so eager to uphold. Water had to be conserved, instead of merely running away uncontrolled and eventually wasting itself in the sea. Irrigation helped to feed the nation, was an insurance against drought, provided stability and flexibility in agriculture for the nation and the farmer, high yields per acre, increased population, closer settlement, and prosperity through increased revenues of state and export incomes. It was an affirmation of the official rationale for irrigation development and the state's policy of the industry's prioritized access to water supplies. And, of course, Coca-Cola was the pinnacle of water refinement.[19]

VIRPO also quickly established itself as a body promoting the extension of irrigation methods. To aid with the improvement of irrigation techniques, it established an extension sub-committee to advise the organization regarding "best ways of getting useful new techniques adopted by irrigators to their advantage" such as "better use of water, fertilizers, machinery, and so on."[20] Its chief public activity in this field was conducting Field Days on new developments in the industry. These events continued to attract public, commercial, and government interest throughout VIRPO's lifespan. They served as a means to convey new developments to irrigation farmers. The first one, named "Irrigation '68," was held at the Shepparton Civic Centre from June 12 to 14 June and boasted an extensive program. On a flyer advertising the convention, the front side invited people to attend panels on subjects such as water policy, agricultural techniques, fertilizers, water storage and re-use, drainage, and product marketing. The other side contained the ladies program, inviting

Victoria, Melbourne, 1; Victorian Irrigation Research and Promotion Organisation, "Minutes of Executive Meeting," August 8, 1969, VPRS 6008/P0001, 68/0629, Public Record Office Victoria, Melbourne, 1.

[19] Frances Leete, *Life, Wealth and Power: The Story of Water in Australia* (Crows Nest: The Coca Cola Export Corporation, 1966(?)), 20–24; 32–35; 54–56; Anonymous, The Coca Cola Export Corporation, Irrigation. Water in Australia Book 3, Crows Nest: The Coca Cola Export Corporation, 1969(?)).

[20] Victorian Irrigation Research and Promotion Organisation, "Notes on the First Meeting of the Sub-Committee on Extension," November 7, 1966, VPRS 6008/P0001, 66/0245, Public Record Office Victoria, Melbourne, 1.

women to panels on "Better Living on the Farm" including tips on cake decorating, fruit preserving, cooking meat, egg, and rice menus.[21]

Apart from being able to attract substantial commercial interest, demonstrated by the high number of exhibiting firms, Field Days such as "Irrigation '68" illustrate that VIRPO was capable of summoning high-level government officers and ministers. The opening of the convention was performed by the Acting Minister for Water Supply, the Honorable J. C. M. Balfour. The Minister for National Development, The Honorable David Fairbairn, addressed the attendees of the "gents' program" on "A National Water Policy," and the Right Honorable J. McEwen, Deputy Prime Minister for Trade and Industry, spoke on the subject of "Will We Sell Our Products?" Other speakers included employees of CSIRO, the Victorian Departments of Agriculture and Mines, and the University of Melbourne.[22] VIRPO was, indeed, able to collocate the parties interested in the welfare of the irrigation community at events regarding its development and future. Or, as Frances Leete put it, the organization's role was "to initiate, inspire and assist its components with the know-how prestige and influence it has shown it commands."[23]

However, despite high ambitions, VIRPO never became a professional organization. Nor did it become self-supporting. Consequently, the Executive was continuously busy fundraising. At the outset, it employed a paid chairperson, executive officer, and secretary to carry out its day-to-day work. It contracted a professional firm to inquire into effective promotional activities.[24] Nevertheless, despite many pledges for support, companies' and state governments' willingness to fund VIRPO was ultimately underwhelming. Salaries for officers were scrapped and replaced

[21] Victorian Irrigation Research and Promotion Organisation, "Irrigation '68. Convention and Exhibition," VPRS 6008/P0001, 68/1934, Public Record Office Victoria, Melbourne.

[22] Victorian Irrigation Research and Promotion Organisation, "Exhibitors at V.I.R.P.O. Field Days/Convention 1968/69, VPRS 6008/P0001, 68/0629, Public Record Office Victoria, Melbourne; "VIRPO, Irrigation '68. Convention and Exhibition."

[23] Francis J. Leete, "VIRPO, Report 1968/69," April 16, 1969, VPRS 6008/P0001, 68/0629, Public Record Office Victoria, Melbourne.

[24] Victorian Irrigation Research and Promotion Organisation, "V.I.R.P.O.—The Next Stage," 1966, VPRS 6008/P0001, 66/0245, Public Record Office Victoria, Melbourne, 1; Leete, "Report 1968/69," 1; Victorian Irrigation Research and Promotion Organisation, "Minutes of Executive Meeting," July 15, 1966, VPRS 6008/P0001, 66/0245, Public Record Office Victoria, Melbourne, 1.

with reimbursements for expenditures and travel, making VIRPO a quasi-volunteer organization.[25] Consequently, it heavily relied on motivated and dedicated individuals. Among these, its executive officer, later secretary, Ernest Menzies Jackson, affectionately nicknamed "Watershed," quickly emerged as a *spiritus rector* and dominant figure. Jackson was not only a zealot and a seemingly tireless worker, who wrote prolifically to newspapers, but he also commanded a large network of personal relationships with news media people and leading politicians. This made him an integral part of VIRPO's activities. Significantly, he also became a key figure in the formation of the Salinity Action Groups.[26]

Based in Albury, Jackson was one of the charismatic leading protagonists that shaped the salinity scene for activists that would fully emerge in the 1970s. Once an irrigation farmer, later a lobbyist for irrigation, he finally turned into a conservationist, without ever relinquishing his fervent advocacy of irrigation. Born in Kew, a suburb of Melbourne, in 1901, he had been an information officer at the SRWSC from 1954 to 1964 when he retired. Far from being a young man, he then became a member of the VIRPO Executive soon after.[27] In his later years, an *Age* reporter described him as "[t]all, if a bit stooped, and gnarled as an old man gum tree."[28] Both a campaigner for much of his life and the author of a "catchment philosophy," he was deemed "Australia's oldest environmental campaigner" and, with a questionable metaphor, hailed as "one

[25] W. A. Borthwick, Minister for Water Supply, to The Honorable Sir Henry Bolte, M.L.A., Premier of Victoria, "VIRPO—Request for Assistance," April 9, 1968, VPRS 6008/P0001, 68/0629, Public Record Office Victoria, Melbourne, 1–2; Victorian Irrigation Research and Promotion Organisation, "Statement of Accounts—September 2 October 16, 1975," VPRS 6008/P0001, 74/2225, Statement of Accounts; Victorian Irrigation Research and Promotion Organisation, "Statement of Receipts and Payments for Year Ended 31st March, 1977," May 12, 1977, VPRS 6008/P0001, 68/0629, Public Record Office Victoria, Melbourne.

[26] Victorian Irrigation Research and Promotion Organisation, "Minutes of Executive Meeting," October 23, 1975, VPRS 6008/P0001, 74/2225, Public Record Office Victoria, Melbourne, 1; Anonymous, "Land Care Crusader 'Watershed' Jackson," *Trees and Natural Resources* 32, no. 4 (1990): 27; Victorian Irrigation Research and Promotion Organisation, "Minutes of Executive Meeting," July 15, 1966, 1.

[27] Geoff Maslen, "A Life Devoted to the Nation's Waterways," *The Age*, October 18, 1990; "Lobbyist With a Missionary Zeal," *The Age*, November 30, 1988; Anonymous, "Land Care Crusader 'Watershed' Jackson," 26.

[28] Maslen, "A Life Devoted to the Nation's Waterways."

of the earliest crusaders for improving Australia's rural environment."[29] Jackson preferred to refer to himself as an "angry realist" and claimed to have adopted his approach to things from Cervantes' Don Quixote: "Only He who Attempts the Ridiculous Can Achieve the Impossible"[30] (Fig. 6.1).

Jackson's father, also named Ernest, was the wool and produce manager of a pastoral house and a farmer, nicknamed "Super" Jackson for his pioneering use and advocacy of superphosphate fertilizer. Jackson, Sr. helped establish the Victorian Pasture Improvement League, making him a role model for his son. Jackson Jr. left school before his intermediate certificate and went jackarooing, an apprenticeship on a sheep station, on Boomanooma Station on the Murray. After this, he and his brother moved to the family's irrigation property at Lockington, Victoria. During the Depression, he quit the farm and began work at the Murray Valley Development League.[31] As an anecdote has it, Jackson did not consider himself to be a good farmer: "The father of a girl he was once friends with told him the two essentials for matrimony were common sense and money. 'He implied I had neither—and he was right!'".[32]

Jackson did make a good organizer, though. In 1944, he was elected organizing secretary of the Goulburn and Waranga Water Users' United League. He campaigned for the enlargement of the Eildon dam, then went on to become an education and film officer of the "Save the Forests" campaign, scripting, filming, and showing the campaign's films. Afterward, Jackson moved on to a job at the Victorian Department of Agriculture where he was the Officer in Charge of the department's mobile extension unit tasked with taking information to rural Victoria. At the ages of 52 and 45, respectively, Ernest Jackson and Vera Margaret

29 Anonymous, "Land Care Crusader 'Watershed' Jackson," 26.

30 Ernest M. Jackson, *Replenish the Earth. Living Resource Conservation for Sustainable Development* (Albury: Catchment Education Trust, 1982), 5; Ernest M. Jackson, "Conservation Philosophies," April 4, 1976, Box, George and Maureen Hardwick, Appin, 1.

31 Anonymous, "Land Care Crusader 'Watershed' Jackson," 26–27; Maslen, "A Life Devoted to the Nation's Waterways," *The Age*, October 18, 1990; "Lobbyist With a Missionary Zeal," *The Age*, November 30, 1988; Wells, Up and Doing, 79–80; Jackson, *Replenish the Earth*, 6; 30; Cliff L. Spowart, "Kerang Irrigation Region Salinity Action Committee, Minutes of Meeting held at the Kerang on the 2nd. of October, 1990," October 2, 1990, VIRPO Folder, George and Maureen Hardwick, Appin, 2.

32 "Lobbyist With a Missionary Zeal," *The Age*, November 30, 1988.

Fig. 6.1 Geoff Maslen, "Lobbyist with a missionary Zeal." The Age, November 30, 1988 (*Source* © The Age, Melbourne. The use of this work has been licensed by Copyright Agency except as permitted by the Copyright Act, you must not re-use this work without the permission of the copyright owner or Copyright Agency)

Haughton married. Vera worked in the Nursing Division of the United Nations. She backed Ernest's activities with patience and understanding, while leaving the center stage to her husband. Together, they set up the Watershed Association (later Catchment Education Trust) in 1960.[33]

"Watershed" Jackson received his nickname from the term used for a catchment in North America, meaning the land area that is drained by a river and its tributaries.[34] In his 80th year, he published a book as a "record of his beliefs and convictions," entitled *Replenish the Earth*, which was dedicated to this concept. What the author described as a "pot-pourri or hopefully, a literate companion," may best be characterized as an idiosyncratic collage of essays and materials that proceeds from a few basic principles and then goes on to support these by an eclectic selection of extracts from writings by others.[35] Ultimately, Jackson wanted to illustrate that the concept of the catchment—"Nature's own natural division"—should be the base of natural resource management.[36]

Water lay at the core of what he termed his "catchment philosophy." The primacy of the element and its gravity flow through the landscape, linking together all components of the environment, was his starting point, even his dogma: "[A]ll living and non-living things exist in run-off or drainage areas. These areas, bounded by their ridges and common water lines, are 'Catchments'. Whatever its size or situation, a catchment, within itself, is a complete physiographic and hydrologic entity in which nature has 'packaged' the inter-related problems of water, land

[33] Anonymous, "Land Care Crusader 'Watershed' Jackson," 26–27; "A Life Devoted to the Nation's Waterways," *The Age*, October 18, 1990; "Lobbyist with a Missionary Zeal," *The Age*, November 30, 1988; Ernest M. Jackson, "Relatives & Friends of the Late Vera Margaret Jackson (Nee Haughton)," January 1988, VPRS 6008/P0001, 84/36649, Public Record Office Victoria, Melbourne; Jackson, *Replenish the Earth*, 6; 30; Spowart, "Minutes," October 2, 1990, 2; Anonymous, "Our Hidden Memorials, 2013," *eSplash eMag* 59 (2013), accessed August 16, 2022, https://ulyssesmelb.com/system/files/artworkz_esplash_059.pdf.

[34] David Lambert, *The Field Guide to Geology* (New York: Facts on File, new edition 2007), 130; "Lobbyist With a Missionary Zeal," *The Age*, November 30, 1988.

[35] Jackson, *Replenish the Earth*, 5; 2.

[36] Ernest M. Jackson, "Australian Assocation for Environment Education. Inaugural Conference—October 19–24. Introduction to the Catchment Philosophy," VPRS 6008/P0001, 80/07411, Public Record Office Victoria, Melbourne, 1.

and people. [...] Indeed Catchments are everywhere and everything to everybody."[37]

Ernest understood himself as a conservationist who emphasized the fundamental unity of land, water, and people, and thus "man's dependence on the elements and his need to toil and protect his earth." He saw his own thinking as fundamentally ecological, understood as "The Science of Everything," embracing all living and non-living things in their relationship with each other: "There are no free lunches in Nature. Everything is dependent on everything else."[38]

With his championing of the idea that the catchment, as a supposedly natural unit, should be the base of resource management, Ernest drew on a long legacy of approaches, both from Australia and other countries. The adoption of river catchments as the basic units for the management of land reclamation and drainage activities was promoted in Britain during the interwar years to replace clumsy, fragmented inherited jurisdictions. In 1930, arrangements were made to subdivide the country into catchment areas, and in 1948, local management bodies were introduced to administer entire catchments. For Australia, J. M. Powell has argued, that "regionalization" according to natural features can be traced back to the squatters who, through practice, established districts whose borders were usually defined by rivers and river basins. These regions were institutionalized by the colonial administration *post factum* and retained during the era of small-scale selection.[39] During the preparations for the Irrigation Act of 1886, a Royal Commission recommended irrigation trusts modeled on "a natural basis, as far as possible, including the entire supply from one watershed." However, by the time the bill was introduced into Parliament, Alfred Deakin had retreated from this position.[40] The River Murray Waters Agreement, governing the interstate use of the river, notoriously showed a marked concentration on the southern part of the MDB with the exclusion of Queensland. However, the Basin-wide purview was not entirely abandoned. During the 1930s and 1940s, the SRWSC,

[37] Jackson, *Replenish the Earth*, 1.

[38] Jackson, *Replenish the Earth*, 2; 4; 15.

[39] Powell, *Bioregionalism*, 48; 56.

[40] Powell, *Garden State*, 113; Deakin, 24 June 1886, Introducing the Water Supply and Irrigation Bill. Victorian Parliamentary Debates: vol. 51, 1886, 428.

reacting to international reports on land degradation and river improvement, became a prime mover in promoting regional management of soil conservation, learning toward the British model of catchment boards.[41]

Jackson's preference for river basins, however, was predominantly shaped during the New Deal era of the 1930s. In the U.S. in 1933, the Roosevelt administration established the Tennessee Valley Authority for the comprehensive development of water resources of an entire river basin, including power generation, flood control, conservation, navigation, and economic and social progress. In 1937, the president proposed a division of the country into principal drainage areas for conservation, planning purposes, and—for the Columbia, Mississippi, and Tennessee Valleys—direct comprehensive management. This plan, however, was rejected due to increasing suspicion of federal involvement.[42]

Another reoccurring reference point was L. E. B. Stretton. The highly regarded judge and Royal Commissioner (1893–1967) had investigated, *inter alia*, the 1939 Black Friday bushfires.[43] He coined the image of an "Inseparable Trinity" of "Forest, Soil, and Water" in his 1946 report into forest grazing along with the formula "None of them can stand alone. [...] If one of them is injured, all three must share the injury."[44] The Trinity became a rallying cry for the "Save the Forests" campaign launched in 1944, for which Jackson worked for some time. The campaign started with forest education and sponsorship of planting of small community forests by volunteers. In 1945 and 1946,

[41] Powell, *Bioregionalism*, 68–69.

[42] Powell, *Bioregionalism*, 48–49. On the TVA see William U. Chandler, *Myth of TVA. Conservation and Development in the Tennessee Valley, 1933–1983* (Cambridge, MA: Ballinger Pub. Co., 1984); North Callahan, *TVA. Bridge Over Troubled Waters* (South Brunswick: A. S. Barnes, 1980); Tim Culvahouse and Richard Barnes, *The Tennessee Valley Authority. Design and Persuasion* (New York: Princeton Architectural Press, 2007); Sybil Thurman, *Tennessee Valley Authority Information Office, A History of the Tennessee Valley Authority* (Knoxville: Tennessee Valley Authority, Information Office, 1983).

[43] Jackson, "Introduction to the Catchment Philosophy," 1; Jackson, *Replenish the Earth*, 51; Powell, *Bioregionalism*, 42–43; Tom Griffiths, Australian Dictionary of Biography, "Stretton, Leonard Edward (Len) (1893–1967)," 2002, accessed August 16, 2022, http://adb.anu.edu.au/biography/stretton-leonard-edward-len-11793/text21097.

[44] Leonard Edward Bishop Stretton, *Report of the Royal Commission to Inquire into Forest Grazing Together with Minutes of Evidence* (Melbourne: Government Printer, 1946), 1.

it distributed 105,000 trees. Powell considers it to be “a forerunner of modern community efforts in the Basin.”[45]

The campaigner Jackson also paid tribute to Louis Bromfield (1896–1956), an American writer, journalist, and farmer-conservationist. Bromfield won a Pulitzer Prize in 1927, but then lost critical acclaim as a novelist. He nonetheless continued to write prolifically and retained a large readership along with best-selling status. Bromfield purchased and restored three exhausted farms from his childhood in Richland County, Ohio, one of which became his Malabar Farm, where he promoted soil conservation. He became famous for his efforts in experimental farming and nature writing.[46]

Like Bromfield, Ernest Jackson was enthralled by the example of “the visionary Muskingum Watershed Conservancy District of Ohio, U.S.A.”[47] This regional coordination organization for natural resource management at the Muskingum River, a tributary of the Ohio River, which itself is a tributary of the Mississippi system, had been inspected by Vera Jackson during her term as a U.N. officer in New York. “The Miracle of Mukingum” was launched in 1933, the same year as the TVA in the U.S. It was prompted by the disastrous floods of 1913 that killed 500 people and gravely damaged property, which were then followed by droughts. Significantly for Jackson, the district covered the catchment of the Muskingum River and facilitated cooperation between local people and government agencies to manage natural resources.[48] To him, this showed that the catchment concept plus community involvement were the base for effective management: “[T]he Muskingum success is an example of effective watershed development” with “skillful co-ordination at the local community, local government levels to create a cohesive unified valley organisation dedicated to its goal of completely integrated river valley development.”[49] He demanded the “same basic

[45] Powell, *Bioregionalism*, 43.

[46] Ohio History Central, “Louis Bromfield,” accessed August 16, 2022, https://ohiohistorycentral.org/w/Louis_Bromfield; Jane Waterman, The Literary Encyclopedia, “Louis Bromfield,” last modified June 3, 2004, https://www.litencyc.com/php/speople.php?rec=true&UID=578; Jackson, “Introduction to the Catchment Philosophy,” 3.

[47] Jackson, *Replenish the Earth*, 38.

[48] Jackson, *Replenish the Earth*, 6; 38–39; Louis Bromfield, *Out of the Earth* (New York: Harper & Brothers, 1950), 147–56.

[49] Jackson, *Replenish the Earth*, 39.

Roosevelt approach" for Australia: "Federal funding of regional conservation projects co-ordinated at the grass-roots by the people of the region themselves."[50]

For all his indebtedness to the New Deal era, Jackson remained interested in the new and emerging discourses about the global environment and ecology throughout his life. He was clearly conscious of the scope of environmental issues, such as the "global population explosion, the widespread exploitation of renewable natural resources and the energy crisis," although he exhibited a certain bewilderment at the momentum of the eco-age[51]: "Decades before the current Environment Consciousness and Public Awakening to the finite limitations of our life-support systems, a small band of enthusiasts struggled to convince the public of the vital importance of safeguarding our Forests, Soil, Water Flora and Fauna. Never in our wildest dreams did we imagine the tremendous current revolution with all, and sundry crowding the 'Conservation Band Wagon', Protest Marches and Angry Writers deploring the ravaging of our priceless Natural Heritage, etc. etc."[52]

While the title of his book is a quote from the Old Testament (Genesis 1:28), the book's subtitle, *Living Resource Conservation for Sustainable Development*, was adopted from the World Conservation Strategy (WCS), published in 1980 by the International Union for Conservation of Nature and Natural Resources (IUCN), the United Nations Environment Programme (UNEP), and the World Wildlife Fund (WWF).[53] This document signified a "fundamental policy change for the international conservation movement." It marked a shift away from the traditional focus on wildlife preservation toward a concern for the wider pressures affecting the natural environment. It confirmed the growing belief that the assimilation of the aims of both conservation *and* development was the key to a sustainable society. By contrast, in the decade before the

[50] Jackson, *Replenish the Earth*, 38.

[51] Jackson, *Replenish the Earth*, 6.

[52] Jackson, *Replenish the Earth*, 5.

[53] *The Holy Bible: King James Version* (Dallas: Brown Books Publishing, 2004), Gen 1:28; International Union for the Conservation of Nature et al., *World Conservation Strategy. Living Resource Conservation for Sustainable Development* (Gland: International Union for the Conservation of Nature, 1980). On the Stockholm Conference and UNEP see Warde et al., *The Environment*, 136–40; McCormick, *The Global Environment Movement*, 107–54.

Stockholm Conference of 1972, Western European and North American environmentalism had generally considered economic growth suspect and inimical to environmental management, which was often equated with undesirable environmental effects like pollution and resource degradation.[54]

The WCS established the original definition of "sustainable development"[55]: "Conservation, like development, is for people; while development aims to achieve human goals largely through use of the biosphere, conservation aims to achieve them by ensuring that such use can continue."[56] This definition was modified in the 1987 "Brundtland Report" to the paradigmatic formula that "Sustainable development is development that meets the needs of the present without compromising the ability of future generations to meet their own needs."[57]

It was this concept of sustainable development that Jackson found most attractive to substantiate his view of a compatibility of conservation and development—to the extent that he used the WCS's definitions for both terms in his book. It was his intention to "promote consensus between 'preservationists' and 'conservationists'," defining "the art of conservation" as the ability to "maintain the productivity of our renewable natural resources while ensuring the maintenance of landscape amenity." He rejected the wilderness idea, which he identified with preservation, that is "non-use or the locking up of renewable natural resources in the belief that such action can protect or maintain any form of life forever," maintaining that "Nature is not static."[58] Jackson's dismissive attitude toward (urban) Greenies, whom he described as "extreme environmentalists that insist that conservation means non-use instead of

[54] John McCormick, "The Origins of the World Conservation Strategy," *Environmental Review* 10 (Autumn 1986), 177–78.

[55] Radkau, *Age of Ecology*, 386.

[56] International Union for the Conservation of Nature et al., *World Conservation Strategy*, 1.

[57] Gro Harlem Brundtland, ed., *Our Common Future. The World Commission on Environment and Development*, Oxford: Oxford University Press, 1987, 43.

[58] Jackson, *Replenish the Earth*, 3–4.

multi-disciplinary management for perpetual use," bordered on disdain—occasionally seasoned with ridicule[59]: "Conservation is much more than planting trees […], saving koalas, kangaroos, and wildflowers. It is more than supporting National Parks or even the very commendable re-cycling efforts of many sincere, dedicated groups."[60] Against this backdrop, a reader of *Replenish the Earth* is somewhat surprised to find a brief chapter on "Alternative Life-Styles and Energy Conservation," which also explores biodynamic and permaculture farming.[61]

For the campaigner and lobbyist Jackson, there were two priorities: Water management reform in the MDB based on the catchment concept and environmental education. Again, quoting the WCS, Jackson was adamant that the latter was to have top priority in schools, since presently, he found that there was "a great lack of public understanding and awareness of even the elementary principles of ecology in relation to the management of our renewable natural resources."[62] This was even more disturbing since ultimately, "the behaviour of entire societies towards the biosphere must be transformed if the achievement of conservation objectives is to be assured. A new ethic, embracing plants and animals as well as people, is required for human societies to live in harmony with the natural world on which they depend for survival and wellbeing."[63]

He understood environmental education as the "practical understanding of the social, economic and ecological implications of the application of the principles of the Catchment Philosophy."[64] Therefore, a great deal of his, and consequently VIRPO's, lobbying efforts went into persuading school boards in Victoria and the Department of Education to make environmental education compulsive in order to heighten environmental awareness.[65] His own book is an example of these

[59] Ernest M. Jackson, "Protecting the Environment. A Conservation Strategy for Victoria," June 15, 1987, VPRS 6008/P0001, 84/36649, Public Record Office Victoria, Melbourne, 2.

[60] Jackson, *Replenish the Earth*, 75.

[61] Jackson, *Replenish the Earth*, 67–70.

[62] Jackson, *Replenish the Earth*, 77.

[63] Jackson, *Replenish the Earth*, 75; International Union for the Conservation of Nature, et al., 30.

[64] Jackson, Introduction to the Catchment Philosophy, 11.

[65] Ernest M. Jackson, "Victorian Irrigation Research and Promotion Organisation, Submission to the Curriculum Revision Inquiry, Dept. of Education," August 4, 1976,

efforts. It was distributed as part of an education kit to municipalities and libraries in Victoria, including material prepared and distributed as a cooperative project of the National Soil Conservation Program, the Soil Conservation Authority, and VIRPO. The idea was to equip geography, environmental science, social science, and science teachers with a kit of materials for secondary school curricula intended to introduce students to the principles of catchment management and wise land use.[66]

Throughout his life, Jackson remained a staunch proponent of public irrigation. Against the critique of the likes of Bruce Davidson, he insisted that its total annual value to the nation exceeded its capital costs. He also defended the ill-fated Ord River project as a "magnificent national concept that will inevitably confound its present short-sighted and anti-irrigation-biased critics."[67] Consequently, Jackson insisted that the salinity problems of the Murray Valley and elsewhere were not to blame on the irrigators, as "uninformed critics" were quick to assert. The "real villain" was the lack of drainage and loss of "evaporative vegetation." Nonetheless, irrigation salinity ***could*** be solved if a way was found to economically dispose of saline effluent. But most of all, it was a question of catchment management. Dryland and irrigation salinity problems stemmed from the fundamental problem of the MDB, which was the lack of "co-ordination"—his other magic word—among all relevant interests. He suggested that the first step toward a solution was the formation of

VPRS 6008/P0001, 74/2225, Public Record Office Victoria, Melbourne; C. Hildebrand, "Victorian Irrigation Research and Promotion Organisation, Chairman's Report to the Eleventh Annual Meeting, College of Advanced Education, Bendigo, August 17th, 1977," August 17, 1977, VPRS 6008/P0001, 74/2225, Public Record Office Victoria, Melbourne, 1; Jackson, "Introduction to the Catchment Philosophy;" Ernest M. Jackson, "VIRPO Achievements," January 26, 1981, VPRS 6008/P0001, 80/07411, Public Record Office Victoria, Melbourne; Ernest M. Jackson, "Catchment Awareness Education," August 13, 1980, VPRS 6008/P0001, 80/07411, Public Record Office Victoria, Melbourne.

[66] Cliff L. Spowart to G. J. Mennie, Town Clerk, City of Swan Hill, "Education Kit," Replenishing the Earth: salinity & water supply resource kit 1985/produced by Victorian Irrigation Research & Promotion Organisation [book], Swan Hill Regional Library, Swan Hill; P. S. Robins, "Catchment Education Project, Feburary 1985," Replenishing the Earth: salinity & water supply resource kit 1985/produced by Victorian Irrigation Research & Promotion Organisation [book], Swan Hill Regional Library, Swan Hill; Chris Wooton, "Catchments. A Resource Kit for Schools, Feburary 1985," Replenishing the Earth: salinity & water supply resource kit 1985/produced by Victorian Irrigation Research & Promotion Organisation [book], Swan Hill Regional Library, Swan Hill.

[67] Jackson, *Replenish the Earth*, 12.

a Ministerial Council of the Commonwealth and four state governments, including Queensland, to finally arrive at an administrative structure that would encompass the whole of the MDB.[68]

He was not the only one who argued that the scope of salinity problems urgently required natural resource management reform in the Basin, shifting toward a way that would ensure that water was not only supplied in reliable quantities but also acceptable quality. The increasing awareness of salinization starting in the late 1960s was a formative period for the institutional reforms of the 1980s and 1990s, in which communities, organizations, and governments pressed for a revision of the increasingly outmoded River Murray Waters Agreement.[69] In August 1975, the MVDL convened a meeting at Swan Hill attended by 51 organizations who were concerned with the river. They demanded governments "Give Us a Valley Authority!" The League wanted to increase the powers of the RMC to effectively manage salinity and drainage, re-afforestation, conservation, and recreation.[70] In the same year, in its final report, the River Murray Working Party recommended that water quality become an overall objective for the Commission.[71]

By the mid-1970s, salinization had also become the fundamental problem for VIRPO.[72] "Nature," Jackson wrote in 1976, "has caught up with our folly of expecting permanent and prosperous irrigated agriculture without drainage and without a practical understanding of environmental realities. In essence the bill for our national and individual neglect and lack of knowledge of basic ecological principles that is now being presented in the form of ever spreading salinity and devastated farm landscapes, and in the ever increasing River Murray salinity that threatens the water supply of Adelaide and much of South Australia."[73]

[68] Jackson, *Replenish the Earth*, 18; 20; 50; 54.

[69] Connell, *Water Politics*, 105–8.

[70] Anonymous, "Give us Murray Valley Authority!," *The Riverlander* 29 (August 1975): 5.

[71] River Murray Working Party, *Report to Steering Committee of Ministers. October 1975*, 3.

[72] Jackson, "Submission to Curriculum Revision Inquiry," 04.08.1976, 1.

[73] Anonymous, "'Survival' Key Word for Local Area," *The Northern Times*, July 2, 1976.

A "mega-problem" of this magnitude required a long-term solution which, in turn, demanded a coordination of economic, social, and ecological interests. This, he argued, would require the cooperative efforts of all four governments involved and a continuation of the communities' self-help philosophy.[74] To accomplish the latter, it was essential that there was full awareness of the problem, its basic causes, and solutions. This, however, required "an understanding of the basic interrelationships of soil, water and plants, an appreciation of basic ecological principles and an understanding of the responsibility of every irrigator within a common drainage area to avoid contributing to his neighbour's and local district drainage problems by poor irrigation practices."[75]

The message was that salinization would not be solved by technical means alone. It required a fundamental transformation of the institutional structure that governed the MDB and an effort by its citizens to educate themselves in order to fight for their future. It was also an acknowledgment that irrigators themselves were partly responsible for their problems and had to become part of the solution so that irrigated agriculture would retain its eroding public and government support and survive. Irrigation communities had to get their "house in order before environmentalists start irrigation bashing in earnest."[76]

Things were not as simple as they used to be.[77] VIRPO, founded with the goal of promoting irrigation against its critics and helping to increase the efficiency of the industry, was now confronted with a more fundamental threat than the economic rationalist critique of Australia's public irrigation systems. It found a new role within the increasing urgency of the salinity issue. Thus, the organization played an important part

[74] Jackson, "Submission to Curriculum Revision Inquiry," 1.

[75] Ernest M. Jackson, "Victorian Irrigation Research and Promotion Organisation, Environmental Studies for Irrigation Area Secondary Schools," 1976(?), VPRS 6008/P0001, 74/2225, Public Record Office Victoria, Melbourne, 2.

[76] Gynlais O. Jones, "V.I.R.P.O. Extension Working Group. Sub-Committee Meeting to Discuss Publicity Film. Held on 31st July, at Department of Agriculture, Melbourne," July 31, 1981(?), VPRS 6008/P0001, 80/07411, Public Record Office Victoria, Melbourne, 1.

[77] Ernest M. Jackson, "VIRPO Objectives," January 23, 1981, VPRS 6008/P0001, 80/07411, Public Record Office Victoria, Melbourne, 1; C. Hildebrand, "Victorian Irrigation Research and Promotion Organisation, Fourtieth Annual Meeting December 12th, 1980," December 12, 1980, VPRS 6008/P0001, 80/07411, Public Record Office Victoria, Melbourne.

in setting up two sister organizations in the Shepparton and Kerang regions, GIRDAC and KIRSAC.[78] The manager at Kerang's community-owned Agricultural Research Farm, Bill Jones, claimed credit for "getting VIRPO to form an action committee in this area."[79] GIRDAC has launched in the wake of the floods of 1973–1974, while KIRSAC, about two years later, was founded as a Kerang equivalent after earlier attempts had failed due to a lack of participation from the community. A public meeting in February 1976, conducted by VIRPO, finally brought success.[80] Jackson became KIRSAC's first secretary and proved influential for the agendas and goals of both organizations.[81] They adopted the catchment concept and became strong proponents of community education about salinity and, more broadly, environmental issues.

In turn, the rise of the community-based Salinity Action Groups put the future of VIRPO in doubt. GIRDAC and KIRSAC drew attention away from it and gradually took over many of its responsibilities.[82] Contributions from irrigator leagues and commercial interests declined, making the organization dependent on public funding.[83] D. R. Tucker, a member of the Extension Working Group, remarked in 1976 that the formation of the Salinity Action Groups may indicate that efforts now put

[78] George E. Hardwick, "Chairman's Report to the First Annual Meeting of the KERANG IRRIGATION REGION SALINITY ACTION COMMITTEE, at Pyramid Hill on Tuesday 31st May, 1977," May 31, 1977, Box, George and Maureen Hardwick, Appin, 1; Hildebrand, "Chairman's Report, Annual Meeting 1977," 1.

[79] Elaine Westblade, "Annual General Meeting of the Kerang Agricultural Research Farm," August 17, 1976, Kerang Agricultural Research Farm, Archive, Agriculture Victoria, Kerang, 3.

[80] Victorian Irrigation Research and Promotion Organisation, "Minutes of Executive Meeting," October 23, 1975, 3; Anonymous, "VIRPO Planning Salt Action," *The Northern Times*, February 10, 1976; Anonymous, "Community Call."

[81] Anonymous, "'Survival' Key Word for Local Area."

[82] E. P. Robinson, "Question of the Future of V.I.R.P.O.," September 30, 1976, VPRS 6008/P0001, 74/2225, Public Record Office Victoria, Melbourne.

[83] Victorian Irrigation Research and Promotion Organisation, "Financial Statement 1/9/77 to 2/11/77," November 8, 1977, VPRS 6008/P0001, 74/2225, Public Record Office Victoria, Melbourne; R. A. Brown, "Victorian Irrigation Research and Promotion Organisation, Statement of Receipts and Payments for the Year Ended 31st March 1978," June 30, 1978, VPRS 6008/P0001, 74/2225, Public Record Office Victoria, Melbourne; C. Hildebrand, "Victorian Irrigation Research and Promotion Organisation, Twelfth Annual Meeting—1977/78," August 11, 1978, VPRS 6008/P0001, 74/2225, Public Record Office Victoria, Melbourne.

into VIRPO may be better directed at them. Furthermore, he argued, that while VIRPO's Field Days had been very successful, they were not well-known or readily recognized in irrigation areas. It had no impact at all in cities and large towns. Farmer and secondary industry representation and involvement had become reduced, and its primary role was the coordination of government departments' activities. Since "its perpetuation essentially rests in the hands of a few public servants several 'die-hard' farmers and the indefatigable and zealous Mr. E.M. Jackson its future should now be carefully considered."[84] Salinization had become the overarching issue.

Disparate Sisters: Community Groups in Kerang and Shepparton

Community action about water issues has a long history in the MDB, starting with agitations in Northern Victoria in the 1870s and 1880s that demanded irrigation development. The River Murray Water Conservation Conference at Corowa in 1902 triggered the sequence of events that resulted in the River Murray Waters Agreement of 1914–1915.[85] In a similar fashion, Northern Victoria had a significant record of self-help related to the salinity problem long before GIRDAC and KIRSAC was established in the mid-1970s. Since 1939, experimental plots on salt-affected land to investigate their improvement had been established in Swan Hill, Kerang, and Cohuna under the supervision of Alan Morgan at the Kerang Agricultural Centre.[86] In Cohuna, a Salt Reclamation and Agricultural Improvement Committee was formed in November 1945.[87] Its goal was to "undertake any necessary steps to assist soil reclamation and agricultural improvement, and [...] recommend investigations into problems associated with pasture improvements and irrigation."[88] Two

[84] D. R. Tucker, "Role of VIRPO 1976," VPRS 6008/P0001, 74/2225, Public Record Office Victoria, Melbourne, 1.

[85] Connell, *Water Politics*, 109; 183–87.

[86] Jones et al., *Fifty Years of Achievement*, 4; 9; State Rivers and Water Supply Commission, "Experimental plots," October 17, 1939; Russ, *The Salt Traders*, 97.

[87] Cohuna Soil Reclamation and Agricultural Improvement Committee to State Rivers and Water Supply Commission, November 23, 1945.

[88] Cohuna Salt Reclamation and Agricultural Improvement Committee to State Rivers and Water Supply Commission, "Constitution," July 1, 1946, 1.

community-owned research farms in Swan Hill and Kerang, founded in 1954 and 1956, were two more steps forward. In August 1954, a proposal to establish an "irrigators' research farm" on the Tyntynder Flats was launched, which led to the establishment of the Swan Hill Irrigators Research Farm (SHIRF).[89] Seventy Kerang irrigators followed suit two years later and adopted a Chamber of Commerce suggestion to establish their own farm, "to be established with local money, and Government departments to work in conjunction with farmers on the district problems, biggest of which is the salting of land because of irrigation"—the Kerang Agricultural Research Farm (KARF).[90]

However, by 1976, with the additions of VIRPO and the Salinity Action Groups, a veritable network of community activists had developed. Significantly, now there was not just a technical, but also a political side to self-help. KARF and SHIRF continued their research and strived to channel their findings to farmers and government departments.[91] GIRDAC and KIRSAC successfully established themselves as lobby organizations with access to ministers, politicians, and officials. They commented on policy and contributed regional expertise. Finally, they worked to educate their communities about salinity and its effects. In the case of Kerang, the local newspaper *The Northern Times* played a significant role in dispersing to the public information about KIRSAC's work and salinity.

All of these groups were linked through personal contacts, correspondence, shared information, finance, and overlapping personnel. Due to their entangled histories—most obvious in the case of VIRPO, GIRDAC, and KIRSAC—many of their leading figures wore multiple hats. A division of labor was established in the 1970s and 1980s. As in the case of VIRPO, the activists sought a close alliance with the state and its relevant departments, for which many of their key protagonists worked. Almost invariably, they were men. However, in many cases, behind these "men on the committees," there were women who played their own significant

[89] "Enthusiastic Support for Research Farm," *The Guardian (Swan Hill)*, August 16, 1954.

[90] "Research Farm at Kerang," *The Age*, April 20, 1956.

[91] Jones et al., *Fifty Years of Achievement*, 12–13.

parts while relinquishing the center stage, such as Vera Jackson. Consequently, their achievements are more difficult to delineate than those of the male protagonists, but they deserve recognition.

Characteristic for both Salinity Action Groups was their focus on governments. Their primary aim was to achieve salinity control through a combination of technical means, allocation of funds, and water management reform. This could only be accomplished with the resources and authority of governments. After GIRDAC was founded, Ernest Jackson, acting as its executive secretary, defined its objectives as follows: In the short term it was "to alleviate the hazardous high-water table situation of the Goulburn Region orchards by ground-water pumping and drainage installations." This was an emergency reaction to the floods in the previous years.[92] Similarly, KIRSAC defined its goal as "to do everything possible [...] to bring about an immediate and progressive amelioration and eventual total solution of the salinity and ground-water-table, and associated problems."[93] They felt that the problem had to be solved at its roots. GIRDAC proclaimed its long-term objectives as "to obtain Federal and State Government finance support for the implementation of the recommendations of the 1970 Murray Valley Salinity Report; and to speed up the Prime Minister—State Premiers River Murray Inquiry as to desirable amendments to the River Murray Waters Agreement to restruct [sic] the River Murray Commission to enable effective interstate co-ordination of water quality, salinity reclamation and the integration of appropriate measures for the management of the natural resources in the Murray Valley in the over-all interest of the total environment."[94]

With the exception of environmental education, this statement included all of the priorities that Jackson himself had proclaimed. KIRSAC's aims had an equally Jacksonite tone. They were: "Support of the total catchment management approach to problems in the Loddon Campaspe and Avoca watershed and flood plains. [...] Pursuing at every opportunity the co-ordinated approach to salinity problems. [...] Pressing Government into providing more finance for important areas like the

[92] Ernest M. Jackson to Minister of Water Supply, June 3, 1974.

[93] Hardwick, "Chairman's Report," May 5, 1977, 1.

[94] Ernest M. Jackson to Minister of Water Supply, June 3, 1974.

Mines Department work on groundwater and the work of environmental scientists. […] Insisting that direct grants and long-term and/or low-interest loans be made available."[95]

KIRSAC's long-term goal was perhaps even more ambitious than that of its sister. It wanted to become a "comprehensive organisation strong enough to become a statutory authority, or at the very least, a state-supported co-ordinating group" and "a clearing house for all information on salinity and associated problems."[96] GIRDAC agreed that there was a "crying need" for the establishment of a Salinity and Groundwater Authority as a coordinating body. However, it did not aspire to be its prototype.[97]

This was the vision of an adaption of the Muskingum example in the Murray Valley. KIRSAC was to develop into a catchment-wide coordination organization, preferably with statutory powers and appropriate financial endowment. George Hardwick, its chairperson, postulated that "a National Body, similar to the now disbanded Snowy Mountains Authority should be conceived embracing Victoria, New South Wales and South Australia" to control the salinity problem in order to weld government departments and community groups together, which presently worked in isolation. The ultimate structure of KIRSAC, Hardwick continued, could be based on similar lines as the Dandenong Valley Authority (DVA), which the committee had inspected in 1977, being a "specialist, Regional Organisation having [jurisdiction] over the complete natural water-shed catchment, […] responsible for flood mitigation and main drainage works as well as the control of stream pollution" whose "works have been planned, designed and constructed based on communication and interplay of ideas with interested Authorities, Citizens and Organisations."[98]

It is likely that this was another idea injected by "Watershed" Jackson, who cited the DVA as a successful example of a realization of his catchment philosophy. A statutory authority responsible for drainage and

[95] Ernest M. Jackson to Secretary, "Re: Press Tour, Kerang Region, 18th–20th July," July 8, 1978, VPRS 6008/P0006, 76/0693/PT.0001, Public Record Office Victoria, Melbourne. Original emphasis. "Aims of KIRSAC" are attached to the letter.

[96] Ernest M. Jackson to Secretary, State Rivers and Water Supply Commission, Press Tour," 6–7.

[97] Parliamentary Public Works Committee at Melbourne, "Heath," 1059.

[98] Hardwick, "Chairman's Report," May 31, 1977, 2; 4; 7.

flood mitigation, it covered the catchment of the Dandenong Creek, which "rises on the south-west slopes of Melbourne's Dandenong Ranges and enters Port Philip Bay through the Carrum Swamp and man-made Patterson River." It was set up in 1964, based on the English River Board System.[99]

Before this grand goal would be achieved, matters were comparatively sobering. Much like VIRPO, neither KIRSAC nor GIRDAC became professional organizations, although, in the beginning, KIRSAC had envisaged employing a full-time chairperson and secretary and setting up a permanent office. Since 1977, however, the Northern Victorian Fruitgrowers Association had given GIRDAC access to the professional secretarial services of Norman Mitchelmore for a nominal fee plus costs, while KIRSAC was able to employ an unsalaried secretary on an expense base.[100] The groups relied on the time and effort of dedicated individuals, who largely worked on a volunteer basis, and they were highly dependent on public funding and occasional donations. Funds from government grants were either channeled to them via VIRPO or obtained from applications on a case-by-case basis. They never established a steady financial basis and did not become self-supporting.[101]

[99] Jackson, *Replenish the Earth*, 34.

[100] Leon G. Heath, "Goulburn Irrigation Region Drainage Action Committee, Chairman's Report 1977–78," October 27, 1978, VPRS 6008/P0006, File No. 74/20650/Pt.0001, Public Record Office Victoria, Melbourne, 1; Cliff L. Spowart, "Kerang Irrigation Region Salinity Action Committee, Minutes of Meeting held at the State Offices, Kerang on 7th August, 1988," July 7, 1988, VIRPO Folder, George and Maureen Hardwick, Appin, 2.

[101] Hildebrand, "Chairman's Report, Annual Meeting 1977," 1; Hildebrand, "Twelfth Annual Meeting—1977/78," 1; C. Hildebrand, "Victorian Irrigation Research and Promotion Organisation, Thirteenth Annual Meeting, 10th October, 1979. Chairman's Report," October 10, 1979, VPRS 6008/P0001, 74/2225, Public Record Office Victoria, Melbourne; Victorian Irrigation Research and Promotion Organisation, "Minutes of Executive Meeting of 22nd June, 1988, held at the State Offices Shepparton, immediatelly following the 21st Annual Meeting of 22nd June, 1988," January 22, 1988, VPRS 6008/P0001, 84/36649, Public Record Office Victoria, Melbourne, 2; P. B. Wighton, "Statement of Income & Expenditure for Year Ended 31st March 1987, 21.12.1987, Public Record Office Victoria, Melbourne, VPRS 6008/P0001, 84/36649; Kerang Irrigation Region Salinity Action Committee, "Minutes of meeting held at the Kerang Shire Offices Tuesday, March 1st 1977," March 1, 1977, Box, George and Maureen Hardwick, Appin; Cliff L. Spowart, "Kerang Irrigation Region Salinity Action Committee, Minutes of Meeting held at the State Offices, Kerang, on 6th December, 1988," December 06, 1988, VIRPO Folder, George and Maureen Hardwick; Cliff L. Spowart, "Kerang Irrigation

Their leading personnel either were public employees or had been in the past. Their membership consisted of bodies such as various shires, cities, chambers of commerce, and irrigator leagues. These facts make the close proximity to government departments involved with irrigation and commercial interests abundantly clear. Of the eight people who presented KIRSAC's evidence to the Salinity Inquiry in Kerang on June 27, 1978, four were either current or former public employees. George Hardwick, a former Fisheries and Wildlife officer; Gyn Jones, irrigation officer at the Department of Agriculture in Kerang; John Girdwood, Officer at the SRWSC; and Ernest Jackson, a former State Rivers employee. In 1978, the Kerang committee's membership included representatives from the shires of Cohuna, Gordon, Kerang, and Swan Hill, the Borough of Kerang, the City of Swan Hill, the Torrumbarry System Irrigators Association, the Goulburn–Waranga Irrigators League, the two research farms, VIRPO, the Kerang Chamber of Commerce, and the Victorian Field and Game Association.[102] GIRDAC's membership was similar, although commercial interest was greater. It was made up of shires, municipalities, fruit and vegetable processing companies, milk processors, and primary producer organizations from the Shepparton region.[103]

This membership profile was the continuation of an idea that VIRPO had pursued. The problems of irrigated agriculture could only be solved by bringing together all relevant public and commercial interests in coordinating bodies—now supplemented with representatives from irrigation communities. As a result, the Salinity Action Groups often found

Region Salinity Action Committee, Minutes of Meeting held at the State Offices, Kerang, on 10.10.1989," October 10, 1989, VIRPO Folder, George and Maureen Hardwick; F.R. Smith, "Kerang Irrigation Region Salinity Action Committee, Annual Report Submitted by Cr. F. R. Smith, Chairman," May 1, 1984, VIRPO Folder, George and Maureen Hardwick, 1; Goulburn Irrigation Region Drainage Action Comittee, "Account of Receipts and Payments for the Year ended 31st July, 1978," July 31, 1978, VPRS 6008/P0006, File No. 74/20650/Pt.0001, Public Record Office Victoria, Melbourne; Goulburn Irrigation Region Drainage Action Comittee, "Account of Receipts and Payments," July 31, 1979, VPRS 6008/P0006, File No. 74/20650/Pt.0001, Public Record Office Victoria, Melbourne; John Dainton, "Goulburn Irrigation Region Drainage Action Comittee, Annual Report 1987, 16.02.1987, VPRS 11493/P0001, 182/1," Public Record Office Victoria, Melbourne.

102 Parliamentary Public Works Committee at Kerang, "Hardwick," 474.

103 Parliamentary Public Works Committee at Shepparton, "Merrigan," 139.

themselves in the ambiguous role of lobbying public representatives and institutions while at the same time being semi-public bodies themselves.

In another anecdote about Ernest Jackson, he would write strongly worded letters in the name of KIRSAC to the SRWSC and then reply to himself in more moderate terms as the Commission's employee. This simply cannot be true because Jackson retired from the SRWSC in 1964, while KIRSAC was not founded until 12 years later. However, the story captures this exact ambivalence of the community activists—they were, so to speak, extensions of Victoria's official public irrigation institutions who commanded regional knowledge and expertise that were needed to reform the irrigation system and effectively control salinity. They supplemented institutional processes and channeled information between communities and state institutions, their representatives often being simultaneously salaried by the latter while having to develop a critical stance toward their policies.[104]

GIRDAC's first move was to press the Victorian government for an immediate emergency response to the 1973–1974 floods. It made representations to the Premier, R. J. Hamer, to lobby for the allocation of funds for drainage and salinity control in the Shepparton region "to ensure that the full economic capacity of the region can continue to be utilised to the fullest extent."[105] It also approached the Minister of Lands, W. A. Borthwick, asking for "long term, low interest finance to enable private landholders to install ground water pumps [...], as an emergency measure to bring under control the rising ground water levels."[106] In response, the government announced a special allocation of $300,000 for the installation of groundwater pumps in the region to protect horticultural blocks suffering from shallow water tables. A further $330,000 were allocated to accelerate the construction of surface drainage extensions, $150,000 to the SRWSC for loans to fruit growers

[104] Anonymous, "Land Care Crusader 'Watershed' Jackson."

[105] Peter Ross-Edwards, "Introduction to the Honorable R. J. Hamer, E.D., M. P., Premier of Victoria by Members of the Drainage Symposium Steering Committee of the Goulburn Irrigation Region on Wednesday 29 May, 1974 at 2–15 P.M.," May 29, 1974, VPRS 6008/P0006, File No. 74/20650/Pt.0001, Public Record Office Victoria, Melbourne, 3.

[106] A. Brown-Graham, "Submission to the Hon. W. A. Borthwick, MP, Minister for Conservation, Minister of Lands and Minister of Soldier Development—To Seek a Further Allocation of Funds," July 1, 1975, VPRS 6008/P0006, File No. 74/20650/Pt.0001, Public Record Office Victoria, Melbourne, 1.

and pump installations, and a final $50,000 to hire private groundwater pumps.[107] Further submissions for additional funds and another long term, low-interest grant to finance private installations of drainage re-use systems were declined, but the initial success was enough to show that approaching governments could be a successful avenue toward obtaining resources for salinity reclamation.[108]

Accordingly, KIRSAC's first move was also directed at the government. The committee's first major commitment was to coordinate submissions from the Kerang region to the Public Works Committee's Inquiry on "Salinity Control and Drainage," which was tasked to assess the SRWSC's "40 Million Dollar Plan" for Northern Victoria's irrigation areas.[109] Interested citizens were invited to help prepare a submission on behalf of the community that would establish the economic and social viability of the region, the inadequacy of the SRWSC's Strategy, propose alternative measures, and delineate the consequences of no action, in order to arrest the region's decline: "The essential need is that the PPWC must be convinced of the total backing of the Kerang region for remedial action."[110] Both Salinity Action Groups presented evidence to the inquiry when the Public Works Committee inspected the Kerang and Shepparton areas on July 13 and 14, 1976. Multiple submissions followed.[111]

[107] State Rivers and Water Supply Commission, "Notes for: The Honorable F.J. Granter, M.L.C., Minister of Water Supply," June 25, 1975, VPRS 6008/P0006, File No. 74/20650/Pt.0001, 1, Public Record Office Victoria, Melbourne; A. Brown-Graham, "Submission to W.A. Borthwick—Further Allocation of Funds," 2; Anonymous, "Community Call."

[108] A. Brown-Graham, "Submission to W.A. Borthwick—Further Allocation of Funds;" A. Brown-Graham, "To Seek a Grant of Long Term, Low Interest Finance," July 1, 1975, VPRS 6008/P0006, File No. 74/20650/Pt.0001, Public Record Office Victoria, Melbourne; State Rivers and Water Supply Commission, "Memorandum to the Chairman," July 2, 1975, Public Record Office Victoria, Melbourne, VPRS 6008/P0006, File No. 74/20650/Pt.0001.

[109] Anonymous, "VIRPO Planning Salt Action," *The Northern Times*, February 10, 1976.

[110] Anonymous, "Be Counted for the KIRSAC," *The Northern Times*, June 8, 1976.

[111] Parliamentary Public Works Committee at Shepparton, "Merrigan"; Parliamentary Public Works Committee at Shepparton, "Leon G. Heath (1)," June 28, 1978, VPRS 11559/P0001, 24.07.27, PT4, Public Record Office Victoria, Melbourne; Parliamentary Public Works Committee at Melbourne, "Heath;" Parliamentary Public Works Committee at Kerang, "Jones."

In 1978, KIRSAC tendered a 158-page brick in which it argued that Murray water quality had to be protected at all costs to avoid further deterioration of the environment and related human well-being.[112] The submission was largely compiled by Ernest Jackson, exhibiting his handwriting with an emphasis on a whole catchment approach complemented by environmental education.[113] The bottom line was that all measures for salinity control had to be considered in a long term, whole valley context in favor of individual solutions that transferred salt from one area into another. Hence, the only solution was diverting saline inflows away from the Murray by evaporation basins, or, ultimately, a pipeline that would transfer them into the sea. Financing for better extension services and long-term loans for on-farm works were required to improve farm layout and introduce re-use schemes, and improve channels and drains in order to prevent accessions to the Murray.[114] The submission included detailed suggestions on the costs for required loans, the measures on which they should be spent, and their conditions for applicants, outlines of possibilities to dispose of saline groundwater, dilution of Barr Creek water and re-use schemes, and groundwater pumping.[115] These suggestions were accompanied by letters of support from shires, boroughs, and organizations concerned with natural resources in the Kerang region in support of KIRSAC's evidence.[116]

The committee kept a close watch on the salinity issue. When the first allocation of low-interest loans to farmers to establish salinity control works on their farms through improved land layout, groundwater pumping, and drainage water re-use, had "dried up," KIRSAC presented a submission to the Premier asking for funds so loans could be offered again. As a result, the government agreed for an additional $750,000 to be made available through the Rural Finance Commission.[117] On another occasion, the committee, among other groups, made representations to the Minister of Business and Consumer Affairs to remove import

[112] Parliamentary Public Works Committee at Kerang, "Hardwick," 474–75.

[113] Parliamentary Public Works Committee at Kerang, "Hardwick," 562.

[114] Parliamentary Public Works Committee at Kerang, "Hardwick," 474–81.

[115] Parliamentary Public Works Committee at Kerang, "Hardwick," 477–95.

[116] Parliamentary Public Works Committee at Kerang, "Hardwick," 501–5.

[117] Anonymous, Salt Loans Dried-Up," *The Northern Times*, November 6, 1981; Anonymous, "Salinity Loans of $3/4m," *The Northern Times*, November 24, 1981.

duties applying to laser control units used on land leveling devices that were becoming increasingly important for improved land layout on farms. Consequently, several purchasers from the district received refunds on import duties.[118]

The Salinity Action Groups established themselves as pressure groups capable of adding regional expertise on environmental conditions and damages who were able to aggregate many interests from their respective communities. Both continued to devote much of their steady activities on commenting policy even after the Salinity Inquiry was completed in 1984, contributing to legislation concerning conservation, irrigation, and salinity.[119]

The second focus was the communities themselves, focusing on awareness and education. KIRSAC set out to gain "the support of the whole valley in having the overall salt load entering the Murray reduced." This required, the committee felt, "[k]eeping abreast of the latest research and informing the public of the likely repercussions of its findings" as well as "[a]ctively promoting environmental education at all levels to ensure that young people will be armed with the knowledge of a co-ordinated regional approach in planning land use in the future."[120] Awareness of

[118] W. E. Bromfield, "Import Duty on Laser Control Units," March 19, 1981, VPRS 6008/P0006, File No. 76/0693/PT.0001, Public Record Office Victoria, Melbourne; Cliff L. Spowart, "Kerang Irrigation Region Salinity Action Committee, Minutes of Meeting held at the State Offices, Kerang, on 6th December, 1983," December 6, 1983, VIRPO Folder, George and Maureen Hardwick, Appin, 3.

[119] Gynlais O. Jones, "Kerang Irrigation Region Salinity Action Committee, Minutes of Meeting held at the State Public Offices, Kerang on 4th October 1988," October 4, 1988, VIRPO Folder, George and Maureen Hardwick, Appin, 1; Kerang Irrigation Region Salinity Action Committee, "Minutes of Meeting held at the State Offices of the R W C, Kerang on 5-2-91," VIRPO Folder, George and Maureen Hardwick, Appin, 1; Cliff L. Spowart, "Kerang Irrigation Region Salinity Action Committee, Minutes of Meeting held at the State Offices, Kerang on 6th Oct, 1987," October 6, 1987, VIRPO Folder, George and Maureen Hardwick, Appin, 1; Cliff L. Spowart, "Kerang Irrigation Region Salinity Action Committee, Minutes of Meeting held at the Shire Offices, Kerang, on 5th March, 1985," March 5, 1985, VIRPO Folder, George and Maureen Hardwick, Appin, 1; Smith, "Annual Report," May 1, 1984, 1–3; Kerang Irrigation Region Salinity Action Committee, "State of the Rivers," April 3, 1984, VIRPO Folder, George and Maureen Hardwick, Appin; Spowart, "Minutes," December 12, 1983, 2; Norman J. Mitchelmore, "Goulburn Irrigation Region Drainage Action Committee, "Salinity Committee Recommendations," May 18, 1985, VPRS 11493/P0001, 182/1, Public Record Office Victoria, Melbourne, 1.

[120] Ernest M. Jackson, "Press Tour," July 8, 1978. Original emphasis.

environmental problems was an "essential first" in any permanent salinity reclamation program. But without some basic understanding of ecological principles, such as the operation of the water cycle, the GMID's headworks, and catchment areas, it was difficult to convey the message of interrelationships of irrigators with their neighbors, district, and region. Some form of specific direction in environmental education related to regional problems was needed in the secondary schools in the irrigation areas. Yet, environmental education could not be taught in classrooms alone. The hands-on experience was needed.[121] GIRDAC agreed; the need of education was "greater than ever." Farmers had to learn more about the difficult soils of the Shepparton region, irrigation practices, and their individual responsibilities.[122] As farmers, KIRSAC's chairperson Russell Smith told a Pyramid Hill meeting in 1986, "[w]e must recognise to a large extent we are responsible. Get advice on what's best for your farm. Work together to make sure no water is allowed to sit. Plant trees on intake areas. Improve irrigation practices."[123] Hence both committees, often in conjunction with VIRPO, conducted field days and seminars on drainage, farming, and irrigation techniques to channel knowledge to irrigators.[124] GIRDAC organized meetings of landholders and industry leaders to bring salinity problems to their attention.[125]

The community work may be best exemplified by two protagonists from Kerang. The first is Gyn Jones (1933–2000). He arrived in Kerang with a Bachelor's degree in Agricultural Science in 1957 as an Irrigation Research Officer and assistant to Ken Garland. When his boss left to become Senior Irrigation Officer at the Department of Agriculture in Melbourne, Jones was his successor as an officer in charge for north-west

[121] Parliamentary Public Works Committee at Kerang, "Hardwick," 560–62.

[122] Heath, "GIRDAC," 3.

[123] Anonymous, "History Has its Salinity Lessons," *The Northern Times*, April 8, 1986.

[124] Murray Valley League, "A Conference on—Land Drainage," August 4, 1989, VPRS 6008/P0006, File No. 74/20650/Pt.0001, Public Record Office Victoria, Melbourne; Dainton, "Annual Report 1987," 2; Victorian Irrigation Research and Promotion Organisation, "20th Annual Meeting. Salinity Control Seminar," January 11, 1987, VIRPO Folder, George and Maureen Hardwick, Appin; Ernest M. Jackson, "Victorian Irrigation Research and Promotion Organisation, Notice of 15th Annual Meeting," November 9, 1981, VPRS 6008/P0001, 80/07411, Public Record Office Victoria, Melbourne.

[125] Dainton, "Annual Report 1987," 2; Norman J. Mitchelmore, "Rural Industry Leaders," October 16, 1980, VPRS 6008/P0006, File No. 74/20650/Pt.0001, Public Record Office Victoria, Melbourne.

Victoria.[126] He was one of the individuals who worked toward setting up a GIRDAC equivalent in the Kerang region.[127] When KIRSAC launched, he was one of the leading figures. Jones became the group's public mouthpiece, which he remained throughout its existence, declaring that its "main objective should be to set the community alight and educate the community on what is happening now and the long-term effects."[128] He represented the committee multiple times before the Salinity Inquiry.[129] Cliff Spowart, then chairperson of KIRSAC, reflected in 1991 that "[o]ne of Gyn Jones's] strong points was his frequent reference to a 'philosophy' of things. Such as a philosophy of irrigated agriculture, or a philosophy of salinity control, and so on."[130] He was lauded as the group's "ideas man,"[131] for "his vision, persistance [sic] and sheer hard work,"[132] or simply for "being Gyn and persisting."[133]

Jones was an example of a public employee who blended his occupation with his activism. On the professional side, he continued the work of his predecessors by developing techniques to reclaim salted land.[134] As an irrigation officer at the Department of Agriculture, he cooperated,

[126] Jones et al., *Fifty Years of Achievement*, 3–4; Plaque for Gynlais O. Jones at the State Offices building, Kerang. Photo by Daniel Rothenburg, August 8, 2017, Agriculture Victoria, Kerang; Russ, *The Salt Traders*, 99; Ray Brindal, "Irrigation Guru Retires," *Salt Force News* 21 (May–June 1991): 13.

[127] Anonymous, "More Drainage Moves by VFU," *The Northern Times*, August 20, 1974; Anonymous, "Salinity Apathy," *The Northern Times*, December 10, 1974; Anonymous, "Who Cares about Drainage Here?," *The Northern Times* November 5, 1974.

[128] Anonymous, "Community Call."

[129] Parliamentary Public Works Committee at Kerang, "Hardwick;" Parliamentary Public Works Committee at Kerang, "Jones."

[130] Kerang Irrigation Region Drainage Action Committee, "Minutes," February 2, 1991, 2.

[131] F. R. Smith, "Annual Report by Cr. Russell Smith, Chairman of K.I.R.S.A.C.," 1985, KIRSAC, Kerang Agricultural Research Farm, Archive, Agriculture Victoria, Kerang, 4.

[132] Kerang Irrigation Region Drainage Action Committee, "Minutes of Annual Meeting held at the State Offices, Kerang, on 1st of August, 1989," August 1, 1989, VIRPO Folder, George and Maureen Hardwick, Appin, 1.

[133] Russell Smith, "Chairman's Report for 1986/87," 1987, KIRSAC, Kerang Agricultural Research Farm, Archive, Agriculture Victoria, Kerang, 2.

[134] Jones et al., *Fifty Years of Achievement*, 10.

inter alia as a technical advisor, with the two research farms in Kerang and Swan Hill.[135] Jones also accompanied and assisted the ACIL consultants and others, who collocated the available information on salinization for the Salinity Committee in 1983.[136] On at least two occasions, he traveled overseas to study irrigation, agriculture, and salinity reclamation techniques. In 1984 and 1986 he went to China.[137] Furthermore, he used his position to promote salinity reclamation measures to the community and educate farmers about them. In 1972, he and his staff set up a night school on farm reconstruction through better land layout and more efficient water use for local farmers. It ran until 1978.[138] He held courses on the geography of the Loddon region at the Kerang Technical High School.[139] Multiple times, he compiled his experiences as an irrigation officer into self-help booklets for farmers of the region. One of them, *Kerang Irrigation Region—Problems, Potential, Productivity*, was

[135] B. L. Milligan, "Annual General Meeting of the Kerang Agricultural Research Farm," August 25, 1965, Kerang Agricultural Research Farm, Archive, Agriculture Victoria, Kerang, 2; F. Murphy, "Annual General Meeting of the Kerang Agricultural Research Farm," August 15, 1968, Kerang Agricultural Research Farm, Archive, Agriculture Victoria, Kerang, 2; Dough Small, "Annual General Meeting of the Kerang Agricultural Research Farm," August 22, 1972, Kerang Agricultural Research Farm, Archive, Agriculture Victoria, Kerang, 2; Dough Small, "Annual General Meeting of the Kerang Agricultural Research Farm," August 10, 1973, Kerang Agricultural Research Farm, Archive, Agriculture Victoria, Kerang, 3; "Westblade," 3–4; Swan Hill Irrigators Research Farm, "Guide Book 1965," Swan Hill Irrigators' Resarch Farm, Swan Hill Irrigators' Research Farm at Tyntynder: a selection of historical documents 1965–1993/[CD-ROM], Swan Hill Regional Library, 2; Swan Hill Irrigators' Research Farm, "Guide Book 1976," Swan Hill Irrigators Resarch Farm, Swan Hill Irrigators' Research Farm at Tyntynder: a selection of historical documents 1965–1993/[CD-ROM], Swan Hill Regional Library, 5.

[136] ACIL Australia et al., *Causes, Extent and Effects of Salinity*, 3.

[137] Cliff L. Spowart, "Kerang Irrigation Region Salinity Action Committee, Minutes of Meeting held at the Shire Offices, Kerang, on 6th Nov, 1984," November 6, 1984, KIRSAC, Kerang Agricultural Research Farm, Archive, Agriculture Victoria, Kerang, 1; Anonymous, "Gun Shots—Food in China," *The Northern Times*, June 24, 1986.

[138] Jones et al., *Fifty Years of Achievement*, 10.

[139] Gynlais O. Jones, "Lecture," 1994, Elaine Jones, Kerang; Gynlais O. Jones, "Modern Techniques in Agriculture," June 1, 1972, Elaine Jones, Kerang; Brindal, "Irrigation Guru Retires," 13.

distributed 10,000-fold by KIRSAC in 1988, covering such basic information as the geological history of the region, the causes for salinity, its main sources, and means to control it.[140]

Another brochure shows that, even though it was Gyn Jones who stood in the center, his wife Elaine and daughter Nicole were also involved. Here, salinity reclamation became a collective family effort. For the booklet *Irrigation Salinity—Attack & Defence* written by Gyn, Elaine helped with the preparation while Nicole contributed the artwork.[141] Elaine's persistent involvement is shown by the fact that both KIRSAC recognized Gyn's and her dedication in 1991.[142] Both of them were honored for their "substantial contribution to the success of Salinity Management" by the Department of Natural Resources and Environment in 2002—which was, in Gyn's case, posthumously.[143]

As the Irrigation Officer, Jones was highly visible in Kerang's public sphere. Much like KIRSAC, he cooperated extensively with the *Northern Times*, for which he regularly wrote, and used his weekly column, "Gun Shots"—"[t]he meaning can be taken several ways but I assure you that our 'shots' will be constructive"—to communicate with the community.[144] It ran from 1973 to 1994.[145] "Gun Shots" was a tool for his activism. Firstly, he promoted research farms' work, the Salinity Action Groups and VIRPO, and communicated their events and meetings.[146]

[140] Gynlais O. Jones, "Kerang Irrigation Region—Problems, Potential, Productivity," 1988, Salinity Kerang, Kerang Irrigation Region, Kerang Agricultural Research Farm, Archive, Agriculture Victoria, Kerang; Kerang Irrigation Region Drainage Action Committee, "Minutes of Meeting held at the State Offices, Kerang on 7th June 1988," June 7, 1988, KIRSAC, Kerang Agricultural Research Farm, Archive, Agriculture Victoria, Kerang, 1.

[141] Gynlais O. Jones and Nicole Jones, "Irrigation Salinity—Attack & Defence," Elaine Jones, Kerang.

[142] Kerang Irrigation Region Drainage Action Committee, "Minutes," February 2, 1991, 2.

[143] Certificates of Recognition for Gyn and Elaine Jones at the "Model Farm," Kerang. Photo by Daniel Rothenburg, August 8, 2017.

[144] Gynlais O. Jones, "Gun Shots—'Gun Shots'," *The Northern Times*, March 27, 1973.

[145] Plaque for Gyn Jones, August 8, 2017.

[146] Gynlais O. Jones, "Gun Shots—What's Involved with a Dairy Inquiry?," *The Northern Times*, August 13, 1974; Gynlais O. Jones, "Gun Shots—It's Still Not Too Late to Apply Nitrogen Fertiliser to Grassy Pastures," *The Northern Times*, August 27,

Fighting against the "can't be bothered" attitude, he tried to raise awareness for Northern Victoria's salinity problems, insisting time and time again that "we're all involved in the prosperity or death of our district."[147] One time, he included a photograph of a salt-encrusted paddock to alert readers about the worrying scale of the problem.[148]

Furthermore, Jones wrote short educational pieces about the nexus between water lying around on the ground and the rising water table.[149] He urged his readers to "[l]et us all do our bit to stop underground water being recharged by getting rid of water lying on the surface."[150]

Ironically referring to himself as "big brother," he occasionally "called out," without naming them, the farmers who handled water carelessly and were caught with "big puddles."[151] "Our family has been tripping to and from Swan Hill and we marvel at the same flow of water issuing from the same gateway on to the side of the highway day after day. Obviously, the farmer is not even going around the water even once each day. It's no wonder our region has a bad name—some tourist attraction!"[152]

On the positive side, Jones provided assistance in the form of short instructions on how to solve farm problems like using less water and shared good practice examples of efficient usage and proper land

1974; Gynlais O. Jones, "Gun Shots—Remarkably Warm Winds," *The Northern Times*, July 29, 1975; Gynlais O. Jones, "Gun Shots—What Do We Do with Ourselves All Day?" *The Northern Times*, August 8, 1975; Gynlais O. Jones, "Gun Shots—Big Dry Is Still with Us," *The Northern Times*, January 10, 1978.

147 Gynlais O. Jones, "Gun Shots—Young Ideas Welcome in North," *The Northern Times*, February 8, 1977; Gynlais O. Jones, "Gun Shots—It Concerns Everyone," *The Northern Times*, May 5, 1980.

148 Gynlais O. Jones, "Gun Shots—To Sow or Not," *The Northern Times*, July 13, 1976.

149 Gynlais O. Jones, "Gun Shots—Facts in Journal," *The Northern Times*, September 10, 1974.

150 Gynlais O. Jones, "Gun Shots—A Question of Salinity," *The Northern Times*, March 28, 1986.

151 Gynlais O. Jones, "Gun Shots—Water Is Flowing," *The Northern Times*, March 2, 1982; Gynlais O. Jones, "Gun Shots—Irrigation Wastage," *The Northern Times*, February 25, 1986; Gynlais O. Jones, "Gun Shots—Raining, Pouring!," *The Northern Times*, April 7, 1981; Gynlais O. Jones, "Gun Shots—Autumn in Summer!," *The Northern Times*, February 2, 1984; Gynlais O. Jones, "Gun Shots—Puddles Show That Rain Caught Some!," *The Northern Times*, April 22, 1986.

152 Gynlais O. Jones, "Gun Shots—New Farmers Tend to Changes," *The Northern Times*, March 30, 1982.

layout.[153] He also made political commentary, usually with regard to how economic and political developments had an impact on rural Australia and the farming community.[154] Lastly, even though he was mostly concerned with matters relating to the Kerang region and agriculture, Jones did occasionally write about environmental issues such as the worldwide destruction of rainforests, urging that "[c]onservation of our environment might be something you tend to leave up to a few radical groups—but really every individual should contribute something to ensure that following generations don't end up trying to eke out an existence in a barren wilderness."[155]

For Gyn Jones, as for Ernest Jackson, salinization was a symptom of "mismanagement of our environment" that could not be solved by technical means alone. What was needed was a combination of improved agricultural techniques, government assistance in the form of funds and drainage installations, awareness, and a sense of stewardship for the environment[156]: "[O]ur rich countryside is a heritage. We have inherited it and should pass it on to future generations."[157]

George Ernest Hardwick agreed with this. Like Gyn Jones, Hardwick came from public service, having been the Chief Wildlife Officer for the Kerang region for 11 years. In 1976, he became KIRSAC's first chairperson. However, he soon stepped down from the position, although remaining an active member, to make more time available for the project

[153] Gynlais O. Jones, "Gun Shots—How to Save It," *The Northern Times*, March 3, 1982; Gynlais O. Jones, "Gun Shots—Guide to Watering," *The Northern Times*, October 5, 1982; Gynlais O. Jones, "Gun Shots—Another Year Opens," *The Northern Times*, January 13, 1976; Gynlais O. Jones, "Gun Shots—Water Budget Essential," *The Northern Times*, July 17, 1984; Gynlais O. Jones, "Gun Shots—Warning on Irrigation Designs," *The Northern Times*, April 28, 1981; Gynlais O. Jones, "Gun Shots—District Driving," *The Northern Times*, October 23, 1984.

[154] Gynlais O. Jones, "Gun Shots—Economy Hits the Country," *The Northern Times*, June 29, 1982; Gynlais O. Jones, "Gun Shots—How Much Dearer?," *The Northern Times*, February 21, 1984.

[155] Gynlais O. Jones, "Gun Shots—Ruination of Evolution," *The Northern Times*, February 1, 1984.

[156] Gynlais O. Jones, "Salinity Spread Is a Reminder to 'Lift Our Performance'," *The Northern Times*, June 27, 1986.

[157] Gynlais O. Jones, "Gun Shots—Don't Miss on Water," *The Northern Times*, January 28, 1981.

that he and his family dedicated much of their lives to—the Kerang Environmental Study Centre (KESC).[158] The Hardwicks took up full-time residence at Macropus Park in Appin near Kerang in October 1975 and began developing the property as a host farm to provide environmental education for primary, secondary, and tertiary students, act as a base for excursions, assist with developing environmental education programs in schools, and provide accommodation facilities and tours. Teachers were offered technical papers, worksheets, lessons for different age groups, and assistance with excursions.[159]

The Centre itself was a mixed-irrigation farm producing beef, cereal crops, summer crops, and vegetables to provide hands-on environmental education. It had facilities for 48 students, plus rooms for teachers, a dining room, a veranda, and a swimming pool. Visitors were offered excursions on various subjects such as rice and citrus growing, grapevines and wineries, dairying, land layout, water recycling, forest management, geology of the region, wildlife, saline wastelands, salt harvesting, and bird watching. The whole family, George, his wife Maureen, their daughter Wendy, and their three sons, Craig, Marshall, and Andrew, ran the Centre together. In 1982, they were recognized for their conservation efforts by the Fisheries and Wildlife Division.[160]

The idea and design of the Hardwick Centre were intimately connected to salinization. In 1991, Hardwick reflected that he "saw the study centre as a base for school and educational groups to come to Kerang and possibly assist Kerang to overcome the salinity problem."[161] He considered it imperative to obtain evidence of the productivity of the Kerang district by re-developing a neglected farm in a saline environment, passing

[158] Gynlais O. Jones, "Pump Salt into the Sea!," *Herald*, October 23, 1978; Anonymous, "First Leader of KIRSAC," *The Northern Times*, April 13, 1976; George E. Hardwick, "Kerang Irrigation Region Salinity Action Committtee, Chairman's Annual Report," May 31, 1977, VIRPO Folder, George and Maureen Hardwick, Appin, 1; George E. Hardwick, "re Resignation as Chairman of K.I.R.S.A.C.," June 24, 1976, Box, George and Maureen Hardwick, Appin.

[159] George E. Hardwick, "Establishing the Kerang Environmental Study Centre," 1976(?), Unnamed Folder 1, George and Maureen Hardwick, Appin, 2; George E. Hardwick, "Kerang Environmental Study Centre. Guide for School Excursions to Northern Victoria," 1980s(?), Unnamed Folder 2, George and Maureen Hardwick, Appin, 4.

[160] Hardwick, "Kerang Environmental Study Centre," I; 3; 8–11.

[161] Anonymous, "Environmental Dream Set to End," *The Northern Times*, July 30, 1991.

on the knowledge to students, teachers, and other landholders,[162] and not least, to "draw attention to the devastating salinity and environmental problems within the Kerang Study Area."[163]

Secondly, they testify to a strong connection to Jackson's thinking and his person. Hardwick wrote of Jackson in 1988 that "Ernest's dedication and moral support to me has been greater than any other person, or Group, including KIRSAC."[164] A strong personal association between the two men is demonstrated by their regular correspondence and sharing of information.[165] The KESC included an Ernest Jackson Field Laboratory (later the Vera and Ernest Jackson Field Laboratory and Library), as a token of appreciation of the work and support by Ernest and Vera, with technical books on environmental studies, which were donated to the Centre by the Jacksons.[166] Environmental education, Hardwick wrote in a booklet advertising the Centre to teachers, should "encourage a sense of responsibility in a world of rapid change, to build a better tomorrow." This required an "understanding of the environment, [...] a set of values and feelings of concern for the environment, and the motivation for participating in environmental protection and improvement," and finally "skills for identifying and solving environmental problems."[167] This was a Jacksonite statement with its emphasis on environmental education as the key component of a sustainable future.[168] The same booklet includes imagery and a verbatim passage from *Replenish the Earth*: "how can people make the right decisions without knowledge and understanding

162 Hardwick, "Kerang Environmental Study Centre," 3.

163 Hardwick, "Kerang Environmental Study Centre," 6.

164 George E. Hardwick, "KIRSAC: re FIELD LABORATORY: KERANG Environmental study centre," December 5, 1988, VIRPO Folder, George and Maureen Hardwick, Appin.

165 Among George Hardwick's documents, I found dozens of letters written to him by Ernest Jackson, although no copies of outgoing ones. It can, however, be reasoned from Jackson's letters that it was a mutual exchange of correspondence. Ernest M. Jackson, "Copy of Evidence," July 22, 1976, Box, George and Maureen Hardwick, Appin; Ernest M. Jackson, "British Empire Medal Award," 1976(?), Box, George and Maureen Hardwick, Appin.

166 Hardwick, "Kerang Environmental Study Centre," 4; Anonymous, "Centre Part of Salinity Fight," *The Northern Times*, September 13, 1988; Hardwick, "KIRSAC: re FIELD LABORATORY: KERANG Environmental study centre."

167 Hardwick, "Kerang Environmental Study Centre," 1.

168 Jackson, *Replenish the Earth*, 77.

of the realities of our environment, of the complex inter-relationships of our renewable natural resources and of their own inter-dependence [on] the total natural environment."[169]

The example of the KESC should, however, not lead one to assume that Ernest Jackson's "philosophy" was unreservedly embraced by the Salinity Action Groups, even though they did adopt many key parts of his thinking. Nor was the KESC a venture of KIRSAC. It remained privately financed by the Hardwicks throughout its existence, although the committee had strongly supported its establishment.[170] KIRSAC's chairperson, Cliff Spowart, did not share Ernest Jackson's "enthusiasm for the sort of hands on experience that students would get at places such as George Hardwicks," worrying that "too many instant experts [...] would be created."[171] It does, however, highlight the staunch conservationism of one of its leading protagonists, who helped to shape the character and aims of the committee.

Nor should the obvious similarities of the committees in Kerang and Shepparton, brought about by their common relation to VIRPO and their place in the web of community groups, overshadow their differences. Notwithstanding Ernest Jackson's gospel of coordination of all interests in the Murray Valley, the committees remained divided by the interests of their own regions. This became apparent in resource conflicts. Leon Heath, GIRDAC's chairperson, dismissed the "hostility from people downstream" toward the SRWSC's "Salinity Control and Drainage" strategy, insisting that the Commission had been "quite objective in deciding that some areas in this Region should receive high priority"[172]: "Some people downstream do not seem to appreciate that their greatest

[169] Hardwick, "Kerang Environmental Study Centre," 7; 32; Jackson, Replenish the Earth, 80.

[170] Hardwick, "Kerang Irrigation Region Salinity Action Committee, Chairman's Report," May 31, 1977, 6; Hardwick, "KIRSAC: re FIELD LABORATORY: KERANG Environmental study centre."

[171] Cliff L. Spowart to Ernest M. Jackson, November 13, 1988, VIRPO Folder, George and Maureen Hardwick, Appin.

[172] Heath, "Chairman's Report," October 27, 1978, 1.

threat from the Shepparton Region will come from a 'no control' situation. [...] If the problems of salinity and groundwater are not tackled in this Region, then the effects downstream will [be] disastrous."[173]

In this regard, the sister-committees reproduced the parochialism which many of their fellow citizens had exhibited during the hearings of the Public Works Committee and the conflicts over the Tyrrell-Tutchewop scheme. KIRSAC, with the Kerang region being disadvantaged by its downstream position, low economic priority, and more difficult environmental conditions, never was in a similar bargaining position as its Shepparton counterpart. Jones, speaking in favor of Kerang, stated: "We feel like the little boy standing beside the railway track who sees the train going from Melbourne to Sydney, but he has no hope of being on that train."[174] This also helps to explain why KIRSAC focused much more on community education and self-help approaches on the farm than GIRDAC because engineering solutions were not in sight.

The two committees discussed here were by far not the only community groups involved with salinization. Of the about 30 Salinity Action Groups which existed in Northern Victoria at the end of the 1970s, some developed into important actors.

Tree Planters, a Pipeline, and Growing Unrest

The United Nations Association of Australia declared 1982 the Year of the Tree. "It is appropriate that such a Year of the Tree be held in Australia," Valerie Swane, the chairperson of the responsible committee, explained, "as this is the world's driest continent and can benefit most from efforts aimed at making the public more aware of the benefits of trees." They were, *The Northern Times* explained to its readers, one of the most fundamental resources of the planet for oxygen, water cycle control, food, timber, paper, and soil conservation. However, since 1950, over half of the world's forests had been lost, more than during the entire previous history of humanity. One important consequence was soil degradation. In

[173] Leon G. Heath, "Goulburn Irrigation Region Drainage Action Committee, Chairman's Report to 1979 Annual Meeting of Goulburn Irrigation Region Drainage Action Committee," October 13, 1980, VPRS 6008/P0006, File No. 74/20650/Pt.0001, Public Record Office Victoria, Melbourne, 1–2.

[174] Parliamentary Public Works Committee at Kerang, Gynlais Oughton Jones, 13.07.1976, 96.

the last 75 years, 75 billion tons of topsoil had been lost to erosion and, in the next 20 years, if present trends continued, many more billion tons would be lost, leading to desertification, encroaching on arable zones of the world's producing areas.[175]

In Northern Victoria, by the 1980s, salinity had become an issue which concerned broad parts of rural communities. Trees were now seen by community groups, individual farmers, and shires as a key factor to halt the spread of salinization and arrest waterlogging and underground seepage of water into streams by retaining water in their roots.[176] KIRSAC was involved, but it was just one of many groups that promoted large-scale tree planting. In 1979, the committee ran a Plant-a-Tree campaign to create awareness of the role trees played for the restoration of the environment, involving some 170 letters, car bumper stickers, and numerous press reports. The campaign was repeated in 1981 with 55,000 trees expected to be planted in the Kerang area.[177] A year later, the Gordon Shire Council announced plans to allow for 1000 trees to be planted as a pilot project against waterlogging and salting. Trees became the Shire Council's top priority for salinity control, also offering incentives to ratepayers to grow trees. In 1981, the shire set out to plant 4500 trees. The Shire of Wycheproof, in the same year, started a planting program for over 15,000 trees annually. The Borough of Kerang followed suit.[178]

It was a collaborative effort. KARF provided expertise by publishing a list of 60 types of trees suitable for the Kerang region in one way or

[175] Anonymous, "Year of the Tree," *The Northern Times*, June 4, 1982. *The Northern Times* used numbers from the Global 2000 Report to U.S. President Jimmy Carter by the Council on Environmental Quality. See Gerald O. Barney, *The Global 2000 [Two Thousand] Report to the President of the U.S.—Entering the 21st Century. A Report* (New York: Pergamon Press, 1980).

[176] Barr and Cary, *Greening a Brown Land*, 61.

[177] Anonymous, "Big Effort to Combat Salt," *The Northern Times*, May 13, 1980; Gynlais O. Jones, "Gun Shots—How Much to Pay?," *The Northern Times*, July 14, 1981.

[178] Anonymous, "Trees Combat Salting," *The Northern Times*, October 3, 1980; Anonymous, "Trees Combat Salt Near Boort," *The Northern Times*, August 12, 1980; Anonymous, "Tree Planting in Shires Salt Fight," *The Northern Times*, May 1, 1981; Anonymous, "2000 Trees for Pyramid," *The Northern Times*, May 19, 1981; Anonymous, "Trees for Borough," *The Northern Times*, September 3, 1982.

another.[179] The SRWSC stepped in with technical advice to tree planting groups to ascertain that the trees had a bearing on the problems they should address. The Forests Commission assisted with advice on suitable types, planting, and future care.[180] Occasionally, they were not just seen as useful, but for beautification. A woman in Tresco, near Swan Hill, applied to the Forests Commission for assistance to plant 50 trees on nature strips, near tennis courts, and public halls.[181] The regional winner for Northern Victoria of the Focus on Farm Trees Awards, organized by the Australian Farm Tree Regeneration Committee in recognition of the Year of the Tree, was the Twigg Family of Serpentine, south of Pyramid Hill, who had planted about 6000 trees over the past ten to twelve years, but had known nothing about the awards until they received notice from Canberra.[182]

Barr Creek caused the greatest agitation. Ultima dryland farmers bitterly opposed the SRWSC's proposal to extend the Lake Tutchewop drainage scheme with a pipeline through adjacent farmland and into evaporation basins. This was a continuation of the controversies of 1968, when Mystic Park and Fish Point farmers had opposed the original scheme that diverted Barr Creek waters into Lakes William, Kelly, and, ultimately, Tutchewop to evaporate, being concerned that this would damage their land but do nothing to aid the salinity problems of their own region. Nonetheless, the scheme was built.[183] In 1975, the SRWSC tabled its proposal to include Lake Tyrrell, a salt lake near Sea Lake, 120 kilometers north-west of Kerang, as the final evaporation basin. This measure had been recommended in the "Gutteridge Report" five years earlier. Dryland farmers from the Kerang region, during a meeting of the Ultima and District Dryland Farmers League, raised money to engage professional help in putting together a case to oppose the scheme. The League argued that it was not intended to solve the drainage problems of the Kerang

[179] Anonymous, "60 Proven Trees in Northern Region," *The Northern Times*, August 19, 1980.

[180] Anonymous, "Trees to Cut Pollution," *The Northern Times*, April 14, 1981.

[181] Anonymous, "Tresco Trees Welcomed," *The Northern Times*, February 3, 1981.

[182] Anonymous, "National Farm Trees Award to Twiggs," *The Northern Times*, July 20, 1984.

[183] Anonymous, "Minister Is Told of Threat to 30 Farms," *The Kerang New Times*, February 6, 1968; Mystic Park Progress Association to M. for National Development, "Barr Creek/Tutchewop Project."

region, but merely to divert existing drainage water from upstream into Lake Tyrrell. They opposed an open channel to transport saline water, arguing that this would be a major threat to farming areas along its route. Such a plan would add to the risk of salting of dryland areas but not reduce the salinity problem of the Kerang region's irrigation areas.[184]

In September 1978, the Water Commission divided the proposed scheme into two and suggested to firstly construct a self-contained extension, known as the "Mineral Reserve Basins Scheme" (MRB). It envisaged an increase of the pumping from Barr Creek and the construction of a pumping station at Lake Tutchewop, in order to divert water into three depressions about ten kilometers west via a pipeline under the Murray Valley Highway, and a ten-kilometer channel. This would allow for an annual diversion of at least 4500 megaliters, or 24%, of Barr Creek's annual flow, containing 16,000 tons of salt. The SRWSC estimated that the $3.7 million scheme would have a life span of 150 years, which could be extended by a later inclusion of Lake Tyrrell and even be "prolonged indefinitely" by disposal into the sea.[185]

The land adjoining the eastern edge of the basin sites was within the Mystic Park Irrigation Area and contained segments which already showed substantial signs of salinization, while the land on the western edge was used for dryland cropping and grazing. A major issue for the Public Works Committee was the effect on these lands regarding possible salinization. It found that the existing drainage pattern would be affected and seepage from the western end of the basins would occur. This would result in a rise of the water table, leading to a salinization of "marginal strips" of about 150–250 meters around the basins, some parts of the eastern irrigated properties, and dryland properties beyond these strips. Therefore, the SRWSC would acquire a portion of the surrounding land and lease it back to interested farmers.[186]

[184] State Rivers and Water Supply Commission, *Salinity Control and Drainage*, 2; Anonymous, "'Dry' Farmers' $3600 for Anti-Salt Case," *The Northern Times*, August 15, 1975; Anonymous, "Ultima Is Opposing Channel," *The Northern Times*, July 16, 1976.

[185] Hugh W. Caffrey, *Progress Report No. 4 from the Parliamentary Public Works Committee on the Salinity Control and Drainage Inquiry with Particular Reference to the Barr Creek Aspect* (Melbourne: Salinity Committee of the Victorian Parliament 1980, 5–6; 10; 16.

[186] Caffrey, *Progress Report No. 4*, 12–14.

The MRB scheme, once again, divided the rural communities of Northern Victoria into upstream and downstream interests. Shepparton farmers favored the scheme, while Kerang, Swan Hill, and Mallee farmers opposed it. Lying further downstream, South Australia's interests in good quality Murray water lingered in the background.[187] When the state government disclosed in late 1980 that it would seriously consider implementing the scheme, KIRSAC announced that it would seek talks with the Premier to receive assurances that all possible safeguards were implemented. In principle, the committee was opposed to the scheme, declaring that it would allow Shepparton irrigators to put more salt into the Kerang area. Shepparton could have its own evaporative basins.[188]

However, the main protagonists of the conflict were not activists from the irrigation areas, but rather dryland farmers from the Mallee who would be principally affected by the scheme: Rex McCann, secretary of the Inland Salinity Action Committee (ISAC); Ray Jewson and Keith Anderson of the Ultima and District Dryland Farmers Protection League (UDDFPL); and Ross Hercott of the "Halt the Salt" campaign. Ross, his brother Olive, and their sons were share farmers near Swan Hill, who mostly grew wheat without the use of fertilizers, insecticides, or herbicides. On weekends, Ross preached Christianity at a local church. In 1981, he and Olive launched their campaign to make academics, government officials, farmer organizations, politicians, and the media more aware of salinity. Ross conducted free tours by land and by air with a Cessna 172, showing people salt pans in the MDB to help them grasp the extent of the problem.[189]

Immediately after the government's announcement, ISAC assembled a deputation to the Victorian Premier, Dick Hamer, in order to convince him to reject the scheme. Rex McCann argued that, while the MRB scheme was inadequate to solve the salinity problem, it would ruin a large proportion of the Mystic Park, Tresco, and Lake Boga districts. In his view, it was simply designed to pacify water users downstream, meaning South Australians. He criticized that the shires of Kerang and Swan Hill had been prevented from submitting further evidence to the

[187] Russ, *The Salt Traders*, 165.

[188] Anonymous, "$3.7m Salt Plan," *The Northern Times*, November 14, 1980; Anonymous, "KIRSAC Opposes Mineral Basins Disposal Scheme," *The Northern Times*, December 16, 1980.

[189] Russ, *The Salt Traders*, 168–70.

PPWC.[190] The deputation, comprising councilors from the shires of Swan Hill, Kerang, and Wycheproof, and representatives of ISAC; the North-west Riding Ratepayers' Association; and the UDDFL, gained a reprieve. The Premier asked the PPWC to reopen the inquiry.[191] Nonetheless, the construction commenced in May 1981 and was expected to be completed within three years.[192]

At a public meeting of about 100, lasting three and a half hours, at the Kerang Town Hall on July 9, "universal condemnation" was expressed. The new Premier, Lindsay Thompson, was asked to halt the work until a reappraisal by the Public Works Committee. Two senior SRWSC officers were unable to convince landholders and local politicians that the proposal had sufficient safeguards. It was an "extraordinary face-to-face meeting between the State Government's technical experts and their vocal opponents" led by ISAC. Commissioner David Constable and research engineer Ken Collett found themselves "placed in the firing line during a torrid period of questioning which saw farmers dispute the commission's scientific evidence." In the heated atmosphere, the crowd turned ugly. One woman broke down crying after stating that the project would take the family farm. Constable reminded the farmers that they were also in the MDB and, therefore, had a common link with other people in the catchment. He criticized that an "overall balanced approach to salinity control" was being denigrated while real benefits to Murray water quality were possible. ISAC members rejected Collett's insistence that waters from the MRB scheme could not seep horizontally through to the Murray. The officers tried to calm the crowd by assuring that in the event of any problem directly attributable to the scheme, it would take preventive or remedial action.[193]

After ISAC had presented further evidence to the Premier, work came to a halt, and the PPWC was once again asked to reconsider the issue. It, however, decided against the further investigation, arguing that ISAC had not raised any new matters. The committee did assure the government that it would dismiss any scheme that would jeopardize the Tresco

[190] Anonymous, "Ruin!" *The Northern Times*, November 18, 1980; Anonymous, "Basin Plan 'Band Aid'," *The Northern Times*, January 13, 1981.

[191] Anonymous, "Basin Plan 'Band Aid'."

[192] Anonymous, "Mineral Basins Start," *The Northern Times*, May 20, 1981.

[193] Anonymous, "Halt Basins Plea Again," *The Northern Times*, July 10, 1981.

Irrigation District. It recommended reinforcing the scheme's safeguards by groundwater monitoring and the installation of further interceptors, if such were found necessary. The cost–benefit ratio was deemed satisfactory and the proposed buffer zone was adequate to protect neighboring dryland farming areas as well as stock and domestic water supplies. ISAC's assertion of technical flaws in the design was dismissed, and the MRB considered a responsible move to lessen the polluting effect of Barr Creek on Murray water quality. Once again, the farmers were reminded of their responsibility. Kerang, Cohuna, Koondrook, and parts of the Tragowel Plains derived large-scale benefit from the use of the creek as a drainage facility, so they clearly had responsibility for the people of Swan Hill, Robinvale, Mildura, Red Cliffs, and cities and towns of South Australia, including Adelaide. In other words, the opposition to the MRB scheme was considered to be founded on parochial grounds, with no base on technical or environmental considerations.[194]

This did little to calm the situation. During an "often emotional" and "potentially explosive" meeting at the Swan Hill Town Hall, over 400 farmers implored the state government to reverse its decision to proceed with the MRB scheme, refusing to accept drainage water from other areas, which was claimed would accelerate the spread of "liquid cancer" in Northern Victoria. The meeting resolved to ask the SRWSC to commission the Soil Conservation Authority to conduct an effectiveness study of the scheme and halt the work until its completion. According to *The Northern Times*'s verdict, Ross Hercott was the most popular speaker, lecturing on the spread of salinization. Rex McCann blamed "Melbourne academics" for turning large areas of Northern Victoria into salt drains, claiming that the amount of soil the government was attempting to save in Shepparton would be destroyed in the Mallee.[195]

The conflict seemed to escalate. In a last-ditch move to stop the scheme, ISAC and Ultima farmers threatened the government to take "direct action" to prevent work on the MRB from proceeding, while McCann refused to indicate whether this action would be violent or not.

[194] Anonymous, "New Evidence on Basins," *The Northern Times*, July 17, 1981; Anonymous, "Basins Halt Order," *The Northern Times*, July 24, 1981; Anonymous, "Basin Decision 'Disgusting'," *The Northern Times*, October 16, 1981; Anonymous, "Salt Scheme Finding: 'No New Matters of Substance'," *The Northern Times*, October 16, 1981.

[195] Anonymous, "Stop the Basins-Farmers," *The Northern Times*, October 20, 1981.

The SRWSC was warned not to continue with substantial works toward the completion of the scheme and thereby provoke "action."[196] ISAC, in conjunction with the UDDFPL, instructed a legal firm to proceed with expeditions "to take whatever action should be necessary to prevent the basins scheme."[197] Cohuna councilor Greg Fehring, who had attended the meeting at Swan Hill, claimed that "mass hysteria" stirred up by "yahoos," was the driving force behind the protests.[198] The Premier met with Ross Hercott to examine his evidence, while the Minister of Water Supply, Glyn Jenkins, warned the farmers not to take any "foolish or intemperate" action to prevent work on the MRB scheme as the government would not tolerate this and "steps would be taken to ensure that law and order prevailed."[199]

The farmers' threats were not substantiated. Instead, the conflict was pacified by being moved into the courtroom, and work was halted again.[200] On December 10, 1981, Ray Jewson and Keith Anderson issued a Supreme Court injunction through their lawyers, Garden and Green in Swan Hill, on the grounds that the SRWSC was acting beyond its powers in implementing the scheme, that it had neglected to carry out adequate research, and the scheme was not cost-effective. They accused the Commission of attempting to acquire land compulsory for a project which had not been assessed properly.[201] The case became the first class-action in Australia, with 51 plaintiffs, which would not be decided until 1986.[202]

Another development supported a calming of the tumultuous MRB issue. At the height of agitation, during the heated confrontations between farmers and the Water Commission at the Kerang meeting, one man sat in the crowd gathered in the Town Hall and said nothing. It was Evan Walker, the Australian Labor Party's (ALP) Shadow Minister for

[196] Anonymous, "Farmers' 'Action' Threat," *The Northern Times*, October 23, 1981.

[197] Anonymous, "Legal Bid to Halt Scheme," *The Northern Times*, October 30, 1981.

[198] Anonymous, "Salt Protest 'Hysteria'-Cr," *The Northern Times*, October 23, 1981.

[199] Anonymous, "Labor Move on Salinity," *The Northern Times*, November 10, 1981.

[200] Anonymous, "Mineral Basin Scheme 'Go'," *The Northern Times*, March 9, 1984.

[201] Anonymous, "Farmers Unite in Legal Fight Against Desalination Plan," *The Age*, November 7, 1985; T. B. Green, *A Country Legal Practice* (Carlton: Koestler Press, 1988), 44; Russ, *The Salt Traders*, 171.

[202] Anonymous, "Salt Basins Apeal Lodged," *The Northern Times*, April 4, 1986.

Conservation and Planning.[203] Walker, in an interview with Peter Russ, later reflected that the meeting convinced him of the widespread unrest in rural communities about salinity and evaporation basins, particularly the MRB scheme, which he felt had to be addressed. Russ has reconstructed the events of the years 1981 and 1982 using interviews with the protagonists of the day, which he conducted in 1990. Based on this, he has argued that, with Victoria facing a state election in 1982, the ALP was in search of a rural issue to wrest seats from the conservatives in order to win the election. Already having supported, for several years, conservation campaigns to stop clear-felling in state reserves and tree planting groups in dryland areas who sought funding for their work, it went to take up salinization as a political issue. The first draft of the ALP's salinity policy was cut on a kitchen table by six people in a farmhouse at Ultima on a property belonging to Ron Davies, a Mallee farmer, ALP member, and good friend of Evan Walker. One attendee of this meeting was Gyn Jones. He contributed the technical expertise on salinity, while Davies provided the political inspiration and direction. The others were mostly young public servants from Swan Hill who were concerned with the salinity problem. The resultant one-page paper became the foundation of the coming ALP government's policy on salinity.[204]

In opposition, the Labor Party had the support of many rural groups. Among them the Hercott brothers and their "Halt the Salt" campaign. Ross Hercott later claimed to have extracted the promise of a fresh start to the deadlocked Public Works Inquiry on salinity from David White, the Shadow Minister for Water Supply, if John Cain won government in 1982. The party also promised to double the funds for salinity control on farms.[205]

In March 1982, a month before the election, the ALP candidates for Bendigo Province and Swan Hill, Fabian Reid and Ian Hardie, made an aerial inspection of Northern Victoria to assess the effects of salinization between Swan Hill and Echuca. Afterward, they toured the region by car to discuss the issue with farmers. The two candidates assured them

[203] Anonymous, Halt Basins Plea Again," *The Northern Times*, July 10, 1981; Russ, *The Salt Traders*, 163.

[204] Russ, *The Salt Traders*, 163–68. Excerpts from the interviews are quoted in the text. On the ALP alliance with anti-logging conservation groups, see Hutton and Connors, *A History of the Australian Environment*, 155–58.

[205] Russ, *The Salt Traders*, 168; 172; "Rural," *The Age*, April 5, 1982.

that they "appreciated the gravity of the problem and regretted that years of Liberal government neglect had resulted in [a] disastrous situation." Adopting the rhetoric of the MRB's opponents, they declared that the evaporation basin approach was a "band-aid" treatment and not a long-term solution. A Labor government, they announced, would immediately set up an expert high-level inquiry drawing upon the Soil Conservation Authority, the SRWSC, the Department of Agriculture, universities, and overseas studies to determine long-term strategies to deal with the problem. Appropriate funding would be provided to conduct the inquiry and to implement its proposals.[206]

A month later, the ALP won the election, claiming 50% of the first preference votes and gaining 17 seats.[207] The new government heeded to its promise. It quickly replaced the Public Works Committee system with specialist parliamentary committees. An all-party Salinity Committee was established in July 1982, tasked to inquire into the social, environmental, and economic effects of salinity, opportunities for cooperative efforts in salinity control, methods of assessing the cost, and administrative arrangements. Particular priority was given to the Barr Creek question and the Mineral Reserve Basins and Lake Tyrrell schemes, on which the committee was to report within six months.[208]

However, the Salinity Committee later advised the Minister of Water Supply that such an inquiry would take much more time. In March 1984, in absence of any conclusion, the minister decided to proceed with work on the scheme as soon as possible, only for it to be stopped again two years later. Chemical deposits were found on pumping test equipment, which, it was feared, could lead to the clogging of sub-surface works.

[206] Anonymous, "ALP Slams Lib. Inaction on Salinity," *The Northern Times*, March 5, 1982.

[207] Australian Politics and Elections Database, "Parliament of Victoria, Legislative Assembly Election. Election of 3 April 1982," University of Western Australia, accessed August 16, 2022, https://elections.uwa.edu.au/elecdetail.lasso?keyvalue=912&summary=false; Anonymous, "Local MPs Returned," *The Northern Times*, April 6, 1982; Anonymous, "State Election Round-Up," *The Northern Times*, April 6, 1982.

[208] Salinity Committee, *Third Report to the Parliament. Salt of the Earth. Final Report on the Causes, Effects and Control of Land and River Salinity in Victoria* (Melbourne: Salinity Committee of the Victorian Parliament, 1984), v; ix–x; Russ, *The Salt Traders*, 172–73.

Furthermore, the government decided that it wanted to wait for the outcome of the legal case.[209]

A day before the case started, the state government introduced a bill to retroactively declare the SRWSC authorized to take all means it considered necessary for salinity control. However, the passing of the amendment was prevented by the farmers with help from Barry Stegall, Legislative Assembly member for Swan Hill. During the legal proceedings, emotions in Northern Victoria once again ran high. According to Peter Russ, this went so far that a fence picket was thrown through the front window of Don Blackmore's home in Swan Hill, who was the local chief project engineer responsible for the construction of the MRB scheme. Rex McCann accused the Victorian government of having deliberately changed the Water Act to ruin the farmers' case.[210]

Justice O'Bryan ruled the Mineral Reserve Basins Scheme lawful on March 19, 1986. The Commission was declared legal in the compulsory acquisition of 11 properties around the three proposed evaporation basins, and the court awarded costs of up to $340,000 against the plaintiffs. However, the justice questioned the economic benefit and water management aspects of the scheme. He criticized the very low rate of return for the now increased cost of $8.6 million and pointed out that it could even turn out to be negative. This made the scheme difficult to justify on economic grounds: "The advantages are minimal, weighted against the interests of the rural community represented by the plaintiffs." It was little consolation to them. Farmers who milled around outside the court building in Melbourne expressed anger and disbelief. Some of their family members were in tears.[211]

On June 18, 1986, at eleven a.m., more than 1000 people lined Victoria Street in Kerang, and businesses closed. Sounds of the fire siren signaled three minutes of silence to protest the commencement of the MRB scheme. Simultaneously, citizens in Lake Boga, Swan Hill, Sea Lake, and Wycheproof did the same. Residents, politicians, farmers, shop

[209] Salinity Committee, *Second Report to the Parliament. The Activities of the Salinity Committee* (Melbourne: Salinity Committee of the Victorian Parliament, 1983), 6; Anonymous, "Mineral Basin Scheme 'Go'," Anonymous, "Chemicals, Court Action Stop MRB Scheme," *The Northern Times*, March 12, 1986.

[210] Russ, *The Salt Traders*, 201–2.

[211] Anonymous, "Salt Basins Doubts Despite Court Ruling," *The Northern Times*, March 21, 1986.

owners, and school children gathered on the streets in all five towns. The silent protest had been hastily organized by local representatives of the three major political parties. In a joint statement, Alan Wood (Liberal), Barry Stegall (National), and Pat Frazer (Labor) declared that a long-term solution to the salinity problem was of such vital importance that it transcended party politics since the salinity question was not a parochial issue, but one that affected everyone would continue to have serious repercussions for future generations. While the rally focused on the MRB, it also addressed the wider issues of salinity control. The three representatives put forward a resolution, demanding that the MRB scheme be abandoned, the legal costs be borne by the state government, regular flushing of Lakes Boga and Charm be done to ensure they did not become salt wastes, the commissioning of a feasibility study into long-term solutions to salinity, and recognition be given that community involvement in planning and decision-making was vital.[212]

It was a tranquil final act to a noisy struggle. In December 1986, after more than ten years of conflict, the Mineral Reserve Basins Scheme was deferred indefinitely following a ministerial tour of the Kerang region. After the Minister for Conservation, Forests, and Lands, Joan Kirner, had met with local community groups and the representatives who had organized the silent protest, she was convinced that the scheme should not proceed any further, and persuaded the rest of the ministers to see it her way. Subsequently, they jettisoned the MRB scheme and agreed to pay some of the farmers considerable legal costs. The Minister for Water Resources, Andrew McCutcheon, released a new cost–benefit analysis showing a negative return for the scheme. Stricter environmental controls had increased the projected annual operating costs to over $200,000, which offset the value of anticipated salt reductions in South Australia. In the end, $7.3 million had been spent on bridges, roads, pumping stations, channels, and land acquisitions.[213]

Those were peanuts compared to the projected cost of another proposal—a pipeline to the sea to dispose of the saline wastes of irrigation. The pipe was a pet project of the opponents of the MRB scheme, routinely praised as the long-term solution to the salinity problem, while

[212] Anonymous, "5 Towns' Salinity Scheme Protest," *The Northern Times*, June 17, 1986; Anonymous, "Silent Show of Solidarity," *The Northern Times*, June 20, 1986.

[213] Russ, *The Salt Traders*, 203–5.

the SRWSC's proposal was vilified as being just a band-aid.[214] Time and time again throughout the decade, pipeline schemes were advocated by activists like Rex McCann and several shire councils in the Kerang area. When Kerang's mayor, Steve Coughlan, supported the silent protest, he did so by declaring that "we should pipe the salt to the sea, where it belongs."[215]

The idea blossomed in the long shadow of the "Supply Channel Scheme," which had been discussed by Gutteridge, Haskins & Davey in 1970 as an option for "a complete and permanent solution to the problem in the Murray valley." The consultants assessed a proposal originally put forward by Simon Pels to the Winston Churchill Memorial Trust in 1967. Pels had advocated a "separation of the irrigation supply from the river system above the points of saline entry" by constructing artificial channels for carrying irrigation water, while the Murray itself would be carrying all saline drainage effluent.[216] They evolved a single-channel solution, which envisaged the delivery of high-quality water to major towns and irrigation areas via a pipeline and an open channel system. The Consultants concluded that this approach would be technically feasible. They did not consider the environmental implications of converting the entire river into an unrestricted drain for the adjacent irrigation areas. The initial costs were calculated to be between $292.2 and $334.6 million ($14.6 to $16.5 billion dollars today), with annual operating costs ranging from $20.3 to $23.4 million dollars ($233.6 million to $269.2 million today). This price tag caused Gutteridge, Haskins & Davey to exclude the scheme from further consideration.[217]

During the hearings of the Public Works Committee, the proposal reappeared in a reversed form. KIRSAC wished to "see a costing carried

214 Anonymous, "Basin Plan 'Band Aid'," *The Northern Times*, January 13, 1981; Anonymous, "'Reprieve' on Basins," *The Northern Times*, January 16, 1981.

215 Anonymous, "5 Towns' Salinity Scheme Protest," *The Northern Times*, June 17, 1986.

216 Gutteridge, Haskins & Davey, *Murray Valley Salinity Investigation—Report*, 208. See Simon Pels, *A Plan to Control Salt Problems in the Murray River. Report Prepared by S. Pels for Submission to the Winston Churchill Memorial Trust, September 1967* (Deniliquin: Water Conservation and Irrigation Commission, 1967).

217 Gutteridge, Haskins & Davey, *Murray Valley Salinity Investigation—Report*, 208–09; 222–25; 428; Gutteridge, Haskins & Davey, *Murray Valley Salinity Investigation. Volume 2. Maps* (Canberra: River Murray Commission, 1970), Map 8. Current values calculated using Hutchinson and Ploeckl, "MeasuringWorth."

out for the ultimate solution to the Murray Valley's salt problems, i.e., a pipeline/drain to the sea carrying concentrated saline drainage from all irrigation regions on both sides of the river."[218] George Hardwick told *The Northern Times* in 1978 that a "pipeline to the sea would be the logical answer" to salinization.[219] Subsequently, however, the committee reverted its policy.[220] Gyn Jones, who was by then convinced that the only feasible solutions for the Kerang region were on the farm rather than in costly large-scale engineering solutions, wrote in 1981: "The big priority for this region is not Mineral Reserve Basin Schemes—it's not a pipeline to the sea—it's the improvement of land surfacing so that water is applied efficiently and equitably. It is the upgrading of farm water application and drainage; it's the establishment of operable re-use schemes both district and on-farm. When we have about 80% adoption of these techniques, then it is time to start putting engineering priorities forward."[221]

However, new bids were repeatedly launched at the government to inquire into the idea. In 1981, the Kerang Borough Council asked the Premier to conduct further investigations into a salt pipe, while it rejected to turn over large areas of land for salt disposal through the MRB scheme.[222] Councilor Wright was quoted saying that the scheme was "only a stop-gap method and is not going to solve the problem."[223] A new attempt was made by Bruce Reid, Legislative Council member for Bendigo, in 1984. He proposed to the Salinity Committee the construction of a pipeline from the Barr Creek-Tutchewop area, along the Swan Hill–Bendigo–Melbourne railway line, to the sea at Port Phillip Bay. The preliminary cost calculation was $149 million to intercept about 150,000 tons of salt per year.[224] The proposal was promoted by the Victorian

[218] Parliamentary Public Works Committee at Kerang, "Hardwick," 484.

[219] Anonymous, "Pump Salt into the Sea!," *Herald*, October 23, 1978.

[220] Anonymous, "Study Pipe Idea," *The Northern Times*, February 17, 1981.

[221] Gynlais O. Jones, "Gun Shots—Planning Pays Off," *The Northern Times*, April 14, 1981.

[222] Anonymous, "New Bid on Salt Pipe," *The Northern Times*, February 6, 1981.

[223] Anonymous, "Study Pipe Idea," *The Northern Times*, February 17, 1981.

[224] Anonymous, "Salt-to-Sea Plan by MLC," *The Northern Times*, February 17, 1984; Anonymous, "Ocean Salt Pipe Plan at Rally," *The Northern Times*, July 31, 1984; Anonymous, "Education on Salt Essential," *The Northern Times*, August 8, 1984.

Farmers and Graziers Association as "a stroke of genius" and supported by Ross Hercott.[225]

In the same year, a "Pipe the Salt Committee," led by Swan Hill shire councilor Jack Larkin, put forward the idea of a pipeline to the sea along the bed of the Murray River using a polyethylene pipe with an anticipated life expectancy of 100 years. The committee was made up of the Kerang, Swan Hill, and Wycheproof shires and claimed to be supported by 21 other municipalities and 20 organizations. Larkin envisioned that the pipe would lend itself to the introduction of town and trade waste disposal for treatment or offshore discharge. He claimed that the project could be completed within three to five years involving all three Basin states and would require no maintenance or running costs except for the required pumping system.[226]

Although the committee maintained that it was expressing the "basic commonsense knowledge of the majority," the idea never enjoyed consensual support in the rural communities and was repeatedly rejected by the government. During a meeting with ISAC representatives, the Premier discounted the idea as not being feasible and too costly at an estimated price of $300 million.[227] Kerang shire councilor Marshall Dawson referred to the projected costs as "fantastic," urging other useful, less ambitious, measures like tree planting.[228] The Gordon Shire Council refused to support the pipeline project, arguing that it was merely an emotional response to the Mineral Reserve Basins Scheme and held little benefit.[229] According to councilor Russell Smith, chairperson of KIRSAC, the proposal was a "pie in the sky."[230]

In 1988, the newly founded Murray–Darling Basin Ministerial Council (MDBMC) attempted to put a lid on the issue by announcing that such a pipeline was not on its agenda. Discarding the project in a polite manner, it was declared to be an option for the future. Council President Neil

[225] Anonymous, "Ocean Salt Pipe Plan at Rally."

[226] Anonymous, "Renewed Effort for Salt Pipe," *The Northern Times*, September 11, 1984; Anonymous, "Saline Effluent Disposal Vital," *The Northern Times*, September 14, 1984; Anonymous, "Salt Pipe 'Furphy' Claim Laughable," *The Northern Times*, October 21, 1988.

[227] Anonymous, "New Bid on Salt Pipe," *The Northern Times*, February 6, 1981.

[228] Anonymous, "'Reprieve' on Basins," *The Northern Times*, January 16, 1981.

[229] Anonymous, "Study Pipe Idea," *The Northern Times*, February 17, 1981.

[230] Anonymous, "Salt Pipeline Is Rejected," *The Northern Times*, April 14, 1981.

Fitzpatrick emphasized that the responsibility for the management of the land and the environment of the Basin ultimately rested with the community.[231]

There would be no Lake Tyrrell Scheme, no pipeline to the sea. There would be no grand-scale engineering solution, period. However, the wish for a technological fix that would "replace that natural drainage system" of the Murray Valley with an artificial one "to protect the [irrigation] environment" and, thereby, permanently retain environmental security in the Murray Valley's irrigation areas, died hard.[232] As late as 1989, Gutteridge, Haskins & Davey were commissioned to conduct a feasibility study, taking account of environmental issues, costs, and a possible time frame. Again, no concrete measures were followed.[233]

It now appeared that the only viable route was to continue living with salinity, using a combination of locally adapted engineering methods—provided they were economically viable—and farming practices, coupled with promoting community awareness and involvement. Decades of self-help by the research farms, Salinity Action Groups, public servants, lobbyists, educators, tree planting volunteers, and concerned individuals had gone far in forging solutions for salinity control. They, however, had not succeeded in arresting the continuing spread of salinization.

Interlude: A Multi-faceted Problem

"This year," GIRDAC chairman John Dainton wrote in 1987, "I believe has been a most significant year in the fight against salinity within the Region. There is a greatly increased awareness of the problem, and salinity issues which once were not discussed to any extent in public are now openly reported and control options debated." By contrast, however, newly released water table maps showed "alarming increases to the level of the water table in the Shepparton Irrigation District. The ever deteriorating situation demonstrated by these maps [is] indisputable."[234]

[231] Anonymous, Sea Pipeline Is Not Being Considered," *The Northern Times*, September 2, 1988.

[232] Anonymous, "Salt Pipe 'Furphy' Claim Laughable," *The Northern Times*, October 21, 1988.

[233] Anonymous, "GH & D Win Pipeline Survey," *Salt Force News* 15 (August–September 1989): 4.

[234] Dainton, "Annual Report 1987," 1–2.

Gyn Jones, in a gloomy prediction, warned that "given the husbandry we've had in the past, the northern plains have 50–70 years of life before they become complete saline deserts."[235]

Indeed, salinity problems were still increasing in scope. In 1984, it was estimated that about 2400 km^2 of land in the state was observably damaged by salt, 1400 of which lay in the GMID. Another 4000 km^2 were found to be salt-prone. Irrigated agriculture, through drainage water from irrigation, seepage from groundwater under irrigation areas, leakage from water channels, and unsuitable agricultural practices, was annually adding about 400,000 tons of salt to the Murray, of which Barr Creek alone contributed 160,000 tons. This was more than half of all saline inflows that could be attributed to Victorian activities, amounting to about 700,000 tons of salt per year. Non-irrigated areas added a mere 10%.[236]

For the Shepparton region, it was estimated that half of the irrigated land had water tables at or near two meters below the surface. Unless further control measures were taken, a major expansion was anticipated within the next decade. As for Kerang, a precarious stability had been achieved. Whereas about 72% of the region was underlain by shallow water tables, the water inputs and outputs appeared to balance, and the water table to be in a state of "dynamic equilibrium," in which evapotranspiration losses balanced accessions, at about one meter below the surface. However, it was believed that this could be upset by very wet years and the area underlain by high water tables could increase into peripheral dryland areas.[237]

This equilibrium was due to the fact that irrigation farmers, partly with assistance from governments, throughout the GMID had invested huge sums into farm reconstruction. In the Kerang region, about a quarter of the irrigated land had been graded to have a steeper slope using laser-leveling equipment, which enabled a more controlled, uniform distribution and drainage of irrigation water and rain. In the Shepparton region, about 22% had been graded. About $70 million had been spent by individuals in the GMID since 1979 on land forming alone. In the

[235] Gynlais O. Jones, "Salinity Spread Is a Reminder to 'Lift Our Performance'," *The Northern Times*, June 27, 1986.

[236] Salinity Committee, *Salt of the Earth*, 22–30; 40.

[237] Salinity Committee, *Salt of the Earth*, xvii; 52; 60; ACIL Australia et al., *Causes, Extent and Effects of Salinity*, xxi.

Nyah, Tresco, and Woorinen irrigation areas near Swan Hill, sub-surface drainage had been installed, predominantly before 1970. Three groundwater control projects for the Shepparton region were completed by the late 1980s, involving almost 100 pumps to dispose of saline groundwater. Farmers, chiefly in horticultural areas, had changed irrigation systems at considerable capital cost. In total, $10 million per year was spent by farmers and governments on salinity control, on interception and disposal schemes for saline groundwater and drainage effluent, land forming, on-farm measures, and tree growing.[238]

In contrast, the adoption of laser grading by landholders in the Kerang area, designed to reduce waterlogging and groundwater accessions while increasing productivity, proved to be disastrous for water quality in Murray. Laser grading was hailed as "a revolution going on in irrigated agriculture" by its advocates.[239] However, there was a catch. Farmers who laser-graded their land were reluctant to invest in re-use systems, despite government loans that were available. With fixed entitlements and cheap irrigation water there was no incentive to save and recycle it. The annual flow and total salt loads in Barr Creek, draining the farmlands of Kerang and Cohuna, which had been decreasing after the installation of the Tutchewop-Tyrrell scheme, climbed after the rapid introduction of laser grading. The pumps on the creek were not large enough to handle the increased volumes of brackish water intended for storage at Lake Tutchewop, which was draining off farmlands. The lake filled rapidly, and the remainder was discharged into the Murray.[240]

Salinization continued to be a strain on the irrigation communities. Kerang Shire President Kevin Inglis told the members of the shire council in 1984 that the region had learned to cope and survive with the effects of salinity "with extreme suffering and heartbreak." For many, he said, it meant a lower standard of living or having second jobs to survive. There was no longer profit from many salt-affected farms, their owners surviving on outside income. Few young people were allowed to or wanted to stay on a farm with the prospect to do so at 30–40% salinity loss. There was

[238] Gynlais O. Jones, "Salinity Spread Is a Reminder to 'Lift our Performance'"; Salinity Committee, *Salt of the Earth*, 82–86; 115–16; Salinity Pilot Program Advisory Council, *Draft Shepparton Land and Water Salinity Management Plan* (Shepparton: Salinity Pilot Program Advisory Council, 1989), 14.

[239] Anonymous, "Revolution Through the Laser," *The Northern Times*, June 8, 1982.

[240] Russ, *The Salt Traders*, 207–9.

a "continuing deterioration of the Kerang region," not just in terms of economic loss, but also of many wetlands, lakes, and forests. Much of this had been caused by engineering works for salinity control, which, to Inglis, were "deliberate works to deteriorate one area, to attempt to keep the status quo in the Murray River."[241]

One did not have to subscribe to Inglis's position that the Kerang region was consciously being allowed to deteriorate for the benefit of other water users in the Murray valley, to acknowledge that salinization was a long-standing burden on the economic and social fabric of the region. The same was true for Shepparton, although only more recently. Perhaps surprisingly, there was still a general paucity of research on these issues and the uncertainties involved were overwhelming. According to a 1983 study by the consultants of ACIL Australia, Gutteridge, Haskins & Davey, and others, salinity's social effects were "perhaps among the least studied of all." Just as in the 1960s, when there had been profound gaps in the knowledge available about the extent and severity of salinization on agricultural land, 20 years later "estimates of the economic costs [...] [were] characterised by imprecise methods of data collection and unverified assumptions."[242]

Overall, the consultants' best approximation was that annual agricultural production losses due to salinity were $6 million in irrigation areas, and $3.1 to $5.5 million in dryland areas. The loss of capital value was deemed to be about $110 million and from $17 to $180 million, respectively. For the Kerang region, it was estimated that more than half of the irrigated area was suffering reduced or negligible agricultural production. Due to lower productivity and prices, the gross income of farmers on salt-affected properties was possibly often less than 50% of comparable non-salt-affected ones.[243] A 1984 report for the Victorian parliament estimated that the average proportion of a farm affected by salinization in the Shepparton region was 30% on horticultural holdings, and 8% on mixed dairy farms. The total cost from damages in the state was approximated to be about $50 million per year (about $433 million dollars today). Kerang's loss of agricultural productivity was approximated to be 25%, Shepparton's to be 7%, resulting in a combined loss of $30 million

[241] Anonymous, "Salinity a Wider Problem," *The Northern Times*, September 18, 1984.

[242] ACIL Australia et al., *Causes, Extent and Effects of Salinity*, 149; xiii.

[243] ACIL Australia et al., *Causes, Extent and Effects of Salinity*, 85; 133.

per year. The report warned that the cost could triple within the next 30 years.[244]

Beneath these numbers lay social repercussions in the irrigation communities. Generally speaking, ACIL, et al., adopted the nexus that the rural sociologist Sharman Stone had put before the Public Works Committee in 1978: As the productivity of the land decreased, younger and more adaptable farmers were likely to leave the area, while older ones stayed on. Those who did stay would either expand their farms or grudgingly accept lower incomes. Farmers would look for off-farm work and leave a district if there were no opportunities available.[245] This resulted in farms becoming absentee-owned and deteriorating due to neglect, giving an area a "general impression of declining fortunes."[246] Finally, the Consultants added, farmers whose productivity was diminished by salinization often commanded insufficient funds to carry out improvements required to ameliorate its effects, like adequate irrigation layout. On the one hand, salinization required them to farm a greater area, while on the other, lower returns made it difficult to finance expansion or development—a fatal situation for many.[247]

However, the consultants objected to the uniformity of the economic and social multiplier effects of lower incomes for farmers that Stone had assumed for all salt-affected areas. Whereas she had accurately assessed the situation of small towns like Pyramid Hill (632 people in 1971), with populations of less than 2000, larger centers might have a broad enough economic base to survive. For example, it could not be assumed that cities like Shepparton, Mildura, and Echuca—which all developed because of irrigation—would suffer a symmetrical depopulation due to salinity problems. Other factors, they argued, such as the terms of trade of the relevant agricultural industries, the mix of agricultural production, and the size of each town, would determine the net effect. Shepparton, a case in point, had continued to grow despite an extended period of declining canning fruit incomes, which were offset by the buoyant conditions for the dairy industry. Shepparton City grew from a population of

244 Salinity Committee, *Salt of the Earth*, xviii; 70–72; 85–86. Current value calculated using Hutchinson and Ploeckl, "MeasuringWorth."

245 ACIL Australia et al., *Causes, Extent and Effects of Salinity*, 149–50.

246 ACIL Australia et al., *Causes, Extent and Effects of Salinity*, 150.

247 ACIL Australia et al., *Causes, Extent and Effects of Salinity*, 134.

10,848 to 17,523 from 1957 to 1966, reaching 24,430 people in 1981. The Shepparton region's growth rate between 1971 and 1986 was similar to the one achieved state-wide, indicating a robust economy based on irrigated agriculture and local processing. Kerang had largely maintained its population of about 4000 from 1970 to 1986, despite the growing extent of salinization. Swan Hill's population had grown from 7700 to 8831 over the same period. All this disputed the causality of salinization for depopulation that had been so forcefully advocated by some witnesses during the Salinity Inquiry.[248]

Furthermore, productivity losses due to salinization were countered by a transformation of the regional economies. Agriculture, while remaining the primary form of non-urban employment, steadily declined. Still, agricultural industries represented 37% of household income and 69% of the value of exports from the Kerang/Swan Hill region in 1986. In the Shepparton region, irrigated agriculture provided 55% of employment. In both regions, substantial links existed between the agricultural sector and related industries, which provided farm machinery, irrigation equipment, feed, fertilizer, sprays, and fruit packing packages, causing flow-on effects of productivity losses in agriculture to the regional economy. However, in Kerang and Swan Hill, the number of agricultural enterprises declined by 28% between 1980 and 1989. State-wide, the trend was a decline of 35% over this period. In 1986, about two-thirds of Kerang's male workforce were employed in agriculture, but this proportion was steadily decreasing while the importance of tourism and hospitality was rising. This was particularly the case in Swan Hill. About 235,000 visitors came to the Murray–Mallee region in 1987–1988. However, ongoing salinization could endanger this potential, as lakes and wetlands in the Kerang Lakes Area, which were valued for recreation and duck hunting, feared becoming brackish or hypersaline and provide reduced habitat for bird life.[249]

[248] ACIL Australia et al., *Causes, Extent and Effects of Salinity*, 150–51; State Rivers and Water Supply Commission, *Salinity Control and Drainage*, 25; John Cary and Neil Barr, "The Kerang Lakes Area Management Project. Social Profile," in *Kerang Lakes Area Draft Salinity Management Plan. Background Papers Volume 1 of 2*, ed. Salt Action Victoria (Melbourne: Salt Action, 1989), 5; 7–8; 11; Salinity Pilot Program Advisory Council, *Draft Shepparton Land and Water Salinity Management Plan*, 18–19.

[249] Cary and Barr, *Greening a Brown Land*, 5; 7–8; 11; Dwyer Leslie Pty. Ltd., "Kerang Lakes Area Working Group. Regional Economic Conditions," in *Kerang Lakes Area Draft Salinity Management Plan. Background Papers Volume 1 of 2*, ed. Salt Action Victoria

Though a causal nexus between salinization, economic depreciation, and the decline of irrigation communities was not established, Inglis and Stone urged that the crisis transcended economics, encompassing the "anxiety associated with a problem that is often not easily explained, is inequitably distributed in the community, and appears to be inexorably growing."[250]

Furthermore, it was extremely difficult, if not impossible, to quantify the real cost of a problem that included such things as future tourist or economic potential, aesthetic judgments, estimations of human and community "worth," levels of mental health, the devaluing of a landscape with dead and dying trees, the loss of freshwater fish, wild flowers, and habitats of fauna. Who, for example, could quantify the mental and physical stress experienced by those who lived with the declining productivity and altered landscape induced by salt?[251]

One final point is significant. One-hundred and thirty years after the first signs of human-induced soil salting appeared in Victoria, the "events set in train with the advent of European man are now both irreversible in the human time scale and have certain inevitabilities."[252] According to engineers of Dwyer Leslie and Maunsell and Partners, "natural" means of controlling the water tables, like major reforestation, cessation of irrigation, etc., would not be successful within the human time scale. For water tables to be controlled in the shorter term, the only feasible means would be mechanical ones, such as pumping.[253] If the Murray basin was a bathtub full of soil that was now almost saturated with water, even turning off the tap would prove insufficient.[254]

(Melbourne: Salt Action, 1990), 5–9; Kerang Lakes Area Working Group, *Kerang Lakes Area Draft Salinity Management Plan*, 91; Salinity Pilot Program Advisory Council, *Draft Shepparton Land and Water Salinity Management Plan*, 21–26.

250 State Rivers and Water Supply Commission, *Salinity Control and Drainage*, 97.

251 Anonymous, "Salinity Multi-Faceted Problem," *The Northern Times*, September 15, 1984.

252 Dwyer Leslie Pty. Ltd. and Maunsell and Partners, *Salinity Control in Northern Victoria. A Strategic Study for the Salinity Committee of the Victorian Parliament* (Melbourne: Salinity Committee of the Victorian Parliament, 1984, 4.

253 Dwyer Leslie Pty. Ltd. and Maunsell and Partners, *Salinity Control*, 5.

254 Salinity Committee, *Salt of the Earth*, 61.

CHAPTER 7

A New World (1979–1994)

Go with the Flow

"There is a tide in the affairs of men / Which, taken at the flood, leads on to fortune." This was Shakespeare's way of saying "go with the flow."[1] Ernest Jackson thought it was a fitting motto for this time of intensified change.[2] A tide was, indeed, coming.

The new parliamentary Salinity Committee, which had been instituted by the ALP state government, was established in July 1982 to provide a basis upon which to develop an effective program for the control of salinity. This objective was later amended to include an inquiry into water allocations in Northern Victoria.[3] It was a mammoth two-year undertaking, drawing on contributions by several hundreds of people from the rural community, government departments, research institutions, industry, and public interest groups, who prepared submissions.

[1] William Shakespeare, *Julius Caesar: Fully Annotated with an Introduction by Burton Raffel. With an Essay by Harold Bloom* (New Haven: Yale University Press, 2006), Act 4, scene 3, 217–18.

[2] Ernest M. Jackson, "Good 'Tidings' for the Murray!" February 1986, VPRS 6008/P0001, 84/36649, Public Record Office Victoria, Melbourne.

[3] Salinity Committee, *Salt of the Earth*, v; xvi.

© The Author(s), under exclusive license to Springer Nature Switzerland AG 2023

D. Rothenburg, *Irrigation, Salinity, and Rural Communities in Australia's Murray-Darling Basin, 1945–2020*, Palgrave Studies in World Environmental History,
https://doi.org/10.1007/978-3-031-18451-2_7

It conducted 104 meetings, of which 16 were public hearings held in Kerang, Mildura, Swan Hill, Bendigo, Hamilton, Shepparton, and Melbourne. Witnesses included representatives of the Australian Conservation Foundation; the shires of Swan Hill, Kerang, and several others; the VFGA; the Murray Valley League (MVL); the UDDFPL; GIRDAC, KIRSAC, and other Salinity Action Groups; VIRPO; the Pipe the Salt Committee; tree planting groups, as well as individuals. The committee inspected salt-affected properties and salinity control projects in Victoria, including KARF; New South Wales; and Western Australia.[4]

It commissioned two further studies into the salinity problem—one led by ACIL Australia and another one by Gutteridge, Haskins & Davey—to assess the success of the techniques that had been implemented so far, and the advantages and disadvantages of implementing possible control techniques.[5] Further, the committee commissioned a strategic study of salinity control, led by the consulting firm Dwyer Leslie in association with Maunsell and Partners, to assess the options and requirements for further protection of agricultural productivity, as well as generation and disposal of saline wastes. Lastly, it obtained information on salinity management in other countries through discussions with overseas specialists and a review commissioned through the International Referral System of UNEP.[6]

Graham Hunter was appointed Director of Research in 1983 and responsible for the driving of the inquiry's research and direction via an interdepartmental steering committee, which consisted of staff from the Land Protection Service at the Department of Conservation, Forests and Lands, the Water Commission, CSIRO, the University of Melbourne, and the Department of Agriculture.[7]

A major difference between the Public Works Inquiry on "Salinity Control and Drainage" and the new Salinity Committee was that the leaders of the new inquiry systematically included community groups concerned about salinization. Previously, they had had the chance to

[4] Salinity Committee, *Salt of the Earth*, x–xi; 267; 269–77.

[5] Salinity Committee, *Salt of the Earth*, xi–xii; ACIL Australia et al., "Causes, Extent and Effects of Salinity;" Gutteridge, Haskins & Davey, et al., *The Application of Salinity Control Techniques in Victoria* (Melbourne: Salinity Committee of the Victorian Parliament, 1983).

[6] Salinity Committee, *Salt of the Earth*, xvi; Dwyer Leslie Pty. Ltd. and Maunsell and Partners.

[7] Salinity Committee, *Salt of the Earth*, vi; Russ, 174.

give evidence, but not been part of the inquiry's working procedures. The community groups' role was upgraded insofar as they were officially included in the workings of the Salinity Committee; however, they were not part of the Steering Committee, but only of a Consultative Group from which their views were fed to the former. Eric Merrigan and Leon Heath of GIRDAC, Kevin Inglis of KIRSAC, plus representatives from other groups, such as the Pipe the Salt Committee, the Australian Conservation Foundation, and the MVL became official participants of the proceedings.[8]

Furthermore, the Salinity Committee issued a newsletter describing its activities and sent questionnaires to the councils of 162 non-metropolitan municipalities in the state to survey the perception of salinity problems. More importantly, it sponsored a community conference on salinity control at the Bendigo College of Advanced Education, held on February 18 and 19, 1984, which was attended by 370 people.[9] The conference was intended as an "informal forum in which people can share ideas" and for committee members to learn more about the views of the community and "the role that each of us might play." John Kerin, the Federal Minister for Primary Industry, delivered a keynote in which he urged a national approach toward salinity control. In five panels, the participants heard lectures on the salinity problem, possible solutions, methods of implementation, and the role of governments by speakers from ACIL Australia; Gutteridge, Haskins & Davey; and academics such as Sharman Stone and Sandford D. Clark of the University of Melbourne. In 13 working groups, they were joined by the committee members to debate the requirements of a salinity control strategy and its components. A forum on priorities for action to governments, farmers, researchers, educators, and communities was held. Finally, the conference and the inquiry were reviewed. The list of participants reveals many a familiar name: George Hardwick, Ross Hercott, Ernest and Vera Jackson, Gyn Jones, Rex McCann, Eric Merrigan, Norman Mitchelmore, and Russell Smith were among the attendants. These usual suspects

[8] Salinity Committee, *Salt of the Earth*, 282–83; Russ, 174.

[9] Salinity Committee, *Salt of the Earth*, xiii–xiv.

were joined by shire representatives, public servants, farmers, teachers, technical experts, politicians, and industry representatives.[10]

The Northern Times called the conference an "outstanding success" and found that there had been a "general consensus" that the problem needed to be tackled on a catchment basis, which involved cooperation between the states and federal government funding.[11] According to Graham Hunter, Bendigo was a "turning point" in the inquiry because it demonstrated the range and depth of community interests and concern to the politicians.[12] It was acknowledged that community organizations had played a vital role in fostering awareness of salinity problems and promoting solutions. The value of the work of KIRSAC, GIRDAC, VIRPO, Pipe the Salt, the VFGA, and Halt the Salt was officially recognized. However, the committee predominantly saw these voluntary groups as useful transmitters between the government and the rural communities, which played "an important part in motivating the community and ensuring that governments are aware of community concerns," but not as equal actors who could have a say in its decisions. Nonetheless, it was a decisive step forward for them.[13]

In October 1984, the Salinity Committee released its final report to the public. It was titled "Salt of the Earth"—both an appeal to bold, enduring steps toward a solution by committed people, and a reminder of the fatal consequences of ignorance and mismanagement, which could turn the natural properties of salt against humanity's endeavors. The report recommended a markedly different approach to salinity control than the SRWSC had taken in 1975 with its "Salinity Control and Drainage" strategy. Salinity control was no longer seen primarily as an engineering feat that would complement the existing water storage and delivery infrastructure with a drainage system, but as a "co-ordinated action by landholders, local communities, and a range of government agencies." This included mechanical measures such as drainage and pumping, but these were not seen as a one-time technological fix. Rather, a strategy

[10] Salinity Committee, *Community Conference on Salinity Control. Bendigo College of Advanced Education. 17–18 February 1984* (Melbourne: Salinity Committee of the Victorian Parliament, 1984).

[11] Anonymous, "Kerin calls for National Assault on Salt," *The Northern Times*, February 21, 1984.

[12] Russ, *The Salt Traders*, 176.

[13] Salinity Committee, *Salt of the Earth*, 171; 203–4; 206.

for salinity control would need to have a long life and the capacity to adapt over time. The previous two decades had been marked by conflicts between governments and communities over the installation of drainage facilities over the possible merit of one area and potential detriment to another, the disposal of drainage water into wetlands, and finally by the bitter struggles about the Tutchewop-Tyrrell and Mineral Reserve Basins schemes. They demonstrated that large-scale engineering solutions could not be politically enforced on the rural communities, even if they were deemed to be effective by experts. The committee declared that the success of salinity control programs depended largely on farmers' voluntary actions. Hence, the future role of governments would have to be "to provide expert advice to assist in the resolution of the problems of the local community" so that landholders would have relevant and reliable information and advice. This was in contrast with the present practice where the affected communities were presented a set of options, followed by a period of public comment, which the agency took into account when making its decision. Another reason why the committee took this view was that the communities had demonstrated capability in solving their own problems and functionality as political actors. Given official support, sufficient funding and coordinated activities, they could be an integral part of the solution.[14]

Furthermore, as ACIL et al. stated, there were still profound uncertainties involved in salinity control. The committee soberly declared that the "basic causes and effect relationships with salinity may take decades, even longer, to fully work out, and remedies may take an equally long time to be fully effective." It noted a lack of reliable information upon which to base the formulation of effective control programs and identified important gaps. While the causes of salinity problems were known in principle, understanding of the precise reasons for specific cases was still often inadequate. There was a reasonable assessment of the distribution of salinity problems but substantial uncertainty about the rate of change in extent and severity. Finally, there was a lack of reliable data on the full range of physical and socioeconomic effects which limited the understanding of

[14] Salinity Committee, *Salt of the Earth*, 157; 164; 174; 199. "Salt of the Earth" is most likely a reference to the New Testament. The meaning of the metaphor is, however, disputed. See *The Holy Bible: King James Version* (Dallas: Brown Books Publishing, 2004), Matt. 5:13.

the problem's full magnitude. Effective salinity control needed sustained local knowledge and involvement.[15]

It was the Committee's view that the salinity problem was a land management, water pollution, environmental, and agricultural problem, as well as one "which can have significant adverse social and economic impacts on individuals and communities." A strategy to "achieve a new balance in the inputs and outputs of the soil–water system at a productive and sustainable level," to protect both agricultural and environmental quality, would require public expenditure of $50 million to $90 million over the next ten years to support farm-based measures and capital works with community education, consultation, and involvement as a major element.[16]

On the farm, land forming, laser grading, water harvesting, micro-irrigation of high-value horticultural crops, and private groundwater pumping and re-use were recommended. A program for improving natural drainage lines was suggested, plus an extension of Shepparton's main drainage system, should this be economically feasible. A groundwater pumping scheme was recommended for Kerang and several for Shepparton. Crucially, the proposed strategy assumed that the suggested management program for Barr Creek, which was under development, would result in effective control of the Creek's salt loads and obviate the need to use Lake Tyrrell as an evaporation basin. In the short term, disposal of saline drainage water had to be through the Murray with appropriate dilution flows, until alternative strategies like re-use, local evaporation basins, and local treatment could be developed. Lake Tutchewop should be restored to more acceptable salinity levels. No further wetlands should be used for evaporation basins, except for naturally saline environments. An improvement in habitat values for wildlife should be a specific aim in the future.[17]

To manage salinity control more effectively, the committee opted for the establishment of a Salinity Control Board, which would be responsible for the development of policy and oversight of central management functions, and a budget for government expenditure on salinity control. Furthermore, it would ensure coordination of government agencies'

[15] Salinity Committee, *Salt of the Earth*, 157; 215–16.

[16] Salinity Committee, *Salt of the Earth*, xix; 160.

[17] Salinity Committee, *Salt of the Earth*, 137–41.

activities, allocate funds for control programs of each agency, and establish guidelines for salinity loans and grants. The board would be composed of state government agencies' staff and community representatives. Communities should be involved in the development of salinity control policies. A framework of state objectives, policies, and plans should be established where a series of regional and local strategies would operate, taking account of the specific and varying needs of each region. The river catchment approach should be incorporated to the extent necessary. A State Salinity Control Strategy should be developed and regional control plans for the major catchments be prepared, which would identify land management practices, methods of encouraging their adoption, and specify disposal methods for saline effluent, if necessary. Following a point from GIRDAC, the report also recommended that landholders linked by a common salinity problem develop sub-regional salinity control plans, while recognizing the relationship between the individual farms' and sub-regions' drainage requirements.[18]

"Salt of the Earth" was a balanced report overall that sought to end the confrontations between upstream and downstream communities, Kerang and Shepparton, irrigation and dryland farmers, government and communities. It was clearly informed by the Salinity Committee's community consultation and involvement, specifically crediting Gyn Jones, Ernest Jackson, Cliff Spowart, VIRPO, GIRDAC, and KIRSAC for their input.[19] Comments, therefore, were favorable. Ernest Jackson noted that the report was "a very realistic and practical assessment of the problem of salinity control. It offers no instant solutions, or reclamation practices of easy, simple, inexpensive application." The committee had listened well to the advice from landholders, scientists, community groups, consultants, and departmental advisers. The report recognized that salinity control "cannot be isolated but is part of the whole environmental problem of the effective management and husbandry of integrated land, water and associated renewable natural resources that include all forms of degradation, water pollution and the flora and fauna habitats." "Watershed" Jackson saw his own ideas endorsed in a document that highlighted "the basic philosophy of renewable natural resource management" and recognized "the need for the spelling out of the principles of renewable natural

[18] Salinity Committee, *Salt of the Earth*, xx; 164–65; 175–80; 189–93.

[19] Salinity Committee, *Salt of the Earth*, 193; 211; 230.

resources management on a catchment basis" as well as "the great need of public awareness and understanding of the salinity problem."[20] The report, he wrote, was "a revolution in public and private acceptance and advocacy of multi-disciplinary co-ordination, integrated land, water and vegetation management on a catchment basis and the realisation of the importance of realistic and practical Environmental Education to bring into common focus the need of departmental and community co-operation at the regional level to inspire remedial action on each and every farm."[21] It implied "that our environment is held in common ownership irrespective of age, sex, wealth or poverty" and acknowledged "the reality of the need to organise ourselves on a communal basis to deal with the problems of the Murray/Darling Catchment."[22]

Shortly after the report's release, Ernest Jackson and his band of enthusiasts jumped into action. GIRDAC declared its support for the appointment of a Salinity Control Board and urged its affiliates to promote the resolutions to the public and the Victorian government so that the recommendations of the committee would be implemented.[23] VIRPO urged that there was a great need for immediate measures "to ensure that the main components of the Salinity Report are given strong support" throughout the Murray Valley. This, the organization felt, had to be done before the forthcoming state election in autumn 1985. Politicians had to be made aware that the time for an immediate resolve on salinity control had come. It called a special meeting at Moama to discuss the contents of the report and reach a consensus on a campaign for political action toward its implementation. The sessions were chaired by VIRPO and its sister organizations. Members of the Salinity Committee attended, including its Research Officer, Graham Hunter, and Legislative Assembly member for Rodney, Eddie Hann who had been deputy chairperson. Other political representatives were also present. The meeting welcomed the Salinity Report, congratulated the

[20] Ernest M. Jackson, "'Salt of the Earth.' Final Report to the Victorian Parliamentary Select Committee into the Causes, Effects and Control of Land and River Salinity, October, 1984," November 10, 1984, VPRS 6008/P0001, 84/36649, Public Record Office Victoria, Melbourne, 1–2.

[21] Ernest M. Jackson, "Notice of Postponed Executive Meeting," February 4, 1985, VPRS 6008/P0001, 84/36649, Public Record Office Victoria, Melbourne.

[22] Jackson, "Notice of Postponed Executive Meeting."

[23] Mitchelmore, "Salinity Committee Recommendations," 1–2.

committee, and supported, in general, its recommendations, urging that "it becomes the first priority of our incoming Victorian State Government to establish a Salinity Control Board with the provision of necessary funds."[24] In late January, the government endorsed the report and committed itself to a funding of $32 million over four years, if re-elected, making measures against salinization the main feature of its agricultural policy.[25]

On the interstate level, salinization was causing a substantial shift in the way the MDB was managed. Conservation issues were gaining profile and attitudes were shifting with the advent of the Age of Ecology. The tumultuous period of the Little Desert and Lake Pedder disputes entailed an "increasing rejection of the simplistic assumptions in the traditional development imperative."[26] The environmental movement scored a victory by preventing the damming of the Franklin River in 1983. With its aid, the Labor Party won office in Canberra in the same year, showing that the movement was not only capable of mobilizing tens of thousands of people but could also influence the outcomes of elections. Few rivers were left to regulate, and the construction of additional dams was becoming difficult to politically enforce. With the arrival of environmentalism on the political agenda the urban majority in Australia was increasingly conscious of concerns that previously had been on its peripheral vision. Land degradation had become a headline in public debate in the late 1970s. This threat had been substantiated by comprehensive studies of society-induced environmental damage. One was "A Basis for Soil Conservation Policy in Australia," published in 1978, which found that 50% of the country's agricultural and pastoral land needed remedial treatment. Another one was the "Gutteridge Report" of 1970, the result of increased visibility of salinity problems in the Murray during the drought of 1967–68,

[24] Ernest M. Jackson, "Salinity Report Recommendations Require Political Action for Implementation," January 7, 1985, 6008, VPRS 6008/P0001, 84/36649, Public Record Office Victoria, Melbourne; Mitchelmore, "Salinity Committee Recommendations," 1; Ernest M. Jackson, "Notes on Meeting of Witness Groups (Non-Government) to the Salinity Inquiry," February 16, 1985, VPRS 6008/P0001, 84/36649, Public Record Office Victoria, Melbourne.

[25] Jackson, "Salinity Report Recommendations"; "Labor Pledges $32m to Fight Salinity," *The Age*, February 1, 1985.

[26] Powell, *Bioregionalism*, 72.

which, for the first time, had raised this issue for the Commonwealth government.[27]

The understanding of the MDB's salinity problems continued to grow during the 1970s and 80s. However, there was a widespread feeling that interjurisdictional water management in the Basin was deficient, particularly in South Australia and Victoria. The MDB was still managed as a "curiously lopsided network" focused on its southern sector, the Murray system, which excluded the Darling and the state of Queensland.[28] Salinization showcased the RMC's primary weaknesses—it was neither empowered to cope with Basin-wide problems nor to control water quality. Accordingly, the Salinity Committee had argued that the Commission's role should be expanded to comprehensively manage the Basin. South Australia, for which salinization had continued to be the primary concern since the 1960s, was pressing the hardest for change. It was felt that water management had to widen its focus from an almost exclusive concentration on quantity and distribution to include an assurance of quality. In a first step, the RMC assumed an in-principle, albeit fairly nominal role, in managing salinity impacts in 1982. However, this made little difference in practice. A full-fledged reform was required—the River Murray Waters Agreement had to be made fit for managing water quality.[29]

Daniel Connell has argued that these developments paved the way for a shift toward Integrated Catchment Management (ICM) in the Basin, a combination of holistic thinking about the biophysical environment and need for community involvement and empowerment. Its advocates urged new interjurisdictional arrangements that would make it easier for government agencies and communities to work together. The coming institutional changes in the MDB reflected an emerging international consensus that sustainable environmental management was essential, and ICM was the best way to achieve it. They key features of a working ICM system include its adaption to the variability and diversity of the area being managed, and the involvement of communities supported by legislation designed to empower and assist rather than unnecessarily restrict. Above

[27] Powell, *Bioregionalism*, 72–75; Barr, *The House on The Hill*, 112; Hutton and Connors, *A History of the Australian Environment Movement*, 165.

[28] Powell, *Bioregionalism*, 76.

[29] Powell, *Bioregionalism*, 76–77; Connell, *Water Politics*, 108–10; Salinity Committee, *Salt of the Earth*, xi.

all, ICM is supposed to encourage participation and cooperation. Meanwhile, soil conservation agencies in the Basin states had moved toward a closer embrace of the ideal of using whole catchments as practical, coordinating units. The management of land degradation was becoming more interdisciplinary in scope and displayed a greater commitment to holistic interpretations.[30]

Another factor for the coming reforms were changing ideas about how public institutions should be organized and operated. The developmental orientation was questioned in favor of a more "mature" approach that de-emphasized big construction. The proponents of change argued that decision-making should no longer be left to small groups of engineers. They criticized that "giant pioneering monoliths" like the RMC had been too closed off from public control and allowed to develop too independently, including their single disciplinary engineering cultures. Instead, a more adaptive administrative structure and more open management systems were favored, that is, ones that were more responsive so that the community and its elected representatives would be enabled to make direct contributions to determination and implementation of policy.[31]

The increasing influence of this style of thinking was reflected in a position paper for a meeting of ministers in Adelaide in November 1985. This meeting was called to discuss the growing controversy about the environmental state of the Murray and its catchment. The authors criticized, inter alia, that there was no single agency responsible for the management of the Basin and that the RMC was only empowered to make recommendations and had limited opportunity to influence land management and only indirect influence over the tributaries of the Murray. Further, they criticized that there were no effective institutional arrangements for intergovernmental cooperation and that general institutional complexity inhibited the integration of water and land-use planning and management. Implicitly, the paper argued that the state of affairs demonstrated the need for a strong organization to manage the MDB as a single unit directly responsible for the catchment and its tributaries. The

[30] Connell, *Water Politics*, 113; Powell, *Bioregionalism*, 76.

[31] Connell, *Water Politics*, 115; Powell, *Bioregionalism*, 77.

author also conveyed that much stronger processes for inter-governmental cooperation were required.[32]

The RMC would not survive the sweeping changes that resulted from this assessment. Two extremes on the continuum of reform options considered were an authority with comprehensive powers and an exchange of letters of understanding between governments. The middle road was taken. Governments would be required to agree on an overall plan with each one responsible for implementation within their own jurisdictions. These new institutional arrangements, called the MDB Initiative, were incorporated into a revised agreement which, for the first time, included Queensland and the Australian Capital Territory, although initially not as fully committed signatories.[33]

The key elements of the new structure were the MDBMC and the MDBC. The former had two or three ministers from each government. The MDBC superseded the RMC and inherited its staff. A Murray–Darling Basin Agreement was signed on October 30, 1987 and came into force at the beginning of 1988. Under these new arrangements, the MDB was to be managed to improve biodiversity and sustainability *as well* as production.[34] "What's been recognised is that if you're going to cope with the most serious environmental threat in Australia it has to be a joint effort and tackle all the facets," the federal Minister for Resources and Energy, Gareth Evans, declared, pledging that the Commission would "treat river, soil and environmental salinity as an integrated problem." The MDBC announced that it would complete a salinity and drainage strategy by the end of the year and develop a broader resource management plan.[35]

Following the civic ideal of ICM, during the new agreement's preparations, the Ministerial Council called for public participation and reached out to volunteer associations, including the MVL, VIRPO, the New South Wales Irrigators' Council, the South Australian Conservation Council, the Australian Conservation Foundation, and various local

[32] Connell, *Water Politics*, 113–14. This account is based on Aynsey Kellow's summary since the original paper is no longer available: Aynsley Kellow, "The Murray-Darling Basin," in *Environmental Politics and Policy Making in Australia*, eds. Timothy Doyle and Aynsley Kellow (South Melbourne: Macmillan Education Australia, 1995), 226–27.

[33] Connell, *Water Politics*, 114–15.

[34] Connell, *Water Politics*, 115; Powell, *Bioregionalism*, 78–79.

[35] "Joint Attack on Rivers Fouled by Salt," *The Age*, March 28, 1987.

government representatives. A Community Advisory Committee to the Ministerial Council formed to distil views and aspirations into advice to the Council made up of select representatives from the major regions and organizations such as the ACF, the MVL, and the National Farmers Federation. The MVL, like VIRPO, had criticized the fractured institutional structure of the MDB for a long time, and both organizations welcomed the changes. Leon Heath of GIRDAC became a member of the first advisory committee in 1986. KIRSAC, however, failed to gain representation. Russell Smith criticized that its membership was not representative of the community of the MDB with several members living outside of the area, and no representative from the Kerang region, despite it being recognized as the single greatest source of saline contamination of the Murray.[36]

Nonetheless, when the "momentous decision" to form the MDBMC was made in Adelaide, Ernest Jackson greeted it with enthusiastic approval. His stance had long been that salinization stemmed from the lack of coordination among all relevant interests in the MDB. He had suggested the formation of a Ministerial Council of the Commonwealth and four state governments to cover the whole catchment of the Basin. Now, after "seventy years of single-disciplinary administration," he cheered, "the decisions of the Adelaide Conference highlight the acceptance of a Philosophy Whose Time Has Come." This was, of course, "the planning and management of the renewable natural resources of the whole natural environment through multi-disciplinary administration on a total catchment basis of all relevant interests." Again, though "with appropriate humility and simplicity," Jackson saw his own philosophy realized in the minister's decisions.[37]

VIRPO used the occasion of its 20th annual meeting to reflect on a decisively new situation, concluding that the catchment philosophy was now officially accepted, though it was "loathe to specifically claim any credit." The organization reflected that "the Salinity Threat brought about the four government[s'] acceptance of ecological realities." However, as encouraging as the formation of the MDBMC was, public recognition of conservation issues was still very sobering: "koalas

[36] Connell, *Water Politics*, 115; Powell, *Bioregionalism*, 78–79; Anonymous, "Murray-Darling Basin Community Advisory Council meets," *Salt Force News* 2 (March 1987): 11; Smith, "Chairman's Report 1986/87," 1.

[37] Jackson, "Good 'Tidings' for the Murray!" Original emphasis.

and kookaburra's [sic!] have more appeal than the more mundane issues of soil and water."[38]

Nonetheless, as Ernest Jackson put it, the MDBMC was "a dream come true."[39] After decades of conflict, inquiries, political haggling, and patient small-scale progress, with the impending launch of the Victorian government's salinity control program and the reform of the MDB's water management institutions enacted, VIRPO and the Salinity Action Groups were now riding the tide.

Winds of Change

Kerang district engineer John Boehm used a different metaphor. The "winds of change," he wrote, "have elements of excitement and apprehension for the future, depending on who you talk to." The winds, however, turned out to be a storm, propelling Northern Victoria's irrigation communities into the future. This storm was called "water trading."[40]

The Dartmouth dam in the Alpine region of Victoria was completed in 1979 and was the last major dam on the Murray River. It marked a turning point in water politics. Instead of allocating the extra water available to irrigators based on the water rights formula, the government auctioned it off to the highest bidder. According to Neil Barr, "[m]ore than anything else, this auction symbolised the new world of irrigation."[41] It was a prelude to the radical changes to the irrigation sector that would take place throughout the next decades. The sacred nexus between the land and water entitlement, in which water was allocated according to an equity-based formula and remained tied to the farm's title, was dissolved. The powerful irrigation supply authorities that had facilitated the development of the Murray Valley for irrigated agriculture were transformed into public companies or government-owned trading enterprises whose

[38] P. S. Robins, "Victorian Irrigation Research and Promotion Organisation,"; "Report of Chairman—20th Annual Meeting," February 12, 1987, VPRS 6008/P0001, 84/36649, Public Record Office Victoria, Melbourne, 1–2.

[39] "Lobbyist with a Missionary Zeal," *The Age*, November 30, 1988.

[40] Anonymous, "Irrigation '82. Kerang's Lifeline," *The Northern Times*, June 8, 1982. See Walter Benjamin, "Theses on the Philosophy of History," in *Illuminations*, ed. Hannah Arendt, trans. Harry Zohn (New York: Schocken Books, 2007).

[41] Barr, *The House on The Hill*, 112.

task was to generate a return on capital. The price for irrigation water gradually rose toward "full cost recovery." Water was transformed into a commodity.[42]

This process signified a fundamental break with the principles that had governed the irrigation sector for almost 75 years. Under the regime of the SRWSC, the rationale for irrigation development had been to establish white, hard-working communities in inland Australia, and to expand economic production. Considered common to Victoria, running water was public property, which could not be owned by a private person. Water rates were compulsory, whether the holder of the title used the water or not. The public was expected to bear the capital costs, while the irrigators only paid for maintenance and operation. This arrangement was now in dissolution.[43]

The state-directed, subsidized water provision for irrigation was part of what Paul Kelly termed the "Australian Settlement," a range of policies designed to achieve as much certainty as possible for Australian businesses and workers. This included the White Australia Policy—a set of policies which forbid non-European immigration—the protection of industry through tariffs, price-fixing and subsidies, centralized wage-fixing, and an active role of government in the economy. It also involved a social contract between city and country: the former valued rural life and accepted that society would support farmers financially, politically, and socially, while the latter would supply the nation with food at reasonable prices in return. By the 1980s, the prosperity resulting from this social contract was under serious threat. Australia's foreign debt rose sharply throughout the decade and into the 1990s, and unemployment and inflation were on the rise. As a result, the federal Labor government, led by Bob Hawke, initiated a radical restructuring of Australia's economy. From 1983 to 1996 a comprehensive program of economic and monetary policy reform was implemented, encompassing the removal of trade barriers, the floating of the Australian dollar, the privatization of public utilities, the deregulation of the banking sector, and the creation of markets for power and water. With a broader acceptance of the limited international demand for irrigation's leading items, chiefly dairy produce

[42] Barr, *The House on The Hill*, 112; Salinity Committee, *Fourth Report to the Parliament. Final Report on Water Allocations in Northern Victoria* (Melbourne: Salinity Committee of the Victorian Parliament, 1984), 63.

[43] Powell, *Garden State*, 144–48; 194–96; 200.

and fruit in Victoria, the rural sector did not escape the revolutionary adjustment that concerned other established economic fields such as steelmaking and the protected clothing and textile industries.[44]

These reforms reflected an ideology of economic rationalism, Australia's form of Neoliberalism—an alternative model of economic regulation which challenged the Keynesian paradigm that had been hegemonic during the Long Boom.[45] Neoliberalism, much like capitalism, has become a container term routinely used in political battles, but it also encompasses a variety of analytical definitions. As a "mutating" political and economic theory with a discernable core, it has spawned a variety of adaptions. As such, it can be understood as either a distinct phase of capitalist expansion characterized by the opening of new sites and the development of new techniques for capital accumulation, a political rationality and a mode of subjectification, or a mode of reconfiguration of categories like class, gender, and sexuality. What connects these different concepts is the premise that markets and the institutions that produce and sustain them are at the core of the project.[46] For the present purpose, it is useful to consider Neoliberalism as a political project designed to expand capital accumulation by opening new sites for markets and expanding private property rights. This project is backed by a theory of political-economic practices that proposes that liberating individual entrepreneurial freedoms within an institutional framework characterized by strong private property rights, free markets, and free trade can best advance human well-being. The state's role is to create and preserve an institutional framework appropriate to such practices and to set parameters ensuring efficient operation of the market, as opposed to directly

[44] Kirsten Henderson, "Dams, Pumps, Pipes and Dreams. Water and the Making of Regional Australia," in *Rural and Regional Futures*, eds. Anthony Hogan, Michelle Young (London; New York: Routledge, 2015), 90–91; Powell, *Garden State*, 287; James Walter, "Growth Resumed, 1983–2000," in *The Cambridge History of Australia*, eds. Alison Bashford and Stuart Macintyre (Cambridge: Cambridge University Press, 2013), 166; Paul Kelly, *The End of Certainty. Power, Politics, and Business in Australia* (Sydney: Allen & Unwin, 1994).

[45] Dyster and Meredith, *Australia in the Global Economy*, 267; Walter, "Growth Resumed," 164–65.

[46] William Callison and Zachary Manfredi, "Introduction: Theorizing Mutant Neoliberalism," in *Mutant Neoliberalism. Market Rule und Political Rupture*, eds. William Callison and Zachary Manfredi (New York: Fordham University Press 2019), 6–14; Quinn Slobodian, *Globalists* (Cambridge, MA: Harvard University Press 2018), 2–7.

intervening. Existing markets must be deregulated[47]: "[I]f markets do not exist (in areas such as land, water, education, health care, social security, or environmental pollution) then they must be created, by state action if necessary."[48] State intervention in markets must be kept to a minimum because the state cannot have enough information to second-guess market signals, and its actions are inevitably distorted by interest groups. Since the absence of clear property rights is seen as one of the greatest barriers to economic development and the improvement of human welfare, the enclosure of commons and the assignment of private property rights are considered the best ways to ensure their rational use. Neoliberalism, therefore, champions the privatization or corporatization of government businesses and the removal of subsidies. This program was enacted by such protagonists as Deng Xiaoping, Margaret Thatcher, and Ronald Reagan in the late 1970s and early 1980s.[49] Hawke's government followed these examples and systematically began expanding the markets' role in the Australian economy.[50]

During this period when the basic tenets of how water was treated were redefined, the sustainability discourse became another driver for change. The advent of the Age of Ecology, with increased attention to water quality and soil degradation, gave impetus to a redefinition of water management's goals, as it had to the formation of the MDBC as a move toward Integrated Catchment Management. This was helped by the emerging paradigm of sustainable development that was injected by the prominence of the Brundtland Commission's findings, "Our Common Future," and the endorsement of the World Conservation Strategy in Australia. Sustainable development attempted to remove the division between economic development and sustainability by defining the use of resources in a way that would not compromise the needs of future generations. In a combination of both philosophies—neoliberalism *and* sustainability—water could no longer be regarded as a public good in unlimited supply, allocated according to protectionist principles.[51]

[47] David Harvey, *A Brief History of Neoliberalism* (Oxford: Oxford University Press, 2005), 1–2; Henderson, "Dams, Pumps, Pipes and Dreams," 90.

[48] Harvey, *A Brief History of Neoliberalism*, 2.

[49] Harvey, *A Brief History of Neoliberalism*, 1–2; 26.

[50] Walter, "Growth Resumed," 166.

[51] John J. Pigram, *Australia's Water Resources: From Use to Management* (Collingwood: CSIRO Publishing, 2006), 64; Henderson, "Dams, Pumps, Pipes and Dreams," 89–90;

The move toward market-based water trading in the MDB, therefore, can be understood as a result of two coinciding factors: Firstly, the crisis of the Keynesian political-economic approach which had been embraced by both major parties but had become dysfunctional throughout the 1970s. Drawing largely on inspiration from the U.S., proponents of neoliberal ideas dominated the political sphere in the 1980s. Neoliberal thought became the hegemonic frame for government action and, as James Walter has argued, for politics itself. Secondly, an increasing appreciation for the MDB's escalating ecological crisis urged a rethinking of water management to achieve what was now termed "sustainable" use. This, together with neoliberal ideas, successfully delegitimized the existing top-down state agencies for natural resource management. Their brief had been to further the common good through state intervention and public ownership of natural resources. Instead, the agencies had privileged vested interests—the irrigators benefited from publicly subsidized water—that wreaked environmental havoc.[52]

Salinization tied both strands together. By now one of the most prominent environmental problems, and certainly the one that irrigation was most notorious for, it served as striking evidence that reform was urgently required. Water could no longer be regarded as an unlimited good to be used in a manner ignorant of the environmental consequences and the costs for the public. The introduction of water trading was meant to ensure that irrigation was no longer a burden on government budgets, to discourage the use of water for low-value crops and encourage the production of more profitable ones, and to benefit the environment by ensuring that less water was wasted. According to the proponents of reform, by transforming water from a public good into private property and allocating its use based on market value, efficiency in production *and* environmental improvements were possible.[53]

This was the rationale for the introduction of Transferable Water Entitlements (TWE) in Victoria. In 1982, amidst the revived deliberations to find a long-term, comprehensive solution to salinity control in the state,

Gro Harlem Brundtland, ed., *Our Common Future*, *The World Commission on Environment and Development* (Oxford: World Commission on Environment and Development, 1987), 43.

[52] James Walter, *What Were They Thinking? The Politics of Ideas in Australia* (Sydney: University of New South Wales Press, 2010), 254–55; 301–02.

[53] Barr, *The House on The Hill*, 112–14.

Bruce Davidson argued that possibly the cheapest method to reduce salt in the Murray was flushing it out with water bought from irrigators rather than by building works designed to reduce the salt level. He argued that any water from new projects should be sold to the highest bidder, for which Dartmouth had set the precedent.[54]

At the same time, the Public Bodies Review Committee conducted an inquiry into Victoria's non-metropolitan water industry.[55] A committee-commissioned report from Monash University contributed a cost–benefit analysis of irrigation development. Its brief was to determine whether the state had received a rate of return from the invested resources that was comparable to one that might have been obtained had the same resources been invested elsewhere. The study team came to a devastating conclusion. The state's irrigation system had been ill-conceived. Its rate of return was low, even negative on the initial outlay to construct it. Its major benefit was increased agricultural production; however, the major beneficiaries were the past irrigators, of whom many had sold their farms and lived elsewhere. Incomes in Victoria as a whole were probably lower than they would have been had the irrigation system not been constructed. All things considered, irrigation had been a poor investment. For Davidson, who had forcefully criticized Australia's public irrigation systems on these exact grounds in the 1960s, this must have been satisfying. However, the report suggested that, rather than abandoning the system, it should be managed more efficiently. This might require the transfer of water from public irrigation to private diverters. The report even considered removing water from agricultural uses altogether, to be used instead for urban consumption, hydroelectricity, and river freshening to ensure the health of riverine ecosystems and control salinity. TWE were, therefore, deemed a useful measure.[56]

[54] Anonymous, "'Irrigation Should Pay Its Way'," *The Northern Times*, February 19, 1982.

[55] Public Bodies Review Committee, *Twelfth Report to the Parliament. Future Structures for Water Management. Volume 4. Final Report: Irrigation and Water Resource Management* (Melbourne: Government Printer, 1984), xiii.

[56] Anonymous, "Irrigation is 'Ill-Conceived'," *The Northern Times*, February 10, 1984. See Centre of Policy Studies, *Monash University. Distribution of Costs and Benefits of Victoria's Irrigation Systems: A Report* (Melbourne: Public Bodies Review Committee of the Parliament of Victoria, 1983).

When the Public Bodies Review Committee released its final report in 1984, its verdict was no more favorable. While the Water Commission was attested to have been effective in pursuing its fundamental objective—the development and maintenance of closer settlement—it had effectively abandoned its second objective—direct financial return to the state. The first objective, however, was deemed irrelevant to contemporary society. It had been appropriate for a pioneering water supply system, but no longer important for a mature one that operated in very different economic and market conditions. The irrigation system was considered an important competitive strength for the state, being a well-developed facility for storage and distribution and capable of reliably delivering large quantities of water. However, there were a number of inhibiting factors that prevented individual irrigators and communities from realizing the possible economic and social benefits to the greatest extent—low levels of efficiency and productivity, both agricultural and economic, considerable modernization needs, and "a serious and growing environmental problem" of which salinity was the most well-known. The committee recommended that pricing policies should recover all operating costs: "Pricing and accounting systems should operate along the lines of a modern commercially based enterprise." Cross-subsidies from one group of irrigators to another should be removed. Compulsory payments for full water rights, whether they were used or not, should be abolished "to introduce greater flexibility into the system." The report strongly endorsed temporary or permanent transferable water rights at market prices.[57]

In the same year, the Salinity Committee released its "Water Allocations" report. The committee had inquired into the water quantities allocated for forest watering and river dilution, and those for existing irrigation districts, as well as changes in the management of water rights in the GMID. It made a plea for flexibility. In the future, important non-irrigation needs, namely water for red gum forests and the protection of both agricultural productivity and the environment through salinity control, would require substantial water allocations—possibly in excess of the extra water available from the Dartmouth dam. Nature conservation should be given highest priority for any surplus stream flows. Therefore, the "irrigation community in constituted irrigation districts [...] should

[57] Public Bodies Review Committee, *Twelfth Report to Parliament*, xv–xvii; 175–76.

not expect further fixed allocations of water." Rather, in the medium- and long-term, the GMID could see a real reduction of water available for each season due to environmental requirements. Water should no longer be added to the existing water rights but flexibly allocated to account for a number of non-irrigation uses whose requirements could not be determined in advance.[58]

Therefore, this committee, too, recommended that the Water Act be amended to include provisions for the transferability of water entitlements. It cited several studies that had recommended TWE, although the findings were ambiguous. These studies anticipated that water's transferability would improve the efficiency of Victoria's irrigation system and save costs and also have a favorable impact on the salinity problem. Increased charges for water and TWE were expected to reduce accessions to the water table in irrigated areas by improving the efficiency of water use, which would also reduce the social costs attributable to salinization. At the very least, the consequences for the water table situation would be insufficient to warrant *not* introducing the policy. TWE would improve the efficiency and equity of the water entitlement system in the GMID and provide a basis for irrigators to adapt to changing circumstances, considering that "[e]conomic conditions coupled with market forces to a large extent will dictate which is the most efficient means of production to maximise returns to the landholder."[59]

The committee was perfectly aware that breaking up the nexus between land and water entitlement was a radical step that would have fear-reaching consequences for the irrigation communities. ACIL Australia, in its study on the possible consequences of TWE, expected that water would move out of Kerang, the Tragowel Plains, and Swan Hill, and

58 Salinity Committee, *Water Allocations*, v; xiii; xiii–xvii; xxiii; 65–66.

59 Salinity Committee, *Water Allocations*, xv; 58; 60–61. The studies referred to here are: Alan Randall, "Resource Allocation in a Maturing Water Economy," in *Irrigation Water: Policies for Its Allocation in Australia. Papers from a Workshop, University of New England*, ed. Australian Rural Adjustment Unit (Armidale, NSW: Australian Rural Adjustment Unit, 1982). The other two studies are only available as summaries. "Financing Salinity Control in Victoria" by Michael Read and Associates is summarized in the papers belonging to the Community Conference on Salinity in Bendigo in 1984: Salinity Committee of the Victorian Parliament, *Community Conference on Salinity Control. Bendigo College of Advanced Education. 17–18 February 1984* (Melbourne: Salinity Committee of the Victorian Parliament, 1984). Finally, ACIL Australia's report, entitled "Transferability of Water Entitlements" is summarized in Appendix VII to the "Water Allocations" report: Salinity Committee, *Water Allocations*.

into Rodney, Tongala/Stanhope, Rochester, and Cohuna, although it was uncertain about the impact on Shepparton. In the case of the Tragowel Plains, the consultants estimated that up to 30% of water rights could be transferred out of the area within just five years. Furthermore, they were acutely aware that introducing transferable water rights would raise another question—would it be desirable if areas with high accessions to the water tables were prohibited the transfer of entitlements inwards, while transfers outwards would be encouraged to achieve more favorable economic and environmental outcomes?[60]

For people in the Kerang region who remembered the original Salinity Inquiry proceedings, this may have had a familiar ring. In 1978, GIRDAC's Leon Heath had argued that a polluting area, meaning Kerang, should not be allowed to irrigate as this threatened the viability of the *whole* of irrigated agriculture. Consequently, many were alarmed.[61]

The Victorian government made it clear that it intended to follow the recommendations in favor of market prices for water. In its Conservation Strategy of 1987, the government declared that the "irrigation system cannot expand to meet increased demand for water—the construction of new storages is unlikely. [...] Unwise allocation and use of irrigation water has contributed to salinity problems in some areas." Meanwhile, Melbourne's water use was increasing at about 3% per year, while irrigation accounted for 80% of water use in the state. In response, the government announced that it would "adjust the price of irrigation water towards the real costs of supply and its target rates-of-return on assets."[62] Irrigation was expected to improve its overall economic and environmental performance. But who would bear the costs?

Ernest Jackson, who could usually read the writing on the wall, bitterly observed that public irrigation "is no longer accepted as 'a good thing' and desirable national investment because it is often blamed for the salinity hazards[.] [...] Public resistance to reclamation and remedial government funding has been triggered off by the extremists of the environmental lobby on the score of why spend good money after bad."[63]

[60] Salinity Committee, *Water Allocations*, 63–64.

[61] Parliamentary Public Works Committee at Melbourne, "Heath," 1069.

[62] Government of Victoria, *Protecting the Environment. A Conservation Strategy for Victoria* (Melbourne: Government Printer, 1987), 57; 62.

[63] Jackson, "VIRPO Objectives," 1. Original emphasis.

He foamed that the State Conservation Strategy was a "drab, dull document" that reflected the "Left Wing policies of the Government and the greenie sympathies of many bureaucrats."[64] VIRPO's chairperson Phil Robins condemned that denigrating irrigation, which had been responsible for thriving cities and towns, had become popular among academics in their proverbial ivory towers. Irrigators should demonstrate their solidarity and refuse to be written off as a backwards industry and a burden to the state.[65]

The reactions in Kerang were no less fierce. The "lifeblood of the Shire of Kerang is under threat," Councilor Ray Hampton warned. The recommendations made to the Public Bodies Review Committee could spell the end of irrigation in the Kerang district: "Once water is put on the open market we will be priced out of it." He speculated that irrigators farming poor land in the area might be the first tempted to sell their water rights, which would leave vast areas of unproductive land in the shire. The family unit was being threatened and irrigators were up in arms over the possibility of transferable water entitlements.[66]

One Dudley J. Walters of Myall complained in *The Northern Times* that the experts' considerations were made purely on economic grounds while political and social aspects were completely left out of sight. Irrigation had created prosperous cities and towns and high agricultural productivity for the state. Selling water to the highest bidder would mean the opposite of closer settlement, namely the concentration on a few large holdings. Walters rejected the notion that waterlogging and salinization could be blamed on the present generation of irrigators—the modern irrigator had learned his lesson.[67]

A. J. Mann and D. A. Leed, of the Tragowel Plains Irrigators' League, questioned the relationship between increasing the price of irrigation water and better management that would preclude salinity problems. The logic that industries would become more efficient if resources became

[64] Jackson, "VIRPO, Protecting the Environment. A Conservation Strategy for Victoria," 2.

[65] Anonymous, "Irrigators' Chance for Solidarity," *The Northern Times*, March 14, 1984.

[66] Anonymous, "Lifeblood 'Theat'," *The Northern Times*, February 23, 1982; Anonymous, "'No Water Rights Auction'," *The Northern Times*, April 6, 1982.

[67] Dudley J. Walters, "Irrigation Problems 'Man-Made' Neglect," *The Northern Times*, March 23, 1982.

more expensive was flawed. They suggested penalization for individual irrigators who mismanaged their water. Furthermore, the Water Commission's "inefficient" delivery system should be examined. More loans should be given to irrigators to re-layout their properties.[68]

I. A. McDonald of Kerang was appalled by what he saw as an academic bias against irrigation, while the food provided by irrigated agriculture was taken for granted. True, a "great many mistakes" had been made, which had "brought discredit to the whole system." However, the verdict that irrigation had been a poor investment was "ridiculous." Lake Hume's expansion had provided food during wartime. A diversity of crops had enabled towns on the Murray to prosper and grow the population. Increased water charges, he dreaded, would throw the whole rural community off balance.[69]

KIRSAC criticized that Monash's report made the "heroic efforts" of those working to overcome the salinity problem more difficult. The report was of more value as an outline of the methods of economic assessors than a useful examination of the irrigation industry. Irrigation had been practiced for 5000 years and was widespread throughout the world, contributing to much of its well-being. The Victorian people had received an ample return on their "modest investment." Irrigation had provided recreational facilities and opportunities for tourism, which otherwise would not have existed due to the variable river flow. It had improved water quality downstream by the sheer amount of water stored and distributed. There were good reasons to be confident that the rate of return would increase as irrigation's potential was fully developed, and the demands for its products would increase with a growing population. The committee warned that "[c]are needs to be exercised that 'economic efficiency' does not become a sacred cow that stands in the way of progress."[70]

At its core, as Gyn Jones's stanza went, the logic that TWE was a solution to salinization, was simple: "raise the price of water and we'll all look after it/not waste a drop/cure the salinity problem." Perhaps,

[68] A. J. Mann and D. A. Leed, "Irrigation Defended," *The Northern Times*, April 10, 1984.

[69] I. A. MacDonald, "Academic Bias," *The Northern Times*, March 9, 1984.

[70] Cliff L. Spowart, "Irrigation Flows for Progress," *The Northern Times*, March 30, 1984.

he continued, "the ball is in our court to come back at these people to remind them of the complexity of the problem."[71]

Irrigators, councilors, and community activists reasserted the belief they had exhibited during the Public Works Committee hearings in the 1970s. In opposition to the emerging neoliberal consensus, public irrigation was defended as a benevolent force for the common good. It had brought prosperity to the state, facilitated settlement, and turned previously dusty regions into a food bowl for the whole nation. They questioned the logic that an increased price for irrigation water would lead to more efficient use and would help solve the salinity problem. It was an appeal to the promise of irrigation which, they felt, had been fulfilled under the auspices of Keynesian policy, while its full potential had still not been realized.

But by then this seemed to be a losing battle. The storm, so to speak, was already underway. When the water from the Dartmouth dam had been auctioned off, some irrigators tried to disrupt the proceedings. Upon announcement of market-based water rates, some Torrumbarry and Kerang irrigators also threatened militant action. But the angry protests from the irrigation communities against the new neoliberal consensus could not prevent a gradual implementation of transferable water entitlements.[72] South Australia and New South Wales introduced the policy in 1983. Victoria followed suit when the government allowed temporary transfers of water entitlements between irrigation farmers in the Goulburn–Murray, Campaspe, and Macalister irrigation districts for an initial trial period in 1987. Permanent trade was allowed in 1991.[73]

It was an overture to further market-based reforms in the 1990s. In February 1994, the Council of Australian Governments (CoAG) adopted a Water Policy Agreement that committed the states and territories to more effective water management policies and practices, as well as an integrated approach to the development and management of water resources. Specifically, the agreement targeted water allocation to the environment,

[71] Gynlais O. Jones, "Gun Shots—Irony With Rain Falls," *The Northern Times*, July 23, 1984.

[72] Barr, *The House on The Hill*, 112; Anonymous, "Water Users Angry," *The Northern Times*, March 27, 1984.

[73] Salinity Committee, *Water Allocations*, 57; Anonymous, "Kerang Hosts Ministers' Meeting," *Salt Force News* 2 (March 1987): 2; Murray-Darling Basin Authority, *Guide to the Proposed Basin Plan*, 783.

a balance of environmental and developmental concerns, the adoption of Integrated Catchment Management, a pricing reform including full cost recovery, the removal of cross-subsidies, and the adoption of TWE. At its meeting in 1995, CoAG also endorsed a National Competition Policy which required competition in the business activities of governments and other economic sectors. The policy's essential aims were to remove unnecessary barriers to trade and competition and to require all businesses, including governments, to operate within the same market rules. Australia was, by then, "firmly engaged in a wide-ranging process of water reform directed towards promoting economic efficiency in water in a competitive environment within the context of ecologically sustainable management of the resource."[74]

In this emerging new world of irrigation, Victoria's super agency for rural development—the glorious SRWSC—was left behind, superseded by the Rural Water Commission (RWC) in 1984, which did not command the same comprehensive powers. A Department of Water Resources was established as the central management agency, while the RWC was responsible for policy and planning.[75]

Against the backdrop of multiple other reforms, this almost seems like a footnote. The decisive outcomes of this eventful period were the establishment of the MDBC as an institutional arrangement which encompassed the whole of the MDB and systematically included the participation of its communities and the commodification of water under the auspices of economic efficiency and sustainability. As for salinization, which had been a factor in all of these developments, a comprehensive attempt at a long-term solution was underway.

A Friendly Takeover

"Salinity is the single greatest threat facing the environment in Victoria."[76] This was the message to Melbournians on May 31, 1988. It was crudely delivered in the form of almost six tons of salt from the Mallee that were dumped on the City Square, representing the amount that entered the Murray every three minutes. What could have appeared like

[74] Pigram, *Australia's Water Resources*, 64–65.

[75] Powell, *Garden State*, 288–89.

[76] Government of Victoria, *Salt Action*, Foreword.

a clandestine art performance was in fact a government-organized event featuring the Minister for Agriculture, Evan Walker, and the Minister for Conservation, Forests and Lands, Joan Kirner. Children from Kew East Primary School, armed with white and blue balloons representing salt and water, burst their white balloons and released the blue ones into the air to symbolize that "joint action is vital if we are to beat salinity." This was the launch of the Victorian government's long-term salinity control strategy entitled "Salt Action—Joint Action," a key component of the State Conservation Strategy. The performance was purposefully held in the metropolis "to raise awareness about the problem amongst the 75 percent of Melbourne people who either have never heard of salinity or don't know what causes it."[77]

Salinization was now officially deemed a top-priority environmental problem for the state government—an indicator of how much political attention had increased since 1967–68, when the drought highlighted the scale of a decidedly low-profile issue by causing slugs of highly saline water to come down the Murray. Phillip Toyne, Director of the Australian Conservation Foundation, called land degradation, including soil erosion and expanding salinization, "Australia's greatest environmental challenge."[78] A timely warning, considering that, at this time, it was estimated that 960 km^2 of irrigated land in the MDB were affected by salinization, and 560 km^2 had water tables within two meters of the surface, which were rising rapidly in many areas.[79]

"Salt Action" also marked a turning point in the approach to salinity control. In line with the Salinity Committee's recommendations, the issue was no longer solved primarily by large scale engineering solutions, but instead, through cooperation between the government and the communities employing a variety of smaller-scale options. As Joan Kirner put it, "[i]nstead of the Government presuming to tell people what's best for them, we are working together to plan future action so that everyone is aware of the programs and why." The communities had shown that

[77] Anonymous, "Launch of Salt Action: Joint Action," *Salt Force News* 9 (June–July 1988): 1.

[78] "Putting our House in Order," *The Age*, December 9, 1988.

[79] Murray-Darling Basin Ministerial Council, *Murray-Darling Basin Environmental Resources Study* (Sydney: State Pollution Control Commission, 1987), 93.

they were capable of solving their own problems. Broader awareness for salinization remained yet to be manufactured.[80]

The Salinity Committee's report, "Salt of the Earth," had attracted many comments. Eight state governments and rural shires, 14 rural community groups, as well as nine universities and private consultants responded to its recommendations. The vast majority of groups and shires had supported the formation of a Salinity Bureau, while the RWC and the Department of Water Resources had opposed it. The latter had criticized the report because the department felt it did not include clearly stated objectives, systematic examination of long-term scenarios, and systematic linkage between the analysis and the recommendations made. Furthermore, it reinforced the idea that costs associated with salinity control programs were socialized while the benefits were capitalized by private landholders.[81]

After Labor won the state election in March 1985, the returning government prepared a response. To this end, Evan Walker, Joan Kirner, and Andrew McCutcheon, the Minister for Water Resources, formed a Ministerial Task Force on Salinity. They adopted 47 of the 49 recommendations that the Salinity Committee had made. Crucially, the Task Force members followed their departments and rejected the establishment of a Salinity Bureau, which would have constituted a quasi-statutory authority on which rural bodies would have had voting rights. Instead, they opted to have community groups develop their own salinity management plans, with the final decision made by the government and implemented with its assistance.[82]

As an initial step, in October 1985, the government established a pilot program to identify the most effective means of addressing salinity control measures to showcase successes. To do so, a catchment was required where community involvement would be strong, including dryland areas, and where the economics of salinity and drainage control measures were favorable. Most importantly, the government looked for conditions in which measures could be planned and implemented within a strong management framework that would enable the community to solve its

[80] Anonymous, "Launch of Salt Action: Joint Action," 2.

[81] Russ, *The Salt Traders*, 193–95.

[82] Russ, *The Salt Traders*, 195–98.

own problems.[83] Eventually, the Goulburn and Broken River catchments were chosen in the Shepparton region. The official rationale was "the high potential losses within the Region if salinity were to continue unabated and because of the irrigation and dryland salinity interactions."[84] Unofficially, the prioritization of the region's fruit and dairy industry promised a better return on investment than other regions prompted the decision. An advisory body to the program was founded, which consisted of representatives from key interest groups and community leaders—the "Salinity Pilot Program Advisory Council" (SPPAC).[85]

The members of SPPAC were appointed after a public call for nominations from the community, most of whom were prominent leading figures from areas such as dryland and irrigation farming, processing industry, education, local media, agricultural organizations, and local government. Among them was Jeremy Gaylard, Mayor of Shepparton. However, the leadership fell to two protagonists from GIRDAC: John Dainton, who became chairperson, and Leon Heath, who served on the Irrigation Sub-Committee with Eric Merrigan, another long-standing member. Dainton was a dairy farmer at Mooroopna, who had been involved with the industry since the 1970s as president of the Victorian Dairy Farmers Association, secretary of the No. 9 District Council of the United Dairyfarmers Victoria, and director, later deputy chairperson, of IBIS Milk Products, a milk processor in Shepparton and Stanhope. Later, he became the director of the industry cooperative Bonlac. Dainton had been nominated to join GIRDAC in 1978 to represent the dairy industry and was elected chairperson in 1984, a post which he held until SPPAC was formed. Leon Heath, who had represented GIRDAC since the days of the Public Works Inquiry on Salinity, was involved with the Victorian Farmers Federation and became a member of the Community Advisory Council of the MDBMC. Shepparton's Salinity Action Group, founded in 1974,

83 Russ, *The Salt Traders*, 199; 224–26; Salinity Pilot Program Advisory Council, *Draft Shepparton Land and Water Salinity Management Plan*, 3.

84 Salinity Pilot Program Advisory Council, *Draft Shepparton Land and Water Salinity Management Plan*, 3.

85 Russ, *The Salt Traders*, 224–26; Salinity Pilot Program Advisory Council, *Draft Shepparton Land and Water Salinity Management Plan*, 3.

had now reached top positions in a government-endorsed community group.[86]

SPPAC's role was to be a forum for the assessment of objectives, policies, and projects of the pilot program and to recommend priorities in salinity control. It was tasked to inform the regional community about the program, enable citizens to participate in discussions about salinity issues, and advise the Ministerial Task Force and the program's management about community attitudes toward salinity. SPPAC had competences and financial support never attained before by a community group; it was granted direct access to the ministers and received millions of dollars in funding. It used the latter to build its own community bureau and pay a service fee to its members. By contrast, the voluntary groups had been in a permanent state of fund-seeking and never commanded more than a few thousand dollars at a time. Since SPPAC was answerable to a committee of ministers instead of just one, each corresponding department became accountable to SPPAC. It soon developed media links, which made it a powerful voice. Its members became public figures who lent credibility to the issue with local faces. In the context of Shepparton's situation, it was their difficult task to convince the public, business people, bankers, and local government representatives that they had a *potential* problem that needed to be acted upon.[87]

However, SPPAC's primary goal was to develop a salinity management plan for the Shepparton region. It was the first of a series of regional plans to be prepared for all of Victoria. To this end, the state was divided into nine Salinity Control Regions, which corresponded to major catchments adjusted to local government boundaries. These regions were self-contained as both causes and effects of any salinity problem were generally within a single region, except salt transfer from east to west in the Murray. Following the work on the Shepparton management plan, working groups were founded for the Campaspe, Barr Creek, Kerang Lakes, Tragowel Plains, and Nangiloc-Colignan areas to develop corresponding plans. For each of these regions, other government-funded groups of community representatives were appointed to develop and

[86] Russ, 226–27; Salinity Pilot Program Advisory Council, *Draft Shepparton Land and Water Salinity Management Plan*, iv; 3–4; John Northage, *The Story of John Dainton's Role in Mending the Goulburn Broken* (Shepparton: Goulburn Broken Catchment Management Authority, 2014), 33; 37–38; 42; 56–57.

[87] Russ, *The Salt Traders*, 228–35; Northage, *Mending the Goulburn Broken*, 103–5.

implement salinity management plans for their area. They received assistance and technical support from consultants, government experts, and agencies. When the government approved of a plan, it was implemented and funded jointly by the government and the community.[88]

On September 25, 1986, the Kerang Lakes Area Working Group (KLAWG) was launched on the shores of Middle Lake by the Premier, John Cain. The group was answerable directly to the government but fiercely attempted to remain independent. As in the case of SPPAC, nominations for water user and special interest representatives had been called through local media. Municipalities and state government organizations had been asked to nominate representatives. Public ballots had been conducted for several positions where there had been more than one nominee. The resulting mixture of members aimed to represent the whole area with its various interests and competences. In the working group, representatives of water users in the region, of social issues, conservation and environment, tourism, recreation, and commerce joined local government and government agency members. The chair fell to Stuart Simms, a leading figure of the Lake Charm Restoration Committee. This community group had been founded in 1976 to halt the deterioration of the lake's water quality by rising salinity levels and worked to have it restored to a freshwater lake. Rex McCann, who had campaigned against the Mineral Reserve Basins Scheme, joined Gyn Jones of the Department of Agriculture and Kerang shire councilor Kevin Inglis, both members of KIRSAC. While the RWC provided secretarial support and a liaison person in Kerang, technical expertise in the form of groundwater models, as well as agricultural, economic, and environmental studies were contributed by government departments. Additionally, KLAWG's work was supported by the consultants of Dwyer Leslie.[89]

"Salt Action" institutionalized an existing cooperation between communities and government institutions. Leading protagonists from the two Salinity Action Groups in Kerang and Shepparton seamlessly transitioned into the newly established, government-funded and endorsed community groups. This development was logical. From the outset,

[88] Government of Victoria, *Salt Action*, 12–17; Russ, *The Salt Traders*, 242.

[89] Kerang Lakes Area Working Group, *Kerang Lakes Area Draft Salinity Management Plan*, iii; xxvi–xxvii; Anonymous, "Lake Charm Fast Dying of Salt," *The Northern Times*, April 9, 1976; Anonymous, "Management Plan for Lakes Region," *The Northern Times*, July 11, 1986.

both groups had been intensely focused on governments because they understood that their salinity problems could only be solved with government authority and resources. Being semi-public bodies themselves, who recruited a large part of their personnel from government agencies and local representatives, they were extensions of Victoria's public irrigation institutions. They had successfully established themselves as respected representatives of their communities in salinity matters who were capable of transmitting opinions to the government. They commanded valuable regional knowledge, as well as technical and environmental expertise. Following an idea championed by VIRPO, they advocated a coordinated approach of bringing all relevant interests together to solve irrigation's problems. KIRSAC's initial goals had been high-flying. The committee wanted to become a "comprehensive organisation strong enough to become a statutory authority, or at the very least, a state-supported co-ordinating group." However, despite this, none of the groups had become a professional, officially endorsed organization. They depended on dedicated individuals such as Gyn Jones, Ernest Jackson, and John Dainton. Not least, the volunteer groups had relied on public financing and goodwill from political representatives. Now, the government launched its own official community organizations, with appropriate funding and backing, and opened it to the well-established community activist scene. It was a friendly takeover, and the Salinity Action Groups did not think twice to accept the deal.[90]

Upon visiting the Kerang Lakes area in April 1986, Joan Kirner was impressed "with the community and the expertise of local conservationists and their willingness to tackle salinity and conservation problems."[91] Likewise, at the Salinity Committee's community conference in Bendigo two years earlier, the communities had convincingly demonstrated their concern for and interest in the issue. The committee had recognized the value of the work of the Salinity Action Groups, VIRPO, and others. Following this successful example, the government conducted a State Salinity Conference in 1987 to assess the draft of its strategy, released in February for public discussion and submissions.[92] When Graham Hunter

[90] Hardwick, "Chairman's Report," May 31, 1977; Ernest M. Jackson, "re: Press tour, Kerang Region, 18th - 20th July," 6–7.

[91] Anonymous, "Lakes Salinity Talks Pledged," *The Northern Times*, April 8, 1986.

[92] Government of Victoria, *Salt Action*, Foreword; Anonymous, "Future Directions for Salinity Control," *Salt Force News* 3 (June 1987): 1–5.

was the guest speaker at KIRSAC's tenth annual meeting, he underlined this trust in communities' capability to help themselves. The goal of "Salt Action," he declared, was "long-term community acceptability, rather than imposing control measures."[93]

For Hunter, the program was an attempt to gain a medium between engineering and land management approaches and develop community structures to help solve salinity problems.[94] The hard-fought and drawn-out conflicts over the Lake Tyrrell and Mineral Reserve Basins Schemes had vividly demonstrated to the government that these types of engineering solutions could not be politically enforced against the will of substantial and well-organized parts of the rural communities—or at least not without paying a high political price. The fact that the government was willing to accept that a radical such as Rex McCann became a salaried member of an officially endorsed working group illustrates how determined it was to restore peace.

Due to all of these factors, "Salt Action," with its emphasis on community involvement and acceptability, contrasted sharply with the strategy that the SRWSC had tabled in 1975. It focused heavily on ambitious engineering solutions, but this did not mean that the government was prepared to upgrade the existing groups, which had tried to reflect a variety of different interests similar to the official working groups, and turn them into autonomous statutory authorities. Instead, it reserved the right to make final decisions for itself and created advisory bodies that the community activists were invited to join. By co-opting their protagonists, the government lent credibility to its pledge to include the communities in decision-making. Put slightly Machiavellian, they not only commanded expertise but were in a good position to sell the program to their fellow citizens.

If a symbol is required for this merger between government and community groups, the opening of the Kerang Environmental Study Centre at Appin, run by the Hardwick family, will suffice. On September 9, 1988, an "army of volunteers, landholders, community groups, scientists and local councils" gathered to rally behind "Salt Action." After 13 years of development, Joan Kirner declared the Centre officially open.

93 Anonymous, "Salinity Focus on the Community," *The Northern Times*, June 11, 1986.

94 Anonymous, "Salinity Focus on the Community".

In her keynote, the minister emphasized that "all parts of the community must understand that awareness and action were of paramount importance in the fight against the degradation of Victoria's land." Kirner paid tribute to the commitment and vision of George Hardwick and his family. Education centers such as the KESC could play a vital role in promoting community awareness. It was a hands-on model for community education, which her department sought to encourage. "This continued high level of Government commitment and strong community support will go a long way toward overcoming the serious land degradation problems," the minister said. Finally, she unveiled a plaque which read: "A vision commenced 1973 and developed by George and Maureen Hardwick and their children, Wendy, Craig, Marshall and Andrew, as a facility for teaching integrated catchment management, and for the passive exchange of experiences between urban and rural people."[95]

On the same occasion, Ernest "Watershed" Jackson officially dedicated the Vera and Ernest Jackson Field Laboratory and Library. The other distinguished *dramatis personae* in this celebration included Kerang shire councilor and KIRSAC member Kevin Inglis; Heather Mitchell, the president of the Victorian Farmers Federation; the notorious Gyn Jones; Don Oberin, who represented the MDBMC; and Eric Merrigan of VIRPO and GIRDAC. "As long as the present strategies of community input and involvement and decision making continued," Inglis said, success lay ahead (Fig. 7.1).[96]

The principal goal of "Salt Action" was to manage the salinity of land and water in order to maintain—and, where possible, improve—the social well-being of communities, the environmental quality, and the productive capacity of the affected regions. To this end, water tables should be stabilized at harmless levels, or their rise at least be significantly retarded. No new evaporation basins or other works would be located in wetlands unless compelling public interest was demonstrated. High-value wetlands and other significant environmental features should be protected or rehabilitated. The government stated frankly that improvements might not be evident for many decades, requiring a long-term political and financial commitment. The price for this was a cost-sharing arrangement between

[95] Anonymous, "Centre Part of Salinity Fight," *The Northern Times*, September 13, 1988.

[96] Anonymous, "Centre Part of Salinity Fight".

Fig. 7.1 "Centre part of salinity fight." The Northern Times, August 13, 1988 (*Source* © Gannawarra Times, Kerang)

the state, regional communities, and landholders. Staying on the road toward market-based water prices, the government was no longer willing to regard irrigation as an inherently beneficial undertaking. Communities were expected to contribute financially to salinity control to the extent that they derived benefits from control measures and local farming, water management, and disposal systems contributed detrimentally to salinization. Harmful practices should be discouraged.[97]

On the other hand, management practices that were cost-effective and not dependent upon continuing subsidy would be encouraged. Each investment in salinity control would be subject to an economic evaluation with a set target economic rate of return, although intangible costs and benefits would be considered. Where an increase in salinity was deemed unavoidable, or the cost of salinity control measures was considered to

[97] Government of Victoria, *Salt Action*, 1; 5; 20; 40–41.

be economically inviable, help would be provided to the communities in question for adapting to saline conditions. For the Kerang region, this meant that the approach that the SRWSC had favored in 1975 would not be revised—it would remain a saline area with permanently high water table conditions and no sub-surface drainage system. Landholders would have to develop saline agricultural practices. Given the decades of work by the two community-owned research farms in Kerang and Swan Hill, this was well underway.[98]

Preparing the salinity management plans for each region would take years. In the meantime, landholders continued to receive assistance for farm planning and the provision of grants and loans at concessional interest rates. Groundwater exploration, community drainage solutions, land forming, groundwater pumping, water harvesting, efficient irrigation methods, and on-farm evaporation basins were eligible for assistance. As a last resort, tax concessions and funds were to be made available to farmers willing or forced to leave the industry. Advisory teams would be set up to operate in all salinity regions in Victoria within ten years.[99]

Within this same timeframe, "a sense of local and regional ownership of the salinity problem" would have to be fostered. According to Joan Kirner, it was imperative "to increase all Victorian's consciousness of salinity problems and to engage country and city people to assume responsibility for the land and its future by acting as custodians." "Salt Action" aimed to create awareness, concern, and responsibility about salinity while providing the communities with the necessary knowledge to take part in salinity control.[100]

The existing non-governmental organizations, specifically the Australian Conservation Foundation, National Farmers Federation, MVL, GIRDAC, and KIRSAC would play a vital role to achieve this. Their work continued to be supported by salinity community grants. Allocation for these was subsequently increased. In 1987, $140,000 were allocated. The next year, $210,000 went to 33 groups, to peak at $290,000 in 1991 when 59 groups received funds through the program. KIRSAC obtained grants for a brochure on salinity control in

[98] Government of Victoria, *Salt Action*, 1; 17; 19; 21–22.

[99] Government of Victoria, *Salt Action*, 29–34.

[100] Government of Victoria, *Salt Action*, 25; 42; Anonymous, "Local Share of Salinity Funding," *The Northern Times*, June 6, 1986.

the Kerang region. The Swan Hill Irrigators Research Farm employed a journalist to revise the farm's guidebook with the support of a grant. Its Kerang counterpart received $3100 to establish an educational trial walk for school children. VIRPO used $5000 for research and to collect material on catchments as the basis for a script for a documentary about Ernest Jackson. GIRDAC raised grants to continue its bus tours through salt-affected areas. Among the many other causes were tree planting projects, newsletters, forums, trials of salt-tolerant pasture species, and in-service training for Victorian Farmers Federation representatives on salinity control. The community initiatives scene, which had already been vibrant in some areas, greatly benefited from the grants scheme. In 1981, a mere three farm tree groups had existed. In 1988, there were 37. Since the scheme had been introduced in 1986, 77 new local Salinity Action Groups had been founded.[101]

Additionally, the government launched a variety of initiatives of its own. In 1986, the first issue of "Salt Force News," a bi-monthly magazine to keep people informed about "what's happening on the salt front," was published.[102] Mary Lou Considine was appointed the first salinity journalist. She was soon succeeded by Vivian McWaters. The magazine acted as a mouthpiece of the government, which informed its readers about new developments in the "Salt Action" program, the state of salinity management plans, and official announcements. It also featured a "Grass Roots" section with success stories in salinity control throughout the state. The magazine ran for ten years and had 49 issues until the last one appeared in June 1996.[103]

As an "action research project" for schools in isolated rural areas, "Saltwatch" was launched in Spring 1987 with over 300 participating schools. Students collected water samples from local streams, rivers, and bores to analyze the salt content in order to gain a local snapshot of salinity

[101] Government of Victoria, *Salt Action*, 42; Anonymous, "Grassroots," *Salt Force News* 1 (December 1986): 7; Anonymous, "1988 Community Salinity Grants," *Salt Force News* 7 (February 1988): 1–2; Anonymous, "Launch of Salt Action: Joint Action," 2; Anonymous, "Salinity Grants Awarded for 1990/91," *Salt Force News* 16 (October 1989–January 1990): 10; Anonymous, "Crabb Announces: Community Salinity Grants," *Salt Force News* 20 (March–April 1991): 3.

[102] Mary Lou Considine, "Editorial," *Salt Force News* 1 (December 1986): 2.

[103] Considine, "Editiorial," 2; 7–8; Vivian McWaters, "From the Editor," *Salt Force News* 3 (June 1987): 2; Carri Tiffany, "Farewell, Salt Force News!" *Salt Force News* 49 (June 1996): 1.

concentrations in surface- and groundwaters. Subsequently, a resource package for schools was made available containing condensed information and classroom activities on salinity. According to the first annual review of "Salt Action," by early 1989 more than 10,000 people had participated in the project.[104]

Another environmental salinity-focused project was inaugurated in Seymour, near Bendigo, on World Environment Day, June 5, 1988. 400 Girl Guides defied the pouring rain and freezing cold to plant over 17,000 trees to mark the start of Greening Victoria. These were supposed to be the first of over one million trees to be planted across the state in known salinity recharge areas. The girls were accompanied by the Lord Mayor of Melbourne, Alexis ("Lecki") Ord, Victorian Guides Commissioner Maren Chandler, Heather Mitchell, and Joan Kirner.[105]

The latter two women were also responsible for what is likely the most prominent initiative that can be linked to the salinity problem—Landcare. Even though its roots can be traced back to farmer-led initiatives in Western Australia and Victoria, which were galvanized by dryland salinity, its breakthrough can be attributed to the alliance between the Minister of Conservation, Forests and Lands and the president of the Victorian Farmers Federation. Joan Kirner and Heather Mitchell collaborated in the formation of this community-based program to reverse land degradation, which would tackle a range of issues like salinization, soil erosion, pest infestation, and loss of native plants and animals. It facilitated community participation, educated farmers about land management, and instilled an ethic of stewardship in their approach to land use. The program was officially launched on November 25, 1986 at Stricta Hill near St. Arnaud alongside the first group at Winjallock, which focused on establishing deep-rooted pastures on the bare hills of the district. Funding for LandCare came from the Victorian government. By 1990 there were 70 groups in the state.[106]

[104] Government of Victoria, *Salt Action*, 42; Anonymous, "SALTWATCH: An Action Research Project for Schools," *Salt Force News* 3 (June 1987): 6; Vera Lubezenko, *Saltwatch. Involve Me and I'll Understand. A Resource Package for Schools* (Melbourne: Salt Action Victoria, 1992); Anonymous, "First Annual Review Highlights Action," *Salt Force News* 12 (January 1989): 1.

[105] Anonymous, "Greening Victoria Launched on World Environment Day" *Salt Force News* 9 (June–July 1988): 8; Salt Force, *Greening Victoria* (Seymour: 1988).

[106] Anonymous, "Minister Launces LandCare," *Salt Force News* 1 (December 1986): 3; Mary Johnson et al., "Landcare in Australia," in *Landcare. Local Action—Global Progress*,

Landcare expanded rapidly during the decade. South Australia took its cue from Victoria and Western Australia in 1989 and formed a State Management Committee for the Year of Landcare. Then, an unlikely alliance between farmers and conservationists led to support on national level. National Farmers Federation executives Rick Farley and Andrew Robb joined forces with Phillip Toyne, director of the Australian Conservation Foundation, and secured bipartisan support for Landcare. Subsequently, the Commonwealth government declared the 1990s the "Decade of Landcare." In 1993, the National Landcare Program was launched. LandCare brought focus, funding, credibility, and national attention to a wide range of conservation issues beyond salinization.[107]

"Salt Action" added a new impetus to the old endeavor of living sustainably in a saline environment. By 1990, about $100 million had been spent on various salinity works and services. For the government, salinization was not so much a single issue, but rather the program was an experimental vehicle that could solve the problem through which other farm services, economic development, infrastructure, and employment could be delivered to the rural community. The program became a focal point for solving many other environmental, farming, economic and social concerns in rural Victoria.[108]

Basin-wide, the MDBC introduced its own Salinity & Drainage Strategy in the same year that "Salt Action" launched. This meant several management changes to reduce evaporation from storages. It allowed for additional saline drainage to flow into the river from new irrigation projects in upper catchment states, Victoria and New South Wales. In return, those states and the Commonwealth invested in groundwater interception works in the middle and lower reaches of the river where the greatest salinity reduction benefits could be obtained. The strategy aimed to significantly reduce the net average salinity levels measured at Morgan in South Australia and to manage river flows in such a way as to avoid

eds. Delia Catacutan et al. (Nairobi: World Agroforerstry Centre, 2009), 14–15; Andrew Campbell, *Landcare: Communities Shaping the Land and the Future: With Case Studies by Greg Siepen* (St. Leonards: Allen & Unwin, 1994), 22–29; Richard Broome et al., *Mallee Country: Land, People, History* (Melbourne: Monash University Publishing, 2019), 331–32.

[107] Johnson et al., "Landcare in Australia," 15–16; Broome et al., *Mallee Country*, 331–32; Campbell, 30–31.

[108] Russ, *The Salt Traders*, 244.

short-term spikes in salinity levels. The MDBC maintained a register to record negative salinity impacts of new irrigation developments, and positive ones of compensating remedial projects. Salinity credits, expressed in Electrical Conductivity (EC), became the currency to measure impacts on the Basin environment. Within ten years, a net average reduction in salinity levels of 80 EC and an average salinity of less than 800 EC at least 95% of the time should be achieved.[109]

"Salt Action," the MDBC, and the introduction of market-based pricing mechanisms for irrigation water, together, can be seen as the conclusion of a phase of the almost steady spread of salinization throughout Northern Victoria, and indeed the MDB, which had begun 135 years prior when the first recorded signs of soil salting were noted by a Western District squatter. The developments around 1988, European Australia's bicentennial, marked the beginning of a new world in rural environmental management. The question remained whether they would succeed in getting the slow catastrophe under control.

[109] Connell, *Water Politics*, 119–21. Electrical conductivity is used to measure the salt content of water. It expresses how much electrical current can flow through a sample of water. The more dissolved salt in the water, the higher the EC. Agriculture Victoria: Measuring the Salinity of Water. See "Measuring Salinity" for details.

CHAPTER 8

Under Control? (1988–2020)

Legacies

In the late 1980s, Kerang was renowned as "the home of salinity."[1] Gyn Jones, nonetheless, claimed credit for the region as one "of the first to experience a major environmental problem, [it] is also the first to successfully tackle that problem, with little outside funding and little outside support until the last five years. The local community has been able to demonstrate the value of co-operation and community action."[2]

Jones pointed to the long history of self-help, commencing with Alan Morgan's reclamation plots in 1939. For him, Kerang's community research farm KARF had been the focus of salt reclamation and productivity increase, not only for the region but for the rest of Australia, which Kerang had "astounded [...] by lifting productivity despite the crippling salt problem." He pointed out that, during the 1940s, the average salt-affected farm had 20 to 40 cows, averaging about 90 kg of butterfat; 20 or so beef cattle; 200 to 300 first cross ewes; and grew some barley or millet. Highways had not been sealed, and not many farmers had cars. Farm roads had been poor, and townships generally reflected a lack of

[1] Jones, "We've Done Something About Salt," *The Northern Times*, March 1, 1988.

[2] Jones, "K.I.R.S.A.C.," 1993(?), 2.

© The Author(s), under exclusive license to Springer Nature Switzerland AG 2023

D. Rothenburg, *Irrigation, Salinity, and Rural Communities in Australia's Murray-Darling Basin, 1945–2020*, Palgrave Studies in World Environmental History,
https://doi.org/10.1007/978-3-031-18451-2_8

vitality. In the late 1980s, an area of two to three times the size of these older farms milked 120 to 150 cows, produced 20 to 30 tons of butterfat, and ran 2000 sheep while producing a variety of crops. Jones admitted that this was also due to increased water availability.[3] He failed to mention the contribution to productivity increases due to mechanization, fertilizer, and chemicals since the 1950s that characterized the agricultural sector in general.[4] Yet there is no denying that the Kerang region was an outstanding example of self-help and organization.

Nevertheless, a lack of knowledge about salinization was still widespread throughout Northern Victoria. When Joan Kirner visited Kerang in 1988, she emphasized that this was one of the area's greatest challenges. The minister quoted a study that concluded that more than half of all urban Victorians—including people who lived in provincial cities such as Kerang and Shepparton—did not know what salinity meant. At the same time, it was almost impossible to drive anywhere in the state without seeing its effects.[5] The study, however, showed that 80% of farmers had some understanding of the salinity problem, while, at the same time, few understood the link between salinization and rising water tables. A minority, 4%, attributed the problem to the use of farm chemicals and fertilizers. Farmers throughout Victoria identified salinization as a major concern for the state's future, and the most difficult question for all participants remained what to do about it.[6]

To further complicate things, farmers' recognition of salinization did not mean that they were prepared to address it. A 1989 survey by the University of Melbourne, inquiring into perceptions of dryland salinity in the upper catchments of the Loddon and Avoca Rivers, concluded that most farmers considered salinization to be a regional- and locality-based problem, not a personal one. While, they recognized that salinization would increase over time, apart from those who already had bare patches

[3] Jones, "We've Done Something About Salt".

[4] Davidson, *European Farming in Australia*, 337–43; McNeill and Engelke, 38–39; 52; 88–89; 144.

[5] Anonymous, "Centre Part of Salinity Fight," *The Northern Times*, September 13, 1988.

[6] Terry White, "80% of Farmers Aware of Salinity," *Salt Force News* 10 (August–September 1988): 1–2.

on their land, farmers generally did not think their property would be affected. Salinization happened to someone else.[7]

This situation appeared to improve in the early 1990s. A 1994 survey found a noticeable increase in awareness and concern about salinization among Melbourne residents. While the average level of distress for popular environmental problems such as pollution, tree clearing, waste, and the extinction of plants and animals had declined since 1989, worry over salinization had increased. By then, it was recognized as a mainstream environmental issue, yet the level of concern increased only marginally among residents in Victoria's major provincial towns and among landholders. However, the level of ignorance about salinity problems had decreased. In 1988, almost one in three Melbourne residents was unaware of salinization's seriousness in the state, but by 1994, ignorance had dropped to just 9%; 59% of farmers knew about their local Salinity Management Plan, and 53% said that they had already begun to implement strategies to reduce salinity on their properties.[8] A 1996 survey by the Goulburn–Murray Landcare Network confirmed these trends: 47% of farmers thought that salinity was increasing and likely to affect them personally in the future.[9]

The environmental legacy of Water Dreaming in the Goulburn–Murray region was profound. Before the intervention of Europeans, the Kerang Lakes Area had been characterized by an abundance of natural lakes and swamps, which were filled with floodwater at irregular intervals. The wetlands, river systems, floodplains, and dryland had supported rich and abundant populations of plant and animal species. Agricultural and commercial development altered the nature and extent of these areas through grazing, clearing, altered flood regimes, salinization, and introduced species. Still, the semi-natural areas that remained continued to be valuable ecosystems that provided broad habitats for a wide diversity of species and abundant populations of native fauna: "The consequence

[7] Anonymous, "Not my Problem. Farmer Perceptions of Dryland Salting: Misplaced Optimism," *Salt Force News* 13 (April–May 1989): 5.

[8] Anonymous, "Melbourne Accepts Salinity as a Major Environmental Issue," *Salt Force News* 39 (October 1994): 15.

[9] Anonymous, "Survey Into Farming Community Response to a Rising Watertable," *Salt Force News* 47 (February 1996): 16.

of these modifications has been extensive and significant environmental degradation."[10]

The main environmental modifications were due to irrigation development, and the activities intended to upgrade streams' water-carrying capacities, including de-snagging, dredging, channelization, and levee-banking, extensively degraded habitat values. The use of streams and wetlands for the conveyance of irrigation water had completely altered water regimes. Contrary to "natural" conditions, streams had low flows during winter, and high ones during summer. Pyramid Creek, formerly a broad, shallow vegetated stream, had been converted into a deep and narrow irrigation channel. The dredging of its bed and other streamlines, coupled with rising groundwater levels, led to the interception of saline groundwater and more saline stream flows. Barr Creek met a similar fate, becoming the main collector drain for irrigation areas in Kerang and Cohuna—infamously the largest individual source of salt in the entire Murray Valley.[11]

Barr Creek water was being diverted into Lakes Little, William, Kelly, and Tutchewop to reduce salt discharge into the Murray. In 1986, 50% of the creek's flow was redirected to Lake Tutchewop for flood mitigation and use as an evaporation basin, turning it into a saline lake, where before it had been a top-rated site for wildlife and birds. In Lake Charm, the water quality deteriorated dramatically when it was used for pondage to provide water for lower reaches in dry years. At this time, the lake contained the worst quality water of all in the Kerang Lakes system.[12] Cullen's Lake became one of the casualties of the high salt load in Pyramid Creek. Previously, its water quality had been considered very good, but it markedly declined. In 1977, over 500,000 water birds were estimated to

[10] Kerang Lakes Area Working Group, *Kerang Lakes Area Draft Salinity Management Plan*, 44.

[11] Kerang Lakes Area Working Group, *Kerang Lakes Area Draft Salinity Management Plan*, 44–45; Russ, 90–91; Hardwick, "A Historical Report of the North Central Catchment Region," 11; Gutteridge, Haskins, & Davey, *Murray Valley Salinity Investigation*—Report, 234.

[12] Anonymous, "Inland Sea Left Legacy of Salt Salinisation," *The Northern Times*, June 6, 1986; Kerang Lakes Area Working Group, *Kerang Lakes Area Draft Salinity Management Plan*, 45.

be on the lake. In March 1988, it was virtually dry because the incoming water from Pyramid Creek was considered too saline.[13]

In all, by the late 1980s, saline intrusions, disposal of saline drainage water, and isolation from natural flushing flows had caused massive salinity increases. In many wetlands, salinity levels were more than 100 times greater than before. This led to the extinction of many freshwater-dependent plant and animal species. Examples were lakes with telling names, such as the Duck Lakes, Dry Lake, Long Lake, and North Drainage Lake, which, while previously fresh or brackish, had been turned into hypersaline environments supporting virtually no flora and fauna.[14]

The Shepparton region, before European settlement, was a rich and diverse ecosystem. Vast stretches of open flood plains of box and red gum forests were dissected by prior streams. Large numbers of open wetlands were scattered throughout the region. Flora and fauna were well adapted to the natural extremes of floods and droughts. Reports by early European settlers confirmed the abundance of wildlife and described the pattern of hunting and gathering by Aboriginal people. Development of the region for European agriculture resulted in the removal of 95% of native forests and woodlands. Extensive clearing in the upstream catchment areas and the introduction of major irrigation schemes had profound effects on surface and groundwater hydrology. The environmental quality of a large proportion of the regional streams, such as the Campaspe River, and wetlands were diminished by clearing, grazing, drainage works, rising water tables, and salinization. During the summer, salinity increased substantially, and white salt inflorescences were common along some bare banks. A number of mammal species had become extinct, and the diversity and abundance of native flora and fauna had been reduced. However, in the late 1980s, some streams and wetlands were still ranked as high-quality habitats.[15]

Another legacy was that of the Salinity Action Groups. With the Victoria government's friendly takeover, many of their leading protagonists transitioned into state-endorsed community groups. This was

[13] Anonymous, "Water is Too Salty for Lake, so Important Wetland is Dry," *The Northern Times*, March 4, 1988.

[14] Kerang Lakes Area Working Group, *Kerang Lakes Area Draft Salinity Management Plan*, 45.

[15] Salinity Pilot Program Advisory Council, *Draft Shepparton Land and Water Salinity Management Plan*, 27–29.

rational since this was the place where they could exercise their influence over the planning and community education they wanted to achieve. Consequently, volunteer organizations had increasingly outlived their usefulness. As Cliff Spowart put it, "[q]uite a lot seems to be happening in salinity control whether or not KIRSAC has a meeting, so in one sense we seem to have worked ourselves out of a job."[16] With "Salt Action," the committee had fulfilled its "major initial objective."[17]

Furthermore, it became obvious that the groups were highly dependent on the committed work of their long-standing figures. By 1989, VIRPO's major problem was a lack of new members, and the present generation was increasingly exhausted. Vera, "the 'gracious lady' of 'Pigeon Bank'," the Jackson's homestead, had passed away in late December 1987. Ernest's health deteriorated, and in early 1989, aged 88, he moved to the Lutheran Rest Home at Albury, no longer able to take an active role as the organization's executive secretary. In October, he was hospitalized. Subsequently, VIRPO's activities were suspended. After 25 years of existence, at the annual meeting in 1990, those present could not agree on the organization's future.[18]

On December 14, 1990, at 3:00 p.m., a ceremony took place at Lake Eildon. In the presence of the Bishop of Wangarratta, the ashes of Ernest Menzies Jackson, who had passed away on October 13, were distributed over the top Mount Pinniger as a tribute to "a great friend and environmental educationalist" who had served for over 50 years in various

[16] Cliff L. Spowart, KIRSAC, "Salinity Activities Round-up. 29/9/87," September 29, 1987, VIRPO Folder, George and Maureen Hardwick, Appin.

[17] Cliff L. Spowart, "Letter to KIRSAC Members," February 21, 1987, KIRSAC, Kerang Agricultural Research Farm, Archive, Agriculture Victoria, Kerang, 2.

[18] Victorian Irrigation Research and Promotion Organisation, "Minutes of Executive Meeting Held at State Offices, Shepparton, on 22 March 1989," March 22, 1989, 6008, 84/36649, Public Record Office Victoria, Melbourne, 2; Jackson, "Relatives & Friends of the Late Vera Margaret Jackson (Nee Haughton);" Eric Leete, "Victorian Irrigation Research and Promotion Organisation, Notice of Meeting," March 1989, VPRS 6008/P0001, 84/36649, Public Record Office Victoria, Melbourne; Daniel M. Ley, "Notice to Executive Members," November 1, 1989, VPRS 6008/P0001, 84/36649, Public Record Office Victoria, Melbourne; Eric Leete, "Victorian Irrigation Research and Promotion Organisation, Chairman's Annual Report 1990," 1990, VPRS 6008/P0001, 84/36649, Public Record Office Victoria, Melbourne; L. Jones, "Note for File 84/36649," March 21, 1990, VPRS 6008/P0001, 84/36649, Public Record Office Victoria, Melbourne.

voluntary organizations concerned with irrigation and conservation.[19] Cliff Spowart wrote in his "Farewell" that Jackson "knew his life's work was done. With his passing this Region has lost a staunch friend and advocate."[20] A cairn and plaque for "Watershed" were unveiled on the same site the next year.[21] VIRPO, unable to find anyone who was willing and capable of filling Jackson's role, disbanded at a meeting in Shepparton in January 1991. Two years later, the remains of the organization were assembled into a fund for the Ernest Jackson Award.[22]

For KIRSAC, the situation was similar. Even though the committee felt that it still had a role as a regional watchdog and to address issues that might fall between the sub-regional working groups, it became obvious that the group was exhausted. Meetings lapsed because of insufficient attendance.[23] Gyn Jones's retirement from the Department of Agriculture in 1991, came as "rather a shock. [They] had thought that Gyn was one of those institutions that just went on and on."[24] Even though, in his retirement, he was still involved with KLAWG in compiling the Kerang Lakes Area Salinity Management Plan, Jones felt forced to resign from his secretarial duties for KIRSAC after 14 years soon after due to his ill

[19] Daniel M. Ley, "Victorian Irrigation Research and Promotion Organisation, Press Release, Eildon Weir—Mount Pinnenger Look-Out, Friday, December 14, 3 p.m. 1990," November 30, 1990, VPRS 6008/P0001, 84/36649, Public Record Office Victoria, Melbourne.

[20] Spowart, "Minutes of Meeting held at the Kerang on the 2nd. of October, 1990," 2.

[21] Kerang Irrigation Region Salinity Action Committee, "Minutes of Meeting held at Kerang on 7th April, 1992," April 7, 1992, VIRPO Folder, George and Maureen Hardwick, Appin, 2. A photo of the cairn and plaque can be found in Anonymous, "Our Hidden Memorials".

[22] Cliff L. Spowart, "KIRSAC, Information & Comments," February 5, 1991, VIRPO Folder, George and Maureen Hardwick, Appin; Gynlais O. Jones, "K.I.R.S.A.C.," 1993(?), VIRPO Folder, George and Maureen Hardwick, Appin.

[23] Kerang Irrigation Region Salinity Action Committee, "Minutes of Annual Meeting held at the State Offices, Kerang, on 1st of August, 1989;" Spowart, "Minutes of Meeting held at the Kerang on the 2nd. of October, 1990," 1; Kerang Irrigation Region Salinity Action Committee, "Minutes of the 1992 Annual Meeting of K.I.R.S.A.C. held in the R.W.C. Meeting Room on Tuesday, 7th July, 1992, at 8.00 p.m.," July 7, 1992, VIRPO Folder, George and Maureen Hardwick, Appin; Kerang Irrigation Region Salinity Action Committee, "1 and 2 March 1993," March 1/2, 1993, VIRPO Folder, George and Maureen Hardwick, Appin, 1.

[24] Spowart, "KIRSAC, Information & Comments".

health.[25] He was honored for "his ability to express the views of KIRSAC clearly and logically," which had "ensured that these views were known and respected by Governmental officers and others."[26] At the occasion of Salt Action's ten-year review, he was acknowledged as "one of the first salinity experts."[27] As for KIRSAC, after 17 years, the group went into recess shortly after Jones's retirement with a last meeting in October 1993.[28] Gordon Shire councilor Doug Miles honored the committee for its "tremendous support to the area."[29]

George and Maureen Hardwick's vision, the Kerang Environmental Study Centre, came to an end in August 1991. Several months behind and unable to pay in the near future, North West Country Credit foreclosed on their loan and ordered that they vacate the property. Creditors had rejected two refinancing proposals. The creation of a public trust fund to ensure the Centre's long-term future would not have been realized before twelve months' time and north-west was unwilling to wait this long. Thus the Hardwicks had nowhere to live and did not know what to do with their belongings. The couple was allowed to stay on the property for an extended period while they prepared for a clearing sale. They received tremendous support from the community, but their dreams were shattered.[30]

Two pioneers of the self-help community, the Kerang Agricultural Research Farm (KARF) and the Swan Hill Irrigators Research Farm, closed their gates, too. KARF, considering that "the project had achieved

[25] Brindal, "Irrigation Guru Retires," 13; Kerang Irrigation Region Salinity Action Committee, "Minutes of Meeting held at Kerang on 7th April, 1992," 1–2.

[26] Kerang Irrigation Region Salinity Action Committee, "Minutes of Meeting held at Kerang on 7th April, 1992," 2.

[27] Anonymous, "Salinity Program 1985 to 1995—A Decade of Achievement," *Salt Force News* 42 (April 1995): 6.

[28] Gynlais O. Jones, "Minutes of Meeting held at the Rural Water Corporation Offices at Kerang, on Tuesday, October 5th, 1993," October 5, 1993, VIRPO Folder, George and Maureen Hardwick, Appin, 1.

[29] Anonymous, "KIRSAC Thanked," *The Northern Times*, January 11, 1993.

[30] Anonymous, "Eviction set to End Environmental Dream," *The Northern Times*, July 30, 1991; Anonymous, "Farm Eviction. NWCC 'More than Fair'," *The Northern Times*, August 2, 1991; Anonymous, "Eviction Deadline Extended," *The Northern Times*, August 8, 1991.

its purpose," sold the property in 1991.[31] SHIRF, which had been struggling with debt, also decided to sell its estate considering the "size and relevance of the industry" and dwindling "[c]ommitment by the community."[32] The Management Committee continued to meet as required to administer windup funds that were progressively awarded to support projects conducted by other community organizations.[33]

GIRDAC, although its leading members had transitioned into SPPAC, continued to operate until the early 1990s, but with markedly decreasing activities.[34] John Dainton, its former chairperson who went on to head SPPAC, continued a distinguished career in water management. From 1996 to 1999, he was chair of the Goulburn Broken Catchment Management Authority and Goulburn–Murray Water. In 1997, he was awarded the Ernest Jackson Award by the Irrigation Association of Australia for his contributions to the improvement of irrigation farm and catchment management.[35]

How successful had the activists been? Russell Smith contended that KIRSAC had been successful in publicizing the Kerang region's salinity problem.[36] The committee felt that it had "endeavoured to put the situation as clearly as possible to the decision makers of our state during the last 10 years."[37] KIRSAC derived great satisfaction from the development

[31] Hardwick, "A Historical Report of the North Central Catchment Region," 40.

[32] Craig Davies, "President's Report 1994," 1994, Swan Hill Irrigators Resarch Farm, Collection of Annual Reports and historical documents by Swan Hill Irrigators' Research Farm Inc: including Minutes of General Meetings from 1954 until 1998/[manuscript], Swan Hill Regional Library, Swan Hill; Craig Davies, "Annual Report—S.H.I.R.F. 1996," 1996, Swan Hill Irrigators Resarch Farm, Collection of Annual Reports and historical documents by Swan Hill Irrigators' Research Farm Inc: including Minutes of General Meetings from 1954 until 1998/[manuscript], Swan Hill Regional Library, Swan Hill.

[33] Swan Hill Regional Library, "Title Control Number: a1182141," March 18, 2017, Collection of Annual Reports and historical documents by Swan Hill Irrigators' Research Farm Inc, Swan Hill Regional Library, Swan Hill.

[34] Eric Merrigan, "Minutes of Meeting of G.I.R.D.A.C. held at the Office of N.V.F.G.A., 21 Nixon Street, Shepparton on Monday 27th April 1992. Commencing at 1 p.m.," April 27, 1992, VPRS 11,493/P0001, 182/1, Public Record Office Victoria, Melbourne, 83–93.

[35] Northage, *Mending the Goulburn Broken*, 35–37.

[36] Anonymous, "KIRSAC Thanked".

[37] Anonymous, "Salinity Focus on the Community," *The Northern Times*, June 11, 1986.

of the "Salt Action" strategy and the establishment of community-based sub-regional working groups, having "achieved a large part of its objectives."[38] They even dared to believe that their organization had been "one of the many factors bringing that about."[39]

This was a sober, but robust, verdict. GIRDAC and KIRSAC had certainly played significant parts in raising public awareness of salinization's threat to their communities. Their members had given public credibility to a cause that was largely abstract and difficult to communicate. With their lobbying efforts, they had scored small victories, which helped to improve the situation. However, ultimately, they did not achieve the aim of becoming the bases for official, state-endorsed organizations in salinity control and resource management, which KIRSAC had championed. Instead, the government preferred to co-opt their members to its own organizations in order to retain the final say in decision-making. To assess the groups' influence on the changes in salinity and water management policy on the state level would mean undue speculation. What seems clear, however, is that without the strong presence and vibrancy of a scene of community activists, the Victorian government would not have entrusted the communities with a role in planning for salinity control measures. The fact that many of the Salinity Action Groups' leading protagonists continued their work there testifies to their acquired expertise and the trust they had achieved in their communities and among government experts and politicians.

Who was to blame for the slow catastrophe? Hilary Susan Howes is surely right when she argues that there was "little benefit in assigning blame for what may, with the benefit of hindsight, be perceived as the mistakes of the past." Rather, it was important to examine the ideas and cultural values; the social, ideological, and economic motives that underlay irrigation development and caused salinization. Already having discussed the latter, here it is worth noting that the question of responsibility did matter to the contemporaries.[40]

One popular narrative was sheer ignorance. A Victorian government brochure explained that "Australia's settlers tried to make a living off the

[38] Kerang Irrigation Region Salinity Action Committee, "Minutes of Annual Meeting held at the State Offices, Kerang, on 1st of August, 1989."

[39] Smith, "Chairman's Report 1986/87," 2.

[40] Howes, "The Spectre at the Feast," 220.

land the only way they knew—by clearing trees and scrub, and farming the way they used to in Europe."[41] Similarly, the general consensus at the 1984 Community Conference in Bendigo was "that salinity had been caused by past and present generations but the problem was created through ignorance."[42] Jones defended the Kerang region, arguing that "an irrigation system had been placed unknowingly in a saline environment."[43] SPPAC agreed: salinization processes were "unknowingly set in train" by the engineers and farmers who introduced irrigated agriculture in the Goulburn Valley.[44]

Perhaps, however, this interpretation is summed up best in the Greening Victoria anthem written by Faye White, "With Our Own Two Hands":

> Well we came from the farms of Europe
> And we cleared the trees from the land
> But the balance of bushland rocks and rain
> We did not understand
> How could we have ever imagined
> When we cut the forests down
> This would cause the water to rise
> From deep below the ground.[45]

The narrative of ignorance has a critical flaw. As Howes has shown, the problem was known before the introduction of irrigation in Australia. Alfred Deakin, during his travels in Egypt and Italy, had noted the development of waterlogging and salting through insufficient drainage, attributed to excessive water use. He advocated for preventive management and education as advanced preparation. Despite this early awareness, drainage had not been a high priority, while the expansion of irrigation had continued unabatedly.[46]

[41] Rural Water Commission, Victoria et al., *Salting Threatens Victoria* (Melbourne: Salt Force, 1988), 2.

[42] Anonymous, "Kerin calls for National Assault on Salt," *The Northern Times*, February 2, 1984.

[43] Brindal, "Irrigation Guru Retires," 13.

[44] Salinity Pilot Porgram Advisory Council, *Draft Shepparton Land and Water Salinity Management Plan*, 1.

[45] Salt Force, *Greening Victoria*, 5.

[46] Howes, "The Spectre at the Feast," 223.

These reflections on responsibility in the face of the previous and continued damage, perhaps inevitably, were about the European colonization of Australia. In Joan Kirner's words, it was "white settlement" that changed the environment "with devastating results."[47] While the appeal to ignorance did not necessarily imply declining the responsibility of the present generation, it often was exculpatory toward the pioneers. Alan Coad, who had been sent to the Tragowel Plains by the SRWSC in the early 1960s to investigate and solve the salinity problem, refused to question this legacy: "Let nobody criticise the pioneer farmers. They made possible the development of our nation."[48]

All of this was a far cry from the triumphant boasts of subjugating "useless nature" and turning it into a wonderland of towns, farms, and orchards, which had prevailed during the heyday of the development imperative. With widespread salinization and many other signs of environmental degradation after just 200 years of European settlement in Australia, the minds and moods were now considerably more sober.

Salinization sparked an equally inevitable and heated debate regarding the mismanagement of irrigation by government planners, engineers, and farmers on the Northern Plains. Given the rearguard position of the industry in the 1980s, this could not easily be separated from the dispute about the legitimacy of irrigation *per se*. Monash University's harsh verdict that the irrigation system had been "ill-conceived" deeply affected the people in the Kerang region.[49]

Some pointed their fingers at the SRWSC. A regional committee meeting of the MVL blamed it for its "lack of foresight," which had caused most of the salinity problems.[50] The Kerang Shire Council rejected the idea that irrigators be blamed for salinization. Rather, they placed responsibility on the governments and their departments in the early years of irrigation who had promoted unwise attitudes toward irrigation and land-clearing. Thus, the irrigator was only a pawn in a scheme to increase production and return a high level of export income to a developing country. In recent years, however, irrigators themselves had led

[47] Anonymous, "Greening Victoria Launched on World Environment Day," 8.

[48] Alan Coad, "The Salinity War – Progress and Problems," *Salt Force News* 10 (August–September 1988): 11; Russ, 113–16.

[49] Anonymous, "Irrigation is 'Ill-Conceived'," *The Northern Times*, February 10, 1984.

[50] Anonymous, "SR&WSC Under Fire," *The Northern Times*, February 28, 1984.

the way toward better practices.[51] In March 1988, a preliminary analysis by the RWC was obtained by the conservationist Des Thomas under the Freedom of Information Act. According to *The Northern Times*, it revealed that Commission's works were to blame for 50,000 tons of salt in the Torrumbarry Irrigation System. This was the result of the deepening of the Box and Pyramid Creeks.[52]

Even though many irrigators who wrote letters to the newspaper accepted that the system had been mismanaged and unsuitable soil types had been farmed with the wrong kinds of crops, they did not readily accept blame.[53] One letter by Drainage Conscious, whose name and address were withheld for good reason, showed himself "amazed at the waste of water in the latter half of the irrigation season." To them, the farmers were their own worst enemy.[54]

The only certain thing was that the situation was a disaster. KIRSAC's chairperson, Gordon Shire Councilor Russell Smith, summed up the status quo in 1986: those "installing irrigation systems did everything wrong as far as the environment was concerned." Natural drains had been blocked, and extra water had nowhere to go. Shires had built roads across the slopes with inadequate cross-drainage and no thoughts for drainage side cuts. The Water Commission had placed channels in sandy ridges, and channel outfalls discharged into water courses. Water application had been chaotic. Land layout had been bad or non-existent. Over-watering had been excessive. All of this had combined to an "environmental disaster." However, Smith said, it was "no use blaming anyone in particular—all have contributed to the situation."[55] The only way was forward.

[51] Anonymous, "Irrigators Not Cause of Salinity," *The Northern Times*, September 14, 1984.

[52] Anonymous, "50,000 Tonnes in Streams," *The Northern Times*, March 4, 1988.

[53] A. J. Mann; Donald A. Leed, "Irrigation Defended," *The Northern Times*, April 10, 1984; I. A. McDonald, "Academic Bias," *The Northern Times*, March 9, 1984.

[54] Anonymous, "Irrigators Wasteful," *The Northern Times*, July 17, 1984.

[55] Anonymous, "History Has its Salinity Lessons," *The Northern Times*, April 8, 1986.

The Way Forward

At the end of the twentieth century, the River Murray and the MDB were "in a state of ecological disaster."[56] The Murray was suffering from declining water quality due to salt, silt, fertilizers, and pesticides from agricultural and pastoral properties.[57] A spectacular blue-green algae bloom in the summer of 1991–1992, which extended along more than 1000 km of the Darling River, gave the ongoing deterioration international prominence. Water diversions continued to grow with previously untapped entitlements that were now being activated. As a response, an audit of water use in the Basin was commissioned. It found that under the level of development which had been reached in 1994, median annual flows from the MDB at the Murray Mouth were only at 28% of what they would have been under pre-development conditions. The percentages of years in which lower reaches of the Murray experienced drought had increased from 5 to 60%. Diversions had grown by 8% since the Salinity & Drainage Strategy had been introduced and were estimated to potentially increase even more in the future. Under the existing entitlements, in most years, more water was approved for diversion than was available. Overall, diversions had tripled in the previous 50 years.[58]

As a response, a cap on further expansion at 1994–95 levels of development was introduced, which aimed to establish a division of water between the environment and production that would be stable over time. Water trading, in combination with restrictions on water use in places where it could cause significant environmental harm, was seen as a useful means to redistribute water to areas where irrigation was of higher value and more environmentally benign.[59]

During this period, the focus of salinity research began to shift. Emphasis was increasingly placed on the influence of biophysical factors in wider catchments rather than just irrigation impacts on major streams. This was prompted by increasing attention to dryland salinity and the impact of dryland agriculture on the catchment's hydrological cycle.[60]

[56] Garden, *Australia, New Zealand and the Pacific*, 113.

[57] Garden, *Australia, New Zealand and the Pacific*, 113.

[58] Connell, *Water Politics*, 123–24.

[59] Connell, *Water Politics*, 124–25.

[60] Connell, *Water Politics*, 134–35.

Even though the Salinity & Drainage Strategy had probably gained the Basin a 20-year reprieve, it was estimated that, without substantial intervention, 30,000 to 50,000 km^2 of land in the Basin would become salinized in the following 100 years. The annual salt load exported to and through rivers would double. In the Lower Murray, the area was expected to increase by about 50% over 50 years. Further deterioration as a result of the increasing impact of salinization from the dryland parts of the catchment would occur.[61] Other visible signs, that "the cumulative level of development had gone beyond of what could be sustained in the long term," were species loss and an overall decline in the health of the Basin's rivers, floodplains, and wetlands.[62]

The MDBMC responded by tabling a new Basin Salinity Management Strategy in 2001. Designed to intercept saline groundwater and excessive revegetation, it was a highly ambitious strategy which, for the first time, covered the whole Basin with the inclusion of Queensland.[63]

Nonetheless, in Northern Victoria, the regional Salinity Management Plans that had been developed in the early 1990s had made decisive steps in the right direction. In 1992, irrigated pasture and cropping areas in the Kerang Lakes Area had been expected to have their annual farm production reduced by 25% in 30 years due to salinization. This would have reduced the value of farm gate production by $3 million per year. In 2002, the gross value of agricultural production was up by 100%. Soil salinity has been surveyed for in 340 km^2 of land. More than 500 whole farm plans had been completed that encompassed the identification of soils, design of channels, drains, laneways, tree belts, bays, and paddocks to achieve an effective irrigation system where water moved quickly over irrigation bays to be drained into a re-use system. Horticultural enterprises received help to convert from flood to micro-irrigation. Almost 90 km of farm drains and 24 re-use systems were constructed. About 46,000 trees were planted. An outfall channel for Lake Charm was completed in February 1997, designed to reduce salinity levels from 5000

[61] Murray-Darling Basin Ministerial Council, *Basin Salinity Management Strategy 2030 BSM2030* (Canberra: Murray-Darling Basin Authority, 2015), 3; Murray-Darling Basin Ministerial Council, *The Salinity Audit of the Murray-Darling Basin. A 100-Year Perspective, 1999* (Canberra: Murray-Darling Basin Commission, 1999), 37; Connell, *Water Politics*, 136.

[62] Murray-Darling Basin Authority, *Basin Plan Evaluation 2017*, 16.

[63] Connell, *Water Politics*, 140–42.

to 2500 EC in 15 years. A groundwater interception scheme for Pyramid Creek was begun, which would intercept 50,000 tons of salt entering the creek annually via groundwater intrusion. A reconfiguration of the Lake Tutchewop salt disposal was underway.[64]

On the Tragowel Plains, the area with a water table of less than a meter below the surface had dropped from 83 to 31%. This was aided by low average rainfall and irrigation deliveries in the late 1990s. Surveys for soil salinity were conducted on 35% of irrigated land. Over 400,000 trees were planted.[65] The Barr Creek Project, initiated in 1987 to decrease the salt loads in the Murray, had achieved a reduction of 4.2 EC in river salinity levels. The evaporative efficiency of the Tutchewop scheme was increased by reducing the amount of fresh off-farm drainage water entering Barr Creek through re-use systems, tree planting, and reducing supply channel leaks and outfall volumes.[66]

These steps were reason enough for cautious optimism. However, in view of the further deterioration of the Basin environment that was expected in the long term, there was a long way to go. In the words of Graeme David, the Manager of the Salinity Bureau, salinity control was "like digging a seemingly endless tunnel. The effort involved is enormous, and it may seem like there isn't an end."[67]

Then, the "independent dynamic," which had made a dramatic appearance in the early 1970s when Northern Victoria experienced widespread flooding and an escalation of salinization, appeared again. A veritable *deus ex machina*, the Millennium Drought completely changed the situation.[68]

[64] Kerang-Swan Hill Community Working Group, *Kerang–Swan Hill Salinity Management Plan 1993–2002* (Melbourne: Department of Natural Resources and Environment, 2002), 2–12.

[65] Julie Brookman and Deidre Stevens, *Tragowel Plains – The Success Story* (Melbourne: Department of Natural Resources and Environment, 2000), 16–17.

[66] Torrumbarry East of Loddon Community Implementation Group, *Turrumbarry East of Loddon Land and Water Management Strategy 1996–2002* (Kerang(?): Torrumbarry East of Loddon Project Team, 2002), 2–3.

[67] Anonymous, "Salinity Program 1985 to 1995—A Decade of Achievement," 5.

[68] Powell, *Bioregionalism*, 7.

Beginning in late 1996 and lasting to mid-2010 in southern Australia, this was "longest and most serious drought in Australian history."[69] In just a decade, the continent's sheep population halved. The particularly water-intensive rice and cotton production collapsed in some years. Irrigators and urban dwellers had to live with frequent water restrictions, leading to the introduction of creative re-use schemes and new industries of water conservation technology. By 2007, kangaroos searching for food and water had become a common sight in southern Australian cities.[70]

The "Big Dry" was particularly grave in the southeast and southwest of the continent, and it severely affected the MDB. Even though the drought prompted record levels of salinity in some areas, overall, it caused groundwater levels to fall to the extent that it reduced many salinity problems in the short term. It halted the spread of salinization in most of the worst-affected regions, especially in south-west Western Australia and Victoria. Still, large areas of New South Wales along the Great Dividing Range, in the Liverpool Plains, the Hunter Valley, and the Greater Sydney Region reported soil salinity as their main issue of concern in 2016.[71]

Furthermore, the mitigation schemes put into place, mostly since 1988, have contributed to the new situation, but it is not clear whether or not they have been the decisive factor. Recent reports from the MDBMC claimed a direct link between the schemes and the success in meeting the salinity targets, irrespective of climatic conditions.[72] In 2009–2010, the MDB's salinity target was reached for the first time, and is still being met at the time of writing. The four reporting sites on the Murray all achieved their targets between July 2014 and June 2019, but due to low flows and a lack of water availability, the target at Burtundy on the Darling has been missed over this period for, on average, almost half the time.[73]

[69] Matthew Heberger, "Australia's Millennium Drought: Impacts and Responses," in *The World's Water. The Biennial Report on Freshwater Resources*, ed. Peter H. Gleick, (Washington, DC: Island Press, 2011), 121.

[70] Heberger, "Australia's Millennium Drought," 97; 101–105.

[71] Australian State of the Environment Committee, *State of the Environment 2016. Land*, 73.

[72] Murray-Darling Basin Ministerial Council, *Basin Salinity Management Strategy 2030*, 3–4; Murray-Darlin Basin Authority, *Basin Salinity Management 2030. Summary Report 2015–16* (Canberra: Murray-Darlin Basin Authority, 2017), 2.

[73] Australian State of the Environment Commitee, *State of the Environment 2011*, 207; 211; Murray-Darling Basin Authority, *Basin Salinity Management 2030. Status Report*

Consequently, the issue has continuously lost its urgency. In 2001, the State of the Environment report warned that by then about 57,000 km^2 of land in Australia had salinity problems and by 2050 that number could rise to about 170,000.[74] The 2006 report still identified loss of vegetation caused by salinity as a major pressure on biodiversity, while the 2016 report did not offer an update on salinization at all.[75] With the end of national-level support for salinity management, the close watch on the issue also came to an end.[76] Accordingly, there is no reliable data on the current extent of the salinization. Dryland salinity occurrence has not been systematically remapped since the Millennium Drought. The best available estimate for Victoria is that about 2470 km^2 of soil were affected in 2013—2% of the total area devoted to dryland agriculture. There is no current data available on irrigation salinity.[77]

What does seem clear, however, is that with each return to wetter conditions, salinization is bound to increase.[78] This explains why the MDBA still considered it a "critical threat to agriculture in the Basin"

2015–16 (Canberra: Murray-Darlin Basin Authority, 2016), 3–4; Murray-Darlin Basin Authority, *Basin Salinity Management 2030. 2018–19 Comprehensive Report. March 2020* (Canberra: Murray-Darlin Basin Authority, 2020), 9–10; Murray-Darlin Basin Authority, *Basin Plan Annual Report 2018–19* (Canberra: Murray-Darlin Basin Authority, 2019), 27–28; Connell, *Water Politics*, 160; Murray-Darling Basin Authority, *2020 Basin Plan Evaluation*, 64–67.

[74] Australian State of the Environment Committee, *Australia State of the Environment 2001*, Canberra: Commonwealth of Australia, 2001), 53.

[75] Australian State of the Environment Committee, *Australia State of the Environment 2006* (Canberra: Commonwealth of Australia, 2006), 38; Australian State of the Environment Commitee, *State of the Environment 2016. Land*, 73.

[76] Australian State of the Environment Commitee, *Australia: State of the Environment 2011* (Canberra: Commonwealth of Australia, 2011), 211.

[77] Commissioner for Environmental Sustainability Victoria, *State of the Environment 2013*, 117; Commissioner for Environmental Sustainability Victoria, *Victorian State of the Environment 2018. Biodiversity (B) Scientific Assessments Part III*, (Melbourne: Government of Victoria, 2019), 226–28.

[78] Australian State of the Environment Commitee, *SoE 2016. Land*, 73; Murray-Darling Basin Authority, *Basin Plan Evaluation 2017*, 17; Commissioner for Environmental Sustainability Victoria, *State of the Environment 2013*, 117; Australian State of the Environment Commitee, *SoE 2011*, 208.

in 2017, even though it believed overall salinity management was successful.[79]

In the wake of a new decade, with water table levels having dramatically fallen and against the backdrop of mounting concerns about the availability of water in the horizon of climate change, which has been intensified by the Millennium Drought, salinization no longer is an urgent, high-profile environmental issue.[80] However, only the future will show whether this slow catastrophe is, indeed, under control.

[79] Murray-Darling Basin Authority, *Basin Plan Evaluation* 2017, 83; Murray-Darling Basin Authority, 2020 *Basin Plan Evaluation*, 66.

[80] Heberger, "Australia's Millennium Drought," 97.

CHAPTER 9

Conclusion

"Salinity control" may appear to be a rather presumptuous term. It evokes the same promise of security which had been implied in the post-war development that was imperative when Australian governments and engineers set out to complete the transformation of the River Murray into a regulated river. Escalating salinization was a by-product of the dialectic of this quest for environmental security. More water, stored and distributed in ever-greater masses, was mobilized for irrigation development. This was deemed necessary for ever-greater production, population growth, and new or expanding inland settlements. It turned into too much water.

The slow catastrophe, however, had started to unfold much earlier, only gaining broad attention during the drought of 1967–1968 when saline slugs in the Murray highlighted a previously mostly disregarded threat. From then on, salinization's priority steadily increased. Faced with declining production, degrading farmland, and decreasing living conditions in a deteriorating environment, the people in the rural communities of the GMID were under considerable stress. Salinization was complex. It spread inconsistently through the district, largely at unknown rates, sometimes appearing after years, sometimes decades. There were profound uncertainties about how to best tackle the problem, how great its effects

© The Author(s), under exclusive license to Springer Nature Switzerland AG 2023

D. Rothenburg, *Irrigation, Salinity, and Rural Communities in Australia's Murray-Darling Basin, 1945–2020*, Palgrave Studies in World Environmental History, https://doi.org/10.1007/978-3-031-18451-2_9

on production and incomes were, and how it affected the social well-being of the communities or the mental constitution of individuals. Decades of degradation disintegrated the fabric of smaller towns when farmers were forced to give up, and young people left the area. The associated stress and anxiety were unquantifiable. The GMID's natural environment was increasingly marked by the impacts of high salinities in streams and wetlands. However, salinization did not cause universal decline of the GMID's communities. It did, however, induce numerous transformations in the district and the whole MDB.

The people in the Kerang and Shepparton areas found that they were linked by their common lifeline—the River Murray. However, they were not only sharing water. Shepparton's more favorable upstream position, along with its economic importance and less saline environment, became the cause of fierce conflicts about rights to pollute the river, resources, and political priorities. With the metropolis of Adelaide all the way downstream receiving saline discharge from their areas, the communities found themselves responsible for the well-being of others in a shared environment. This, however, did not always result in cooperation to safeguard a common natural resource. Instead, Kerang and Shepparton became opposing political camps which fiercely defended their economic interests, united in their mutual determination to shield their access to water from urban consumers.

Cornered by an insidious threat, they organized themselves. The community-owned agricultural research farms in Kerang and Swan Hill were self-help efforts of learning to live in a saline environment. In the 1970s, the Salinity Action Groups added a political aspect to their efforts. Public servants, local government representatives, and farmers turned into activists. Grounded in the tradition of conservationism and closely aligned with governments, they persuaded politicians and government agencies of their case. Decades before urban environmental activists discovered the problems of the MDB, these community-based initiatives fought against the apathy and ignorance of many of their fellow citizens toward the peculiarities of their environment and the damage of ordinary day-to-day farming. A vibrant scene of rural activists thrived throughout the 1970s and 1980s. When in doubt, these activists were determined: dryland farmers from the Kerang region militantly opposed the state government's intention to build a series of evaporation basins throughout their farmland.

During the same period, the spirit of the Age of Ecology spread. Salinization became a significant factor for a shift away from an understanding of the Murray Valley as an unlimited resource toward one of quantitative restraints. It required stewardship. The history of these local initiatives, although inextricably linked to governments, is part of that paradigm shift around 1970, adding a decidedly rural perspective to it, which so far has rarely been examined.[1] The GMID's Salinity Action Groups, despite their heterogenous character, are usefully understood within the context of environmentalism. This, in turn, shows that the focus of Australian historiography needs to be widened beyond the narrow view of urban movements that correspond to the idea of environmentalism as civil society, campaigning to force governments to take environmental protection seriously. Rather, environmentalism must be understood as a hybrid in which state and civil society cannot easily be disentangled. In this respect, impulses from European histories of the environmental movement provide inspiration for new perspectives.[2] Not least, the Salinity Action Groups show their lasting significance as key actors in the previously untold pre-history of Landcare in Australia. Rob Youl acknowledges Ernest Jackson's contribution to Landcare's formation, he only considers Jackson as one of a few "dogged individuals." The adjective is fitting. However, Jackson, as this story has shown, was part of a movement that paved the way for the emergence of Landcare, and not just a solitary individual whose ideas were then adopted.[3]

[1] See Timothy Doyle, *Green Power. The Environment Movement in Australia* (Sydney: UNSW Press, 2000); Timothy Doyle and Sherilyn MacGregor, *Environmental Movements Around the World: Shades of Green in Politics and Culture* (Santa Barbara, California: ABC-CLIO, 2013); Hutton and Connors, *A History of the Australian Environment Movement*; Robin, *Little Desert*. There are two notable exceptions to this. J.M. Powell devotes some attention to Ernest Jackson as influential for the "emergence of bioregionalism" in the MDB. See Powell, *Bioregionalism*, 68. Peter Russ recognizes the role of the Salinity Action Groups and other community initatives in the context of the management reform processes related to the "Salt Action" program. See Russ, 140–43; 149–53; 163–72; 223–39. However, both authors do not understand these actors within the context of conservationism/environmentalism.

[2] Frank Uekötter, *The Greenest Nation? A New History of German Environmentalism*, Cambridge, MA: The MIT Press, 2014, 86–91; Radkau, *Ära der Ökologie*, 19–20; Frank Uekötter, "Myths, Big Myths and Global Environmentalism," in *The History of Social Movements in Global Perspective: A Survey*, ed. Stefan Berger and Holger Nehring (Basingstoke: Palgrave Macmillan, 2017), 419–23.

[3] Johnson et al., *Landcare in Australia*, 5.

However, notwithstanding the rise of ecological consciousness, the promise of development was never fundamentally put into question in the GMID's communities. Unfazed by the condemnation of irrigation as a cavalier industry in both economic and environmental terms, the irrigation dream was not easily abandoned. Community leaders valiantly defended irrigation's privileged position and their own access to the Murray's water. Many favored an engineering approach that would fix the deficient natural drainage system of the Murray Valley and thereby solve the salinity problem. The wish-image of a pipeline to the sea was the most obvious plea to continue the water dreamers' technological transformation of Australia's ecological conditions. With rural livelihoods and self-conceptions tied to irrigation farming, it was not accepted as just business, despite much evidence to the contrary.

These findings run contrary to Michael Cathcart's suggestion that the utopian impetus of water dreaming had been exhausted around 1970.[4] Its appeal continued among many of its main carriers—rural Australians—well into the 1990s. This also supports Richard Waterhouse's thesis that the idea of farming as a lifestyle lost its appeal in the decades after World War II.[5] True, the ideology of water dreaming, was, to some extent, a self-serving device to safeguard irrigator interests against the demands of sprawling metropolises and the environment. However, it should not be discarded as rhetoric. On the contrary, that it was consistently invoked may help explain the emotionally charged debates about irrigation in contemporary Australia beyond pure economic terms and resource conflicts. Where access to water is negotiated, the old divide between city and country is never far away.

By the mid-1980s, it became obvious that there would be no large-scale technological fix for salinity control, requiring a more nuanced approach. This meant a farewell to dreams about permanently transforming the Murray Valley's ecosystems for irrigation purposes. This, however, was only partly the result of spreading environmental awareness and a growing skepticism about technological interventions into ecological systems. The pillars of High Modernism, a belief in progress and the limitless transformation of nature, stood strong *despite* decades of environmental catastrophe. Rather, the grand designs to perfect Northern

[4] Cathcart, *The Water Dreamers*, 246–59.

[5] Waterhouse, *The Vision Splendid*, 209.

Victoria's irrigation environment failed because they were impossible to politically enforce against a multitude of interests unless decision-makers were willing to pay a high political price. Furthermore, the costs involved were deemed prohibitive when weighed against the economic gains in a time when irrigation was no longer seen as a beneficent undertaking in itself. The only solution appeared to be to continue living with salinity in a dynamic natural environment, adapting with a combination of agricultural and incremental technical approaches. The promise of environmental security remained unrealized.

Through the combined forces of Neoliberalism and sustainability, a political constellation developed in which salinization became a decisive factor for wide-ranging reforms in water management in the MDB. As such, the present history is also a part of another shift—the end of the Australian Settlement and the implementation of neoliberal policies as part of a worldwide trend that began after the end of the Long Boom. However, contrary to what James Walter has argued, the passionate rural opposition to water market reform in the 1980s suggests that Neoliberalism did not fill an intellectual vacuum in Australian politics. Rather, it seems the passing world of protectionism did not go quietly or easily.[6]

The coincidence of neoliberal adjustment and the rise of sustainability, which salinization fused, in turn raises further questions. Firstly, one may inquire whether there is a systematic link between neoliberal ideas and the idea of sustainability. The emerging hegemony of Neoliberalism in the West, which entailed privatization, economization, and the creation of markets, coincided with the rise of environmental politics, environmental economics, and international forums and institutions to coordinate efforts at environmental protection. New instruments were developed to bring economic categories into environmental management, such as indicating the ecological effects of production in prices, taxes, fees, and pollution certificates. The idea of ecosystem services aimed to quantify the manifold uses of the environment for human society. This changing relationship of economy and ecology during this period begs further investigation.[7]

[6] Walter, *What Were They Thinking*, 254–55.

[7] Rüdiger Graf, "Die Ökonomisierung der Umwelt und die Ökologisierung der Wirtschaft seit den 1970er Jahren," in *Ökonomisierung. Debatten und Praktiken in der Zeitgeschichte*, ed. Rüdiger Graf (Göttingen: Wallstein, 2019), 189–99; McCormick, *Global Environmental Movement*, 107–78; Frank Zelko, "The Politics of Nature," in *The Oxford*

Secondly, the impact of neoliberal policies on the environment must be examined further. David Harvey has observed that the era of neoliberalization coincides with an accelerating negative trend of environmental degradation since about 1970. Therefore, he regards the unrestrained application of neoliberal principles as the cause of this environmental loss, suggesting that protection can only be achieved by suspending these. Why, then, did environmental protection, during the same period, became a task *sui generis* for national governments and supranational organizations? The "environmental management state" with its national environment programs and authorities emerged at the same time as neoliberals forcefully (and successfully) demanded that states should curtail their interference into markets. Furthermore, Australia shows that the equation between neoliberal reform and environmental degradation is too simple. The damage caused by salinization in the MDB, for a long time, has been done under the premises of Keynesianist approaches, whereas Neoliberalism's balance sheet has not been fully examined, since the long-term effects of a fully emerged water market are still unclear.[8]

Overall, environmental history has much to gain from systematically reflecting on the role of political-economic regimes. In an age where carbon emissions trading is a central tool to reducing greenhouse gas emissions in the European Union, there is reason to question Christof Dipper's assumption that the relationship between the two central post-modern patterns of order, Neoliberalism and sustainability, is indeed a binary opposition, prompting a radical choice between hope and fear.[9]

The high-level move toward managing the MDB as a single unit, taken up in the 1980s, was also catalyzed by a growing awareness of salinization. Previously, it has been described as a top-down endeavor, instigated by state institutions, politicians, and experts.[10] The history of salinization in the GMID shows that it had its grassroots complement, the reform campaigns by the community activists. However, they fiercely opposed

Handbook of Environmental History, ed. Andrew C. Isenberg (Oxford: Oxford University Press, 2014), 718.

[8] Harvey, *A Brief History of Neoliberalism*, 172–75.

[9] For an outline, see "EU Emissions Trading System (EU ETS)," European Commission, accessed June 3, 2022, https://ec.europa.eu/clima/eu-action/eu-emissions-trading-system-eu-ets_en; Dipper, "Docupedia-Zeitgeschichte: Moderne," 20.

[10] See Henderson, "Dams, Pumps, Pipes and Dreams;" Connell, *Water Politics*; Powell, *Bioregionalism*; Pigram, *Australia's Water Resources*; Sinclair, *The Murray*.

the related market-based reforms undertaken in the name of efficiency and adaptability, denying that these would improve economic and environmental outcomes. Yet, market environmentalism ultimately proved to be the stronger force.[11]

The Victorian government's long-term strategy "Salt Action – Joint Action" was the final keystone at the end of the tumultuous period between 1968 and 1988. While it undoubtedly reflected the urban majority's changing attitudes toward the environment and development policies, it would not have occurred without the decades of self-help and political campaigning by the rural communities affected by salinization. The program added fresh impetus, resources, and political will to the old search for sustainable land use in inland Australia.

The program's initial successes, however, have been eclipsed by the onset of the Millennium Drought in the mid-1990s. Since then, salinization has decreased in extent and severity and lost much of its urgency as a public issue in Australia. While its environmental legacies are still visible and the water management institutions which emerged from it continue to operate, the local pillars of community self-help that it spawned have been consigned to history. In the twenty-first century, the GMID and the MDB face other challenges.

[11] Sue Jackson and Leslie Head, "Australia's Mass Fish Kills as a Crisis of Modern Water: Understanding Hydrosocial Change in the Murray-Darling Basin," *Geoforum* 109 (2020): 54.

CHAPTER 10

Epilogue

In the winter of 2017, Lake Tutchewop lay tranquil on a mild afternoon. Some water birds were around. Saltbush, bead bush, and pig's face thrived in the former recreation reserve, now the site of a barely filled evaporation basin for saline drainage water. To think I had imagined its environment to be lifeless! Instead, it had been colonized by salt-tolerant vegetation. Glen, the irrigation officer who took me out here, dared me to try some of them. I ripped out a leaf from a batch of pig's face and took a bite—it tasted as salty as sea water.

As we toured the Kerang Lakes Area, Glen showed me how to map land using an EM38 meter in order to determine the electrical conductivity of the soil, and hence its salinity. He patiently answered all my questions, teaching with obvious joy. As we passed field after field on the Murray Valley Highway, he prompted me: "Saline or not?" More often than not, I replied, "Saline." With a bit of practice, I soon saw salinity everywhere. Many fields were salted out, and revegetated to prevent erosion and further salinization. "What do you do with them?", I asked. "Nothing," Glen replied. "This used to be heaven for Aboriginals," he

© The Author(s), under exclusive license to Springer Nature Switzerland AG 2023

D. Rothenburg, *Irrigation, Salinity, and Rural Communities in Australia's Murray-Darling Basin, 1945–2020*, Palgrave Studies in World Environmental History,
https://doi.org/10.1007/978-3-031-18451-2_10

Fig. 10.1 Lake Tutchewop, August 8, 2017 (*Source* Daniel Rothenburg. Licensed under a Creative Commons Attribution-NonCommercial 4.0 International license)

mused, "until we came and screwed everything up," though he used a slightly more colorful word than "screwed." It is "a haunted country"[1] (Figs. 10.1 and 10.2).

Everyone in the area had a story to tell as soon as I said the key word. "Salinity" seems to have become part of the local lore. It is engraved in biographies and formed memories. Meeting George and Maureen Hardwick who became farmers at Appin, and Elaine Jones, who still lived on the Model Farm, the family's home in Kerang, I could not help but feel a deep respect for these people who have dedicated decades of their lives to a cause which they found worthy. Sadly, Gyn Jones had passed away 17 years before the outsider came to Kerang. I did find him though. In

[1] Judith Wright, *Born of the Conquerors* (Canberra: Aboriginal Studies Press, 1991), 30.

Fig. 10.2 Dead trees from salinization near Bears Lagoon, Victoria, July 21, 2017 (*Source* Daniel Rothenburg. Licensed under a Creative Commons Attribution-NonCommercial 4.0 International license)

the State Offices building, now home to Agriculture Victoria, there was a plaque in his honor. To me, this book is very much their story (Fig. 10.3).

One could tell that the times had changed, too. “Salinity,” one farmer told me, “is a thing of the past.” And, indeed, what I saw in the Lakes Area were only the remnants of almost 100 years of slow catastrophe. While its signs can still be read in the landscape, other issues have since assumed priority.

Climate change has become the central concern about the future of the MDB’s environment and communities. As Royal Commissioner Bret Walker, in his 2019 report on the Basin, has made unequivocally clear, even though there is a range of projections about the future impact of anthropogenic climate change on the Basin, ultimately, the information is clear. He pointedly added that “[o]nly a change to the laws of physics will prevent the general direction of climate change projections eventuating. No political party, ideological persuasion, illogical position, nor

Fig. 10.3 Plaque for Gyn Jones at Agriculture Victoria, August 8, 2017 (*Source* Daniel Rothenburg. Licensed under a Creative Commons Attribution-NonCommercial 4.0 International license)

the attitude taken by the MDBA will, it is expected, change the laws of physics."[2]

In 2007, it was projected that increased daily average temperatures, particularly in the southern Basin, and thus increased aridity were likely. Evidence suggested that water availability would decline, resulting in less run-off both for the environment and human consumption. Dry conditions were likely to persist and intensify in the future.[3] In 2008, the CSIRO's Sustainable Yields Project research projected that, by 2030,

[2] Walker, *Royal Commission Report*, 245.

[3] Walker, *Royal Commission Report*, 255; PMSEIC Independent Working Group 2007, *Climate Change in Australia: Regional Impacts and Adaptation—Managing the Risk for Australia, Report prepared for the Prime Minister's Science, Engineering and Innovation Council (PMSEIC)* (Canberra: Prime Ministers' Science, Engineering and Innovation Council, 2007), 14.

water availability could decline by about 11% in the whole Basin, and by about 17% in its southern part, even though this is bound to vary substantially between regions. In drier years, it was estimated that diversions would fall by more than 10% in most of New South Wales, by 20% in the Murrumbidgee and Murray regions, and between 35 and 50% in Victoria. These projections, however, are likely underestimates.[4] CSIRO concluded that "ecological thresholds may be crossed and the resulting changes may well be largely irreversible."[5] Additionally, research indicates that the most severe impacts of climate change will be due to more extreme events, like drought, flood, heatwaves, and bushfires, rather than the gradual rise in temperature.[6] The Intergovernmental Panel on Climate Change, in its 2014 report, warned that, if future water predictions and scenarios of severe drying were realized, there would be a significant reduction in agricultural production in the Basin, even with comprehensive adaptations including "more efficient water use, allocation, and trading."[7]

Against this backdrop, the future availability of water in the MDB has become an even more pressing problem. In the past, there have been numerous attempts to recover water for the environment and reduce water use to sustainable levels. One of these was the Living Murray Initiative in 2004. This, however, proved to be insufficient. In 2007, the Commonwealth government took control of high-level policy for the Basin. A new Water Act was signed, and the MDBA was formed in 2008. It was tasked with implementing the Murray–Darling Basin Plan, introduced in 2012 to "achieve long-term sustainability by rebalancing water use to sustainable levels and introducing new measures to use water more efficiently and effectively."[8] At its core are Sustainable Diversion Limits on the average amount of water that can be extracted from the Basin. In

[4] Walker, *Royal Commission Report*, 255; Commonwealth Scientitic and Industrial Research Organisation, "Water Availability in the Murray–Darling Basin: A Report to the Australian Government, 2008–09," accessed August 16, 2022, http://hdl.handle.net/102.100.100/116603?index=1, 5; 8; 10; 26; 28; 34; 43; 45.

[5] CSIRO, 56.

[6] Walker, *Royal Commission Report*, 256.

[7] Andy Reisinger et al., "Australasia," in *Intergovernmental Panel on Climate Change, Climate Change 2014: Impacts, Adaptation, and Vulnerability*, eds. Christopher B. Field, Vicente R. Barros, David Jon Dokken et al. (Cambridge: Cambridge University Press, 2014), 1376.

[8] Murray–Darling Basin Authority, *Basin Plan Evaluation 2017*, 7.

other words, the Plan envisaged to achieve nothing less than "to reverse over a century of development-induced decline and protect the MDB from the predicted impacts of climate change."[9] This was backed by $5.8 billion for on-farm infrastructure projects and a further $3.2 billion for the purchase of water for the environment. The level of water extractions is allowed to vary from year to year depending on factors such as storage levels, expected inflows, groundwater levels, rates of recharge, and interception activities such as new forest plantations.[10]

Water trading, which had been fiercely opposed in the GMID's irrigation communities, is now a central, Basin-wide instrument to redistribute water to the environment and to maximize the social and economic benefits from water use. Daniel Connell, in an assessment of the policy, has argued that water markets have made it easier to move water back to the river in order to achieve environmental outcomes than in the past. This is because the leverage of opponents concerned about the impact of water leaving their district is now minimized compared to the previous, consensus-oriented administrative process. Now, all it takes is a buyer. Markets have provided new incentives to extract a maximum financial benefit from water, which has encouraged innovation in crops and irrigation practices. Trading has also helped reduce waterlogging and salinization. During the Millennium Drought, economic returns for irrigation activities stayed at around pre-drought levels because the producers willing to pay higher prices caused water to move from low- to high-value crops. However, the markets have also created new incentives to sell water that would otherwise not have been used and thus would have remained in the river.[11]

According to Connell, there have been significant benefits for the environment, at least in the short term. The government's buyback of licenses has been the first major success in returning water to the environment. In the name of sustainability, public money has been used to buy back water that had been handed out almost for free to irrigators in past decades. In 2020, more than 2.1 billion m^3 of water was held by the Basin's environmental water holders. While the environmental outcomes

[9] Daniel Connell, "Irrigation, Water Markets and Sustainability in Australia's Murray–Darling Basin," *Agriculture and Agricultural Science Procedia* 4 (2015): 133.

[10] Connell, "Irrigation, Water Markets, Sustainability," 134–36.

[11] Connell, "Irrigation, Water Markets, Sustainability," 138–39.

of market solutions are still unclear and subject to continued recalibration, the legacy of the old world of irrigation, when development had been reason of state, will stay with the MDB for a long time.[12]

Meanwhile, the old irrigation settlements in the southern Basin, such as those in the GMID, are in a profound process of transformation. The district has experienced extensive restructuring since the early 2000s due to the Millennium Drought, changes in the agricultural sector, and the availability of water. Low commodity prices for dairy, with a dramatic drop in milk prices in 2008 due to the Global Financial Crisis, as well as for wine grapes, which were made worse by high prices for grain feed, have placed increasing pressure on farm businesses. Since water is now a commodity, water shortage has led to allocations, temporarily or permanently, moving out of some parts of the district, particularly Pyramid Hill and Boort. As a result, a new dryland has developed in these areas. Furthermore, water allocations have increasingly been diverted away from mixed farming into dairy and horticulture. The combined forces of drought and rationalization of agricultural industries have led to a declining number of farms. While tourism is considered an opportunity along the River Murray and in the Kerang Lakes area, the sector will not be able to match the contribution of irrigated agriculture and is also partly very dependent on the availability of water.[13]

These developments have resulted in major shifts in the workforce throughout the region that reflect long-term trends of reduced labor demand related to improvements in productivity that have been felt throughout the MDB. In all communities, the agricultural workforce has decreased between 2001 and 2016 by 31% in Kerang–Cohuna and by 50% in Shepparton. This trend has been even more pronounced in irrigated agriculture, where the workforce has decreased by 43% in Kerang–Cohuna and by an enormous 61% in Shepparton over the same period.[14] Whereas the population of the large towns is growing, small towns are in decline,

[12] Connell, "Irrigation, Water Markets, Sustainability," 139; Murray–Darling Basin Authority, *2020 Basin Plan Evaluation*, xi; Neal Hughes, "Water Markets are not Perfect, but vital to the Future of the Murray–Darling Basin, The Conversation," March 1, 2021, last modified March 1, 2021, https://theconversation.com/water-markets-are-not-perfect-but-vital-to-the-future-of-the-murray-darling-basin-155880.

[13] Murray–Darling Basin Authority, *Guide to the Proposed Basin Plan*, 780–81; 805–06.

[14] Murray–Darling Basin Authority, *Kerang–Cohuna Community 2018*, 1; Murray–Darling Basin Authority, *Shepparton Irrigation Area Community 2018*, 1.

especially in the Gannawarra and Loddon shires. All communities have been structurally aging, and young people tend to leave the area.[15]

Given the unequivocal warnings of Royal Commissioner Bret Walker, it seems clear that climate change and unsustainable water use in the Basin will continue to add severely to these transformations unless forceful measures are taken. In an age of environmental insecurity, the MDB and the GMID are likely to face further pronounced change in the future.

[15] Murray–Darling Basin Authority, *Guide to the Proposed Basin Plan*, 786; 802; Murray–Darling Basin Authority, *Basin Plan Evaluation 2017*, 21.

Glossary

Measuring Salinity Salinity refers to the concentration of soluble salts in soil or water. All natural waters contain some dissolved salts such as sodium, magnesium, and calcium. Sodium chloride (table salt) is the most common of all the salts. It is the main constituent of seawater.

There are two main methods of determining the salt content of water: Electrical Conductivity (EC), and Total Dissolved Solids (TDS). EC is a measure of how much electrical current can flow through a sample of water, while TDS is a measure of the relative weight of dissolved materials in a sample of water.

EC is measured by passing an electric current between two metal plates (electrodes) in the water sample and measuring how readily current flows (conductance) between the plates. The more dissolved salt in the water, the stronger the current flow and the higher the EC. An EC meter measures conductance through a solution in a container of specific dimensions. The standard EC unit used by the Victorian Salinity Program and the Murray Darling Basin Authority is microSiemens per centimeter (μS/cm) at 25 degrees Celsius.

© The Editor(s) (if applicable) and The Author(s), under exclusive license to Springer Nature Switzerland AG 2023

D. Rothenburg, *Irrigation, Salinity, and Rural Communities in Australia's Murray-Darling Basin, 1945–2020*,
Palgrave Studies in World Environmental History,
https://doi.org/10.1007/978-3-031-18451-2

TDS is measured by evaporating a known volume of water to dryness, then weighing the solid residue remaining. It is recorded in milligrams of dissolved solids in one liter of water (mg/L) or parts per million (ppm); 1 ppm is equal to 1.56 μS/cm.[1]

[1] Agriculture Victoria, "Water Quality for Farm Water Supplies."

Bibliography

Sources Cited

Published Works

ACIL Australia, et al. *Causes, Extent and Effects of Salinity in Victoria*. Melbourne: Salinity Committee of the Victorian Parliament, 1983.

Australian Publicity Council. *Liquid Gold Australia*. Melbourne: Australian Publicity Council 1960(?).

Anonymous. "New Game Shooters' Association." *Australian Clay Target Shooting News* 12, no. 1 (1958): 9–11.

Anonymous. New Organisation Promote Irrigation. *Pyramid Hill Advertiser*, January 12, 1966.

Anonymous. "It's Called VIRPO." *Kyabram Free Press*, March 23, 1966.

Anonymous. *Irrigation: Water in Australia Book 3*. Crows Nest: The Coca-Cola Export Corporation, 1969(?).

Anonymous. "Pump Salt into the Sea!" *Herald*, October 23, 1978.

Anonymous. "Land Care Crusader 'Watershed' Jackson." *Trees and Natural Resources* 32, no. 4 (1990): 26–27.

Anonymous. "Our Hidden Memorials." *eSplash eMag* 59 (2013): 27.

Brady, Edwin J. *Australia Unlimited*. Melbourne: G. Robertson, 1918.

Bromfield, Louis. *Out of the Earth*. New York: Harper & Brothers, 1950.

Brookman, Julie, and Deidre Stevens. *Tragowel Plains—The Success Story*. Melbourne: Department of Natural Resources and Environment, 2000.

© The Editor(s) (if applicable) and The Author(s), under exclusive license to Springer Nature Switzerland AG 2023

D. Rothenburg, *Irrigation, Salinity, and Rural Communities in Australia's Murray-Darling Basin, 1945–2020*, Palgrave Studies in World Environmental History, https://doi.org/10.1007/978-3-031-18451-2

Brundtland, Gro Harlem, ed. *Our Common Future: The World Commission on Environment and Development*. Oxford: World Commission on Environment and Development, 1987.

Caffrey, Hugh W. *Progress Report No. 4 from the Parliamentary Public Works Committee on the Salinity Control and Drainage Inquiry with Particular Reference to the Barr Creek Aspect*. Melbourne: Salinity Committee of the Victorian Parliament, 1980.

Cameron, R. J. *Official Year Book of Australia. No. 61, 1975 and 1976*. Canberra: Australian Bureau of Statistics, 1976.

Campbell, K. O. "An Assessment of the Case for Irrigation Development in Australia." In *Water Resource Use and Management: Proceedings of a Symposium held at Canberra by the Australian Academy of Science, 9–13 September 1963*, edited by the Australian Academy of Science, 450–457. Melbourne: Australian Academy of Science, 1964.

Cary, John, and Neil Barr. "The Kerang Lakes Area Management Project. Social Profile." In *Kerang Lakes Area Draft Salinity Management Plan: Background Papers Volume 1 of 2*, edited by Salt Action Victoria. Melbourne: Salt Action, 1989.

Centre of Policy Studies, Monash University. *Distribution of Costs and Benefits of Victoria's Irrigation Systems: A Report*. Melbourne: Public Bodies Review Committee of the Parliament of Victoria, 1983.

Clarke, Ernest E. (Mick). *River in Danger*. Albury-Wodonga: Border Morning Mail, 1981.

Commonwealth Scientitic and Industrial Research Organisation. "Water Availability in the Murray-Darling Basin: A Report to the Australian Government, 2008–09." Accessed August 16, 2022. http://hdl.handle.net/102.100.100/116603?index=1.

Deakin, Alfred. "Introducing the Water Supply and Irrigation Bill." *Victoria: Parliamentary Debates* 51 (1886): 415–447.

Davidson, Bruce R. *Australia Wet or Dry? The Physical and Economic Limits to the Expansion of Irrigation*. Melbourne: Bruce Robinson, 1969.

Davidson, Bruce R. *The Northern Myth: A Study of the Physical and Economic Limits to Agricultural and Pastoral Development in Tropical Australia*. Melbourne: Bruce Robinson, 1972.

Department of Agriculture. *Victoria: Soils and Land Use Near Swan Hill*. Victoria. Melbourne, 1966.

Dunk, W. P. *Report on Goulburn-Murray Irrigation District Water Tables and Salting: Part III, Causes and Prevention of Salting in the Goulburn-Murray Irrigation District*. Melbourne: State Rivers and Water Supply Commission, 1961.

Dwyer Leslie Pty. Ltd. "Kerang Lakes Area Working Group. Regional Economic Conditions." In *Kerang Lakes Area Draft Salinity Management Plan: Background Papers Volume 1 of 2*, edited by Salt Action Victoria. Melbourne: Salt Action, 1990.

Dwyer Leslie Pty. Ltd., and Maunsell and Partners Pty. Ltd. *Salinity Control in Northern Victoria: A Strategic Study for the Salinity Committee of the Victorian Parliament*. Melbourne: Salinity Committee of the Victorian Parliament, 1984.

Fairbairn, David. "Victoria Grant (River Murray Salinity) Bill 1968. Second Reading. Speech. Wednesday, 1 May 1968." In *Parliamentary Debates: House of Representatives*, edited by the Commonwealth of Australia.

Goldsmith, Edward, and Robert Allen. *A Blueprint for Survival*. London: Tom Stacey, 1972.

Government of Victoria. *Protecting the Environment: A Conservation Strategy for Victoria*. Melbourne: Government Printer, 1987.

Government of Victoria. *Salt Action: Joint Action: Victoria's Strategy for Managing Land and Water Salinity*. Melbourne: Salt Force, 1987.

Gutteridge, Haskins & Davey. *Murray Valley Salinity Investigation: A Summary*. Canberra: River Murray Commission, 1970.

Gutteridge, Haskins & Davey. *Murray Valley Salinity Investigation: Volume 1: The Report*. Canberra: River Murray Commission, 1970.

Gutteridge, Haskins & Davey. *Murray Valley Salinity Investigation: Volume 2: Maps*. Canberra: River Murray Commission, 1970.

Gutteridge, Haskins & Davey, et al. *The Application of Salinity Control Techniques in Victoria*. Melbourne: Salinity Committee of the Victorian Parliament, 1983.

Green, T. B. *A Country Legal Practice*. Carlton: Koestler Press, 1988.

Hill, Ernestine. *Water into Gold*. Melbourne: Robertson & Mullins, 1937.

International Union for the Conservation of Nature, et al. *World Conservation Strategy: Living Resource Conservation for Sustainable Development*. Gland: International Union for the Conservation of Nature, 1980.

Jackson, Ernest M. *Replenish the Earth: Living Resource Conservation for Sustainable Development*. Albury: Catchment Education Trust, 1982.

Johnstone, John Glass, et al. *Final Report from the Royal Commission on Closer Settlement as to the Working of the Closer Settlement Acts in the Irrigable Districts and a General Review of the Finances of Closer Settlement together with Appendices*. Melbourne: Government Printer, 1916.

Jones, Gynlais O. *Kerang Irrigation Region—Problems, Potential, Productivity*. Kerang: Salt Force, 1988.

Jones, Gynlais O., et al. *Fifty Years of Achievement. Department of Agriculture and Rural Affairs Kerang 1989*. Kerang: Department of Agriculture and Rural Affairs, 1989.

Kerang-Swan Hill Community Working Group. *Kerang—Swan Hill Salinity Management Plan 1993–2002*. Melbourne: Department of Natural Resources and Environment, 2002.

Kerang Lakes Area Working Group. *Communities in Partnership: Kerang Lakes Area Draft Salinity Management Plan*. Kerang: Kerang Lakes Area Working Group, 1992.

Leete, Frances. *Life, Wealth and Power: The Story of Water in Australia*. Crows Nest: The Coca Cola Export Corporation, 1966(?).

Lubezenko, Vera. *Saltwatch: Involve Me and I'll Understand: A Resource Package for Schools*. Melbourne: Salt Action, 1992.

Mitchell, Thomas L. "Account of the Recent Exploring Expedition to the Interior of Australia." *Journal of the Royal Geographical Society of London* 7 (1837): 271–285.

Mitchell, Thomas L. *Three Expeditions into the Interior of Eastern Australia, with Descriptions of the Recently Explored Region of Australia Felix, and of the Present Colony of New South Wales, Second Edition, Carefully Revised*. 2 Vols. London: T & W Boone, 1839.

Moore, J. W., and A.M.I.E. Aust. *Report on Goulburn-Murray Irrigation District Water Tables and Salting: Part II, Drainage Requirements due to Irrigation Development*. Melbourne: State Rivers and Water Supply Commission, 1961.

Pels, Simon. *A Plan to Control Salt Problems in the Murray River: Report Prepared by S. Pels for Submission to the Winston Churchill Memorial Trust, September 1967*. Deniliquin: Water Conservation and Irrigation Commission, 1967.

PMSEIC Independent Working Group 2007. *Climate Change in Australia: Regional Impacts and Adaptation—Managing the Risk for Australia: Report Prepared for the Prime Minister's Science, Engineering and Innovation Council (PMSEIC)*. Canberra: Prime Ministers' Science, Engineering and Innovation Council, 2007.

Public Bodies Review Committee. *Twelfth Report to the Parliament: Future Structures for Water Management: Volume 4: Final Report: Irrigation and Water Resource Management*. Melbourne: Government Printer, 1984.

Randall, Alan. "Resource Allocation in a Maturing Water Economy." In *Irrigation Water: Policies for Its Allocation in Australia: Papers from a Workshop. University of New England*, edited by the Australian Rural Adjustment Unit. Armidale: Australian Rural Adjustment Unit, 1982.

Reisinger, Andy, et al. "Australasia." In *Intergovernmental Panel on Climate Change: Climate Change 2014: Impacts, Adaptation, and Vulnerability*, edited by Christopher B. Field, Vicente R. Barros, David Jon Dokken, et al., 1371–1438. Cambridge: Cambridge University Press, 2014.

River Murray Working Party. *Report to Steering Committee of Ministers: October 1975*. Canberra: Australian Government Publication Service, 1977.

Robertson, A. G. *Report on Goulburn Murray Irrigation District Inquiry*. Melbourne: State Rivers and Water Supply Commission, 1977.
Rural Water Commission, Victoria, et al. *Salting Threatens Victoria*. Melbourne: Salt Force, 1988.
Salinity Committee of the Victorian Parliament. *Second Report to the Parliament: The Activities of the Salinity Committee*. Melbourne: Salinity Committee of the Victorian Parliament, 1983.
Salinity Committee of the Victorian Parliament. *Community Conference on Salinity Control: Bendigo College of Advanced Education: 17–18 February 1984*. Melbourne: Salinity Committee of the Victorian Parliament, 1984.
Salinity Committee of the Victorian Parliament. *Third Report to the Parliament: Salt of the Earth: Final Report on the Causes, Effects and Control of Land and River Salinity in Victoria*. Melbourne: Salinity Committee of the Victorian Parliament, 1984.
Salinity Committee of the Victorian Parliament. *Fourth Report to the Parliament: Final Report on Water Allocations in Northern Victoria*. Melbourne: Salinity Committee of the Victorian Parliament, 1984.
Salinity Committee of the Victorian Parliament. *Fourth Report to the Parliament: Final Report on Water Allocations in Northern Victoria*. Melbourne: Salinity Committee of the Victorian Parliament, 1984.
Salt Force. *Greening Victoria*. Seymour: Salt Force, 1988.
Soil Conservation Authority. "Soil Salting of Non-Irrigated Soils." *S.C.A. Farmers' Bulletin* 10 (1958).
Salinity Pilot Program Advisory Council. *Draft Shepparton Land and Water Salinity Management Plan*. Shepparton: Salinity Pilot Program Advisory Council, 1989.
State Rivers and Water Supply Commission. *Fifty-Second Annual Report: 1956–57*. Melbourne: Government Printer, 1957.
State Rivers and Water Supply Commission. *Fifty-Third Annual Report: 1957–58*. Melbourne: Government Printer, 1958.
State Rivers and Water Supply Commission. *Fifty-Fourth Annual Report: 1958–59*. Melbourne: Government Printer, 1959.
State Rivers and Water Supply Commission. *Fifty-Sixth Annual Report: 1960–61*. Melbourne: Government Printer, 1961.
State Rivers and Water Supply Commission. *Fifty-Eight Annual Report: 1962–63*. Melbourne: Government Printer, 1963.
State Rivers and Water Supply Commission. *Fifty-Ninth Annual Report: 1963–64*. Melbourne: Government Printer, 1964.
State Rivers and Water Supply Commission. *Sixty-First Annual Report: 1965–66*. Melbourne: Government Printer, 1965.
State Rivers and Water Supply Commission. *Sixty-Second Annual Report: 1966–67*. Melbourne: Government Printer, 1967.

State Rivers and Water Supply Commission. *Sixty-Third Annual Report: 1967–68*. Melbourne: Government Printer, 1968.

State Rivers and Water Supply Commission. *67th Annual Report: 1972–73*. Melbourne: Government Printer, 1974.

State Rivers and Water Supply Commission. *68th Annual Report: 1973–74*. Melbourne: Government Printer, 1974.

State Rivers and Water Supply Commission. *70th Annual Report: 1974–75*. Melbourne: Government Printer, 1975.

State Rivers and Water Supply Commission. *The Development of Victoria Depends on Water*. Melbourne: State Rivers and Water Supply Commission, 1964.

State Rivers and Water Supply Commission. *Salinity Control and Drainage: A Strategy for Northern Victorian Irrigation and River Murray Quality*. Melbourne: State Rivers and Water Supply Commission, 1975.

Stretton, Leonard Edward Bishop. *Report of the Royal Commission to Inquire into Forest Grazing Together with Minutes of Evidence*. Melbourne: Government Printer, 1946.

Tisdall, A. L. "Irrigation in Victoria." In *Water Resource Use and Management. Proceedings of a Symposium held at Canberra by the Australian Academy of Science, 9–13 September 1963*, edited by the Australian Academy of Science, 458–464. Melbourne: Australian Academy of Science, 1964.

Torrumbarry East of Loddon Community Implementation Group. *Turrumbarry East of Loddon Land and Water Management Strategy 1996–2002*. Kerang(?): Torrumbarry East of Loddon Project Team, 2002.

Turnbull, James, et al. *Report of the Royal Commission on Soldier Settlement Together with Appendices*. Melbourne: Government Printer, 1925.

Webster, A. *Report on Goulburn-Murray Irrigation District Water Tables and Salting. Part V, Current Investigations on the Control of Underground Water and Land Reclamation*. Melbourne: State Rivers and Water Supply Commission, 1963.

Wood, W. E. "Increase of Salt in Soil and Streams Following the Destruction of the Native Vegetation." *Journal of the Royal Society of Western Australia* 10, no. 7 (1924): 35–47.

Archival Sources

Public Record Office Victoria, Melbourne

VPRS 241/P0000, pp. 64–131, Salinity Control and Drainage (Joint Select Committee); Kerang; Public Works Committee.

VPRS 6008/P0006, File No.57/20441 / Pt.0001, [UNTITLED—KERANG AGRICULTURAL RESEARCH FARM] * *.

VPRS 6008/P0001, 66/0245, VICTORIAN IRRIGATORS RESEARCH AND PROMOTION ORGANISATION.
VPRS 6008/P0001, 68/0629, V.I.R.P.O.—VICTORIAN IRRIGATORS RESEARCH & PROMOTION ORGANISATION.
VPRS 6008/P0001, 68/1934, OPENING CEREMONY OF V.I.R.P.O. IRRIGATION CONVENTION—SHEPPARTON.
VPRS 6008/P0006, File No.57/20441/Pt.0001, [UNTITLED—KERANG AGRICULTURAL RESEARCH FARM] * *.
VPRS 6008/P0006, File No.74/20650/Pt.0001, GOULBURN IRRIGATION REGION DRAINAGE ACTION COMMITTEE.
VPRS 6008/P0001, 74/2225, V.I.R.P.O.—GEN.
VPRS 6008/P0006, File No.76/0693 / PT.0001, KERANG IRRIGATION REGION SALINITY ACTION COMMITTEE (K.I.R.S.A.C).
VPRS 6008/P0001, 80/07411, VICTORIAN IRRIGATION RESEARCH & PROMOTION ORGANISATION.
VPRS 6008/P0001, 84/36649, VICTORIAN IRRIGATION RESEARCH AND PROMOTION ORGANISATION.
VPRS 11493/P0001, 182/1, G.I.R.D.A.C.—General 83–93.
VPRS 11559/P0001, 24/07/27, THE SALINITY CONTROL & DRAINAGE INQUIRY PT 1.
VPRS 11559/P0001, 24/07/27, THE SALINITY CONTROL & DRAINAGE INQUIRY PT 2.
VPRS 11559/P0001, 24/07/27, THE SALINITY CONTROL & DRAINAGE INQUIRY PART 4.
VPRS 11559/P0001, 24/07/27, THE SALINITY CONTROL & DRAINAGE INQUIRY PART 5.
VPRS 11559/P0001, 24/07/27, THE SALINITY CONTROL & DRAINAGE INQUIRY PART 7.

National Archives of Australia, Canberra

A571, 1968/2238, National Water Resources Development Programme—River Murray Salinity Reduction Projects.
A1690, A1969/742, River Murray Commission—Study of Salinity in the River Murray.
A1960, 1973/774, Sunraysia and Kerang Salinity Reduction Scheme—VIC NWRDP—Department of National Development.
AA1976/642, 192, River Murray Commission—Policy Part V.
AA1976/642, 198, Salinity of Murray River—Part I.

Agriculture Victoria, Kerang

Kerang Agricultural Research Farm, Archive.

Swan Hill Regional Library, Swan Hill

Swan Hill Irrigators Resarch Farm, Collection of Annual Reports and Historical Documents by Swan Hill Irrigators' Research Farm Inc: Including Minutes of General Meetings from 1954 Until 1998/[manuscript].

Swan Hill Irrigators Resarch Farm, Swan Hill Irrigators' Research Farm at Tyntynder: A Selection of Historical Documents 1965–1993/[CD-ROM].

Victorian Irrigation Research and Promotion Organisation, Replenishing the Earth: Salinity & Water Supply Resource Kit 1985/Produced by Victorian Irrigation Research & Promotion Organisation [Book].

Newspaper and Magazine Articles

The Age (Melbourne), 1956, 1963–1964, 1967–1968, 1982, 1985–1988, 1990, State Library of Victoria, Melbourne.

The Guardian (Swan Hill), 1954, 1973–1974, Swan Hill Regional Library, Swan Hill.

The Kerang New Times (Kerang), 1945, 1956, 1968, 1970–71, State Library of Victoria, Melbourne.

Northern Times (Kerang), 1974–1978, 1980–1986, 1988, 1991, 1993, State Library of Victoria, Melbourne.

Gannawarra Times (Kerang), 2017, State Library of Victoria, Melbourne.

The Riverlander (Albury), 1970–1972, 1974–1981, 1983–1985, National Library of Australia, Canberra.

Salt Force News (Melbourne), 1986–1996, National Library of Australia, Canberra.

Private Documents

George and Maureen Hardwick, Appin.

Elaine Jones, Kerang.

Works Cited

Agriculture Victoria. "Water Quality for Farm Water Supplies." Last modified June 9, 2020. https://agriculture.vic.gov.au/farm-management/water/managing-dams/water-quality-for-farm-water-supplies.

Anonymous. "Pollutant, Pollution." In *A Dictionary of Ecology*, 4th ed., edited by Michael Allaby, 300. Oxford: Oxford University Press, 2010.

Anonymous. "Sharman Stone: Liberal MP for Murray Retires from Politics After 20 Years." *ABC News*. Last modified March 26, 2016. https://www.abc.net.au/news/2016-03-26/sharman-stone-murray-victoria-standing-down-from-politics/7277610.

Anonymous. "Explosive Four Corners Investigation into the Barwon-Darling River Systems Must Trigger ICAC Investigation." Australian Conservation Foundation. Last modified July 25, 2017. https://www.acf.org.au/barwon_darling_icac_investigation.

Anonymous. "District Profile. Goulburn-Murray Irrigation District." *Irrigation Leader Magazine*. Accessed June 10, 2022. https://irrigationleadermagazine.com/goulburn-murray-irrigation-district.

Anonymous. "Pumped: Who's Benefiting from the Billions Spent on the Murray-Darling?" *ABC TV*. Accesed 10 June 2022. http://www.abc.net.au/4corners/pumped/8727826.

Australian Bureau of Statistics. "Australian Social Trends 2012: Australian Farming and Farmers." Last modified December 12, 2012. http://www.abs.gov.au/AUSSTATS/abs@.nsf/Lookup/4102.0Main+Features10Dec+2012.

Australian Bureau of Statistics. "Water Account, Australia, 2012–13." Last modified November 27, 2014. http://www.abs.gov.au/AUSSTATS/abs@.nsf/Lookup/4610.0Main+Features12012-13?OpenDocument.

Australian Bureau of Statistics. "Water Account, Australia, 2015–16." Last modified November 23, 2017. http://www.abs.gov.au/ausstats/abs@.nsf/mf/4610.0.

Australian Bureau of Statistics. "2016 Census Quick Stats: Kerang." Accessed June 10, 2022. http://www.censusdata.abs.gov.au/census_services/getproduct/census/2016/quickstat/SSC21320.

Australian State of the Environment Committee. *Australia State of the Environment 2001.* Canberra: Commonwealth of Australia, 2001.

Australian State of the Environment Committee. *Australia State of the Environment 2006.* Canberra: Commonwealth of Australia, 2006.

Australian State of the Environment Committee. *Australia: State of the Environment 2011.* Canberra: Commonwealth of Australia, 2011.

Australian State of the Environment Committee. *State of the Environment 2016: Inland Water*. Canberra: Commonwealth of Australia, 2016.

Australian State of the Environment Committee. *State of the Environment 2016: Land*. Canberra: Commonwealth of Australia, 2016.

Australian State of the Environment Committee. *State of the Environment 2016: Overview*. Canberra: Commonwealth of Australia, 2016.

Abott, Ian. "Aboriginal Fire Regimes in South-West Western Australia: Evidence from Historical Documents." In *Fire in Ecosystems of South-West Western*

Australia: Impacts and Management, edited by Ian Abbott and Neil Burrows, 119–146. Leiden: Backhuys Publishers, 2003.

Aitkin, Don. "'Countrymindedness': The Spread of an Idea." *Australian Cultural History* 4 (1985): 34–41.

Aitkin, Don. "Return to 'Countrymindedness'." In *Struggle Country: The Rural Ideal in Twentieth Century Australia*, edited by Graeme Davison, Marc Brodie, 11.11–11.16. Melbourne: Monash University Press, 2005.

Arnold, V. H. *Victorian Year Book 1965*. Melbourne: Commonwealth Bureau of Census and Statistics, Victorian Office, 1965.

Arnold, V. H. *Victorian Year Book 1968*. Melbourne: Commonwealth Bureau of Census and Statistics, Victorian Office, 1968.

Arthur, Jay Mary. *The Default Country*. Sydney: Unversity of New South Wales Press, 2003.

Ballinger, Robyn. *An Inch of Rain: A Water History of Northern Victoria*. North Melbourne: Australian Scholarly, 2012.

Barney, Gerald O. *The Global 2000 [Two Thousand] Report to the President of the U.S.—Entering the 21st Century: A Report*. New York: Pergamon Press, 1980.

Barr, Neil. *The House on the Hill: The Transformation of Australia's Farming Communities*. Canberra: Land & Water Australia, 2009.

Barr, Neil, and John Cary. *Greening a Brown Land: The Australian Search for Sustainable Land Use*. Melbourne: Macmillan Education Australia, 1992.

Benjamin, Walter. "Theses on the Philosophy of History." In *Illuminations*, translated by Harry Zohn, edited by Hannah Arendt, 253–264. New York: Schocken Books, 2007.

Beresford, Quentin. "Developmentalism and Its Environmental Legacy: The Western Australian Wheatbelt 1900–1990s." *Australian Journal of Politics and History* 47, no. 3 (2001): 403–415.

Beresford, Quentin et al. *The Salinity Crisis. Landscapes, Communities and Politics*. Crawley: University of Western Australia Press, 2004.

The Holy Bible. King James Version. Dallas: Brown Books Publishing, 2004.

Bonneuil, Christophe, and Jean-Baptiste Fressoz. *The Shock of the Anthropocene: The Earth, History and Us*. London: Verso, 2016.

Botteril, Linda. "From Black Jack McEwen to the Cairns Group: Reform in Australian Agricultural Policy." *National Europe Centre Paper No. 86*, Australian National University, 2003.

Bowden, N. *Victorian Year Book 1978*. Melbourne: Commonwealth Bureau of Census and Statistics, Victorian Office, 1978.

Bowler, J. M. "Environmental and Salinity History of the Murray Basin in the Last 500,000 Years." In *Abstracts Murray Basin 88 Conference, Canberra, 23–26 May 1988*, edited by Bureau of Mineral Resources Department of Primary Industries & Energy, Geology & Geophysics, 19–20. Canberra: Department of Primary Industries & Energy, Bureau of Mineral Resources, Geology & Geophysics, 1988.

Broome, Richard. *Aboriginal Victorians: A History since 1800*. Sydney: Allen & Unwin, 2005.

Broome. Richard, et al. *Mallee Country: Land, People, History*. Melbourne: Monash University Publishing, 2019.

Brown, Sarah, et al. "Can Environmental History Save The World?" *History Australia* 5, no. 1 (2008): 3.1–3.24.

Butlin, Noel. *Our Original Aggression: Aboriginal Populations of Southeastern Australia 1788–1850*. Sydney: Allen & Unwin, 1983.

Butzer, Karl W., and David M. Helgren. "Livestock, Land Cover, and Environmental History: The Tablelands of New South Wales, Australia, 1820–1920." *Annals of the Association of American Geographers* 95, no. 1 (2005): 97–99.

Callahan, North. *TVA. Bridge Over Troubled Waters*. South Brunswick: A. S. Barnes, 1980.

Callison, William, and Zachary Manfredi. "Introduction: Theorizing Mutant Neoliberalism." In *Mutant Neoliberalism: Market Rule und Political Rupture*, edited by William Callison and Zachary Manfredi, 1–37. New York: Fordham University Press, 2019.

Campbell, Andrew. *Landcare: Communities Shaping the Land and the Future: With Case Studies by Greg Siepen*. St Leonards: Allen & Unwin, 1994.

Campbell, Judy. *Invisible Invaders: Smallpox and Other Diseases in Aboriginal Australia 1780–1880*. Melbourne: Melbourne University Press, 2002.

Carson, Rachel. *Silent Spring*. New York: Ballantine, 1962.

Cathcart, Michael. *The Water Dreamers: The Remarkable History of Our Dry Continent*. Melbourne: Text Publishing, 2009.

Chakrabarty, Dipesh. "The Climate of History: Four Theses." *Critical Inquiry* 35 (Winter 2009): 197–222.

Chan, Gabrielle. "Murray-Darling Basin: Allegations of Water Theft Spark Calls for Judicial Inquiry." *The Guardian*, July 25, 2017.

Chandler, William U. *Myth of TVA: Conservation and Development in the Tennessee Valley, 1933–1983*. Cambridge, MA: Ballinger Pub. Co., 1984.

Commissioner for Environmental Sustainability Victoria. *Victorian State of the Environment*. Melbourne: Government of Victoria, 2013.

Commissioner for Environmental Sustainability Victoria. *Victorian State of the Environment 2018: Biodiversity (B) Scientific Assessments Part III*. Melbourne: Government of Victoria, 2019.

Connell, Daniel. *Water Politics in the Murray-Darling Basin*. Leichhardt: Federation Press, 2007.

Connell, Daniel. "Irrigation, Water Markets and Sustainability in Australia's Murray-Darling Basin." *Agriculture and Agricultural Science Procedia* 4 (2015): 133–139.

Courtney, Chris, and Fiona Williamson. "Disasters and the Making of Asian History." *Environment and History* 26 (2020): 1–5.
Crabb, Peter. *Murray-Darling Basin Resources*. Canberra: Murray-Darling Basin Commission, 1997.
Crutzen, Paul J. "Geology of Mankind." *Nature* 3 (January 2002): 23.
Crutzen, Paul J., and Eugene F. Stoermer. "The Anthropocene." *IGBP Newsletter* 41 (2000): 17–18.
Culvahouse, Tim, and Richard Barnes. *The Tennessee Valley Authority: Design and Persuasion*. New York: Princeton Architectural Press, 2007.
Davidson, Bruce R. *European Farming in Australia: An Economic History of Australian Farming*. Amsterdam: Elsevier, 1981.
Davies, Anne, and Lisa Martin. "Menindee Fish Kill: Another Mass Death on Darling River 'Worse than Last Time'." *The Guardian*, January 27, 2019.
Davies, Anne. "Murray-Darling Basin Royal Commission Report Finds Gross Maladministration." *The Guardian*, January 31, 2019.
Davies, Anne. "Barwon-Darling River Ecosystem on Path to Collapse, Review Warns." *The Guardian*, July 24, 2019.
Davies, Anne. "Water Wars: Will Politics Destroy the Murray-Darling Basin Plan—And the River System Itself?" *The Guardian*, December 13, 2019.
Davies, Peter, and Susan Lawrence. "Engineered Landscapes of the Southern Murray-Darling Basin: Anthropocene Archaeology in Australia." *The Anthropocene Review* 6, no. 3 (2019): 179–206.
Davison, Graeme. "Decentralisation." In *The Oxford Companion to Australian History*, edited by Graeme Davison, et al., 176–177. Melbourne: Oxford University Press, 1999.
Davison, Graeme. "Country Life: The Rise and Decline of an Australian Ideal." In *Struggle Country: The Rural Ideal in Twentieth Century Australia*, edited by Graeme Davison and Marc Brodie, 01.01.–01.15. Melbourne: Monash University Press, 2005.
Delanty, Gerard. *Community*, 2nd ed. London: Routledge, 2010.
Dipper, Christof. Docupedia-Zeitgeschichte. "Moderne (English Version), Version: 2.0." Last modified November 28, 2018. http://docupedia.de/zg/Dipper_moderne_v2_en_2018.
Doyle, Timothy. *Green Power: The Environment Movement in Australia*. Sydney: UNSW Press, 2000.
Doyle, Timothy, and Sherilyn MacGregor. *Environmental Movements Around the World: Shades of Green in Politics and Culture*. Santa Barbara, California: ABC-CLIO, 2013.
Dyster, Barrie, and David Meredith. *Australia in the Global Economy. Continuity and Change*, 2nd ed., Cambride: Cambridge University Press, 2012.

Elkins, Paul, Joyeeta Gupta, and Pierre Boileau, eds. *Global Environmental Outlook—GEO6: Healthy Planet, Healthy People*. Cambridge: Cambridge University Press, 2019.

European Commission. "EU Emissions Trading System (EU ETS)." Accessed June 6, 2022. https://ec.europa.eu/clima/eu-action/eu-emissions-trading-system-eu-ets_en.

Flanagan, Richard. *A Terrible Beauty: History of the Gordon River Country*. Richmond: Greenhouse, 1985.

Foucault, Michel. "On the Genealogy of Ethics: An Overview of Work in Progress." In *The Foucault Reader*, edited by Paul Rabinow, 340–372. New York: Pantheon Books, 1984.

Foucault, Michel. "Le souci de la verité (1984)." In *Dits et Écrits 1954–1988. IV. 1980–1988*, edited by Daniel Defert and François Ewald, 823–836. Paris 1994.

Gammage, Bill. *The Biggest Estate on Earth: How Aborigines Made Australia*. Sydney: Allen & Unwin, 2011.

Garden, Don. *Australia, New Zealand and the Pacific: An Environmental History*. Santa Barbara: ABC-CLIO, 2005.

Garden, Don. *Droughts, Floods and Cyclones: El Niños That Shaped Our Colonial Past*. North Melbourne: Australian Scholarly Publishing, 2009.

Gascoigne, John. *The Enlightenment and the Origins of European Australia*. Cambridge: Cambridge University Press, 2002.

Gaynor, Andrea, et al. "Looking Forward, Looking Back: Toward an Environmental History of Salinity and Erosion in the Eastern Wheatbelt of Western Australia." In *Country: Visions of Land and People in Western Australia*, edited by Andrea Gaynor, et al., 105–123. Perth: Western Australia Museum, 2002.

Gestwa, Klaus. *Die Stalinschen Großbauten des Kommunismus: Sowjetische Technik- und Umweltgeschichte, 1948–1967*. München: Oldenbourg 2010.

Gestwa, Klaus. "Ökologischer Notstand und sozialer Protest." *Archiv für Sozialgeschichte* 43 (2003): 349–383.

Graf, Rüdiger. "Die Ökonomisierung der Umwelt und die Ökologisierung der Wirtschaft seit den 1970er Jahren." In *Ökonomisierung. Debatten und Praktiken in der Zeitgeschichte*, edited by Rüdiger Graf, 189–2013. Göttingen: Wallstein, 2019.

Griffin, Grahame. "Selling the Snowy: The Snowy Mountains Scheme and National Mythmaking." *Journal of Australian Studies* 79 (2003): 39–52.

Griffiths, Tom. "History and Natural History. Conservation Movements in Conflict?" In *The Humanities and the Australian Environment*, edited by Derek John Mulvaney, 87–109. Canberra: Australian Academy of the Humanities, 1991.

Griffiths, Tom. "Ecology and Empire: Towards an Australian History of the World." In *Ecology & Empire: Environmental History of Settler Societies*, edited by Tom Griffiths and Libby Robin, 1–16. Edinburgh: Edinburgh University Press, 1997.

Griffiths, Tom. *Forests of Ash: An Environmental History*. Cambridge: Cambridge University Press, 2001.

Griffiths, Tom. "Environmental History, Australian Style." *Australian Historical Studies* 46, no. 2 (2015): 157–173.

Griffiths, Tom. "The Transformative Craft of Environmental History: Perspectives on Australian Scholarship." *RCC Perspectives: Transformations in Environment and Society 2017*, no. 2 (2017): 115–124.

Griffiths, Tom. ABC. "Radical Histories for Uncanny Times." Last modified February 27, 2018. http://www.abc.net.au/radionational/programs/bigideas/radical-histories-for-uncanny-times/9478670.

Griffiths, Tom. Australian Dictionary of Biography. "Stretton, Leonard Edward (Len) (1893–1967)." Accessed August 16, 2022. http://adb.anu.edu.au/biography/stretton-leonard-edward-len-11793/text21097.

Guha, Ramachandra. *Environmentalism: A Global History*. Oxford and New York: Oxford University Press, 2000.

Hamilton, Clive. "Human Destiny in the Anthropocene." In *The Anthropocene and the Global Environmental Crisis: Rethinking Modernity in a New Epoch*, edited by Clive Hamilton, et al., 32–43. London; New York: Routledge, 2015.

Hamilton, Clive, et al. "Thinking the Anthropocene." In *The Anthropocene and the Global Environmental Crisis: Rethinking Modernity in a New Epoch*, edited by Clive Hamilton, et al., 1–13. London and New York: Routledge, 2015.

Harris, Edwyna. "Development and Damage: Water and Landscape Evolution in Victoria, Australia." *Landscape Research* 31, no. 2 (2006): 169–181.

Harvey, David. *A Brief History of Neoliberalism*. Oxford: Oxford University Press, 2005.

Heberger, Matthew. "Australia's Millennium Drought: Impacts and Responses." In *The World's Water: The Biennial Report on Freshwater Resources*, edited by Peter H. Gleick, 97–125. Washington, DC: Island Press, 2011.

Henderson, Kirsten. "Dams, Pumps, Pipes and Dreams. Water and the Making of Regional Australia." *Rural and Regional Futures*, edited by Anthony Hogan and Michelle Young, 84–97. London and New York: Routledge, 2015.

Henzell, Ted. *Australian Agriculture: Its History and Challenges*. Collingwood: CSIRO Publishing, 2007.

Herbert, Ulrich. "Europe in High Modernity. Reflections on a Theory of the 20th Century." *Journal of Modern European History* 5 (2007): 5–20.

Hiscock, Peter. *Archaeology of Ancient Australia*. London: Routledge, 2008.

Hughes, Neal. "Water Markets Are Not Perfect, but Vital to the Future of the Murray-Darling Basin." *The Conversation*. Last modified March 1,

2021. https://theconversation.com/water-markets-are-not-perfect-but-vital-to-the-future-of-the-murray-darling-basin-155880.

Howes, Hilary Susan. "The Spectre at the Feast: The Emergence of Salt in Northern Victoria's Irrigated Districts." *Environment and History* 14, no. 2 (2008): 217–239.

Hutchinson, Diane, and Florian Ploeckl. "MeasuringWorth: Five Ways to Compute the Relative Value of Australian Amounts, 1828 to the Present." Accessed August 16, 2022. http://www.measuringworth.com/australiacompare.

Hutton, Drew, and Libby Connors. *A History of the Australian Environment Movement*. Melbourne: Cambridge University Press, 1999.

Isenberg, Andrew C. "Introduction: A New Environmental History." In *The Oxford Handbook of Environmental History*, edited by Andrew C. Isenberg, 1–20. Oxford: Oxford University Press, 2014.

Jackson, Sue, and Leslie Head. "Australia's Mass Fish Kills as a Crisis of Modern water: Understanding Hydrosocial Change in the Murray-Darling Basin." *Geoforum* 109 (2020): 44–56.

Johnson, Mary, et al. "Landcare in Australia." In *Landcare: Local Action—Global Progress*, edited by Delia Catacutan, et al., 13–30. Nairobi: World Agroforestry Centre, 2009.

Kaika, Maria. "Dams as Symbols of Modernization: The Urbanization of Nature Between Geographical Imagination and Materiality." *Annals of the Association of American Geographers* 96, no. 2 (2006): 276–301.

Keating, Jenny. *The Drought Walked Through: A History of Water Shortage in Victoria*. Melbourne: Department of Water Resources Victoria, 1992.

Kellow, Aynsley. "The Murray-Darling Basin." In *Environmental Politics and Policy Making in Australia*, edited by Timothy Doyle and Aynsley Kellow, 220–238. South Melbourne: Macmillan Education Australia, 1995.

Kelly, Paul. *The End of Certainty: Power, Politics, and Business in Australia*. Sydney: Allen & Unwin, 1994.

Lake, Marylin. *The Limits of Hope: Soldier Settlement in Victoria 1915–1938*. Melbourne: Oxford University Press, 1987.

Lambert, David. *The Field Guide to Geology*. New York: Facts on File, 2007.

La Trobe University Melbourne. "Graduate Profiles. Dr Sharman Stone MP, Master of Arts 1978." Accessed August 16, 2022. https://web.archive.org/web/20060920215533/https://www.latrobe.edu.au/alumni/profiles/stone.html (archived website).

Le Feuvre, Juliette. Environment Victoria. "Stop Sucking Rivers Dry." Last modified August 8, 2017. https://environmentvictoria.org.au/2017/08/08/stop-sucking-rivers-dry.

Linn, Rob. *Battling the Land: 200 Years of Rural Australia*. Sydney: Allen & Unwin, 1999.

Lu, Donna. "'Everything Is Saturated': What's Driving the Latest Floods in Eastern Australia." *The Guardian*, October 17, 2022.

Lloyd, Clement J. *Either Drought or Plenty: Water Development and Management in New South Wales*. Parramatta: Department of Water Resources, New South Wales, 1988.

McCormick, John. *The Global Environmental Movement*, 2nd ed. Chichester: Wiley, 1995.

McCormick, John. "The Origins of the World Conservation Strategy." *Environmental Review* 10 (Autumn 1986): 177–187.

McNeill, John R. *Something New Under the Sun: An Environmental History of the Twentieth-Century World*. New York and London: W. W. Norton & Company, 2000.

McNeill, John R., and Peter Engelke. *The Great Acceleration: An Environmental History of the Antropocene Since 1945*. Cambridge, MA: Harvard University Press, 2016.

McNeill, John R., and Verena Winiwarter. "Soils, Soil Knowledge and Environmental History: An Introduction." In *Soils and Societies: Perspectives from Environmental History*, edited by John R. McNeill and Verena Winiwarter, 1–4. Isle of Harris: The White Horse Press, 2006.

Murray-Darling Basin Authority. *Guide to the Proposed Basin Plan. Appendix C: Goulburn-Murray Community Profile*. Canberra: Murray-Darling Basin Authority, 2010.

Murray-Darling Basin Authority. *Pyramid Creek Salt Interception Scheme*. Canberra: Murray-Darling Basin Authority, 2011.

Murray-Darling Basin Authority. *Basin Plan Annual Report 2015–16*. Canberra: Murray-Darling Basin Authority, 2016.

Murray-Darling Basin Authority. *Basin Salinity Management 2030: Status Report 2015–16*. Canberra: Murray-Darling Basin Authority, 2016.

Murray-Darling Basin Authority. *Basin Salinity Management 2030: Summary Report 2015–16*. Canberra: Murray-Darling Basin Authority, 2017.

Murray-Darling Basin Authority. *Kerang-Cohuna Community*. Canberra: Murray-Darling Basin Authority, 2018.

Murray-Darling Basin Authority. *Shepparton Irrigation Area Community*. Canberra: Murray-Darling Basin Authority, 2018.

Murray-Darling Basin Authority. "A Path to Water Reform: Timeline of Water Resource Management in the Murray-Darling Basin." Accessed August 16, 2022. https://www.mdba.gov.au/publications/products/river-management-timeline-poster.

Murray-Darling Basin Authority. "Governance." Accessed August 16, 2022. https://www.mdba.gov.au/about-us/governance.

Murray-Darling Basin Authority. *Basin Plan Annual Report 2018–19*. Canberra: Murray-Darling Basin Authority, 2019.

Murray-Darling Basin Authority. *Basin Salinity Management 2030: 2018–19 Comprehensive Report. March 2020.* Canberra: Murray-Darling Basin Authority, 2020.

Murray-Darling Basin Authority. *The 2020 Basin Plan Evaluation*. Canberra: Murray-Darling Basin Authority, 2020.

Murray-Darling Basin Ministerial Council. *Murray-Darling Basin Environmental Resources Study*. Sydney: State Pollution Control, Commission, 1987.

Murray-Darling Basin Ministerial Council. *The Salinity Audit of the Murray-Darling Basin: A 100-Year Perspective, 1999*. Canberra: Murray-Darling Basin Commission, 1999.

Murray-Darling Basin Ministerial Council. *Basin Salinity Management Strategy 2030 BSM2030*. Canberra: Murray-Darling Basin Authority, 2015.

Meyer, William. *Human Impact on the Earth*. Cambridge: Cambridge University Press, 1996.

Morgan, Ruth A. "Histories for an Uncertain Future: Environmental History and Climate Change." *Australian Historical Studies* 44, no. 3 (2013): 350–360.

Muir, Cameron. *The Broken Promise of Agricultural Progress: An Environmental History*. London: Routledge, 2014.

Mulligan, Martin, and Stuart Hill. *Ecological Pioneers: A Social History of Australian Ecological Thought and Action*. Cambridge: Cambridge University Press, 2001.

National Library of Australia. "Stone, Sharman Nancy, (the Hon.), (Dr) (1951–)." Accessed August 16, 2022. https://nla.gov.au/nla.party-727075.

Nixon, Rob. *Slow Violence and the Environmentalism of the Poor*. Cambridge, MA: Harvard University Press, 2011.

Northage, John. *The Story of John Dainton's Role in Mending the Goulburn Broken*. Shepparton: Goulburn-Broken Catchment Management Authority, 2014.

Ohio History Central. "Louis Bromfield." Accessed August 16, 2022. https://ohiohistorycentral.org/w/Louis_Bromfield.

O'Gorman, Emily. "Unnatural River, Unnatural Floods? Regulation and Responsibility on the Murray River in the 1950s." *Australian Humanities Review* 48 (2010): 67–88.

O'Gorman, Emily. *Flood Country: An Environmental History of the Murray-Darling Basin*. Collingwood: CSIRO Publishing, 2012.

Pardoe, Colin. "Riverine, Biological and Cultural Evolution in Southeastern Australia." *Antiquity* 69 (1995): 969–713.

Pascoe, Bruce. *Dark Emu: Black Seeds: Agriculture or Accident?* Broome: Magabala Books, 2014.

Pigram, John J. *Australia's Water Resources: From Use to Management*. Collingwood: CSIRO Publishing, 2006.

Powell, Joseph M. "'Action Analysis' of Resource Conflicts: The Little Desert Dispute, Victoria, 1963–72." In *The Making of Rural Australia. Environment, Society and Economy: Geographical Readings*, edited by Joseph M. Powell, 161–179. Melbourne: Sorrett Publishing, 1974.
Powell, Joseph M. *An Historical Geography of Modern Australia: The Restive Fringe*. Cambridge, et al. Cambridge University Press, 1988.
Powell, Joseph M. *Watering the Garden State: Water, Land and Community in Victoria 1834–1988*. Sydney: Allen & Unwin, 1989.
Powell, Joseph M. *The Emergence of Bioregionalism in the Murray-Darling Basin*. Canberra: Murray-Darling Basin Commission, 1993.
Powell, Joseph M. "Snakes and Cannons: Water Management and the Geographical Imagination in Australia." In *Environmental History and Policy: Still Settling Australia*, edited by Stephen Dovers, 47–71. Melbourne: Oxford University Press, 2000.
Radkau, Joachim. *Die Ära der Ökologie: Eine Weltgeschichte*. Bonn: Bundeszentrale für politische Bildung, 2011.
Radkau, Joachim. *The Age of Ecology: A Global History*, translated by Patrick Camiller. Cambridge, et al.: Polity Press, 2014.
Raphael, Lutz. "Ordnungsmuster der 'Hochmoderne?' Die Theorie der Moderne und die Geschichte der europäischen Gesellschaften im 20. Jahrhundert." In *Dimensionen der Moderne: Festschrift für Christoph Dipper*, edited by Lutz Raphael and Ute Schneider, 73–91. Frankfurt am Main: Peter Lang, 2008.
Reynolds, Henry. *Frontier: Aborigines, Settlers and Land*. Sydney: Allen & Unwin, 1996.
Robin, Libby. "Of Desert and Watershed. The Rise of Ecological Consciousness in Victoria, Australia." In *Science and Nature: Essays in the History of the Environmental Sciences*, edited by Michael Shortland, 114–149. Oxford: British Academ of Sciences, 1993.
Robin, Libby. *Defending the Little Desert: The Rise of Ecological Consciousness in Australia*. Melbourne: Melbourne University Press, 1998.
Rolls, Eric. *A Million Wild Acres: 200 Hundred Years of Man and an Australian Forest*. Melbourne: Thomas Nelson, 1981.
Daniel Rothenburg. "'The Majesty of Concrete.' Hume Dam and Australian Modernity." *Australian Studies Journal* 31 (2017): 101–113.
Russ, Peter. *The Salt Traders: A History of Salinity in Victoria*. East Melbourne: The Department of the Premier and Cabinet, State of Victoria, 1995.
Schönfelder, Timm. *Roter Fluss auf schwarzer Erde. Der Kuban und der agromeliorative Komplex. Eine sowjetische Umwelt- und Technikkgeschichte, 1929–1991*. Paderborn: Brill Schöningh, 2022.
Scott, James C. *Seeing Like a State: How Certain Schemes to Improve the Human Condition Have Failed*. New Haven and London: Yale University Press, 1998.

Scott, James C. "High Modernist Social Engineering. The Case of The Tennessee Valley Authority." In *Experiencing the State*, edited by Lloyd I. Rudolph and John Kurt Jacobsen, 3–52. Oxford and New York: Oxford University Press, 2006.

Seddon, George Seddon. *Searching for the Snowy: An Environmental History*. St. Leonards: Allen & Unwin, 1994.

William Shakespeare. *Julius Caesar: Fully Annotated with an Introduction by Burton Raffel: With an Essay by Harold Bloom*. New Haven: Yale University Press, 2006.

Sinclair, Paul. *The Murray: A River and Its People*. Melbourne: Melbourne University Press, 2001.

Slobodian, Quinn. *Globalists*. Cambridge, MA: Harvard University Press, 2018.

Stone, Sharman Nancy. "Rural Communities. Facility Development, Attitudes and Self-Determination in the Gordon Shire." Unpublished M.A. Thesis, School of Social Sciences, La Trobe University Melbourne, 1977.

Summons, Martin. *Water: The Vital Element: 150 Years of Shepparton's Growth*. Shepparton: Shepparton Heritage Centre, 2010.

Thurman, Sybil. *A History of the Tennessee Valley Authority*. Knoxville: Tennessee Valley Authority Information Office, 1983.

Uekötter, Frank. "Myths, Big Myths and Global Environmentalism." In *The History of Social Movements in Global Perspective: A Survey*, edited by Stefan Berger and Holger Nehring, 419–447. Basingstoke: Palgrave Macmillan, 2017.

Uekötter, Frank. *The Greenest Nation? A New History of German Environmentalism*. Cambridge, MA: The MIT Press, 2014.

Australian Politics and Elections Database. "Parliament of Victoria, Legislative Assembly Election. Election of 3 April 1982," University of Western Australia. Accessed August 16, 2022. https://elections.uwa.edu.au/elecdetail.lasso?keyvalue=912&summary=false.

Veth, Peter and Sue O'Connor. "The Past 50,000 Years: An Archaeological View." In *The Cambridge History of Australia*, edited by Alison Bashford and Stuart Macintyre, 17–42. Cambridge: Cambridge University Press, 2013.

Walker, Bret. *Murray-Darling Basin Royal Commission Report*. Adelaide: Government of South Australia, 2019.

Walter, James. "Growth Resumed, 1983–2000." In *The Cambridge History of Australia*, edited by Alison Bashford and Stuart Macintyre, 162–186. Cambridge: Cambridge University Press, 2013.

Walter, James. *What Were They Thinking? The Politics of Ideas in Australia*. Sydney: University of New South Wales Press, 2010.

Warde, Paul, Libby Robin, and Sörlin Sverker. *The Environment: A History of the Idea*. Baltimore: John Hopkins University Press, 2018.

Waterhouse, Richard. *The Vision Splendid: A Social and Cultural History of Rural Australia*. Fremantle: Curtin University Books, 2005.

Jane Waterman. "Louis Bromfield." The Literary Encyclopedia. Last modified June 3, 2004. https://www.litencyc.com/php/speople.php?rec=true&UID=578.

Wells, Adrian. *Up and Doing: A Brief History of the Murray Valley Development League and the Murray Darling Association from 1944 to 2014*. Albury: Murray Darling Association, 2014.

Wigmore, Lionel. *Struggle for the Snowy: The Background of the Snowy Mountains Scheme*. Oxford: Oxford University Press, 1968.

Williamson, Fiona, and Chris Courtney. "Disasters Fast and Slow: The Temporality of Hazards in Environmental History." *International Review of Environmental History* 4, no. 2 (2018): 5–11.

Worster, Donald. *Nature's Economy: A History of Ecological Ideas*. Cambridge: Cambridge University Press, 1994.

Worster, Donald. "Appendix: Doing Environmental History." In *The Ends of the Earth: Perspectives in Modern Environmental History*, edited by Donald Worster, 289–307. Cambridge: Cambridge University Press, 1998.

Wright, Don. "The River Murray. Microcosm of Australian Federal History." In *Federalism in Canada and Australia: The Early Years*, edited by Bruce W. Hodgins, et al., 227–286. Waterloo, ON: Wilfrid Laurier University Press, 1978.

Wright, Judith. *Born of the Conquerors*. Canberra: Aboriginal Studies Press, 1991.

Young, Ann R.M. *Environmental Change in Australia Since 1788*, 2nd ed. Melbourne: Oxford University Press, 2004.

Zelko, Frank. "The Politics of Nature." In *The Oxford Handbook of Environmental History*, edited by Andrew C. Isenberg, 717–742. Oxford: Oxford University Press, 2014.

Index

© The Editor(s) (if applicable) and The Author(s), under exclusive license to Springer Nature Switzerland AG 2023

D. Rothenburg, *Irrigation, Salinity, and Rural Communities in Australia's Murray-Darling Basin, 1945–2020*, Palgrave Studies in World Environmental History,
https://doi.org/10.1007/978-3-031-18451-2

Printed in the USA
CPSIA information can be obtained
at www.ICGtesting.com
LVHW021448221123
764620LV00001B/17

Principles and Advances in Supramolecular Catalysis

Jubaraj Bikash Baruah

CRC Press
Taylor & Francis Group
Boca Raton London New York

CRC Press is an imprint of the
Taylor & Francis Group, an **informa** business

CRC Press
Taylor & Francis Group
6000 Broken Sound Parkway NW, Suite 300
Boca Raton, FL 33487-2742

© 2019 by Taylor & Francis Group, LLC
CRC Press is an imprint of Taylor & Francis Group, an Informa business

No claim to original U.S. Government works

Printed on acid-free paper

International Standard Book Number-13: 978-0-367-11164-9 (Hardback)

This book contains information obtained from authentic and highly regarded sources. Reasonable efforts have been made to publish reliable data and information, but the author and publisher cannot assume responsibility for the validity of all materials or the consequences of their use. The authors and publishers have attempted to trace the copyright holders of all material reproduced in this publication and apologize to copyright holders if permission to publish in this form has not been obtained. If any copyright material has not been acknowledged please write and let us know so we may rectify in any future reprint.

Except as permitted under U.S. Copyright Law, no part of this book may be reprinted, reproduced, transmitted, or utilised in any form by any electronic, mechanical, or other means, now known or hereafter invented, including photocopying, microfilming, and recording, or in any information storage or retrieval system, without written permission from the publishers.

For permission to photocopy or use material electronically from this work, please access www.copyright.com (http://www.copyright.com/) or contact the Copyright Clearance Center, Inc. (CCC), 222 Rosewood Drive, Danvers, MA 01923, 978-750-8400. CCC is a not-for-profit organization that provides licenses and registration for a variety of users. For organizations that have been granted a photocopy license by the CCC, a separate system of payment has been arranged.

Trademark Notice: Product or corporate names may be trademarks or registered trademarks, and are used only for identification and explanation without intent to infringe.

Library of Congress Cataloging-in-Publication Data

Names: Baruah, Jubaraj Bikash, author.
Title: Principles and advances in supramolecular catalysis / by Jubaraj Bikash Baruah.
Description: Boca Raton : CRC Press, Taylor & Francis Group, 2019. | Includes bibliographical references.
Identifiers: LCCN 2019000217| ISBN 9780367111649 (hardback : alk. paper) | ISBN 9780429059063 (ebook)
Subjects: LCSH: Supramolecular chemistry. | Catalysis.
Classification: LCC QD878 .P75 2019 | DDC 541/.226--dc23
LC record available at https://lccn.loc.gov/2019000217

Visit the Taylor & Francis Web site at
http://www.taylorandfrancis.com

and the CRC Press Web site at
http://www.crcpress.com

Contents

Preface

In the context of our general life, the word *catalyst* is related to describe the facilitation of everyday activities. Similarly, the term *supramolecular catalyst* is like working under an umbrella to do everyday works in a protective and controlled manner to bring about a speedy, desired and specific output. In a chemical sense, supramolecular chemistry has emerged as a subject forecast to improve fuel, energy, healthcare and the environment. Nature performs reactions in self-made enclosures and has its own carrier systems to mix reagents. Supramolecular catalysis utilises complementary weak interactions through molecular recognition to control and increase the efficacy of chemical reactions. The redox and photochemical properties of any reacting components guide the chemical reactivities of many supramolecular catalysts, and thereby provide scopes to tune catalytic reactivities. Self-assemblies, cages, surface modifications, molecular recognition, control of particle sizes and multicomponent catalyst systems have enabled the design of new supramolecular catalysts. Such catalysts have found applications to tune reactivity and, importantly, in many selective syntheses. Many natural processes such as self-replication and self-assembly performed through supramolecular interactions have increased the importance of the subject. Solid-state assemblies are useful in carrying out cycloaddition reactions and mechano-chemical reactions. Metal organic frameworks have provided many artificial metalloenzymes. Foldamers, atropisomers, capsules, bowl-like structures and molecular flasks have provided scopes to stabilise, modify reactive intermediates and increase the efficacy of reactions. This book is an attempt to agglomerate those interdisciplinary topics in a systematic way and to decipher a new subject which will have long-term impact for general concerns.

Jubaraj Bikash Baruah

Acknowledgments

The desire to systematically present an emerging subject like supramolecular catalysis in the form of a book has been a concern of the author's for quite some time. The author accepted Renu Upadhyay of Taylor & Francis's invitation to write this book. I wish to sincerely thank Renu Upadhyay for this opportunity. I also wish to thank Shikha Garg for handling the production tasks.

A newly growing subject has much to absorb from very recent as well as past original literature. In this book, an attempt is made to articulate the relevant contributions as a reference section in each chapter. I acknowledge all the scientists who contributed to the subject through acknowledged original works. It is not possible in a textbook to highlight certain literature; I acknowledge the contributors if their work has not appeared in reference sections.

Finally, one requires support from family for such a work; the support received from my wife, Helen, and my son, Jnanbikash, is immense. I also thank the Indian Institute of Technology, Guwahati for the use of facilities related to the literature.

Author

Jubaraj Bikash Baruah hails from Assam, India. He obtained his PhD from the Indian Institute of Science, Bangalore, from research under the supervision of Prof. A. G. Samuelson. After his PhD, he did postdoctoral study under Prof. Kohtaro Osakada and Prof. Takakazu Yamamoto of the Tokyo Institute of Technology, Japan. He served Gauhati University as a lecturer before moving to the Indian Institute of Technology, Guwahati. He was the founding member of the Department of Chemistry of the Indian Institute of Technology, Guwahati, where he serves as a senior professor. He served as founder Head in-Charge of Chemistry, Head of the Chemistry Department and Dean of Research and Development of the Indian Institute of Technology, Guwahati. His major research interest is in the novel properties associated with supramolecular systems. He has published over 250 research articles in reputed journals and has written two books.

1 Introduction

1.1 DEFINITION AND APPROACHES

A catalyst is a chemical compound that changes the rate of chemical reactions but is not consumed. Jons Berzelius introduced the term *catalysis*, which originated from the Greek words *kata* and *lyein*. In Greek, *kata* means *down* and *lyein* means *loosen*. Commonly, the word *catalysis* is extensively used in expressing general matters of discussion in every sphere of life, and it generally refers to difficult tasks done by an object in an easier way, where the object under consideration is not affected after performing the job. The study on the chemical aspects of catalysis started from the latter part of the 1700s. The literature on catalysis started with some of the early works of Sir Humphry Davy, who observed an increase in the rate of certain organic chemical reactions by platinum. At that time, Michael Faraday also observed that the rate of hydrogen and oxygen produced by electrolytic splitting of water was accelerated by platinum. Russian chemist Gottlieb Kirchhof, in the early part of the eighteenth century, noticed that a suspension of starch in boiled water remained suspended without decomposition, but in the presence of concentrated acid, the starch transformed to glucose. This reaction was facilitated by acid, and at the end of the reaction, the acid remained unchanged. Continued research on these systems led to the development of many homogeneous and heterogeneous catalytic reactions. In the later part of the eighteenth century, mercury catalyzing reactions to produce indigo dye was discovered. During this period of the century, nickel catalysed hydrogenation reactions of oil to prepare edible fats was also discovered, which made gateways for several new catalytic reactions. Discovery of these reactions provided a timely impetus to explore catalysis and helped to establish the facts about catalytic reactions in a systematic and formulated manner. Another discovery in the early part of the nineteenth century was on the use of a catalyst to combine nitrogen and hydrogen under pressure to form ammonia, this reaction attracted large industrial interest to catalytic reactions. In the middle part of the nineteenth century, independent discoveries of new catalysts for synthesis of polymers by Ziegler and Natta made a great impact on the industrial production of polymeric materials. Accompanied by progressive growth on the topic of catalysis, the logistical aspects of selective products with stereo- and regio-chemical control of products emerged. In these topics, weak interactions have contributed significantly, and the topic of supramolecular catalysis has taken a stand on its own as a subject area for research and applications.[1–4] Generally, a recognition unit or unit with prominent supramolecular features is incorporated into the scaffold of a conventional catalyst to design a supramolecular catalyst. By doing that, a specific molecule gets along with such a catalyst. For example, urea derivatives are useful in molecular recognition. They catalyse or mediate the reactivity of various nucleophiles and activate substrates for catalytic reactions. A catalyst with supramolecular features does not necessarily significantly improve catalytic ability with respect to another catalyst analogue without a supramolecular feature; such compounds are suitable as catalysts but do not function as supramolecular catalysts.

It is a concern for all of us to keep the environment clean and safe. To do so, control of the toxic substances generated from automobile engines is of the utmost concern. It is necessary to destroy or convert hazardous substances such as gases like carbon monoxide and sulphur dioxide that come out as exhaust to less harmful substances. Catalytic converters do so to a limited extent. Catalytic converters are composed of regenerable materials accelerating the decomposition of specific hazardous materials. For the safety of future life in the world, there is also a compulsion on us to keep the atmosphere protected from any hazardous matter. The ozone layer protects living beings from exposure to ultraviolet radiation by absorbing it in the atmosphere. Exposure to the ultraviolet radiation of the sun by living beings is dangerous. One threat to the environment immediately is from the depletion of the ozone layer in the atmosphere. The depletion of ozone is caused by the chlorofluoro hydrocarbons produced on the earth, which penetrate to the ozone layer and decompose ozone molecules. The chlorine radicals produced from chlorofluoro hydrocarbons are converted to chlorine. These chlorine molecules act as catalysts to deplete ozone by transforming it to oxygen. The catalytic activity for such a conversion by chlorine is so high that about a million ozone molecules can be destroyed by a single chlorine molecule in a second. Hence, the utilization of chlorofluoro hydrocarbons is banned and replacements have come to be used. These are two examples of simple catalytic reactions that concern our environment, and the examples show the important role of catalytic reactions in day-to-day life.

1.2 WHAT BIOLOGY TELLS US ABOUT CATALYSIS: SPECIFIC CASES

Biological conversion of starch to sugar was a stage setter to reveal an independent class of catalyst systems called enzyme catalysts. Enzymes are biological macromolecules composed of numbers of amino acids linked together through amide bonds. The enzymes are uniquely able to enhance or retard the rates of many biological reactions. Some such rate-enhanced reactions do not take place in water and/or at ambient conditions without an enzyme catalyst. Enzyme catalysts find use in industrial processes involving preparation of fermented products, baking, brewing, detergents, pharmaceuticals and textiles. Some common processes encountered in biology are urea conversion to ammonia, which is done by an enzyme called urease; glucose to alcohol done by zymase; inversion of cane sugar is catalysed by invertase; and starch conversion to maltose in the presence of an enzyme called diatase. According to the types of catalytic reactions caused by enzymes, they are classified and categorised into six main enzyme classes, which are listed in Table 1.1.

In-depth knowledge of catalytic intermediates, mechanistic paths and catalytic cycles of catalytic reactions provides insights into the sustainability and utility of catalysts. New efficient design of a catalyst requires understanding of the energetics as well as kinetics of the catalytic reactions. The fundamental principles that rule a common catalytic reaction are valid for natural catalytic reactions caused by enzymes. Due to the superior ability of nature in the design aspects to manipulate reactivity and selectivity, an enzyme performs catalytic reactions at an enormously

TABLE 1.1
Different Enzymes and Their Functions

Name of Enzyme	Performing Reactions
Oxidoreductases	Redox reactions, especially involving C—O, C—C, C—N bonds. Reductive amination of C—O bonds; oxidation of C—H bonds.
Transferases	Reactions that transfer functional groups such as methyl, acyl, amino, phosphoryl, glycosyl, nitro, thiolate groups.
Hydrolases	Reversible and irreversible hydrolysis of functional groups such as esters, amides, nitriles, strained rings.
Lyases or synthases	Addition of small molecules to double bonds.
Isomerases	Racemizations, epimerizations and rearrangement reactions involving isomers.
Ligases	Enzymatically active only when combined with another counterpart such as adenosine triphosphate.

accelerated rate as compared to a synthetic catalyst. Nature has evolved through natural processes over a vast period of time and provided a robust system; hence, design of catalysts in a similar manner to nature by artificial means is an uphill task. Despite the complicated structures of enzymes and the reaction of complicated intermediates involved in an enzyme-catalysed reaction, enzyme catalysis is understood to an extent just sufficient to develop biological mimics through synthetic means. Enzyme-catalysed reactions are guided by four principal factors: (a) templates provided by nature through complementary interactions to confine substrates, and the active sites those may be referred to as catalytic pockets; (b) generation of hydrophobic pockets within an enzyme catalyst by noncovalent interactions in water as medium; (c) association of each enzyme with self-replication properties to decide the type and course of a reaction; and, finally, (d) the allosteric properties associated with enzymes due to cofactors. Lowering the free energy of the transition state to provide new path(s) leading to product(s) is the common referral issue in synthetic as well as natural catalysis. Enzymes in general are more effective catalysts due to the specific substrate binding ability and the presence of logical active sites for substrate-specific reactions to provide specific products. Broad ranges of organic reactions, such as redox reactions, carbon—carbon bond-forming reactions and hydrolytic reactions, are catalysed by enzymes. Nature provides the required architecture for enzyme catalysis to specifically bind stereo-isomers. It makes way for stereo-selective products. It is interesting to note that enzyme catalysis initially was not a tasty cup of tea for organic chemists. It was due to studies on the reactivities that were restricted within a narrow window of substrates, limiting applicability of the subject. Moreover, the isolation and stability of enzymes were also concerns to scientists. But with the progress of time, economically viable procedures for the preparation of enzymes through high cell density fermentation has helped enzyme catalysis to emerge as one of the prominent catalytic process. The three-dimensional active sites of an enzyme recognise molecules through confinement or complementing weak interactions, hence imbuing selectivity in reactions and products. Besides this, high chemo- and stereo-selectivity of products in enzyme catalysis have established the topic at the crest of the sciences dealing with the catalytic reactions. Many difficult regio-, enantio-, and diastereo-selective reactions are made easy by using enzyme-catalysed reactions.[5] Biological chemical reactions are guided by supramolecular interactions and self-assemblies. In a real sense, nature has no reaction vessel to mix components; neither are there hands to mix the components or control the mixing process. Biological reactions conventionally proceed in aerobic conditions in the presence of water and also do not use drastic reaction conditions to synthesise a product. In biology, the rate of mixing and other thermodynamic parameters, including transport of the waste as well as useful products, occur in a controlled manner.

The kinetic approach to studying enzyme catalysis banks on the formation of activated species known as an enzyme—substrate complex. The activated species formed between an enzyme and reactant transforms reactants into products. A subsequent stage is required in the case where there is interaction between the product and enzyme; in such a situation, a product is bound to the enzyme. In general, the products are dismantled from the enzyme-product adduct, and thereby the enzyme gets ready to cause the next catalytic cycle by interacting again with a substrate molecule, forming an enzyme—substrate complex (reactive intermediate). Enzyme-catalysed reactions are formulated from different intermediate equilibriums to set up equations known as the Michaelis and Menten equations.

For the success of an enzyme-catalysed reaction, the hydrogen ion concentration and temperature are prime parameters. Enzyme-catalysed reactions are green by virtue of definition; in biological chemistry, substrates, enzymes and products are in water. Many enzyme-catalysed reactions are also performed in organic solvents that are not environmentally benign. In many cases, the use of solvents other than water helps to overcome the problem of insolubility of many organic compounds in water. This enhances the viability of using enzyme catalysts in organic synthesis as a common practice.

Lock-and-key theory is one way to understand the catalytic action of enzymes. This theory uses the analogy of having a relation between a substrate and enzyme as a process of fitting a key into a lock. It also suggests that when a substrate binds to an active site by a lock-and-key process (**1.1a**) there is no change in the shape of the active site. Such complementary fitting of a substrate to an enzyme (Figure 1.1) guides enzyme-catalysed reactions. There is also another model called the induced-fit model (**1.1b**) for understanding enzyme activities. According to this model, there exist some amounts of noncomplementarity to bind, but the host has sufficient flexibility and shape to have compatible interactions through reorganizing itself to adopt the appropriate geometry. This model explains the catalysis of certain enzymes that catalyse reactions of multiple substrates.

Enzyme-catalysed reactions are keys in light production as well as in certain protective measures. Some living organisms convert chemical energy into light through a process called bioluminescence. There are many living organisms such as bacteria, fungi, crustaceans, molluscs, fishes and insects that glow when light falls on them. In general, emission of light arises from the oxidised product of compound luciferin, caused by the catalytic reaction of an enzyme luciferase. Two insects using catalytic reactions for light generation and protective measures are shown in Figure 1.2.

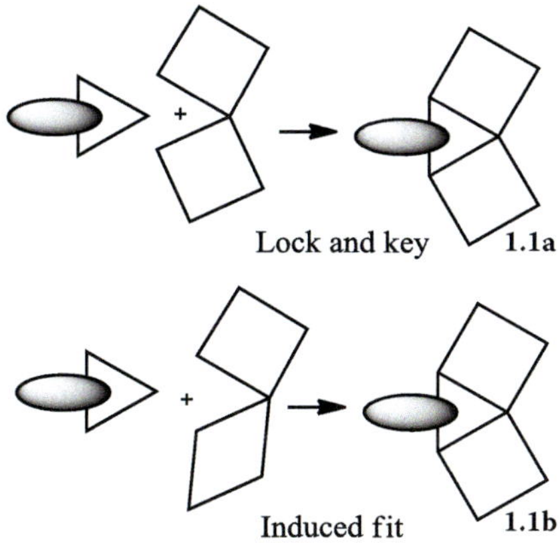

FIGURE 1.1 Lock-and-key and induced-fit mechanism.

1.2a 1.2b

FIGURE 1.2 Firefly (**1.2a**) and beetles (**1.2b**). Insects using enzyme catalysis to generate (a) light and (b) defence.

Firefly bioluminescence is fuelled by conversion of adenosine triphosphate to inorganic pyrophosphate to cause conversion of firefly luciferase to the oxidised form adduct of luciferase with oxyluciferin (Figure 1.3). Firefly luciferase specifically forms an adduct with the nucleotide triphosphate. The adenylate formed in this way is the true substrate of the subsequent oxidative chemistry. In the first step, luciferase adds to D-luciferin to form enzyme-bound D-luciferin—adenosine mono-phosphate. Formation of this adduct is facilitated by magnesium ions. The adduct reacts with oxygen to produce an oxidised form in an excited state. These excited luciferin-oxyluciferin adducts decay to a ground state through emission of light, with quantum yield ranging from 0.4–0.6. In the course of the reaction, carbon dioxide gas evolves. The reactions are shown in Scheme 1.1. The luciferase enzyme acts as a mono-oxygenase to cause the catalytic oxidation.[6]

For protective measures of certain insects, they release gases, and the release of gas is very fast, comparable to an explosion at a micro level caused by such insects. An insect named the bombardier beetle, under a panic situation or in apprehension of danger, sprays an irritating substance through a sudden blast from the prostate part of the body (circled in Figure 1.2), which it uses as a biologically built-in weapon. The spray is as hot as 100°C and spreads to a 10–20 cm distance with a velocity of about

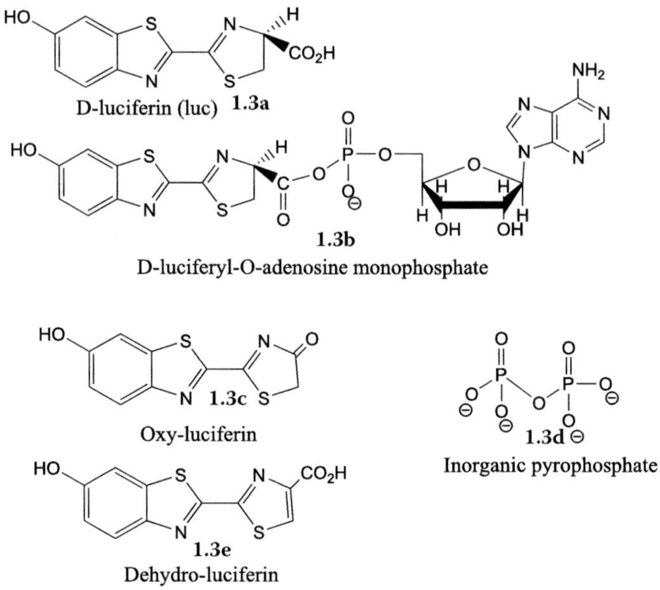

FIGURE 1.3 Structures of different components involved in biophotoluminescence of firefly.

$$\text{Luciferin} + \text{Luciferase} + \text{ATP} \xrightleftharpoons{Mg^{2+}} \text{Luciferase.Luciferine.AMP} + \text{inorganic polyphsphate}$$

ATP = Adenosine triphosphate, AMP = Adenosine monophosphate

$$\text{Luciferase.Luciferin.AMP} + O_2 \longrightarrow \text{Luciferase.Oxyluciferin*} + \overset{*}{\text{AMP}} + CO_2$$

* = excited state

$$\text{Luciferase.Oxyluciferin*} \longrightarrow \text{Luciferase.Oxyluciferin} + \text{Light}$$

SCHEME 1.1 The reactions resulting in emission by oxidation of luciferin.

10 m/sec. The entire process happens as a result of the ability of the pygidial glands of the beetles to control the reactants of the process. The gland acts as a reservoir to provide (a) a reaction chamber and also (b) the exit path to discharge the products. The reservoir stores an aqueous solution of hydrogen peroxide, 1,4-dihydroxybenzene and alkanes. The reaction chamber contains peroxidase and catalase enzymes that perform catalysis to cause very fast reactions among the reactants. When beetles spray the hot, irritating gas for their self-defence, there is a biological process that transfers the components of the reactants in the reservoir to the enzyme-containing chamber. By this process, a molecule of 1,4-dihydroxybenzene is oxidised to 1,4-benzoquinone and, as a consequence of this oxidation, oxygen also evolves by the decomposition of hydrogen peroxide. The 1,4-benzoquinone molecule being a component of the spray, it is the root cause of the irritation caused by the dispersed spray by the beetle. Whereas the exothermic nature of the reaction releases hot oxygen and provides

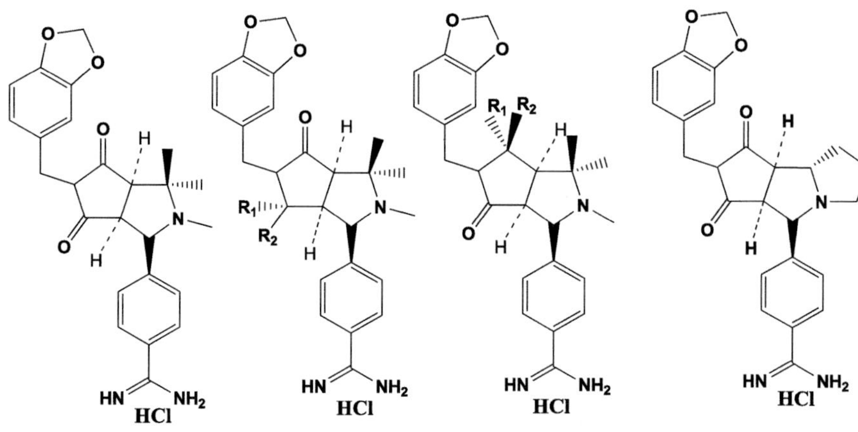

1.4a

when $R_1 = R_2 = H$ **1.4b**
When $R_1 = H$ $R_2 = CH_3$ **1.4c**

when $R_1 = R_2 = H$ **1.4d**
When $R_1 = H$ $R_2 = CH_3$ **1.4e**

1.4f

FIGURE 1.4 Examples of molecules recognised to different extent by thrombin and trypotosin enzymes. (U. Obstl et al. *Chem. Bio.* 1997, 4, 287–295.[7])

the heat to the spraying liquid/gas mixture. During these courses of action, water vaporises, pressure builds up and the spray explodes from the exit channel.

The natural blood-clotting process is controlled by certain enzymes; thrombin is an enzyme which facilitates the process of blood clotting. This enzyme is also used as part of drug discovery to find selective binding with inhibitors for the process.

The lead compounds **1.4a** to **1.4f** shown in Figure 1.4 are suitable for such a purpose. They fit into the hydrophobic pockets in which the residual functional groups located at different sites of a protein can interact through complementary hydrogen bonds. The crystal structure of the thrombin binding to **1.4a** has the phenolic O–H group of tyrosine of the enzyme interacts with an oxygen atom of the ether part of the compound. Whereas the glycine residues interact with the N–H and C=O bonds of the compound. The amidinium part has complementarities with carboxylic acid of aspartate residue. The compound **1.4f** is enclosed in hydrophobic pockets of thrombin, as shown in Figure 1.5. Each pocket has a characteristic name; for example, the D-pocket of thrombin is named because of its location at a 'distal' from the catalytic Ser195. The S_1 pocket is a deep hydrophobic pocket of the enzymes like trypsin and thrombin that contain an aspartate moiety, and this pocket dictates the selective cleavage of peptide bonds. The aspartate part of this pocket binds to the amidinium cation. The relative binding abilities of these compounds are determined by the ratios of their binding constants with two enzymes listed in Table 1.2. These results suggest the ability of thrombin to recognise them differs from the closely related enzyme trypsine. Thrombin has high selectivity for the compound **1.4c**, whereas trypsin recognises the compound **1.4b** in a better way than the others.

Certain enzymes require additional nonprotein components called cofactors to perform catalysis.[8] Cofactors are either linked to the enzyme through covalent linkage or through noncovalent interactions. The covalently linked ones are known as

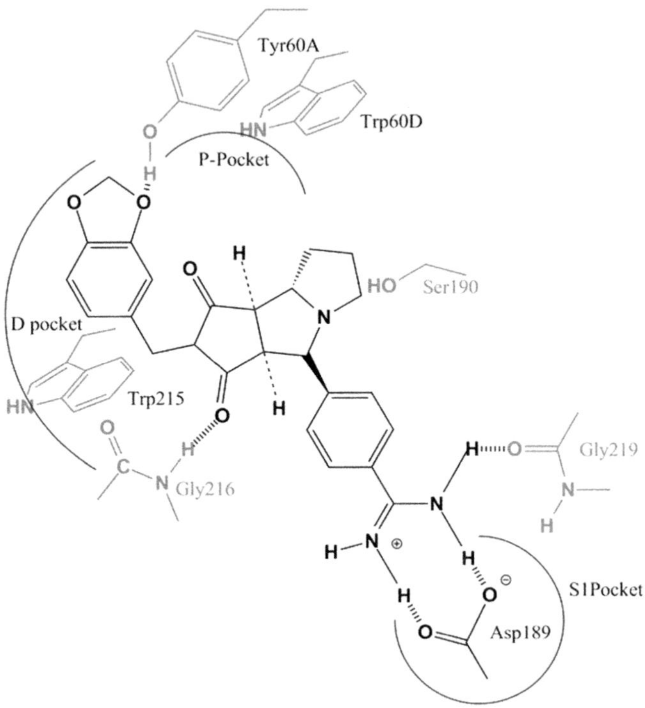

FIGURE 1.5 The environment around the molecule **1.4f** embedded in the hydrophobic pockets of thrombin. (U. Obstl et al. *Chem. Bio.* 1997, 4, 287–295.[7])

TABLE 1.2
Comparison of the Binding Abilities of Different Compounds with Trypsin and Thrombin

	Binding Constant (K in μM)		Selectivity
Compound No	**Thrombin**	**Trypsin**	($K_{trypsin}/K_{thrombin}$)
1.4a	0.5	1.1	2.2
1.4b	2.0	64.0	32.0
1.4c	9.2	40.0	4.34
1.4d	6.2	31.0	5.0

prosthetic groups, while those noncovalently linked are called coenzymes. Biological catalytic reactions involve weak interactions, and several of them involve changes in the redox properties at intermediate steps. Due to such facts, coenzymes get modified during the course of reactions, making it less suitable to use cofactors in catalytic amounts. To avoid such operational difficulties of coenzymes, a stoichiometric amount of an additional reactant is required to keep the cofactor active for multiple catalytic cycles.

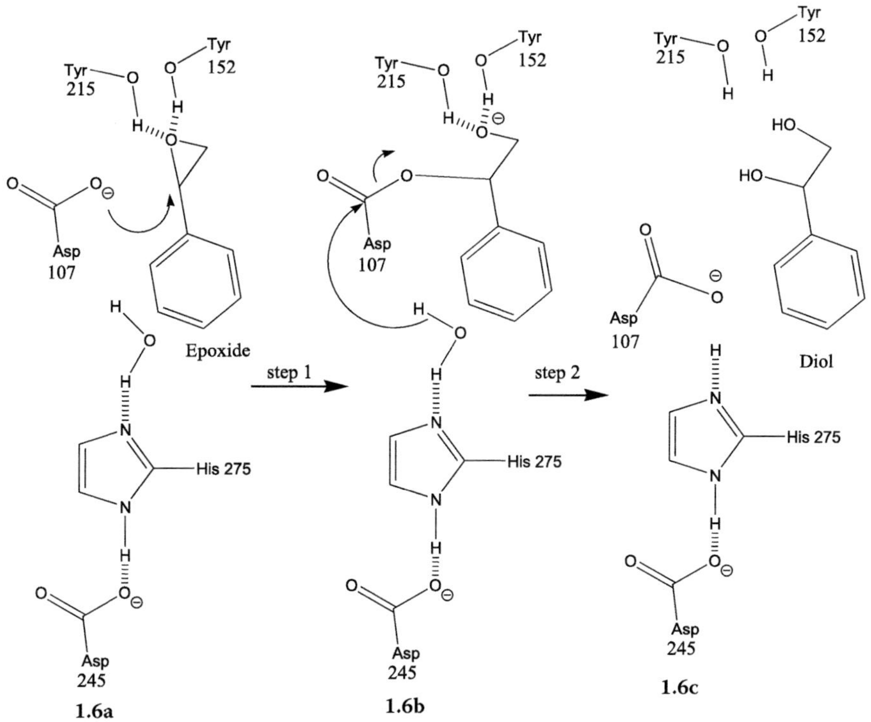

FIGURE 1.6 Epoxide hydrolase-catalysed conversion of epoxide-converting enzymes through covalently linked intermediate. (K. Drauz et al. *Enzyme Catalysis in Organic Synthesis*, 3rd ed. 2012, Wiley-VCH Verlag GmbH & Co. KGaA.[8])

In biological systems, hydrolytic ring-opening reactions of epoxides to form the corresponding diol are caused by epoxide hydrolases. The intermediates of such hydrolysis have covalent links between enzyme and substrate. Accordingly, these types of enzyme-catalysed reactions are classified as covalent enzyme catalysis. In this hydrolysis process, the nucleophilic oxygen atom of aspartate attacks the electrophilic carbon atom of the epoxide ring (Figure 1.6). This results in ring opening as well as ester formation, producing a covalent intermediate. Hydrolysis of this intermediate by water is facilitated by a histidine residue. The reaction sequences are shown in Figure 1.6 where, in the first step, an ester bond is formed at Asp107. In the second step, hydrolysis of the covalent intermediate occurs; this hydrolysis process involves an extended hydrogen-bonded intermediate which has a His275 residue anchoring water and the reactant. The enzyme is regenerated after the hydrolysis and eventually a diol is formed. Thus, during formation of the diol, in fact, epoxide takes up an oxygen atom from the enzyme and water compensates for this in the enzyme.

On the other hand, leukotriene A4 epoxide hydrolase contains a zinc ion, and hydrolysis by this enzyme involves a noncovalent catalytic mechanism. The zinc ion of the active site is coordinated to two nitrogen atoms of histidine residues, and the other two coordination sites have a water molecule and a glutamate residue (**1.7a**).

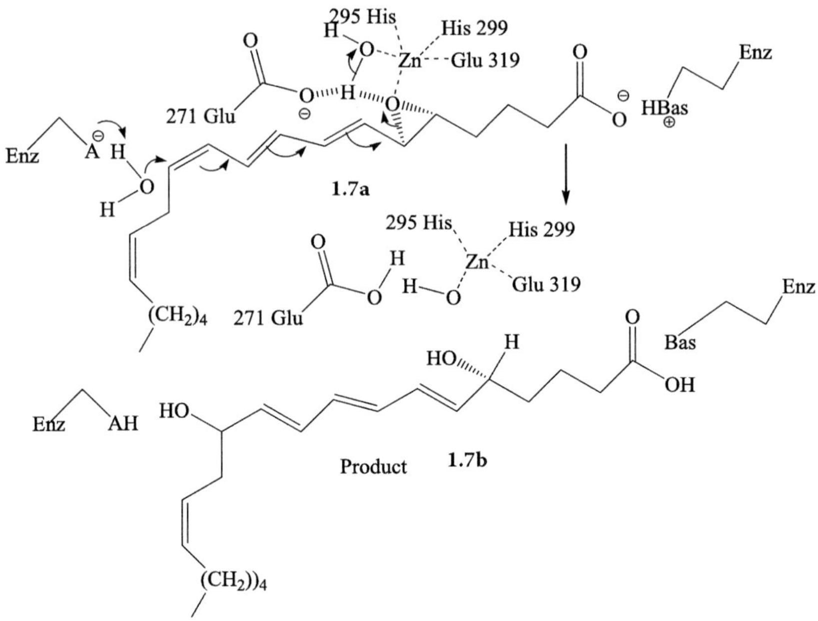

FIGURE 1.7 A noncovalent enzyme catalysed ring opening reaction. (K. Drauz et al. *Enzyme Catalysis in Organic Synthesis*, 3rd ed. 2012, Wiley-VCH Verlag GmbH & Co. KGaA.[8])

The ring opening of epoxide is facilitated by the water molecule held to the zinc ion. The example of substrate shown in Figure 1.7 has a conjugated olefin part. A water molecule attacks a carbon located at the distal end of the conjugated double bonds. This is due to the Lewis acid property of the zinc ion and to draw electron density towards the coordination site through resonance by opening the epoxide ring. The carboxylic acid group of the substrate forms an ion pair with the enzyme to act as the lock. There are many examples of metal ion-catalysed reactions in biology where modification of intermediates of hydrolytic reactions shifts the respective equilibrium towards the product side. There are also many catalytic electron-transfer biological reactions involving metal ions, so metalla-enzymes are useful in hydrolytic as well as redox reactions of biological interest.

Radiobacter halo-alcohol dehalogenase enzymes cause halo-hydroxylation reactions of epoxides. In such enzyme-catalysed reactions, the halide ions are the nucleophiles to open the epoxide ring. The example shown in Figure 1.8 has arginine—tyrosine residues which bind to the epoxide group of the reactant and activate the epoxide. The halide ions held in the selective pockets of the enzymes react with the activated epoxide. Specific pockets of the enzymes provide the required spaces to confine the nucleophiles. They also confer geometry and orientation to the reactant appropriate for the nucleophilic attack of halide ions (in this case, it is chloride). This catalytic reaction belongs to a category of enzyme-catalysed reactions called electrostatic catalysis. The term *electrostatic* is used for these reactions as the electrostatic interaction is the predominant factor in their transition states.

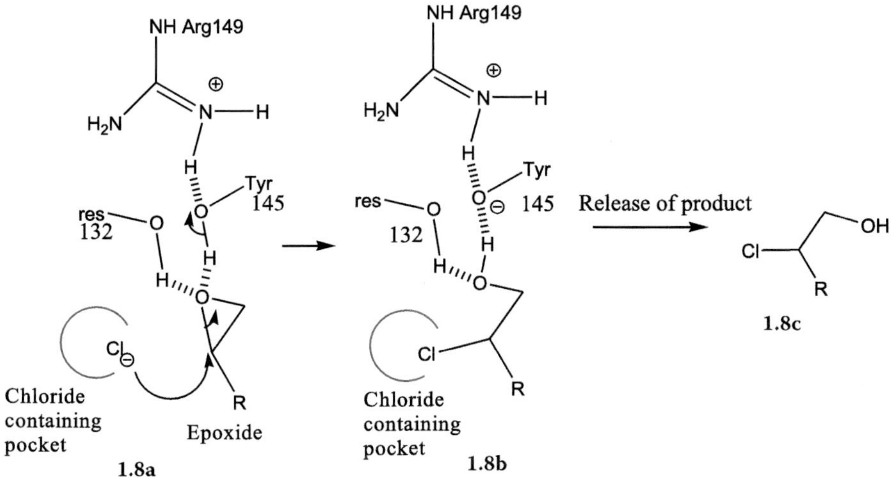

FIGURE 1.8 Chlorohydroxylation of epoxide by radiobacter haloalcohol dehalogenase. (K. Drauz et al. *Enzyme Catalysis in Organic Synthesis*, 3rd Ed. 2012. Wiley-VCH Verlag GmbH & Co. KGaA.[8])

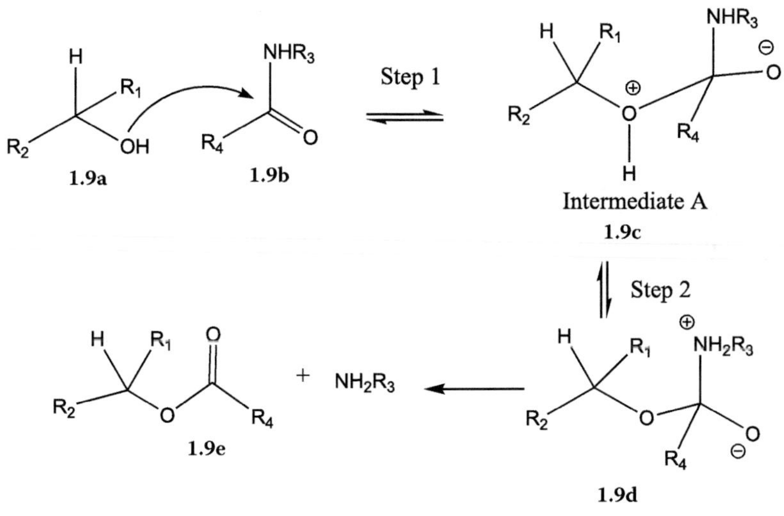

FIGURE 1.9 A reaction undergoing proton transfer to form product.

There are many proton-transfer reactions in biology that are catalysed by enzymes. These reactions provide several interesting useful insights to design new enzymes mimicking supramolecular catalysis. The intermediate **1.9c** (Figure 1.9) is not stable without a supramolecular environment. Such intermediates are responsible for the formation of esters in the enzyme-catalysed reactions of alcohol with amide. In those reactions, the intermediate **1.9c** remains in equilibrium with **1.9d**. The intermediate **1.9d** reacts with a hydronium ion to form ester **1.9e**. If the exchange between **1.9c** and

1.9d is fast, the ester formation rate is not influenced by a functional group. An acid or base counterpart in the enzyme increases the rate of the reaction of the intermediate **1.9c**. When the reaction of water with the intermediate **1.9c** is slow, then a fraction of the intermediate formed at a particular time is converted to the product. This is prominently caused by the enzyme while performing such catalytic reactions.

Many routine catalytic reactions perform at a close capacity to that of a corresponding biological reaction when the characteristic features of noncovalent interactions are improvised. The subject of supramolecular catalysis has emerged from thorough knowledge of host—guest chemistry, complexity of ions, self-assembly and encapsulation together with the fundamental aspects of conventional catalysis. The supramolecular chemistry introduced in design principles of a catalyst or efforts to understand special supramolecular aspects to justify specific features in conventional catalysis has strengthened the subject of supramolecular catalysis. The concepts of molecular recognition on rate enhancement, selective product formation, discrimination between reactants and preference to release a product from a supramolecular system coupled with reactivity study encompass supramolecular catalysis. Chemo-, stereo- and regio-selective products with high turnover and atom economy have demarcated supramolecular catalysis as a pioneering new subject area with future prospects. Thus, this category of catalysis is not different from conventional catalysis but involves supramolecular interactions as extra features. In general, due to the interplay among different noncovalent interactions and the necessity of kinetically controlled assemblies in the catalytic process, it is difficult to forecast the design of a multifunctional assembly as an intermediate in supramolecular catalysis. Thus, it is a difficult task to design a supramolecular catalyst *a priori*. There are many weak interactions such as cation-π, C–H-π, π–π interactions and steric repulsions responsible for stabilizing intermediates of catalytic reactions, and these interactions also facilitate enantioselective catalysis and control rates.

The supramolecular features of many inorganic coordination complexes significantly change the reactivity, and many of those serve as catalysts due to such features associated with them. This provides scope to describe a supramolecular catalytic process from the first principle of catalysts. Alternatively, supramolecular catalysis includes recognition of substrates followed by organization of constructing assemblies as a reaction intermediate or of new molecules. The foregoing general descriptions may be summarised by limiting supramolecular catalysis to those reactions that involve supramolecular interactions where weak interaction schemes are independent of the basic catalytic reaction.

1.3 COMPLEMENTARY WEAK INTERACTIONS AND MOLECULAR RECOGNITION

Depending on the nature of the host, there are several ways host molecules retain guest molecules. The spatial relationship between host and guest categorises them based on the process of retaining guests. The processes are entrapping guests in a capsular structure, nesting, perching, polar surface interactions, sandwiching and wrapping. In capsular interaction, the guest molecule gets into the interior void present within a host, whereas nesting involves fitting the guest into a curved space available on the host molecule, similar to a nest. In the perching process, the guest is held on the edge of a polyhedral host. The sandwiching process involves engulfing the guest between sandwich types of structures, whereas warping is a process in which host molecules form enclosure assemblies or adopt a shape to wrap up a guest.

Recognition of a molecule means specific and selective interactions between two or more molecules to form a stable host—guest complex over other similar analogous combinations. The complementary effects of weak interactions or additional weak interactions over normal interactions are of prime concern in molecular recognition. To have molecular recognition in supramolecular chemistry, two or more molecules bind to one another in a specific geometry. This is especially the case when a central molecule under consideration prefers to interact with one among several alternative molecules. To execute molecular recognition, a set of molecules utilise specific noncovalent interactions such as hydrogen bonds, metal coordination, hydrophobic forces, van der Waals forces, π-π interactions, ion-π interactions and electrostatic interactions. Generally, a relatively bigger or comparably sized partner molecule among a set of interacting molecules, which is called a guest molecule, accommodates or binds the other partner molecules called guest molecules. The interactions between such molecules are referred to as host—guest interactions. However, host—guest interactions or molecular recognition require complementing sizes and shapes inside a cavity, vacant spaces in container-like molecules or suitable space on a surface. The lock-and-key interaction between enzyme and substrate is an example of a molecular recognition process.

Molecular recognition may be an intermolecular or an intramolecular phenomenon. For example, protein folding takes place through an intramolecular recognition process. Among self-assemblies of biological systems with intermolecular hydrogen bonds, protein—protein interactions, protein—substrate interactions, base—pair interactions in DNA, DNA—substrate interactions and so on are commonly observed. In the biological context, there is always a necessity to explore the role of noncovalent interactions for molecular understanding of biochemical processes. Such processes with biological molecules provide practical applications and pave the way to extrapolate the concepts to artificial synthetic systems. In biological or rather in any noncovalent assembly, a final self-assembled structure is obtained as a consequence of the interplay of weak interactions. Hence, the magnitudes of weak interactions contributing to a self-assembly are very important. Some such common interactions influencing supramolecular catalysis are listed in Table 1.3. Depending on the magnitude of hydrogen bond interactions, there are three different types of hydrogen-bonds: strong, moderate and weak hydrogen bonds.

Moderate hydrogen bonds are of energy <20 kJ/mol, and weak hydrogen bonds have energy less than 10 KJ/mol in nonpolar environments. A conformational change requiring conversion of the C—C bond to transform an *anti*-conformer to *gauche* is about 3–5 kJ/mol. Thus, the torsional strain can be utilised either in a constructing or opposing way to guide a particular self-assembly. In a biological system, the torsional factors competing with other weak interactions become prominent; thus, it becomes very difficult to predict how a biological self-assembly will be formed. Intuition and extensive optimization or experimentation are required

TABLE 1.3
Strengths of Some Supramolecular Interactions

Noncovalent Interactions	(kJ/mol)
Strong hydrogen bond	60–120
Moderate hydrogen bond	16–60
Weak hydrogen bond	<10
Ionic interaction	100–350
Dipole-dipole interaction	5–50
Van-der Waals interaction	<5
π-π interaction	0–50
Cation-π interaction	5–80

to establish a structure. There are also complications in simple noncovalently linked self-assemblies to suggest predesign synthesis. Establishing the interplay of weak interactions requires a thorough understanding of the energy of each competing interaction. In many instances, one may predict new noncovalent assemblies based on rational complementing interactions. In general, without considering the energetic aspects of weak interactions within the assembly and without in-depth information on solute—solvent interactions, it is difficult to predict self-assemblies in solution with multiple competing interactions.

The orientations of interacting functional groups are the deciding factors to have maximum interactions between them. The numbers and energy of interactions govern the stability of noncovalent assemblies. For example, carboxylic acid forms intermolecular hydrogen bonds with or without the assistance of a solvent or guest molecule. The formation of hydrogen-bonded dimer of benzoic acid in an aromatic solvent is a common phenomenon. It is observed in solvents such as benzene, whereas it remains as a monomer in solvents like chloroform and carbon-tetrachloride. A dimeric structure of carboxylic acid **1.10a** and the discrete hydrogen-bonded **1.10b** structure is illustrated in Figure 1.10. The self-assemblies of carboxylic acids are influenced by the electronic steric effect and position of other functional groups in the molecules. For example, 2-hydroxybenzoic acid (salicyclic acid) adopts an intramolecular hydrogen-bonded structure **1.10c**.

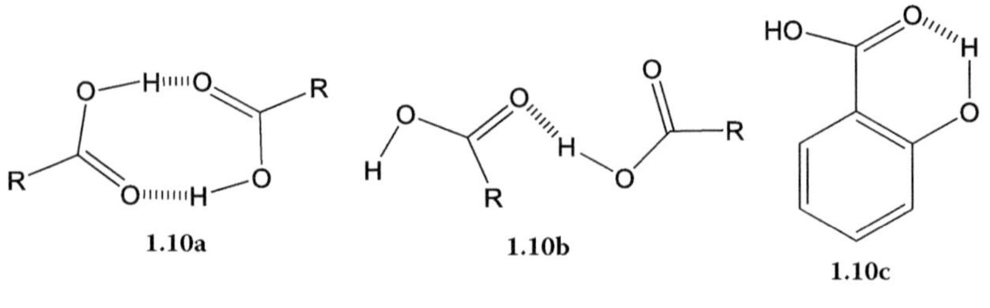

1.10a **1.10b** **1.10c**

FIGURE 1.10 Hydrogen-bonded cyclic dimer of carboxylic acid or discrete hydrogen-bonded unit and intramolecular hydrogen-bonded 2-hydroxybenzoic acid.

Though the self-complementing properties of two hydrogen bond partners makes it possible to have strong hydrogen bonds, the intrinsic acidity or basicity plays a major role in hydrogen bond formation. When an acid and a base interact, they either form hydrogen-bonded cocrystal, supramolecular adduct or salt; in the last case, the base may abstract the proton, generating cation and leave aside other part as counter anion. In such situations, the pKa values of the acids predict the formation of solid adducts as cocrystal or salt. The latter is formed by protonation-deprotonation. Experiments that are generally carried out in solution provide high flexibility to have equilibrium for the exchange of protons between an acid and base. In such a condition, distinction between a cocrystal and salt may not be possible. There is an empirical way to predict salt and cocrystal formation. Salts are formed between an acid and base when ΔpKa (pKa_{base}–pKa_{acid}) is greater than 2. This relationship is not universal, as there are other contributions from the lattice to affect the validity of such a prediction. There are examples of salts forming further adducts with the acid or base partners through weak interactions. In the context of supramolecular catalysis, such interactions are very important in ionic liquids, surfactants, solid catalysts, nanocatalysts and acid-base catalysis. The complementing weak interactions other than hydrogen bonds guide the stability of many self-assemblies. For example, one may consider the interactions between a quinoline molecule with a carboxylic acid. There are three ways **1.11a–c** to form hydrogen-bonded units by quinoline with a carboxylic acid (Figure 1.11) without forming the corresponding salt. One simple way is the discrete N—H···O interaction. Other ways to represent different hydrogen-bonded motifs include cyclic units of hydrogen-bonded structures having two hydrogen bond donors and two acceptors forming an eight-membered unit or formation of a hydrogen-bonded cyclic unit with two hydrogen bond donors and two hydrogen bond acceptors containing seven atoms. The other form of salt **1.11d** is unlikely to form which is reflected in high energy of this form in theoretical calculation. The three units **1.11a–c** are found in self-assemblies of quinoline interacting with carboxylic acids, depending on crystallization conditions and the other functional group attached on the carboxylic acid.

The size and shape of guest molecules decide the required dimension for a molecular recognition to be made available by a host and also decide the compatibility for host—guest binding. This requirement is applicable at the nano-scale as well

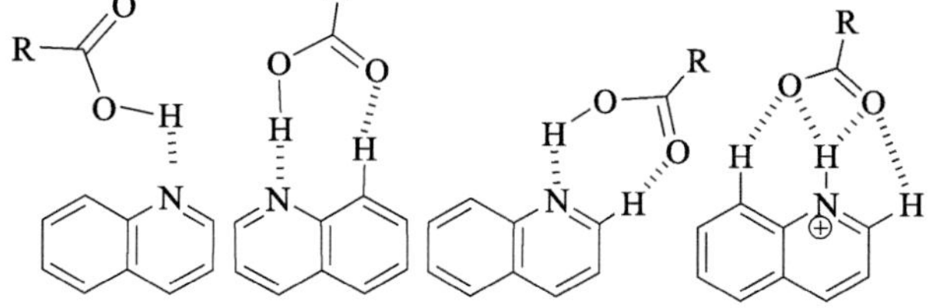

1.11a **1.11b** **1.11c** **1.11d**

FIGURE 1.11 Four possible hydrogen-bonded units between quinoline and carboxylic acid. (D. Singh et al. *Cryst. Growth Des.* 2010, 10, 348–356.[9])

as at the meso-scale. For example, protein recognition by a host molecule is a conventional meso-scale molecular recognition process. There are also large numbers of examples of compounds such as lariat ethers as well as dendrimers that specifically bind cytochrome c protein. Though cytochrome c is a relatively small protein, a molecule must have multiple binding sites to satisfy to recognise this. There are examples of lariat ether complexes with cytochrome c that are useful asymmetric catalysts for oxidation reactions of naphthylmethyl sulfoxide. The rational drug design requires search for selective performance of catalytically active sites for new catalytic reaction is an important issue, but it is a near-impossible task to design enzymes or biological molecules from first principles that will recognise a specific small molecule. The superstructure of noncovalent assemblies generated by multiple numbers of small molecules through self-recognition through weak intermolecular forces to encapsulate a substrate has drawn interest to identify templates for catalytic reactions. Such dynamic assemblies available in nature are easily dismountable. The stability of such assemblies varies and is reflected in their lifetimes. Some of these assemblies are reversibly formed in solution. There are also certain assemblies whose formations are guided by the guest molecules in appropriate solvents. In such situations, additional guest molecules are encapsulated into the assembly formed between the host and guest. Capsulelike assemblies serve as reaction chambers for different reactants. The space created within capsulelike assembly may be such that specific stereoisomer fits into the cavity. Self-assembled capsulelike structures also help to stabilise solvated guest molecules; those solvated molecules residing inside the cavity act as an activated complex for certain solvolytic reaction. An activated complex of a catalytic reaction may also stabilised within such assemblies, facilitating catalytic reactions. Encapsulation provides selectivity and controls on the reaction rates significantly.

The hydrolysis of triphosphate to monophosphate is an example of catalysis involving a molecular recognition (Figure 1.12). The protonated form of aza-crown **1.12a** is a catalyst for the reaction which provides suitable hydrogen-bonding sites to bind to the triphosphate anion, exposing a terminal NH in appropriate geometry to form an N—P bond. This activation process facilitates the selective cleavage of P—O bonds and results in the formation of diphosphate. The P—N bonded intermediate reacts with water to regenerate the original triphosphate. Thus, equilibrium between the degradation of the triphosphate to mono- and diphosphate and their subsequent recombination is established.

Supramolecular features of carboxylic acid to form complementary hydrogen bonds with the carboxylic acid functional group of a reactant are utilised in a manganese-based catalyst (**1.13a**) to perform selective C—H bond activation of iboprufen. The catalyst has terpyridine ligand substituted by Kemp's triacid (Figure 1.13). Both Kemp's triacids form complementary hydrogen bonds with carboxylic acid groups of the two independent iboprufen molecules, as illustrated by intermediate species **1.13b**. Between these two molecules, one is recognised for oxidation, as it is in near proximity to the active manganese site. So only one of the bound molecules is a recognition site to be easily oxidised by the manganese-oxo species. Iboprufen has two oxidizable benzylic groups; one such group is located next to the carboxylic group. This is away from the reactive manganese-oxy site and

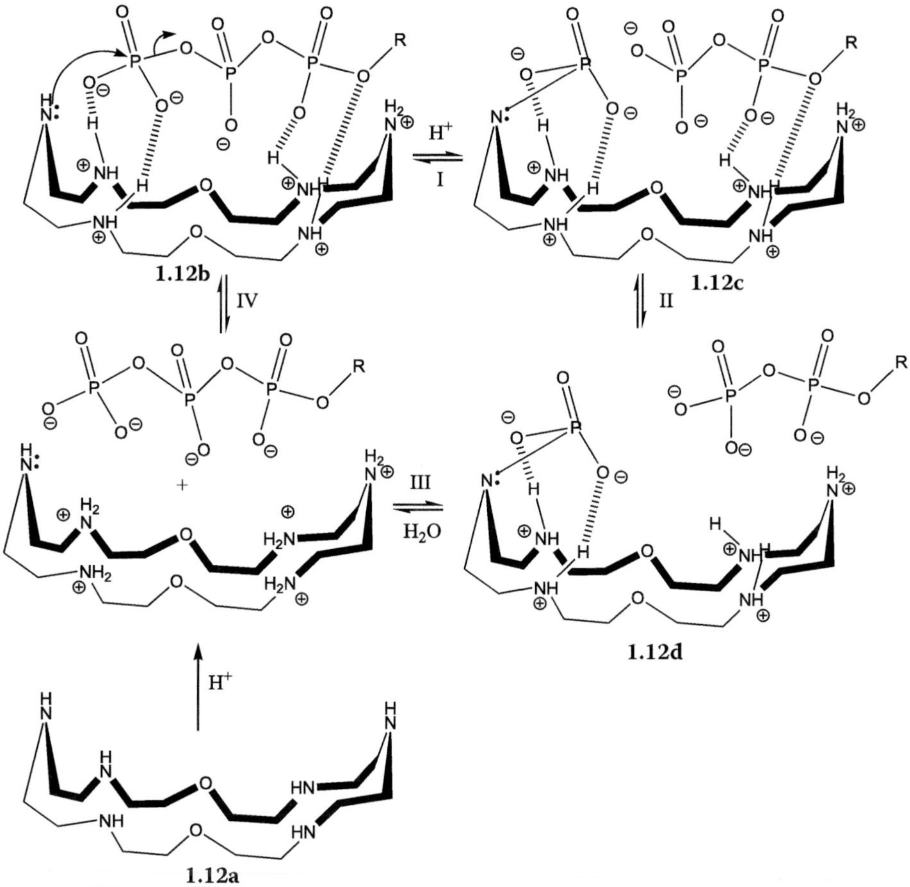

FIGURE 1.12 Triphosphate recognition and subsequent hydrolysis. (Yohannes et al. *J. Am. Chem. Soc.* 1985, 107, 8288–8289.[10])

does not get activated. The oxidation of a secondary benzylic position of iboprufen is also useful, as it generates a stable oxidised compound, whereas if the oxidation next to carboxylic acid takes place, it generates a compound that would easily undergo decarboxylation.

Having control to obtain stereo-selective products from catalytic reactions is a major concern. Supramolecular catalysis provides a solution for stereo-selective synthesis in conventional and photochemical reactions. For example, a catalyst with a triplet sensitiser linked to Kemp's triacid derivative **1.14b** helps to bring the reactant 4-(allyloxy)quinolin-2(1H)-one to close contact with the sensitising group. The intermediate formed between the substrate with a catalyst is due to a complementary hydrogen bond in the form of adduct **1.14c**. The reaction shown in Figure 1.14 results in the formation of specific stereoisomer (3aS,4aR,10bR)-3,3a,4,4a-tetrahydro-2H-2-oxa-bicyclo[3.2.0]hept-1(7)-eno[7,1-c] quinolin-5(6H)-one.

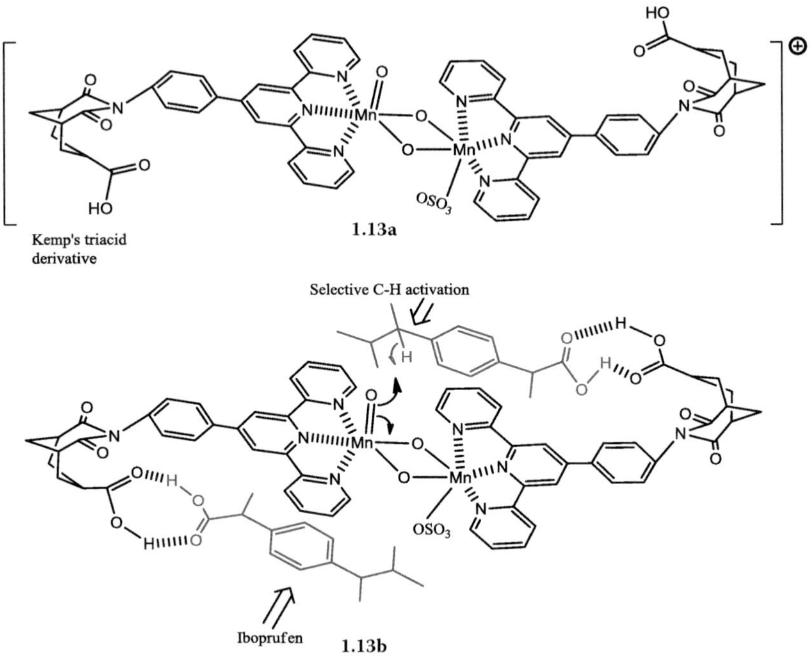

FIGURE 1.13 A manganese catalyst for C—H oxidation of iboprufen. (D. Balcells et al. *Dalton Trans.* 2009, 5989–6000.[11])

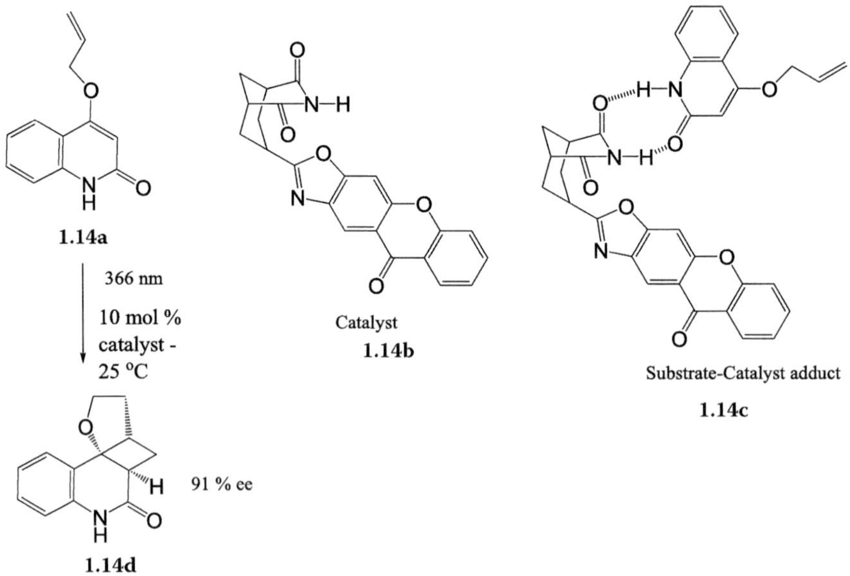

FIGURE 1.14 A photo-cyclisation reaction catalysed by hydrogen-bonded intermediate with catalyst. (M. M. Maturi, T. Bach, *Angew. Chem. Int. Ed.* 2014, 53, 7661–7664.[12])

1.4 SELF-ASSEMBLY OF MULTICOMPONENTS SUITABLE FOR CATALYTIC REACTIONS

Efficient and selective catalysts are developed by introducing supramolecular features to a conventional catalyst. For example, 1-octene reacts with carbon monoxide and hydrogen to undergo hydroformylation reaction as well as an isomerization reaction in the presence of catalytic amount of the complex $[Rh(H)(PR_3)_2(CO)_2]$ (R=aryl, alkyl) to yield nonan-1-al, nonan-2-al and 2-octene. The $[Rh(H)(PR_3)_2(CO)_2]$ complex adopts two different isomeric structures in solution. A similar complex to this rhodium complex but with a different composition, $[Rh(H)(CO)_3(PR_3)]$ shows catalytic reactivity for the above reaction differently. In terms of the rate of the formation of product, this compound has better catalytic activity. The reaction is nonselective for the formation of nonan-1-al. The catalytic hydroformylation reaction by the catalyst $[Rh(H)(PR_3)_2(CO)_2]$ is a relatively slow reaction but provides nonan-1-al as a major product over the others. Based on such observations, new catalysts for hydroformylation reactions are developed by introducing a pyridine-based phosphine-capped ligand **1.15b**. This ligand has nitrogen atom/s of pyridine/s anchored to zinc-porphyrin complexes to make highly hydrophobic environments. The ligand has also a large cone angle; hence, limited amounts of **Zntpp** can go into the coordination sphere of palladium. Thus, only one **Zntpp** ligand (Figure 1.15) binds to the palladium ion. Accordingly, the phosphorus site of **tpp** and **Zntpp** independently ligates to rhodium (I) ion and *in-situ* complexes [Rh(acac)(**tpp**)(CO)] and [Rh(acac)(**Zntpp**)(CO)] (where acac=acetylacetonate anion), respectively, are generated for the catalytic hydroformylation reaction. Both the complexes are active catalysts for the hydroformylation reaction of 1-octene at 80°C and 20 bar pressure. The rhodium complexes[Rh(acac)(**tpp**)(CO)] and [Rh(acac)(**Zntpp**)(CO)] drastically influence the catalytic hydroformylation process. With the catalytic amounts of these ligands, together with the $[Rh(CO)_2(acac)]$ catalyst, the catalytic turnover is doubled compared to using tpp than by using the ligand **Zntpp**. The percentage ratio of isomers relative to each other is also nearly doubled when **Zntpp** is used. The ratio of linear to branched aldehyde formed in the case of a catalytic reaction using **tpp** is 2.8, and for **Zntpp** it is 1.5. Thus, the extent of supramolecular features through hydrophobic sites and steric factors guides the ratios of linear and branched chain products as well as the enantiomers.

Self-assembling and complementing hydrogen bonds help to accomplish highly selective and difficult transformations. For example, functionalised C_{60} barbutric acid **1.16a** has a barbiturate unit inclined at about 45° to the C_{60} surface (Figure 1.16). Advantage of this point is taken to release a catalyst that facilitates the 2 + 2 photochemical cyclisation reaction. Two molecules of birbiturate derivative are assembled by complementing interactions with ureidopyrimidinone derivative to form the adduct **1.16c**. This dimer, upon irradiation at 337 nm in dilute solution, undergoes a photo-dimerisation reaction to give a dimer **1.16d** of C_{60}.

The thiourea derivative 2-ureidoisocytosine **1.17a** forms a dimeric noncovalent assembly **1.17b**. This assembly is formed as a result of complementing hydrogen bonds of the isocytosine unit in the compound. This assembly brings together thiol and epoxide molecules to react by acting as a template (Figure 1.17). The phosphonium

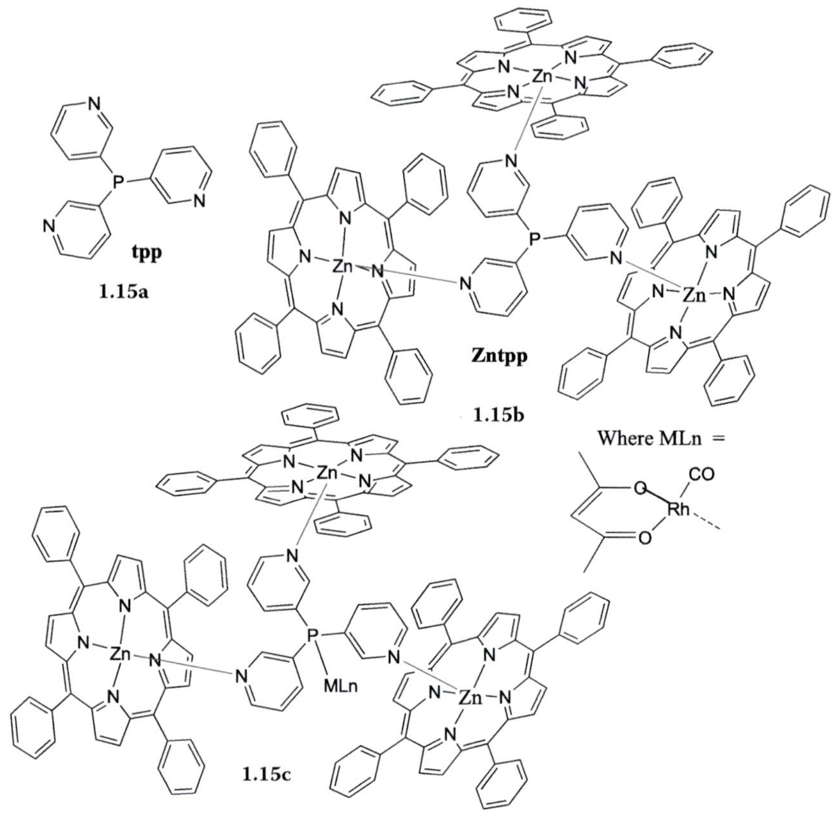

FIGURE 1.15 Ligand decoration by hydrophobic bulky units to improve catalytic efficacy. (V. F. Slagt et al. *J. Am. Chem. Soc.* 2004, 126, 1526–1536.[13])

ion present at the remote end of the parent compound favours self-assembly of the two molecules of **1.17a** to have an organization between them in a head-to-tail manner. Each counterpart of the parent molecule in the dimer has intramolecular hydrogen bonds. The carbonyl groups of the isocytosine form a hydrogen bond with a thiol molecule, and the phosphorus atom interacts with the oxygen atom of an epoxide to activate them. The self-assembly through hydrogen bond formation brings these reactants to close proximity for the reaction. The activated epoxide reacts with the assistance of the self-assembled 2-ureidoisocytosine derivative. The catalytic reactions take place simultaneously at the two ends of the dimeric self-assembled catalyst.

Supramolecular catalysis provides avenues to use multiple compounds to act as catalysts. Proline and an optically active thiourea derivative together act as catalyst (**1.18d**) for an enantioselective reaction between ethyl methyl ketone and aryl nitro-olefin compound **1.18a** to form the product **1.18c**. Proline reacts with ethyl methyl ketone and forms (S)-1-((E)-but-2-en-2-yl)pyrrolidine-2-carboxylic acid **1.18c** by a C—C bond formation reaction (Figure 1.18). This intermediate serves as a precursor

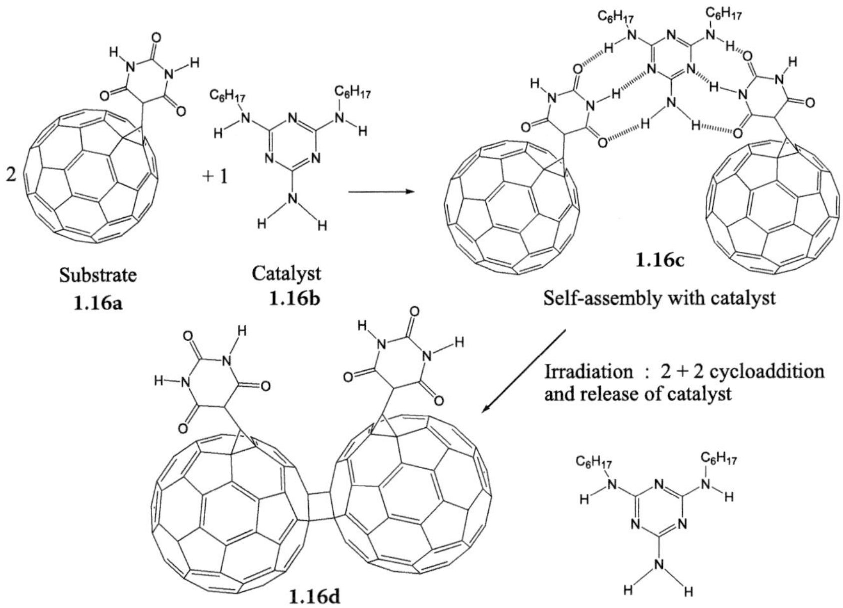

FIGURE 1.16 Hydrogen-bonded self-assembly and subsequent cleavage to provide catalytic means for a reaction. (N. D. McClenaghan et al. *J. Am. Chem. Soc.* 2003, 125, 13004–13005.[14])

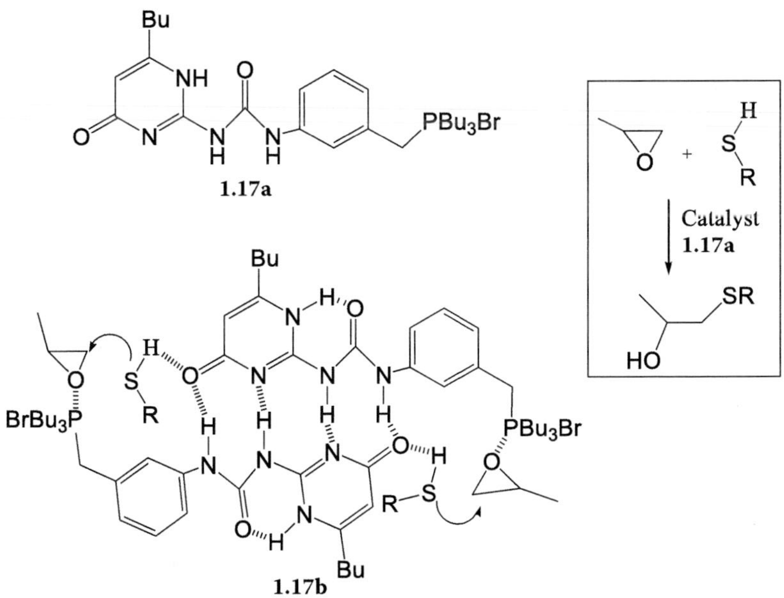

FIGURE 1.17 A urea-based catalyst and the self-assembly activating epoxide and thiol molecules. (K. Ohsaki et al. *Chem. Commun.* 2002, 1690–1691.[15])

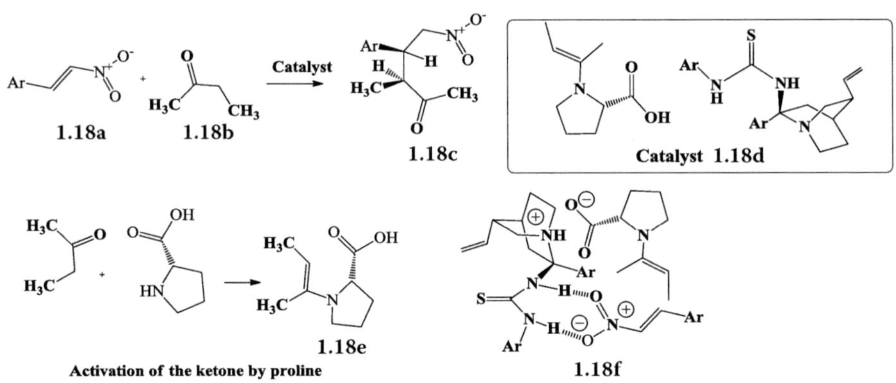

FIGURE 1.18 Two component catalysts for enantioselective Michael addition reaction. (B. List et al. *Org. Lett.* 2001, 3, 2423–2425.[16])

for an enantioselective C—C bond reaction by reacting with aryl-substituted nitro olefins in the presence of chiral thiourea catalyst. The chiral induction and activation process involves the intermediate **1.18f**. The nitro group of the one reactant hydrogen bonds with two N—H bonds of the thiourea part of the catalyst, whereas the proline-based precursor carboxylic acid forms an ion pair by protonating the amine part of the thiourea. The intermediate thus formed gives the product, and the overall reaction involves release of the product from the proline and thiourea adduct, to perform the next catalytic cycle.

Certain compounds are difficult to oxidise, but once such compounds form an assembly or complex, they undergo oxidation reactions. Assembling, complexation or functional modification relocates the positions of the functional groups for activation. Such changes provide the appropriate geometry for a reaction to occur. This is equivalent to post-modification of certain reactive sites to enhance the reactivity of the original compounds, leading to unusual reactivity. For example, the compound **1.19a** is not oxidised in ambient conditions to form the corresponding sulphoxide, but the particular sulphur site is easily oxidised by hydrogen peroxide through preparation of a zinc(II) Schiff base complex **1.19b**. During this conversion, the aldehyde groups are transformed to corresponding imines and the hydrogen atoms on the hydroxyl-groups are lost due to complexation. These two factors help the zinc complex to selectively and effectively oxidise the thioether to sulfone **1.19c** illustrated in Figure 1.19.

Host molecules are designed to offer concave surfaces to accommodate guests with convex shape. This commonly occurs in cation recognitions in macrocyclic polyethers like crown ethers. Based on cation recognition, the reactivity of a cation is triggered or the reactivity of a reactive site at a remote position is regulated. Using the recognition process, new catalysts are generated. The recognition process activates substrates by bringing them together, providing selectivity and modifying reaction intermediates. The recognition of alkali metal ions by crown ethers enhances the rate of the reactions of various water-soluble oxidants to function in nonaqueous media. For example, the oxidation activity of potassium permanganate in organic solvent suffers due to poor solubility, but the problem is partly overcome with the aid of

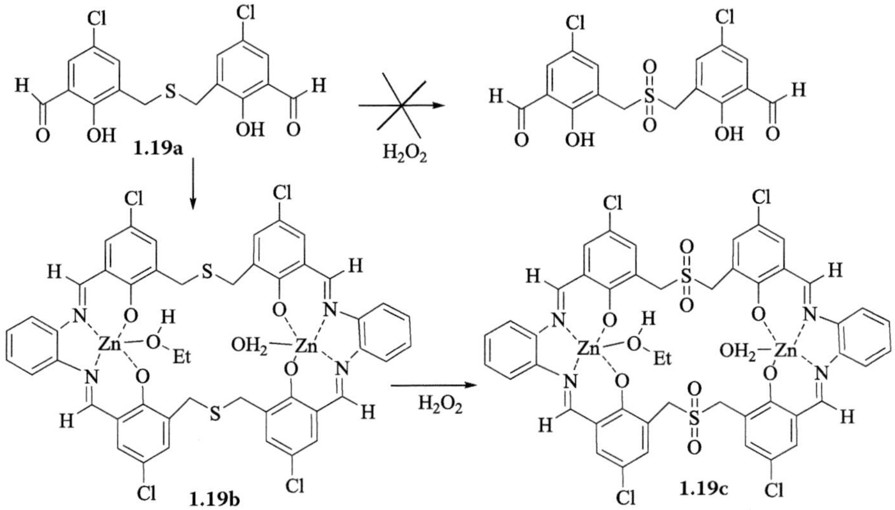

FIGURE 1.19 Oxidation facilitated by cyclic structure. (G. Feng et al. *Scientific Reports* 2017, 7, 15881.[17])

crown ethers, which trap potassium ions and facilitate oxidation reactions by taking the oxidant to a nonaqueous solvent. Crown ethers are commonly used to enhance the reactivity of a reagent in phase transfer catalysis where alkali cations are recognised by crown ethers and brought to interphase. The crown ethers connected as side arms of a compound bring an ionic reactant into close proximity to a reactive functional group. For example, the functionalised crown **1.20b** has the ability to recognise the ammonium cation. It binds to the protonated **1.20a** to form **1.20c**. The adduct provides suitable geometry to the substrate to undergo reaction with the functional group and catalytically attach on the side arm (Figure 1.20).

Crown ether combined with a ligating part modulates the reactivity of metal complexes. For example, the pincer-crown-ether ligand complex of iridium complex

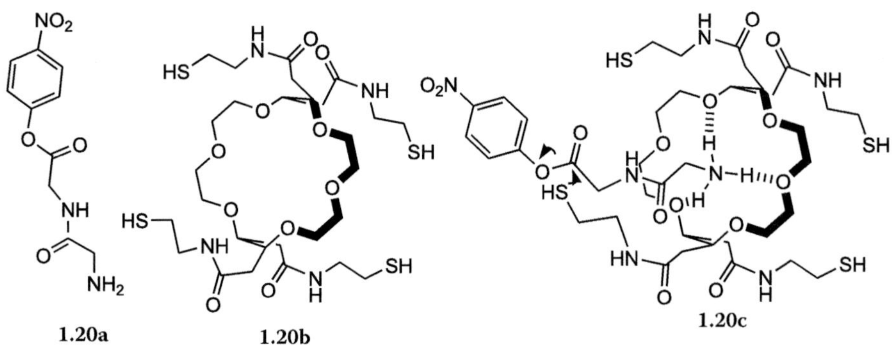

FIGURE 1.20 Crown ether helping in binding and activating ester group for attack by thiol unit. (K. Koga, S. Sasaki, *Pure Appl. Chem.* 988, 60, 539–543.[18])

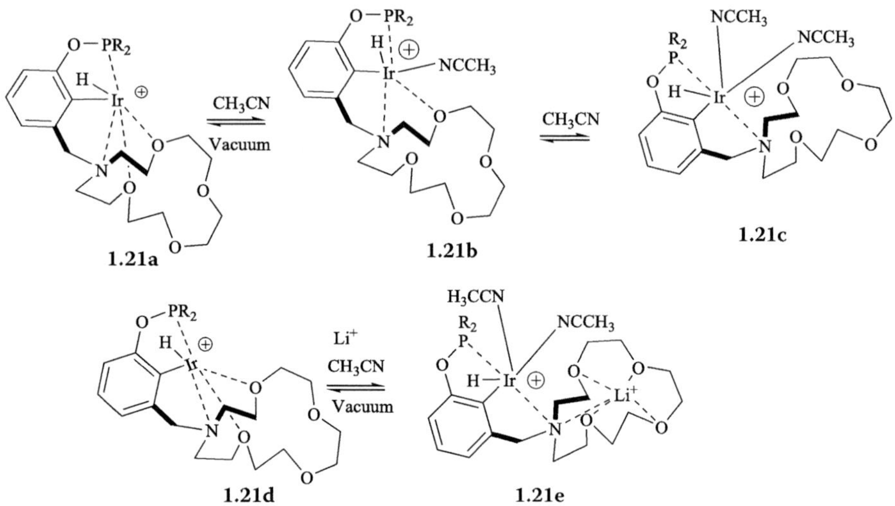

FIGURE 1.21 Manipulation of coordination effect through crown ether alkali metal ion binding. (M. R. Kita, A. J. M. Miller, *J. Am. Chem. Soc.* 2014, 136, 14519–14529.[19])

1.21a adopts different coordination modes (Figure 1.21). Depending on the number of coordinated acetonitrile molecules attached to the central iridium ion, the ability of the crown ether to coordinate iridium ion changes. Accordingly, the ligand adopts tri-dentate, tetra-dentate or penta-dentate binding to adopt different structures as shown through different equilibriums between structures **1.21b–e**. Reaction of the complex with an excess amount of acetonitrile forms a *bis*-acetonitrile complex, but this process is more easily achieved when the reaction is carried out in the presence of alkali metal cations. The alkali metal cations bind to the crown, facilitating the incorporation of acetonitrile as ligand. This happens as the ligating atoms on the crown are engaged to bind to the alkali metal cations. This generates vacant coordination sites on the iridium ion making coordination unsaturation. Such effects lower the activation energy of a reaction and help to create a suitable coordination environment at the iridium centre for catalysis.

Trapping cations by macrocycles is extensively utilised in phase transfer catalysis. But similar anion-guided catalysis is less known. One such example of an anion-guided reaction is the reaction between (E)-N-benzylidenebenzenamine (**1.22b**) and 2,3-dihydrofuran to form 2,3,3a,4,5,9b-hexahydro-4-phenylfuro[3,2-c]quinoline. This reaction is specifically catalysed by the dianion of ethane-1,2-disulfonic acid in the presence of cyclic *bis*-thiourea derivative **1.22e**. This reaction is catalysed by ethane-1,2-disulfonante, but the same catalytic reaction is accelerated by the macrocycle **1.22e**. The host macrocycle **1.22e** is an hydrogen-bonded adduct of acetone. The acetones are dislodged upon reaction with ethane-1,2-disulfonic acid, and the dianion is recognised by the crown effect to form **1.22f**. The catalytic path involves anion recognition process by the macrocycle and further participation of the **1.22f** in the catalytic reaction. Initially (E)-N-benzylidenebenzenamine reacts with ethane-1,2-disulfonic acid to form protonated imine and the bis-sulphite anion.

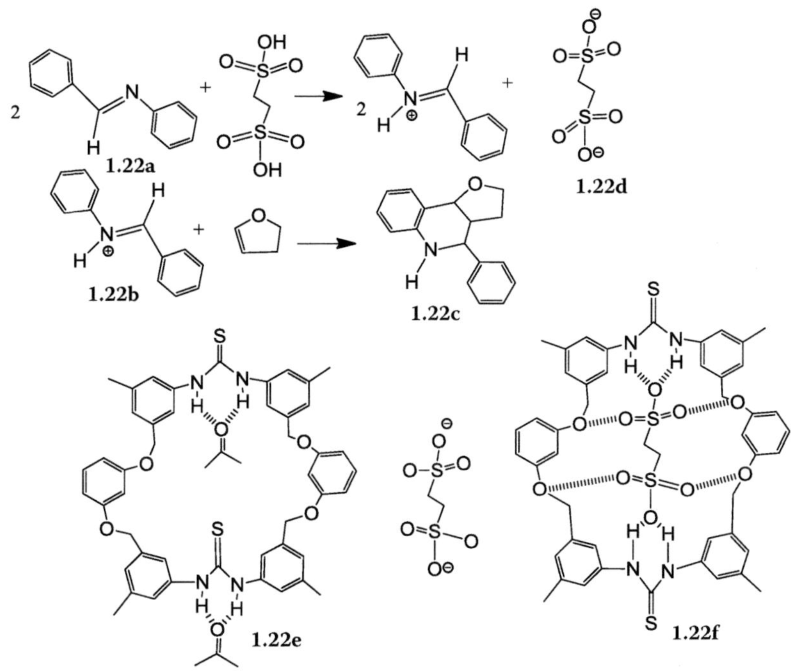

FIGURE 1.22 Anion binding enhancing reactivity of a reaction. (R. Ning et al. *Chem. Eur. J.* 2018, 24, 1–6.[20])

The bis-sulphite anion is recognised by the macrocyclic host (Figure 1.22). To do so the host molecule reorganises to selectively bind the dianion **1.22d**. The tight binding of the anion to the macrocycle facilitates the protonation of the imine. Using 0.25 mol % macrocycle in the reaction mixture, the reaction between imine with 2,3-dihydrofuran caused by the diacid turns out to be a catalytic reaction.

1.5 ENERGETIC ASPECTS OF SUPRAMOLECULAR CATALYSIS

There are two major paths for noncovalent activation of a substrate in supramolecular catalysis. In one case, the substrate provides space and a surface to anchor the catalyst by adopting or having complementing units in it. Alternatively, the catalyst accommodates the substrate through supramolecular features associated with the catalyst.

As described in the foregoing section, the supramolecular catalysis of conventional acid-base or substitution and elimination reactions may occur in many ways, some of which are shown in Figure 1.23. In equation (i) of the figure, it is shown that a substrate provides a template for weak interactions to bring the catalyst to close proximity and provide a correct geometry for undergoing a reaction. On the other hand, a part of the catalyst may also act as a template to activate and provide a correct geometry. These two types of intermediate processes are comparable to the formation of transient host—guest interactions where the host in the first case is the substrate,

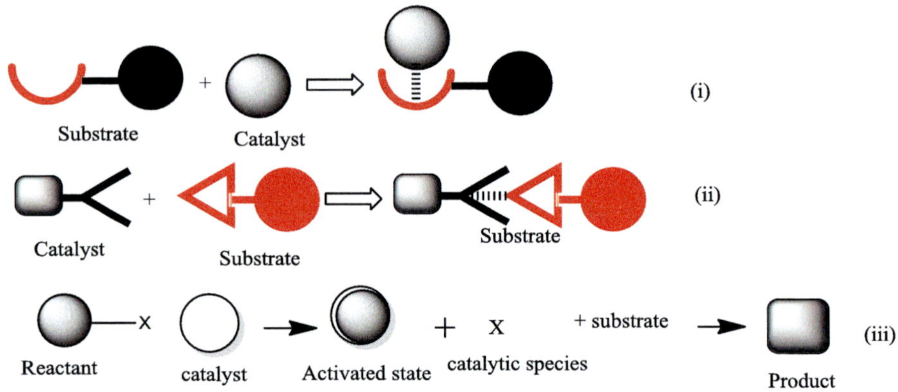

FIGURE 1.23 Some ways interactions facilitate catalytic activities.

whereas in the second case, the host is the catalyst. Accordingly, a reactant may react with the activated species by binding or without binding. There are situations in which two or more reactants are assembled by a catalyst, whereas a reactant may hold another reactant together with a catalyst through noncovalent interactions. There are also reactions, shown in a schematic way in equation (iii) of Figure 1.23, which involve activation of a selectively trapped reactant within a host molecule where the host acts as a catalyst. The reactive catalytic species generated at ease through the principles shown in the equations in Figure 1.23 cause catalytic reactions more effectively than that would have taken place ineffectively without these modifications.

In supramolecular systems, reactants get together; hence, a bimolecular or higher molecular reaction turns out to be like an intramolecular process of a unified entity. Such a process at times causes faster reactions than one where such interactions do not occur. For example, reactants may have direct interacting sites on its skeleton or have interacting sites tethered to them at a distal end. The assembling process in such systems helps a reaction to proceed faster. The increase in reactivity in such cases occurs as the entropy of the reaction decreases due to the formation of orderly arrangements. Hence, the binding equilibriums between two reactants through supramolecular systems have high priorities for accomplishing catalytic activity. A simple scheme of reactions differentiating a bimolecular reaction with or without supramolecular assistance is shown in Figure 1.24. It is necessary to find out the reactivity difference of a reaction carried out under an identical reaction that is assisted by supramolecular interactions with the one that has no such interactions

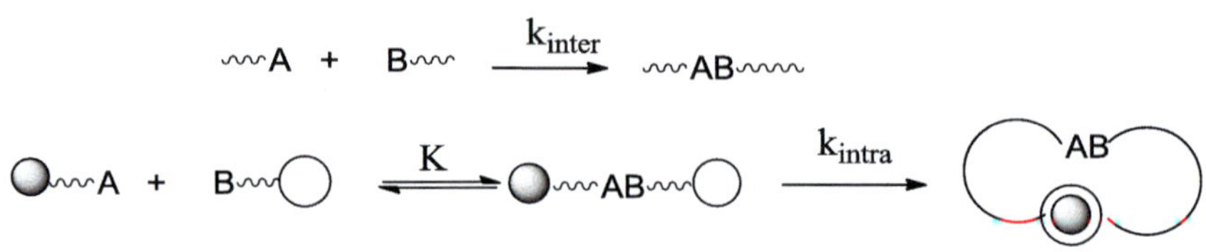

FIGURE 1.24 Formation of supramolecular adduct in supramolecular catalytic reactions.

but resembles the reactive intermediate. A reaction may proceed by association at the reactive site through two paths, shown in Figure 1.24. Such reactions without supramolecular features are generally used as references to determine the kinetic parameter called effective molarity. The effective molarity is the ratio between the rates of a supramolecular-assisted reaction and a similar reaction devoid of supramolecular assistance.[21] The effective molarity of a reaction is independent of the concentration of reactants and products. It is determined by comparing the pseudo—first order rate constants. On several occasions, the concentration of products is referred to as the parameter for comparisons.

There is no clear rationalization for the observed differences in the effective molarity in absolute terms to compare them in a generalised manner, but they provide information for comparisons of individual examples with analogous standards. It is customary to include strain and probability factors in the rate change effects caused by intramolecular reactions, shown in Figure 1.23. It is also a fact that for long chain tethering, a supramolecular site at a distal end will form an activated complex. The formation of which will effect the strain energy of the cyclic transition state and formation of it will be highly influenced by the probability of getting the ends together to the closest proximity to form cyclic motifs. Thus, there is a necessity of relationships that accounts for such features in addition to the rates.

To accommodate those two parameters, the effective molarity (EM) is expressed as

$$EM = \left[\left(\exp\left\{-\left(\Delta H^{\#}_{intra} - \Delta H^{\#}_{inter}\right)\right\}/RT\right] \cdot \left[\exp\left(\Delta S^{\#}_{intra} - \Delta S^{\#}_{inter}\right)/R\right]$$

Alternatively, effective molarity is expressed as the product of the two parameters EM_H and EM_S, one relating to enthalpy and other to entropy. In the equation, the quantity $\Delta H^{\#}_{intra} - \Delta H^{\#}_{inter}$ provides the extent of the strain in the ring-shaped transition state depicted in Figure 1.23. The effects from substituents and steric factors are accommodated in this parameter. The EM_H value is less when there is more ring strain. In a situation where the enthalpy of the ring formation and, without a ring, the enthalpy changes are comparable, the EM_H value tends to unity. This happens when the reactions are solely guided by the entropy parameter EM_S. The value of EM_S is decided by the number of single bonds in the tether that are utilised in the formation of the cyclic geometry. In solution, there are many opposing actions other than the scheme shown in Figure 1.24 that arise from the interactions of solvent molecules and functional groups. The entropy parameters from these effects are nullified.

In general, the effective molarity values are useful to understand the factors that affect intracomplex reactions and also help differentiate a stoichiometric reaction from a catalytic reaction. In the example given in Figure 1.25, the crown ether **1.25a** with redox active arms easily transfers hydride to the acceptor-activated aromatic unit of the reactant and thereby forms the oxidised form. The reactants **1.25b** and **1.25c** can both independently oxidise the compound **1.25a**. Both compounds independently cause a hydride-transfer reaction through the 1,4-dihydronicotinamide ring. The compound **1.25c** has a pendant ammonium cation; this cation of the reactant binds to the crown ether to form a supramolecular adduct **1.25d**. Due to formation of the adduct, the redox reaction is accelerated as compared to the similar

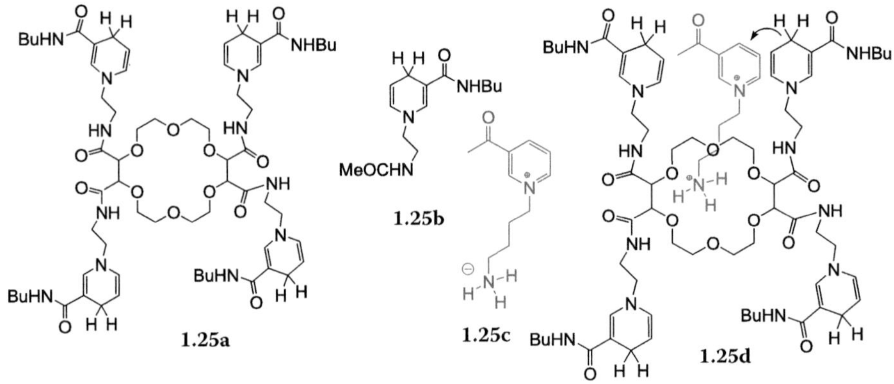

FIGURE 1.25 Crown ether as template for hydride transfer reaction. (J.-P. Behr, J.-M. Lehn, *J. Chem. Soc., Chem. Commun.* 1978, 143–146.[22])

redox reaction caused by the compound **1.25a**. The compound **2.15b** is devoid of the ammonium cationic unit to be recognised by the crown ether. The compound **1.25c** has an association constant of 10^4 M^{-1} in acetonitrile at 23°C. By comparing the rate of the reaction in the presence of the compound **1.25c** with the rate of the reaction in the presence of **1.25b** in acetonitrile, the effective molarity is found to be 0.0005 M.

Kemp's triacid-based receptor **1.26c** anchors the amine derivative **1.26a** and activates it for amide bond formation between them. The same activation is not possible by the N-methylated derivative of the receptor, as it does not have an anchoring ability. The self-assembled molecules formed by the hydrogen bond bring the amine unit attached through the methylene group to the five-member ring into close proximity to the ester linkage. This change provides the correct geometry for the nucleophilic substitution reaction to form an amide bond; as a result compound **1.26b** is selectively formed (Figure 1.26). The effective molarity in chloroform at about 23°C for this reaction is 0.16 M.

Reactions involving metal complexes embedded in a supramolecular assembly may have similar pathways to operate in catalysis, but the noncovalent interactions on the exterior of the catalytic site help increase efficacy and selectivity. Catalyst modifications in inorganic catalysis may occur several ways (Figure 1.27): (a) weak interactions such as hydrogen bonds, π-interactions and ion-pair interactions among the ligands to influence the catalytic path; (b) functional groups on the exterior ligands connected through or without tethers interact with another metal ion—generating polynuclear species to change catalytic activity; (c) the ligand is anchored to a template like self-assembled units, polymers, dendrimers and so on; and (d) confinement is provided by the ligand(s) to provide hydrophobic or hydrophilic pocket(s), influencing the reactivity.

However a catalyst is modified, the fundamental aspects of the principle of functioning of the catalyst remain unchanged. The activation barrier in the energy profile is modified. During the course of the conversion of a reactant to a product, the energy profile generally follows an uphill path, followed by a downhill energy

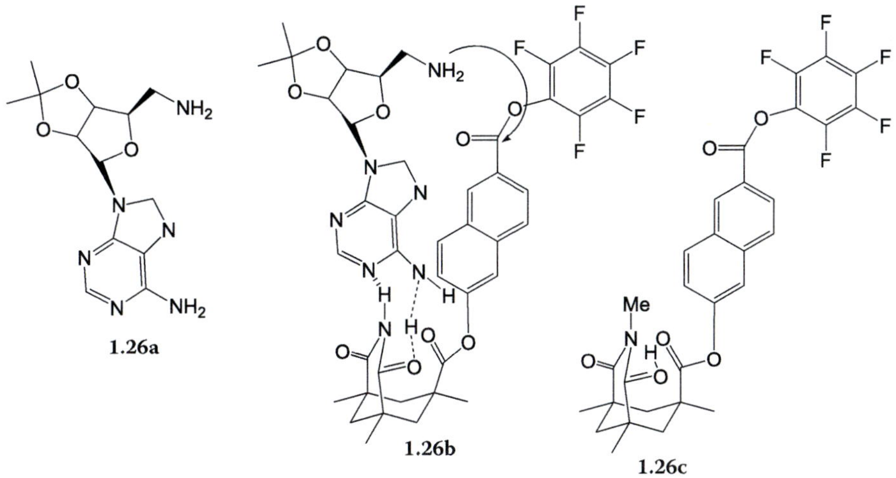

FIGURE 1.26 Functionalised Kemp's acid as template for amide bond formation. (J. S. Nowik et al. *J. Am. Chem. Soc.* 1991, 113, 8831–8839.[23])

path. The job of a catalyst is to reduce such an activation barrier. An energy profile involving a catalytic reaction may involve a single maximum or multiple maxima from intermediary steps. An energy profile depicting multiple numbers of intermediate steps in a catalytic reaction is shown in Figure 1.28.

In a catalytic reaction, substrates A and B, reacting to form product P by catalyst C, may pass through forming an intermediate AC, which may in turn react with B to yield product P. It may also be possible that BC, ABC or some other combination involving C forms a product. To understand such a process, let us consider the acid-catalysed substitution reaction of the chloro-pentaaquo-chromium(III) complex to

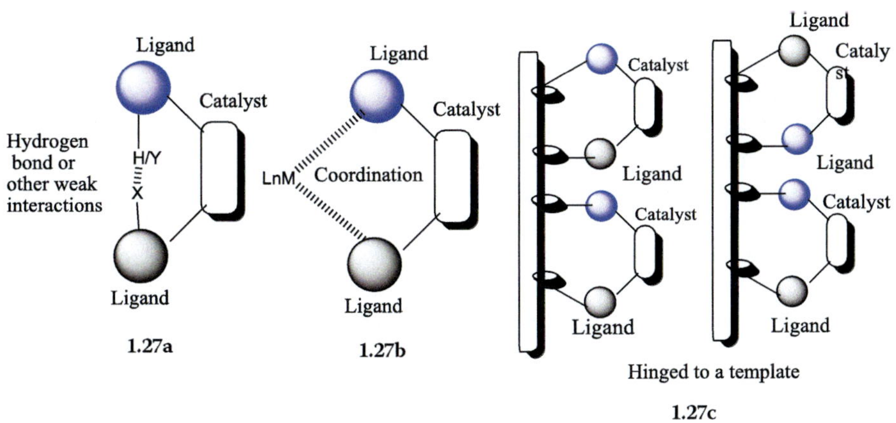

FIGURE 1.27 Weak interactions of ligands that result in modification of catalytic activity.

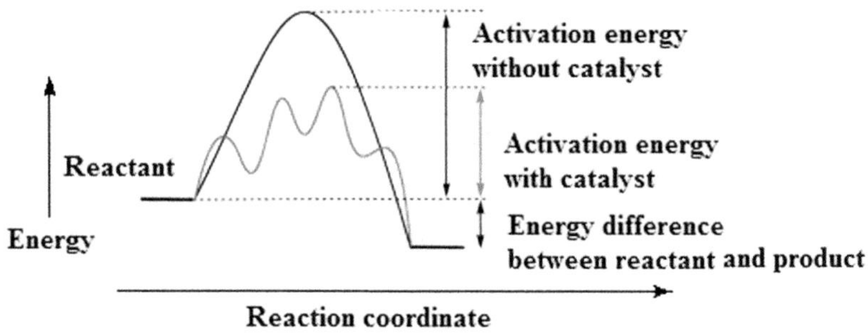

FIGURE 1.28 The activation energy profile for a catalytic reaction involving multiple intermediates as compared to an energy profile reaction without a catalyst.

form the hexa-aquo-chromium(III) complex. This reaction passes through multiple steps

$$[Cr(H_2O)_5Cl]^{2+} + H_2O \rightarrow [Cr(H_2O)_6]^{3+} + Cl^-$$

$$[Cr(H_2O)_5Cl]^{2+} \rightleftharpoons [Cr(H_2O)_5(OH)Cl]^{+} + H^{+} \qquad \text{fast}$$

$$[Cr(H_2O)_4(OH)CI]^{+} + [Cr(H_2O)_6]^{2+} \rightleftharpoons [Cr(H_2O)_5OH]^{2+} + [Cr(H_2O)_5Cl]^{+} \qquad \text{slow}$$

$$[Cr(H_2O)_5(OH)]^{2+} + H^{+} \rightleftharpoons [Cr(H_2O)_6]^{3+} \qquad \text{fast}$$

$$[Cr(H_2O)_5Cl]^{+} + H_2O \rightleftharpoons [Cr(H_2O)_6]^{3+} + Cl^{-} \qquad \text{fast}$$

The reaction is accelerated by the addition of a catalytic amount of $[Cr(H_2O)_6]^{2+}$. The rate of the reaction has direct proportionality to the concentrations of $[Cr(H_2O)_5Cl]^{2+}$ and $[Cr(H_2O)_6]^{2+}$ and is inversely proportional to the hydrogen ion concentration. The labiality associated with the chromium (II) ions and chemical inert behaviour of Cr (III) ions controls the overall reductive substitution reaction. The rate-determining step upon addition of $[Cr(H_2O)_6]^{2+}$ involves a hydroxy-bridged dinuclear species $[Cr(H_2O)_5(\mu\text{-}OH)Cr(H_2O)_5Cl]^{3+}$ that facilitates a rapid electron transfer process between the two chromium ions at different oxidation states; it alters their oxidation states thorough mononuclear complex intermediates. Due to the formation of mononuclear intermediates in the absence of $[Cr(H_2O)_6]^{2+}$, the reaction is relatively slow. This example suggests that the single activation barrier path may not suit to explain the catalytic activity of a catalyst, and the rate law governing unanalysed reactions may vary widely from the rate law of a catalytic reaction.

In supramolecular systems, the binding constant of a substrate or catalyst to a reacting molecule or template make great differences in catalysis. Such a difference in the bindings arises due to the complementing binding of neutral adducts as well as due to proton transfer or ionic interactions. In the examples shown in Figure 1.29, a dicarboxylic acid binds to a host through electrostatically guided hydrogen bonds, forming a salt **1.29a**, whereas in another case, it interacts through hydrogen bonds to form a cocrystal **1.29b**. The binding between two components may result

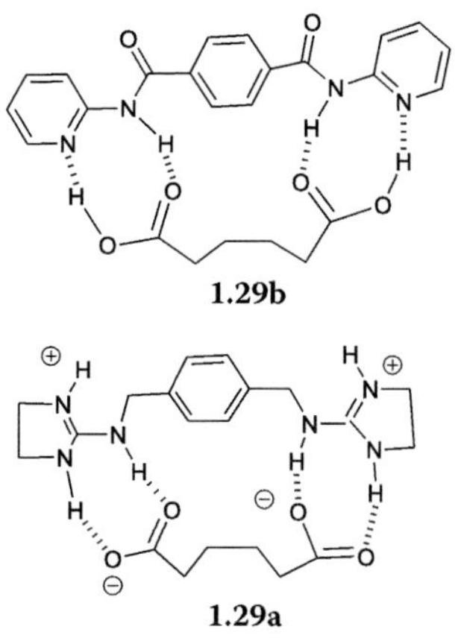

1.29b

1.29a

FIGURE 1.29 Recognition of dicarboxylic acid through salt formation or cocrystal formation. (E. Fan et al. *J. Am. Chem. Soc.* 1993, 115, 369–370.[24])

in different equilibrating self-assembly to adopt chainlike, ringlike or interlocked self-assembling, and each form will have differentiable interactions with a substrate. Due to such factors, the activation processes are reaction conditions and are substrate specific in supramolecular catalysis.

The stereochemistry or steric factors of intermediate species guide the formation of selective enantiomers. Enantioselective products from such reactions generally occur when one or more interacting species among substrate, catalyst or intermediate is asymmetric. Furthermore, asymmetric induction in a product is modulated thorough specific interactions at a remote site from the chiral centre. As an example, the crown ether-tethered guanidinium ion linked to a naphthyl group recognises L-tryptophan **1.30a**. Similar host—guest interactions are useful in asymmetric inductions in different catalytic reactions to provide stereo-selectivity. The π-interactions between the rings and the interactions of guanidinium-carboxylates together with the recognition of the ammonium cation by the crown ether provide adequate geometry to the assembly that suits binding L-tryptophan.

In another example, the host compound **1.30b** is composed of a structure which is like a functionalised calix-4-arene unit. It has two phenolic rings at alternating positions linked to an ethereal unit, and another two phenyl groups are functionalised to link carboxylic acids through rigid tethers. The crown-etherlike portion of the compound binds sodium ion to form **1.30c**. The flexibility of the crown etherlike portion of the compound (**1.30b**) is lost upon binding to the sodium ion. Due to this binding two carboxylic acid ends to relocate them to have a fixed separation from each other. Hence, the geometries of the sodium-bound complex are different from that without a sodium ion. Both form hydrogen bonds to tetrahydropyrimidin-2(1H)-one, but the binding constant in the absence of sodium for formation of the compound **1.30d** is 2200 M^{-1}, whereas in the presence of the sodium ion, it is 21,000 M^{-1}. This suggests that molecular recognition takes place when the sodium ion is present. Such concepts are important in supramolecular catalysis to guide substrate binding and influence the activated state.

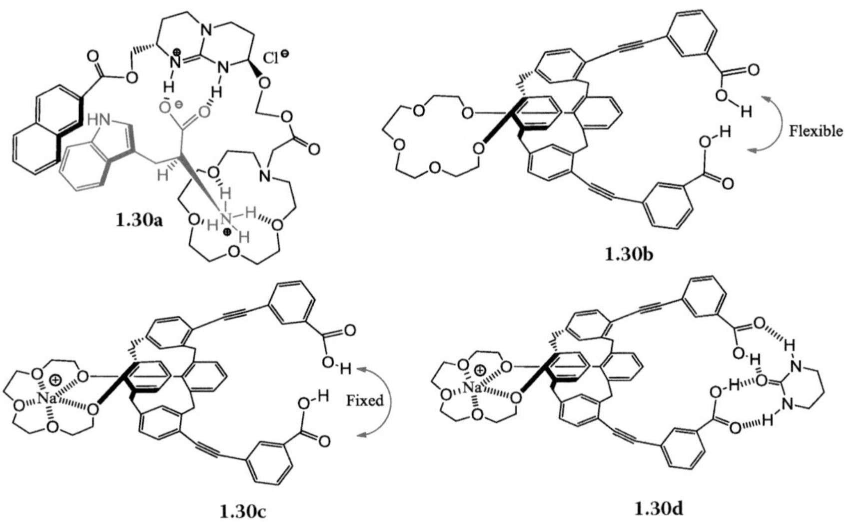

FIGURE 1.30 The stereo-electronic effects controlling selective binding. (A. Galan et al. *J. Am. Chem. Soc.* 1992, 114, 1511–1512.[25])

Cyclodextrins are an interesting class of water-soluble compounds that provide hydrophobic and hydrophilic sites to guest molecules. The compound **1.31a** is a cyclodextrin derivative in which the C2 position of one of the glucose units is functionalised with a flavin moiety that is a derivative of 10-methylbenzo[g]pteridine-2,4(3H,10H)-dione (Figure 1.31). This compound is suitable to form a host—guest complex with phenylmethanethiol (**1.31b**). In such a host—guest complex, phenylmethanethiol resides in the interior of hydrophobic pockets. Due to such encapsulation, exposed thiol groups are oxidised to disulphide in aqueous methanol under basic conditions. Under the same reaction conditions, a similar oxidation reaction of phenylmethanethiol occurs with riboflavin (**1.31c**). Comparisons of both reactions provide an effective molarity of 0.10 M for the reaction mediated by cyclodextrin. These reactions are catalytic in the presence of air. Oxygen present in air reconverts the flavin formed during the reaction to riboflavin, and the catalytic process continues.

Supramolecular catalysis provides a scope to have designs based on large numbers of combinatorial approaches to supramolecular interactions. Contact catalysis is one such approach, in which a neutral molecule is separated as cation and anion by interacting at two different locations of a host molecule. By doing such a separation, the molecule is activated and performs reactions that conventionally can not be achieved. For example, squaramide **1.32a** reacts with trialkylsilylfluorosulphonate (R_3SiSO_3F) to separately stabilise the anion and cation parts at different locations of the squaramide derivative. The SiR_3^+ species is held by two carbonyls of the squaramide, whereas the trifluoromethylsulphonate part is stabilised by hydrogen bonding with the N—H bonds to form an intermediate **1.32b**. This intermediate reacts with **1.32d** and forms intermediate **1.32c**, which reacts with furan to form the product (2R)-2-methoxy-8-oxabicyclo[3.2.1]oct-6-en-3-one **1.32e** and regenerates the catalyst for next cycle of reaction, as illustrated in Figure 1.32.

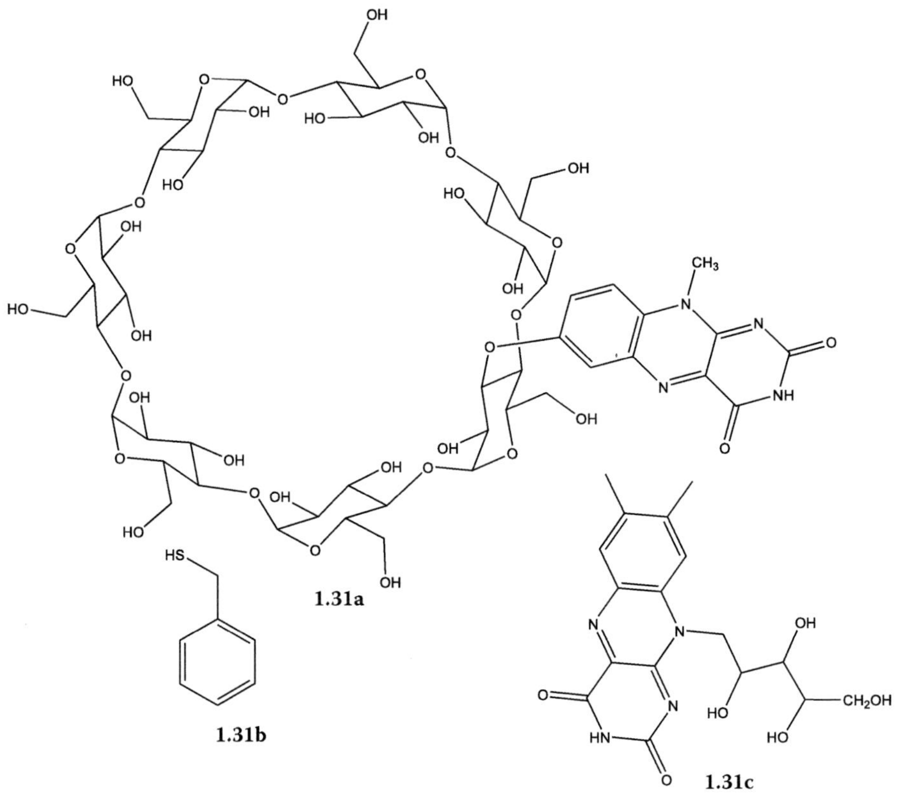

FIGURE 1.31 Cyclodextrin-tethered flavin for oxidation of phenylmethanethiol. (M. Kojima et al. *J. Chem. Soc. Perkin Trans.* 1981, 1, 1647–1651.[26])

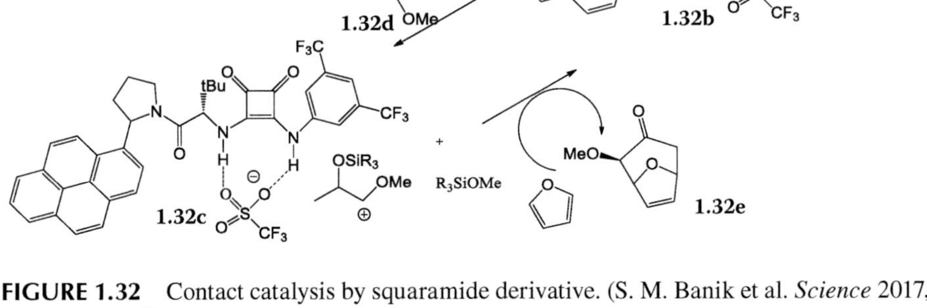

FIGURE 1.32 Contact catalysis by squaramide derivative. (S. M. Banik et al. *Science* 2017, 358, 761–764.[27])

1.6 ADVANTAGES OF SUPRAMOLECULAR CATALYSIS

The foregoing discussions clearly suggest that kinetic study coupled to thermodynamics provides an understanding of supramolecular catalysis as well as the activation process of a reaction that utilises supramolecular effects. Significant progress in understanding self-assemblies and the knowledge gathered for predesigning supramolecular assemblies has helped in the development of new supramolecular catalyst systems. The self-assemblies formed by the combined effects of the dative bonds with other weak interactions have enabled the preparation of new container molecules, porous materials, capsules, nanovessels, clefts and noncovalently linked interlocked systems. These systems have opened new dimensions in the area to carry out catalytic reactions. The characteristic features associated with these container molecules attract interest and provide unique examples of molecular assemblies. Self-assembling in different ways to recognise molecules, equilibrate and provide flexible structures through induced fitting methods helps to systematically develop catalysts that are understood at the molecular level. Such catalysts have stability, robustness and specific functions. They are able to generate polymers and imprinted materials in designed manner. Selective substrate binding, trapping of reactive intermediates or stabilisation of unstable species within cavities of hosts provide avenues for efficient and green methods to mimic biological reactions. In inorganic chemistry, cages and clusters have taken forward catalytic reactions to guide desired products. There are many complexes or supramolecular assemblies of metal complexes which possess kinetically labile metal—ligand interactions; these interactions are useful to enforce the correct equilibrium to push a reaction towards a specific goal. The stability of supramolecular assemblies depends on the concentration; lower molecular weight species are favourably formed in dilute solution. Thus, reactivity based on concentration-dependent reactions has emerged. The catalytic reactions of helicates provide enantiomeric products.

The self-assembly of the compound **1.33a** has a shape resembling a tennis ball (Figure 1.33). The compound serves as a container to bring together dienophile and diene into its void. It acts as a catalyst for a 2 + 2 cycloaddition reaction. The reaction shown in Figure 1.33 involves the inclusion of two molecules of 1,4-benzoquinone into the tennis ball-like enclosure to form **1.33b**. One molecule of benzoquinone from this inclusion compound is replaced by 2,5-dimethylthiophene 1,1-dioxide, which acts as a diene. In the next step, the diene and dienophile (benzoquinone) undergo a cyclisation reaction within the assembly (**1.33d**) and form (1S,4S)-1,4-dimethyl-1,4-epithionaphthalene-5,8(1H,4H)-dione 9,9-dioxide as a cycloaddition product. The product is finally released from the cavity and the next catalytic cycle begins by the inclusion of diene and dienophile.

Size- and shape-selective spaces are created for specific reactions through assembling inorganic or organic components. One way to do this is by organizing different similar or dissimilar molecules as barricades to make different geometries surrounding the active site. The reactants are included into voids and the efficiency of a reaction taking place within the void is enhanced. For example, the porphyrin-based manganese complex **1.34** provides space for size- and shape-selective olefin in the stacks between the porphyrin rings. It is an efficient catalyst for size-selective

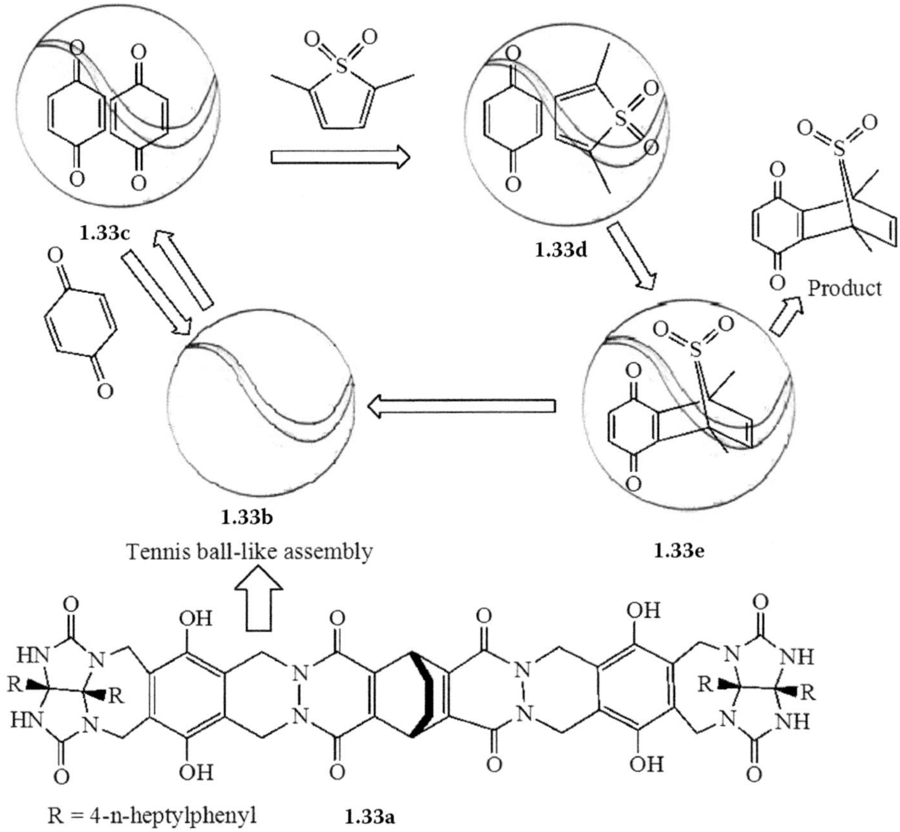

FIGURE 1.33 Tennis ball-like assembly for catalytic 2 + 2 cycloaddition reaction. (J. Kang et al. *J. Am. Chem. Soc.* 1998, 120, 3650–3656.[28])

olefin epoxidation reactions. The manganese ion coordinated by the porphyrin ring is located centrally (Figure 1.34), this is the active site for this catalytic epoxidation reactions of olefin in the presence of tertiary butyl hydro peroxide.

Metallic surfaces provide support to immobilise a catalyst as well as to perform catalytic reactions differently from the catalyst without the support. In the example illustrated in Figure 1.35, the gold surface provides the anchoring site to the thiol-bearing organo-catalyst **1.35b**. The catalyst anchored on the gold surface acts as a bifunctional catalyst for the intramolecular cyclisation reaction of phosphate ester, namely 2-hydroxypropyl (4-nitrophenyl) phosphate (**1.35d**). The phosphate ester has an alcohol functional group on one side which is recognised to form the intermediate **1.35c**. This intermediate facilitates the cyclisation reaction to yield the product **1.35e** and 4-nitrophenol. The intermediate **1.35c** has two imidamide units forming hydrogen bonds with one molecule of the reactant; this hydrogen-bonded assembly activates the reactant by increasing the nucleophilicity of the hydroxyl group and providing a suitable geometry for the catalytic reaction.

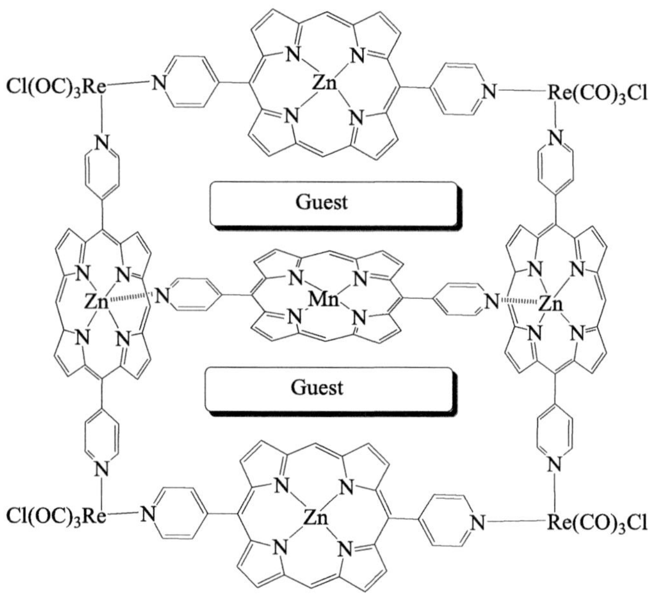

1.34

FIGURE 1.34 A multiporphyrin complex with suitable space for guests to cause catalytic epoxidation. (M. Merlau et al. *Angew. Chem. Int. Ed.* 2001, 40, 4239–4242.[29])

Selective acetylation reactions on carbohydrate molecules are done by using protein-based catalysts. For example, the acetylation reaction of carbohydrate **1.36c** is catalysed by N-methyl imidazole **1.36a** as well as an oligopeptide-based catalyst **1.36b**. The compound **1.36b** is a superior catalyst for the acetylation reaction at the selective position of the carbohydrate with a higher efficiency than the N-methyl imidazole. The carbohydrate has two distinguishable OH in the molecules; these groups are marked as O_aH and O_bH in Figure 1.36. The acetylation of the carbohydrate at these positions caused by the catalysts differ and so does the yield of the reaction. The peptide provides high mono-acylation at the O_aH position (90%), whereas N-methyl imidazole gives a mixture of acyl-products through mono-acylation as well as diacylation. Protein templates are used to hold multiple catalysts to perform independent site-selective catalytic reactions.

The self-assembly of cyclodextrin-based molecules improves the catalytic efficiency of organometallic reactions under biphasic conditions. Triglycerides are natural compounds with double bonds. The conversion of the C=C of triglycerides to aldehydes provides a source for preparing various functional groups such as alcohols, amines or carboxylic acids. A hydroformylation reaction by using an organometallic catalyst is promising to functionalise them, for which the other reactants are carbon monoxide and hydrogen. However, in such a catalytic process, solubility is the major concern. Triglycerides are hydroformylated through supramolecular means by using cyclodextrins along with a water-soluble rhodium catalyst.[32] In this reaction, a transient supramolecular complex is formed between triglyceride alkyl chains and

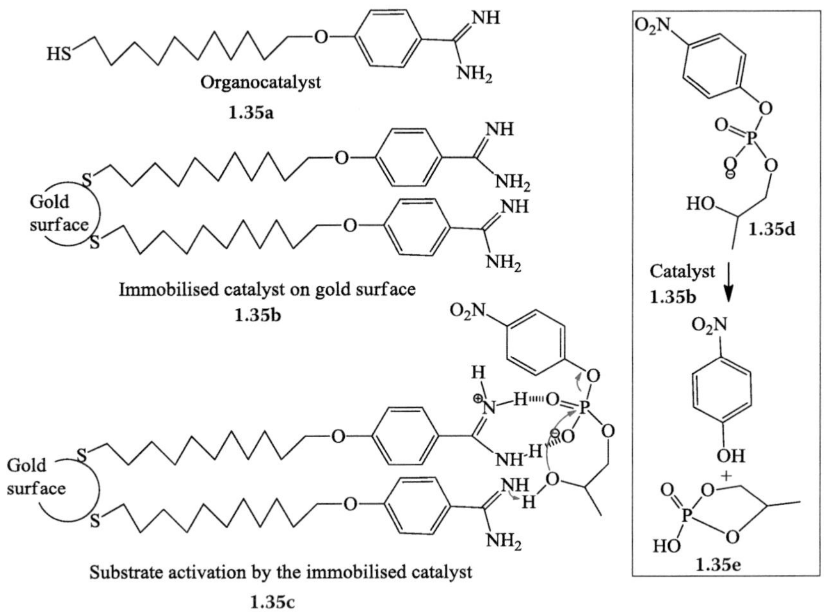

FIGURE 1.35 Surface of metal controlling self-assembling in catalysis. (R. Salvio, A. Cincotti, *RSC Adv.* 2014, 4, 28678–28682.[30])

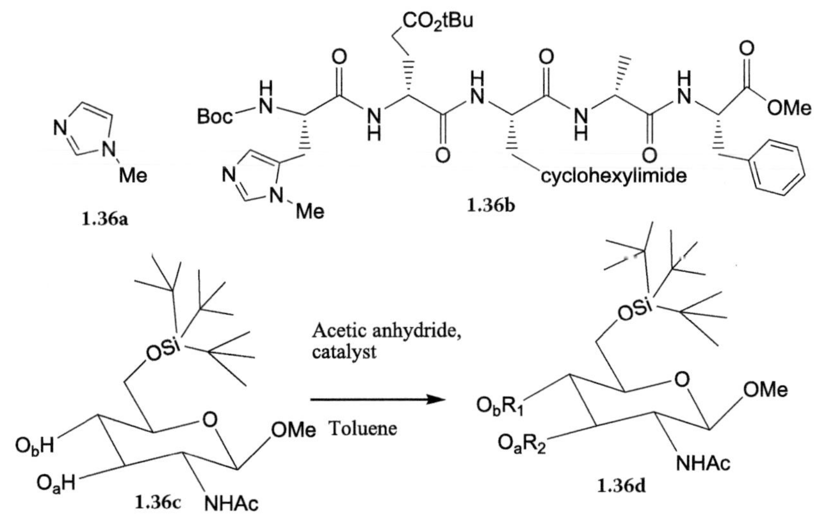

When Catalyst is **1.36a** R_1 = H, R_2 = $-COCH_3$ is 50 %
R_1 = $-COCH_3$ R_2 = H is 22 % and 28 % diacetylated product

Catalyst is **1.36b** R_1 = H, R_2 = $-COCH_3$ is 90 %
R_1 = $-COCH_3$ R_2 = H is 3 % and No diacetylated product

FIGURE 1.36 Selective acylation catalysed by peptide. (K. S. Griswold, S. J. Miller, *Tetrahedron* 2003, 59, 8869–8875.[31])

cyclodextrin at specific concentrations. At such concentrations, the interphase of the triglyceride/water is extended to favour contacts between the double bonds and the rhodium catalyst. Cyclodextrin molecules also control the second coordination sphere of the metal catalyst to influence the observed selectivity.

Immobilised catalysts are of general concern due to their high efficacy and reusability. Immobilised polyoxometalate complexes by surfactants provide more scopes to modify a catalyst for selective transformations than the original polyoxometallate. A hydrophobic nanodimensional environment is created in such catalysis. The substrate binding with polyoxometallate complexes controls and releases the product from a catalytic site. Sol-gel techniques to immobilise reverse-micellelike supramolecular catalysts are in practice to develop new catalysts. The advantages of such catalysts are due to the easy manipulation for modification as well as efficacy. Stimuli-guided catalytic processes through interlocked systems are used for synthesis peptides and also to cause selective reactions through ON and OFF catalytic processes.

There is a bright future ahead with zeolite-based transition metal catalysts needing an understanding of functional behaviour correlating with structures on metal—zeolite composites. Furthermore, nanometal ensembles trapped in zeolites or dendrimerlike structures as well as the use of oxides and metal-chalcogenides would provide active supported catalysts, and such catalysts are expected to perform catalysis at the conjunction of homogeneous and heterogeneous catalysis. Hence, rational design of active and selective heterogeneous catalysts is desirable. Greater attention is being given to supramolecular catalysis in material synthesis through biological processes with biological and biodegradable components, and this will provide superior artificial models to compete with natural processes through self-construction of vessels and performing reactions. The fundamental basis for these aspects are dealt with in the other chapters of this book.

REFERENCES

1. J. M. Lehn, Supramolecular reactivity and catalysis. *Appl. Catal. A: General* 1994, 113, 105–114.
2. M. Raynal, P. Ballester, A. Vidal-Ferran, P. W. N. M. van Leeuwen, Supramolecular catalysis. Part 1: Non-covalent interactions as a tool for building and modifying homogeneous catalysts. *Chem. Soc. Rev.* 2014, 43, 1660–1733.
3. H. J. Davis, R. J. Phipps, Harnessing non-covalent interactions to exert control over regioselectivity and site-selectivity in catalytic reactions. *Chem. Sci.*, 2017, 8, 864–877.
4. J. Meeuwissen, J. N. H. Reek, Supramolecular catalysis beyond enzyme mimics, *Nature Chem.* 2010, 2, 615–621.
5. R. Breslow, Biomimetic chemistry and artificial enzymes: Catalysis by design. *Acc. Chem. Res.* 1995, 28, 146–153.
6. S. M. Marques, J. C. G. E. da Silva, Firefly bioluminescence: A mechanistic approach of luciferase catalyzed reactions. *IUBMB Life* 2009, 61, 6–17.
7. U. Obstl, D. W. Banner, L. Weber, F. Diederich, Molecular recognition at the thrombin active site: Structure based design and synthesis of potent and selective thrombin inhibitors and the x-ray crystal structures of two thrombin-inhibitor complexes. *Chem. Bio.* 1997, 4, 287–295.
8. K. Drauz, H. Groger, O. May (Eds.) *Enzyme Catalysis in Organic Synthesis*, 3rd Ed. 2012. Wiley-VCH Verlag GmbH & Co. KGaA.

9. D. Singh, P. Bhattacharyya, J. B. Baruah, Structural studies on solvates of cyclic imide tethered carboxylic acids with pyridine and quinoline. *Cryst. Growth Des.* 2010, 10, 348–356.
10. P. G. Yohannes, M. P. Merles, K. B. Merles, Pyrophosphate formation via a phosphoramidate intermediate in polyammonium macrocycle/metal ion-catalyzed hydrolysis of ATP. *J. Am. Chem. Soc.* 1985, 107, 8288–8289.
11. D. Balcells, P. Moles, J. D. Blakemore, C. Raynaud, G. W. Brudvig, R. H. Crabtree, O. Eisenstein, Molecular recognition in Mn-catalyzed C–H oxidation. Reaction mechanism and origin of selectivity from a DFT perspective. *Dalton Trans.* 2009, 5989–6000.
12. M. M. Maturi, T. Bach, Enantioselective catalysis of the intermolecular [2 + 2] photocycloaddition between 2-pyridones and acetylene dicarboxylates. *Angew. Chem. Int. Ed.* 2014, 53, 7661–7664.
13. V. F. Slagt, P. C. J. Kamer, P. W. N. M. van Leeuwen, J. N. H. Reek, Encapsulation of transition metal catalysts by ligand-template directed assembly. *J. Am. Chem. Soc.* 2004, 126, 1526–1536.
14. N. D. McClenaghan, C. Absalon, D. M. Bassani, Facile synthesis of a fullerene-barbituric acid derivative and supramolecular catalysis of its photoinduced dimerization. *J. Am. Chem. Soc.* 2003, 125, 13004–13005.
15. K. Ohsaki, K. Konishi, T. Aida, Supramolecular acid/base catalysis via multiple hydrogen bonding interaction. *Chem. Commun.* 2002, 1690–1691.
16. B. List, P. Pojarliev, H. J. Martin, Efficient proline-catalyzed Michael additions of unmodified ketones to nitro olefins. *Org. Lett.* 2001, 3, 2423–2425.
17. G. Feng, Y. Shi, L. Zhang, R. Shi, W. Huang, R. Wang, Air-oxidation from sulfur to sulfone bridged Schiff-base macrocyclic complexes showing enhanced antimicrobial activities. *Scientific Reports* 2017, 7, 15881.
18. K. Koga, S. Sasaki, Functionalization of crown ethers: An approach to the enzyme model for peptide synthesis. *Pure Appl. Chem.* 1988, 60, 539–543.
19. M. R. Kita, A. J. M. Miller, Cation-modulated reactivity of iridium hydride pincer-crown ether complexes. *J. Am. Chem. Soc.* 2014, 136, 14519–14529.
20. R. Ning, Y.-F. Ao, D.-X. Wang, Q. -Q. Wang, Macrocycle-enabled counter anion trapping for improved catalytic efficiency. *Chem. Eur. J.* 2018, 24, 1–6.
21. S. D. Stefano, R. Cacciapaglia, L. Mandolini, Supramolecular control of reactivity and catalysis—effective molarities of recognition-mediated bimolecular reactions. *Eur. J. Org. Chem.* 2014, 7304–7315.
22. J.-P. Behr, J. -M. Lehn, Enhanced rates of dihydropyridine to pyridinium hydrogen transfer in complexes of an active macrocyclic receptor molecule. *J. Chem. Soc., Chem. Commun.* 1978, 143–146.
23. J. S. Nowik, Q. Feng, T. Tjivikua, P. Ballester, J. Rebek Jr., Kinetic studies and modeling of a self-replicating system. *J. Am. Chem. Soc.* 1991, 113, 8831–8839.
24. E. Fan, S. A. Van Arman, S. Kincaid, A. D. Hamilton, Molecular recognition: Hydrogen-bonding receptors that function in highly competitive solvents. *J. Am. Chem. Soc.* 1993, 115, 369–370.
25. A. Galan, D. Andreu, A. M. Echavarren, P. Prados, J. de Mendoza, A receptor for the enantioselective recognition of phenylalanine and tryptophan under neutral conditions. *J. Am. Chem. Soc.* 1992, 114, 1511–1512.
26. M. Kojima, F. Toda, K. Hattori, β-cyclodextrin–nicotinamide as a model for NADH dependent enzymes. *J. Chem. Soc. Perkin Trans.* 1981, 1, 1647–1651.
27. S. M. Banik, A. Levina, A. M. Hyde, E. N. Jacobsen, Lewis acid enhancement by hydrogen-bond donors for asymmetric catalysis. *Science* 2017, 358, 761–764.
28. J. Kang, G. Hilmersson, J. Santamaria, J. Rebek Jr., Diels-Alder reactions through reversible encapsulation. *J. Am. Chem. Soc.* 1998, 120, 3650–3656.

29. M. Merlau, M. Mejia, S. Nguyen, J. Hupp, Artificial enzymes formed through directed assembly of molecular square encapsulated epoxidation catalysts. *Angew. Chem. Int. Ed.* 2001, 40, 4239–4242.
30. R. Salvio, A. Cincotti, Guanidine based self-assembled monolayers on Au nanoparticles as artificial phosphodiesterases. *RSC Adv.* 2014, 4, 28678–28682.
31. K. S. Griswold, S. J. Miller, A peptide-based catalyst approach to regioselective functionalization of carbohydrates. *Tetrahedron* 2003, 59, 8869–8875.
32. T. Vanbesien, E. Monflier, F. Hapiot, Supramolecular emulsifiers in biphasic catalysis: The substrate drives its own transformation. *ACS Catal.* 2015, 5, 4288–4292.

2 Organic and Inorganic Supramolecular Catalysts
Design and Function

2.1 SALIENT FEATURES OF SUPRAMOLECULAR CATALYSIS

The inspiration to develop the subject of supramolecular catalysis stems from natural enzymatic systems. At the initial state of the inception of this subject, interest in the subject was in the ambit of synthetic chemistry. At that stage, the main interest was in understanding the mechanisms of enzyme-catalysed reactions and also preparing artificial enzyme catalysts. As the chemical understanding on such subjects was enriched, interest in developing new synthetic methods for enantiopure products by using such concepts in regular catalytic reactions emerged. Biocompatible, environmental, benign supramolecular catalytic reactions for efficient preparation of enantiopure compounds requires understanding the fundamental principles of organo-catalytic reactions. To achieve such goals, there are ample scopes to introduce supramolecular features to conventional homogeneous reactions. The discovery of homogeneous catalytic reactions mimicking enzymatic reactions has made new avenues to develop many artificial metalloenzymes.[1] Hybrid catalysts of two or more components with integrated catalytic effects provide selectivity and control of reactivity and create avenues for autocatalytic reactions. Catalysts prepared by combining a metal-containing component with a host protein through covalent or noncovalent links are developed. Interactions among a metal, organic or bioorganic part with a host molecule occur through supramolecular, dative and covalent linkages. The vital point of supramolecular catalysis is to utilise those interactions in developing supramolecular catalysts through substrate—catalyst interactions.

A supramolecular catalyst performs catalytic reactions with or without confinement, but in either situation, supramolecular interactions increase efficacy in the rate of reactions and in the formation of selective products. The supramolecular effects may be intramolecular, which provides different reactivity than the situation where such interactions are not possible. Intramolecular interactions provide scopes to modify geometries under specific reaction conditions so as to change reactivity and product selectively. In supramolecular catalysis, the synergic effect of the local environment and the surrounding environment of active site dictates the rate and selectivity. In many reactions, a slight variation in environment may stabilise a particular oxidation state of a metal ion. The redox potential of a metal centre is also highly influenced by the environment. Thus, the redox behaviour of a metal ion varies with the environment to influence or decide whether a reversible reaction will take place. It may be easily understood from the redox property associated with iron (II) and iron (III) of ferrocene. The compound shows a reversible redox couple under an ordinary situation, but the redox potential involved in the oxidation and reduction process changes in confined media. On the other hand, commonly observed redox couples of copper (II) complexes are quasireversible. This is attributed to the fact that a copper (I) ion would prefer to adopt four-coordinate geometry, whereas the copper (II) ion often forms a stable coordination complex with five or six coordination numbers. The copper (II) complexes with four coordination numbers are less common; in such complexes, a reversible Cu (I/II) redox couple may be expected. Hence, a redox process involving geometrical changes around a copper ion would require good amounts of reorganization of bond parameters during a redox process; hence, quasireversible redox processes are often observed from copper (II) complexes. Thus, the coordination environment, nature of ligands and central metal ions decide the faith of a redox process of a metal ion. Such effects provide scopes to utilise supramolecular effects to change the course of a reaction that would have occurred as a stoichiometric reaction to a catalytic reaction. A reversible redox process may occur in organic molecules with or without a reagent. Such a property an organic compound may be utilised for catalytic or stoichiometric redox reactions. The redox behaviour of a catalyst in which the central redox active unit has functional groups suitable to form hydrogen-bonded self-assembly or a redox catalyst with a redox active site confined in a containerlike molecule changes the overall redox properties associated with a similar catalyst where supramolecular features are absent. On the other hand, a complex at a particular oxidation state may act as catalyst, whereas at another oxidation state, it will not; thus, the redox properties of such complexes may be controlled to cause catalytic reactions. For example, the complexes **2.1a** and **2.1c** are at lowers oxidation states; which are easily oxidised to a higher oxidation state by electrochemical oxidations. The respective oxidised forms of these complexes, namely the complexes **2.1b** and **2.1d**, are catalytically active for the ring-opening polymerization reaction of 3,6-dimethyl-1,4-dioxane-2,5-dione (Figure 2.1). Similarly, the copper (I) complex **2.2b** undergoes an oxidation reaction to form copper (II) complex **2.2c** on application of potential amounting to −400 mV in the presence of 2-bromopropanoate (**2.2a**). The oxidised complex picks up bromine from the 2-bromopropanoate and generates a radical species by bromide extraction, forming the copper(II) complex (**2.2b**) (Figure 2.2). This electrochemical reaction initiates generation of radical to catalyse the polymerization of methacrylate to form polymethylacrylate.

2.1.1 Fundamental Basis

The fundamental aspect of supramolecular catalysis is the understanding of the inherent activation process of a substrate by supramolecular interactions to enhance the rate by a catalyst. Thus, to arrive at a systematic understanding, it is necessary to classify catalytic reactions in terms of performing an act leading to the products or by the stimuli used for performing the reactions. For example, substitution, addition and elimination reactions belong to the first category and reactions such as thermochemical, electrochemical and mechanochemical reactions fall under the second category. Regardless of the classifications of the reactions, each class

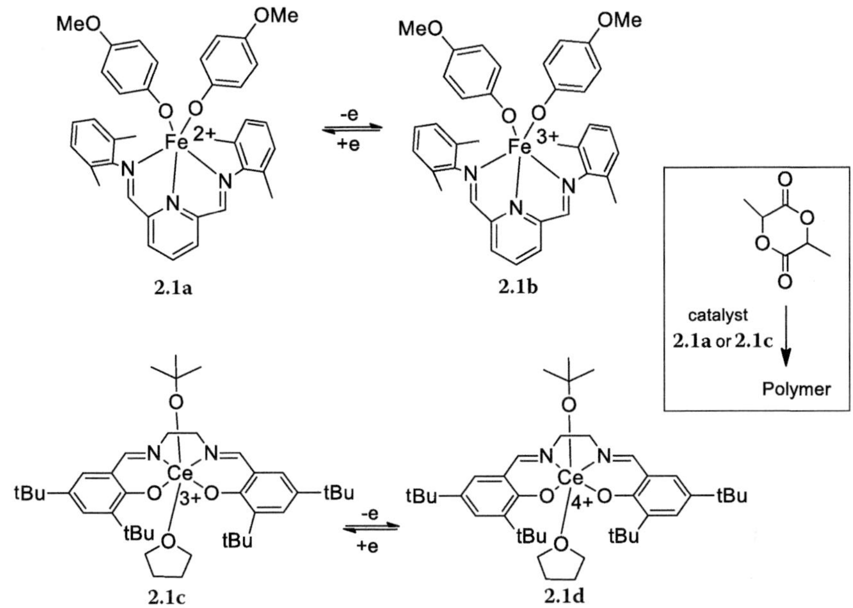

FIGURE 2.1 The redox equilibrium of two complexes and active catalyst for ring-opening polymerization. (From E. M. Broderick et al. *Chem. Commun.* 2011, 47, 9897–9899.[2])

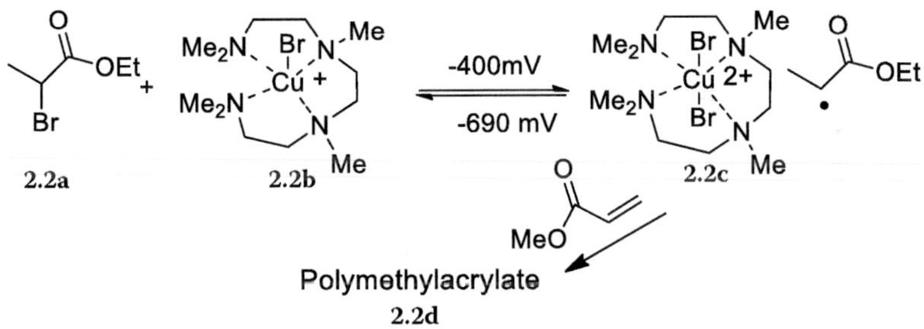

FIGURE 2.2 Catalytic radical polymerization under electrochemical conditions. (From T. G. Ribelli et al. *J. Am. Chem. Soc.* 2018, 140, 1525–1534.[3])

of reaction requires a mechanistic path and needs analysis of intermediate species, and, in supramolecular catalysis, the supramolecular features associated with all the components, including the products. Hence, intermediate characterization is a prime concern not only to understand a process but also to design new catalysts based on those understandings.

Supramolecular assemblies in the solid state help the photodimerisation reactions of olefin compounds. In the solid state, molecules are generally in fixed geometries. To cause a reaction between two double bonds of two independent compounds (intermolecular) or two double bonds separated by spacers within a molecule (intramolecular), the double bonds must be at adequate positions to cyclise. The separation distance between

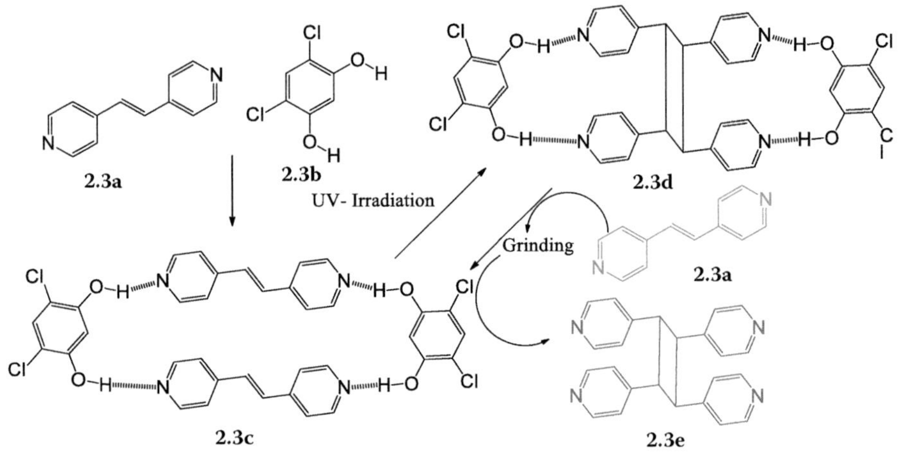

FIGURE 2.3 Solid-state assembly formed through cocrystal by photochemical 2 + 2 cycloaddition. (From L. R. MacGillivray et al. *Topics in Current Chem.* 2004, 248, 201–221.[4] D.-Y. Ma, K. Warnmark, *Chem. Cat. Chem.* 2010, 2, 1059–1060.[76])

the double bonds should be conducive for constructive overlap. The required distance of separation between two double bonds to react together should be in the region of 3.5–4.2 Å. Such an intermolecular arrangement of double bonds suitable to perform 2 + 2 photochemical reactions is achieved by forming cocrystals. A suitably designed cocrystal of olefin compounds brings double bonds close enough for such a reaction. For example, the cocrystal of trans-1,2-bis(4-pyridyl)ethylene (**2.3a**) with 2,5-dichloro-resorcinol (**2.3b**) undergoes a quantitative photodimerisation reaction (Figure 2.3). The role of resorcinol derivative in this reaction is to align the double bonds at the appropriate distance for a photodimerisation reaction. Grinding a solid mixture of about 20 mol % of resorcinol derivative together with the cocrystal **2.3a** generates a cocrystal **2.3c**. The cocrystals generated are composed of a four-component hydrogen-bonded self-assembly where the distance of separation between the double bond is 3.71 Å. Upon exposure to UV radiation, two molecules of the cocrystal **2.3a** undergo [2+2] photo-cycloaddition reaction to form the product **2.3d**. The reaction requires an optimum volume change. Thus, such reactions carried out by adopting the cocrystal technique have a high possibility of undergoing crystal-to-crystal transformation and are found in many related examples. Substrates binding to catalysts and the release of product from catalysts in these reactions are achieved by simple grinding of the reaction mixture. Due to this, no product inhibition occurs in these reactions. Several repeated grindings followed by irradiation after each grinding yield highly selective stereoisomers from this class of reactions. The reactions work at catalyst concentrations as low as 1%. To carry out such reactions requires repeated grinding during the course of the reaction, which may be avoided if continuous ball milling of the samples is carried out. Thus, such supramolecular-catalysed reactions are performed under mechanochemical conditions.

A conventional catalytic reaction may be carried out by including supramolecular features in the catalyst or by adding substrate, which will add up with the catalyst to provide supramolecular features. Typical examples of such reactions are catalytic

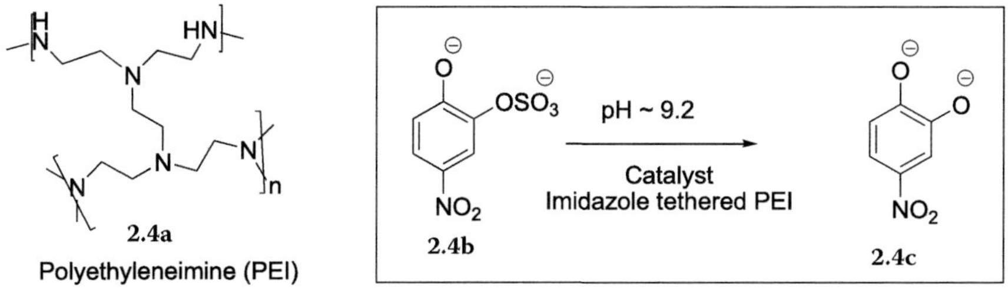

FIGURE 2.4 Modified PEI as catalyst for hydrolysis reaction. (From H. C. Kiefer et al. *Proc. Natl. Acad. Sci. U.S.A.* 1972, 69, 2155–2159.[5])

reactions of polyethyleneimine, commonly known as PEI (**2.4a**). The hydrophobicity of PEI is generally increased by attaching hydrocarbon chains such as dodecyl chains and also by adding basic imidazole side chains. Such polymers enhance the hydrolysis reaction of 2-hydroxy-5-nitrophenylsulfate (**2.4b**) to 4-nitrocatechol (**2.4c**) (Figure 2.4), and this particular reaction is enhanced by 10^{12} times over the reaction carried out without a catalyst.

In certain cases, a series of sequential reactions occur by using the same catalyst. Such reactions may be compared to reactions where the reaction intermediate of a catalytic reaction is modified to another form by the reaction of the catalyst. The same catalyst acts as the catalyst for the transformed form to convert it into another form. Thus, based on reaction conditions, the final product may differ. Such reaction sequences are observed in the amination reaction of α-keto-carboxylic acid. The sequences of reactions with different intermediates involved in the course of forming products are given in Figure 2.5. The catalytic transamination reactions catalysed by the base involve modifications of various intermediate products or transient isolable compounds (**2.5c–e**). The intermediate compounds finally transform to the desired

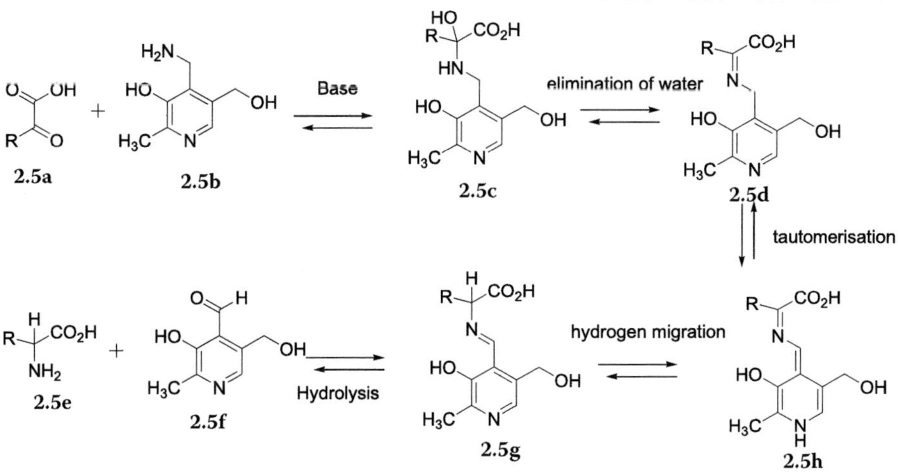

FIGURE 2.5 Several intermediate steps in a catalytic process. (From L. Liu et al. *J. Am. Chem. Soc.* 2004, 126, 8136–8137.[6])

products through multiple steps. The intermediate products formed in each step are stable enough to establish their identity by spectroscopic study. Thus, by varying reaction conditions, the product may be varied with the same catalyst.

Confining a reactant or part of a reactant into a cavity or self-assembly of a host confers selective products with an enhanced rates of reaction. From such a point of view, cyclodextrin molecules are commonly used as host molecules as well as catalysts. Cyclodextrins are cyclic oligomers of sugar molecules where sugar molecules are connected through ether linkages. The cyclodextrin are of three forms, namely α-, β- and γ-cyclodextrin. These forms differ in the number of sugar unit in their respective rings. The α-form is hexa-amylose (**2.6a**), the β-form is hepta-amylose (**2.6b**) and the γ-form is an octa-amylose (**2.6c**). The structures of the three forms are shown in Figure 2.6.

Each form has hydrophobic and hydrophilic rims in its respective structure. The hydrophobic part in each case provides conelike spaces differing in dimensions. Due to differences in the size of the rim; size- and shape-selective encapsulation of molecules take place. For example, α-cyclodextrin is suitable to encapsulate *p*-methoxybenzene (anisole) and, due to encapsulation of anisole into the cavity, the chlorination of the anisole by hypochlorous acid (HOCl) in the presence of α-cyclodextrin specifically takes place at the *para*-position to form 1-chloro-4-methoxybenzene (Figure 2.7a). On the other hand, the selective cyclic product 2-phenylquinoxaline is obtained from the reaction between 2-bromo-1-phenylethanone and 1,2-phenylenediamine in the presence of catalytic amounts of β-cyclodextrin (Figure 2.7b).

Confinement also helps in multicomponent reactions to enhance selectivity and rate. Two- or three-component reactions are specifically guided by such encapsulation, concealing portions of the reactive sites of a reactant to provide specific products. One such example is shown in Figure 2.7c. For example, heterocycle ethyl 2,5-dihydro-5-oxo-2-phenyl-4-(phenylamino) furan-3-carboxylate (**2.7c**) is formed in 25% yield from the reactions among the three components benzaldehyde, aniline and diethyl and but-2-ynedioate is formed in the absence of a catalyst, but this cyclised product is formed in 75% when the reaction is carried out with β-cyclodextrin as the catalyst. The possible reason is that a reaction between the acetylenic derivative with amine is facilitated by cyclodextrin, and the phenyl group of the aniline component of the intermediate product remains encapsulated and helps in further reaction with the aldehyde to form the heterocycle, ethyl 5-oxo-2-phenyl-4-(phenylamino)-2,5-dihydrofuran-3-carboxylate (**2.7c**).

The intrinsic acidity of a supramolecular catalyst is a major influencing factor in performing catalytic activity. The increasing acid or base strength benefits them to have higher catalytic activity. A series of catalytically active thiourea molecules are listed in Figure 2.8 with their respective pKa values. The urea **2.8a** has higher pKa than the corresponding thiourea derivative **2.8b**, but the latter is more suitable for organo-catalysis. This is due to the lower stability of the self-aggregated state of the thiourea derivative with respect to the urea derivative, as the relative bond strength of the N—H···O bond is greater than that of the N—H···S bond. Thus, knowledge of the pKa values of such substrates is helpful in designing supramolecular catalysts. For example, the isomeric thiourea derivatives **2.8b** and **2.8c** show similar pKa values in dimethylsulphoxide solvent, suggesting that both aromatic rings on these two thiourea derivatives are involved in the stabilization of the anion in dimethylsulphoxide solvent. The slight difference arises due to dynamic proton exchange between the

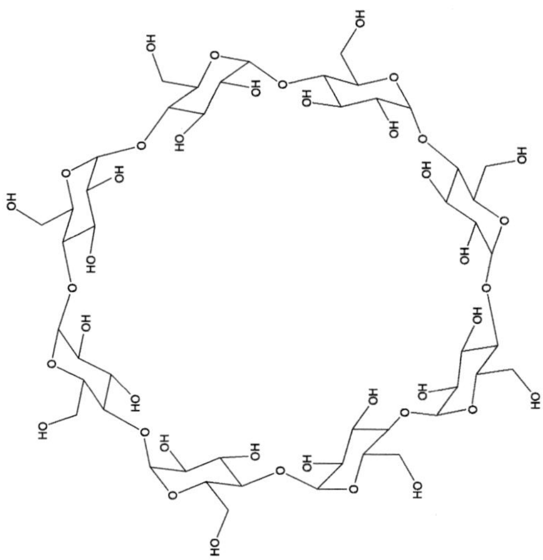

2.6c

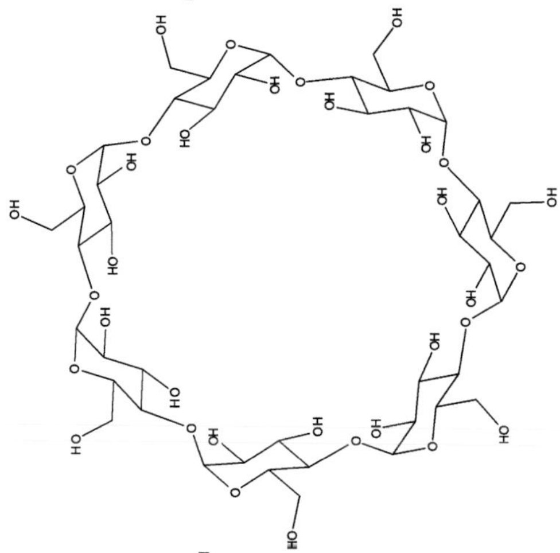

2.6b

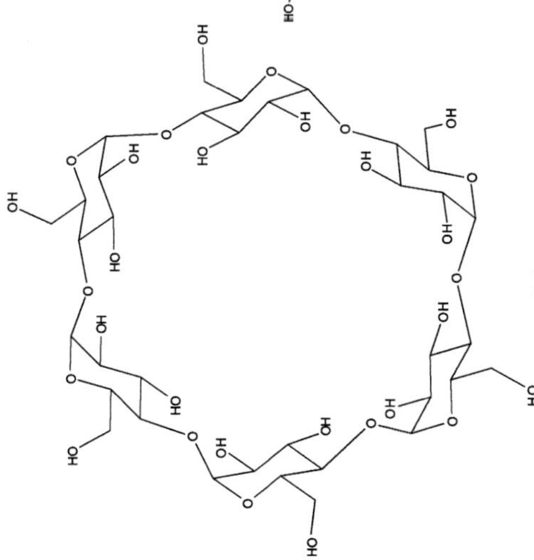

2.6a

FIGURE 2.6 Structures of three different cyclodextrins.

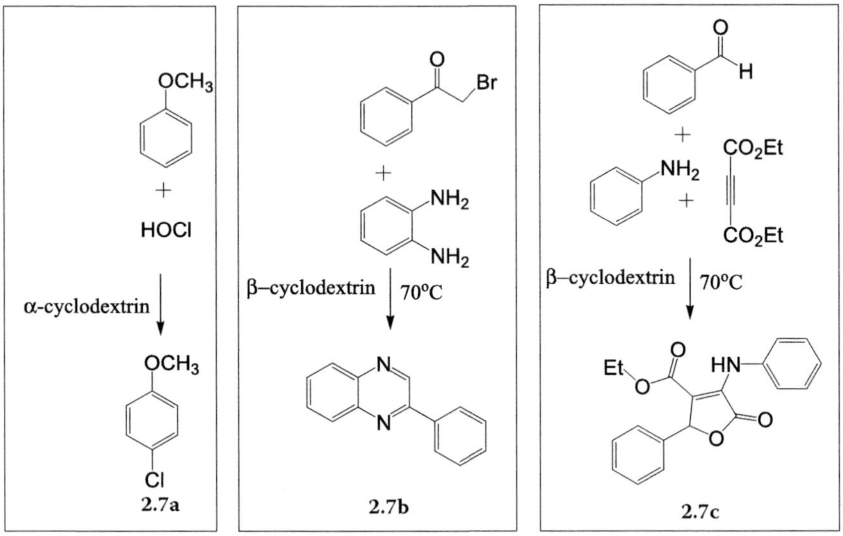

FIGURE 2.7 (a) Chemoselective chlorination in presence of α-cyclodextrin; (b) a selective cyclisation reaction and (c) a multicomponent catalytic reaction of β-cyclodextrin. (From C. C. Bai et al. *Molecules.* 2017, 22, 1475.[7])

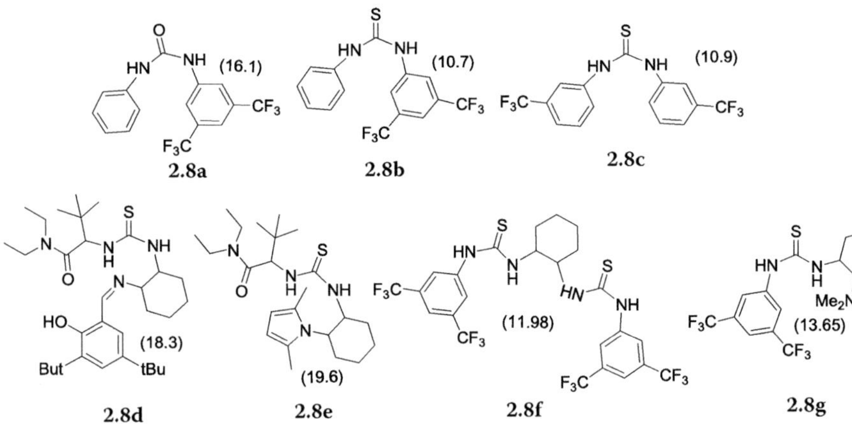

FIGURE 2.8 Different thiourea catalysts and their pKa (shown within brackets).

two nitrogen atoms of thiourea to create a partial negative charge on each site. The aryl group reduces the pKa values of N-arylthioureas, and this effect helps to sort the arylthioureas to be better catalysts to activate organic substrates where acidity is one of the issues. The intramolecular hydrogen bond influences pKa values; for example, the compounds **2.8d** and **2.8e** have pKa values 18.3 and 19.4, respectively. These two compounds are active catalysts for many organic transformations. Though certain thiourea catalysts **2.8f** and **2.8g** have the same structural backbones, they are different

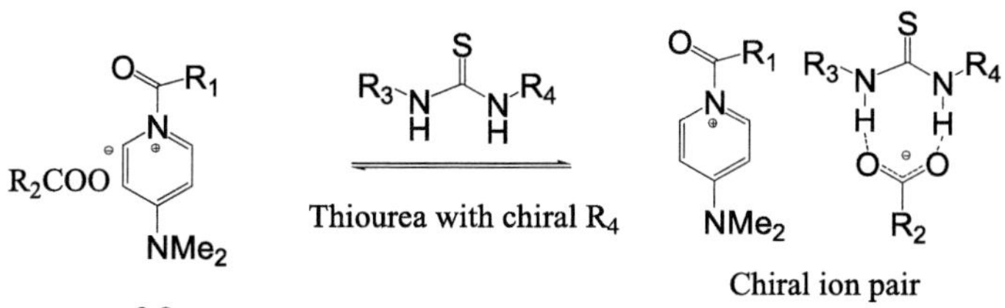

2.9a

2.9b

FIGURE 2.9 Chiral ion formed by hydrogen bonds with chiral thiourea derivative. (From Z. Zhang and P. R. Schreiner, *Chem. Soc. Rev.* 2009, 38, 1187–1198.[8])

in having a second thiourea group, as in **2.8f**. In the case of **2.8f**, the second thiourea group is in relatively close proximity to provide higher stability to a hydrogen-bonded anion containing an assembly and due to which it confers a lower pKa value over the compound **2.8g**.

Cocrystals of chiral ion pairs formed between chiral host molecules with the anion of a salt is an important factor to control the supramolecular activity of a catalyst. In such a situation, the reactivity of the chiral host changes through host—guest complex formation between the ion pair due to complementary hydrogen bonds. Hence, one can modulate reactivity of a host by varying the anion and by locking a particular form of a chiral host. For example, carboxylate salt **2.9a** reacts with chiral thiourea derivative to form a hydrogen-bonded assembly **2.9b**, where the conformation of thiourea is locked as *syn-syn* geometry (Figure 2.9). This effect not only stabilises the conformer due to self-assembling, the carboxylate anion blocks the facile attack of a nucleophile at the C=S site.

Such a principle is utilised in the reaction between N-methylindole with α,β-unsaturated aldehyde such as cinnamaldehyde to form (S)-3-(1-methyl-1H-indol-3-yl)-3-phenylpropanal in the presence of a thiourea derivative and trifluoroacetate salt of a heterocyclic compound (Figure 2.10). The salt **2.10a** is a poor catalyst to couple

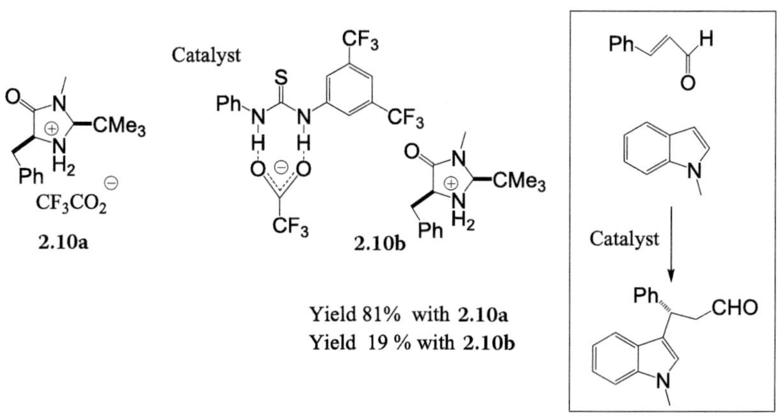

2.10a

2.10b

FIGURE 2.10 Catalytic C—C bond formations by ionic cocrystal. (From M. Mahlau and B. List, *Angew. Chem. Int. Ed.* 2013, 52, 518–533.[9])

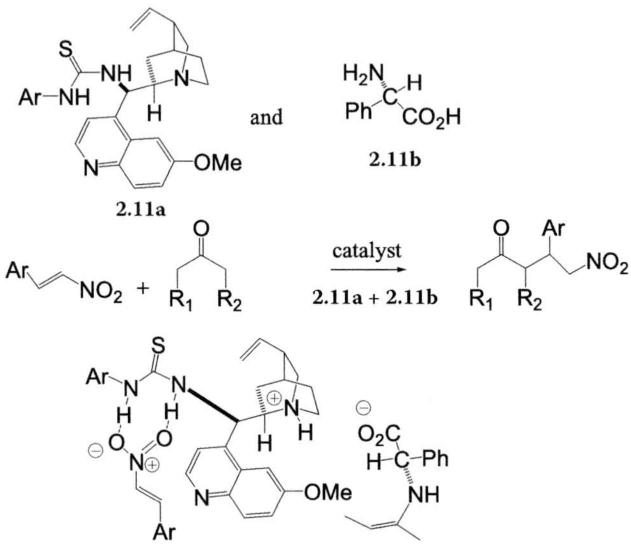

2.11a

2.11b

2.11a + 2.11b

Activated complex 2.11c

FIGURE 2.11 A reaction by a bifunctional catalyst with ionic hydrogen bonds. (From J.-R. Chen et al. *Tetrahedron* 2010, 66, 5367–5372.[10])

N-methylindole with cinnamaldehyde. The cocrystal **2.10b** is a superior catalyst, and upon use of this catalyst, the yield of the reaction increases up to 81%. In this example, the cocrystal results in comparable chirality but high yield. The use of fluorinated salt in such reactions generally increases the solubility of the catalyst in nonpolar solvents to help in the homogeneous catalytic C—C bond formation reaction.

A bifunctional catalyst may be generated by self-assembling through charge-assisted hydrogen bonds. Figure 2.11 shows such a reaction in which the two components of a bifunctional catalyst are a thiourea containing quinidine derivative **2.11a** and an amino acid **2.11b**. These reactants together form a salt, and the salt provides a binding site to the nitro group of nitro-olefin through a complementary hydrogen bond with N—H bonds of the thiourea derivative. The nitro group of the reactant is anchored to the salt by hydrogen bonds and is activated by forming the activated complex **2.11c**. This activation lowers the activation barrier for the nitro-olefin coupling with the ketone and helps undergo the catalytic C—C bond formation reaction.

Protonation of a catalytic species sometimes causes deactivation of a catalyst. For example, rhodium-carbene—based organometallic complex **2.12a** is a good catalyst for ring-opening polymerization reactions. This complex has tertiary amine groups at two ends of the ligand that provide hydrophobic confinement to the metal catalytic site and allow organic unsaturated compounds to be activated at the metal centre (Figure 2.12). But upon protonation of these two sites, it forms the cationic complex **2.12b**, which provides a hydrophilic environment that prevents the approach of olefin compounds towards the active site. The protonation causes loss of conjugation of the amino group with the ring, as the lone pairs of the electrons at the nitrogen atoms are no

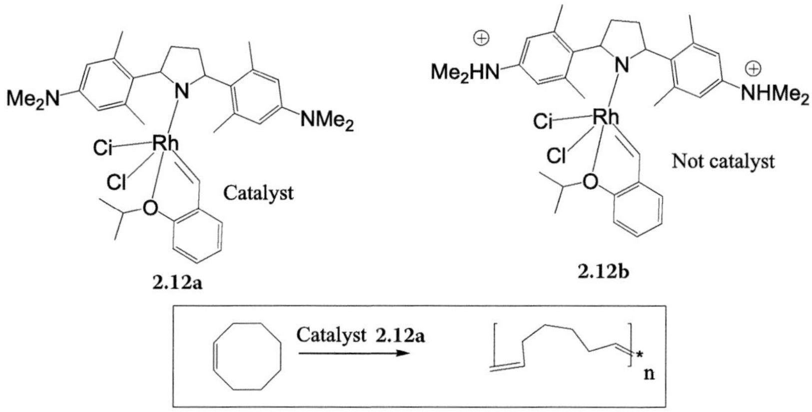

FIGURE 2.12 Deactivation of catalyst by protonation at exterior of the coordination site. (From S. L. Balof et al. *Dalton Trans.* 2008, 5791–5799.[11])

longer available for conjugation. With the hydrophilicity introduced by modifications of the ligands, along with the changed electronic factors at the rhodium centre by the loss of conjugation, the catalytic activity of the protonated complex is reduced.

Intramolecular hydrogen bonds influence or stabilise a particular conformer and geometry of a catalyst to influence the catalytic activity of a catalyst. For example, the *cis*-isomer of the compound **2.13b** has an appropriate geometry to have intramolecular hydrogen bonds (Figure 2.13a). As the N—H bonds of urea are involved in the intramolecular hydrogen bonds in the compound **2.13b**, an interacting substrate has to overcome the effect of such hydrogen bonds in order to bind to the urea unit, whereas the *trans*-isomer **2.13a** does not have intramolecular hydrogen bonds. Due to these reasons, a catalytic reaction caused by the *cis*-isomer is slower, but it is relatively faster when, for a similar reaction, when the *trans*-isomer is used as a catalyst. Advantage of this fact can be taken to control the reactivity of such a catalytic reaction. The reaction between nitro-olefin compound (E)-(2-nitrovinyl)4-bromobenzene with 2,3-pentanedione to form 3-(1-(4-bromophenyl)-2-nitroethyl) pentane-2,4-dione is rapid when the *cis*-form of the catalyst is used. The *cis*-form of the catalyst is generated *in situ* by UV irradiation. The *cis*-form of the compound transforms to the *trans*-form in dark and the rate of the reaction of the *cis*-form is reduced on heating the reaction mixture. The *cis-trans* isomerization also changes the catalytic activity of a reaction. For example, the rearrangement reaction shown in Figure 2.13b is slow in the presence of the *cis*-isomer **2.13c**, whereas it is fast in the presence of the *trans*-isomer **2.13d**. This is due to the steric crowd in the *cis*-isomer, which reduces the interactions of the catalytic active site, namely the pyridine part of the molecule, thereby influence catalytic activity. In the case of the *trans*-isomer, the substituent groups are away from the pyridine; hence, the crowding effect at the active centre is less.

The redox properties associated with a complex influence the chemical reactivity. Generally, rhodium phosphine complexes are good hydrogenating catalysts, the complex **2.14a** is such an example, which is a rhodium phosphine complex with a cobaltocene unit attached to it. The cobaltocene part is easily oxidisable by

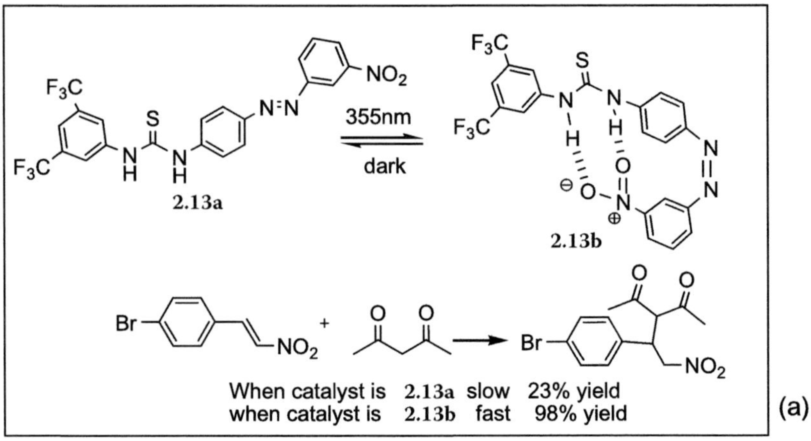

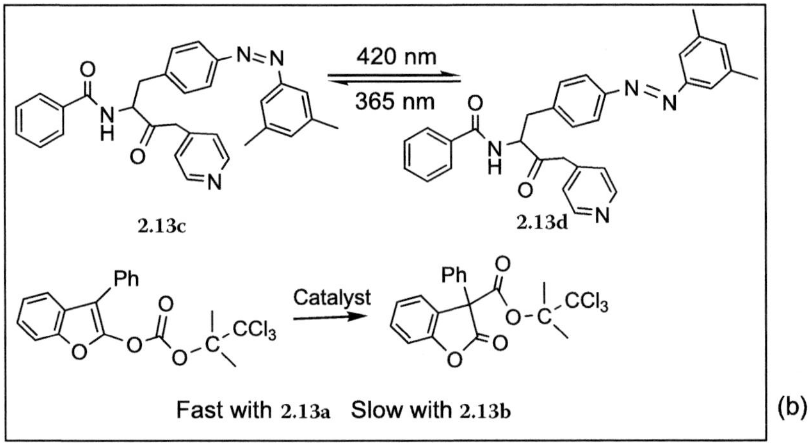

FIGURE 2.13 Different catalytic activities due to (a) intramolecular hydrogen bond and geometry (From R. S. Stoll. et al. *J. Am. Chem. Soc.* 2009, 131, 357–367[12]); (b) geometrical isomers. (From L. Osorio-Planes et al. *Org. Lett.* 2014, 16, 1704–1707.[13])

electrochemical means. Once it is oxidised to the complex **2.14b**, this complex is a poor hydrogenating catalyst for cyclohexene to cyclohexane (Figure 2.14).

There are examples where multiple components act together as a catalyst system. For example, the equimolar mixture of the compounds **2.15a–c** is a good catalyst for polymerization of oxepan-2-one (Figure 2.15) under a nitrogen atmosphere. In this reaction, the benzyl alcohol acts as an initiator of polymerization, whereas the base used in the reaction accelerates the rate by deprotonating benzyl alcohol. In an atmosphere of carbon dioxide, the same composition of the catalytic components does not catalyse the polymerization reaction. This is due to the passivation of the catalyst by carboxylation of the benzyl alcohol and subsequent protonation of the amines.

The structural geometry of a catalyst also influences the rate of a catalytic reaction. For example, the racemisation of chiral amino acid D-alanine is caused by the cationic species **2.16a**, whereas the isomeric form of the compound **2.16b**

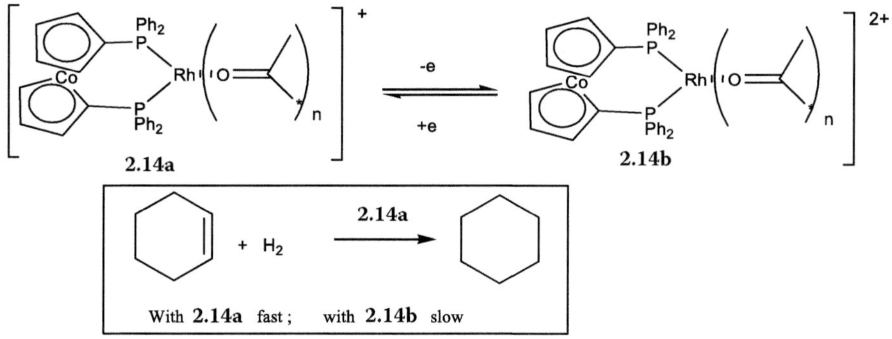

FIGURE 2.14 A rhodium catalyst whose activity is tuned by change of oxidation state of another metal ion present as part of the catalyst. (From I. M. Lorkovic et al. *J. Am. Chem. Soc.* 1995, 117, 3617–3618.[14])

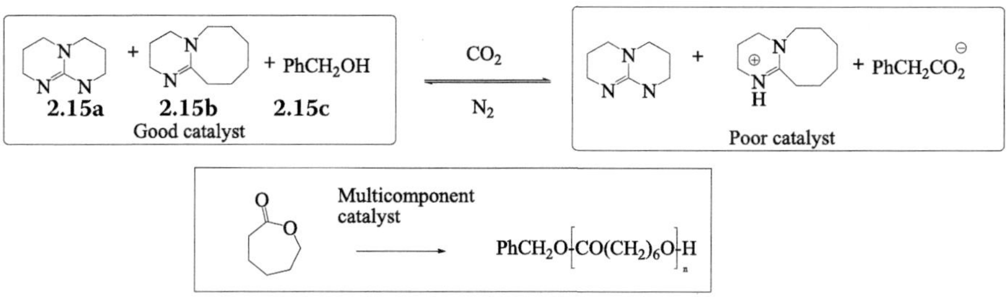

FIGURE 2.15 Multicomponent catalyst changing reactivity on change of medium. (From O. Coulembier et al. *Macromolecules* 2014, 47, 486–491.[15])

does the exchange in a very slow process. These two isomeric cationic species are equilibrated between them by irradiating with appropriate wavelengths, as illustrated in Figure 2.16a. Thus, the deuterium exchange can be modulated to have a fast or slow catalytic reaction by changing the wavelength of irradiation. Similarly, the hydroboration reaction of styrene is catalysed by the rhodium-carbene complex **2.16d** in an efficient manner. The catalytic activity of the complex **2.16d** is drastically reduced when it is converted to the isomer **2.16c** (Figure 2.16b). The ring closing and opening of the dithienylethene scaffold changes the environment in the vicinity of the rhodium, and the conjugation in the ligand scaffold also changes. Due to these effects, the N-heterocyclic carbene-rhodium complex is electronically modulated. The effective path of the conjugated π-system is changed and the electron density of the ligating atoms changed, directly influencing the carbenoid centre. The ability to activate the boron—hydrogen bond of the hydroborating agent by these two rhodium complexes differs and in turn influences the catalytic ability of the complexes in the hydroboration reaction by the two forms.

The organization of molecules at the exterior of an active site enhances or reduces the activity of a catalytic reaction. For example, the aromatization of the amide derivative 1-benzyl-1,4-dihydropyridine-3-carboxamide (**2.17a**) by the potassium ion—encapsulated crown ether (**2.17b**) with an oxidisable part is approximately

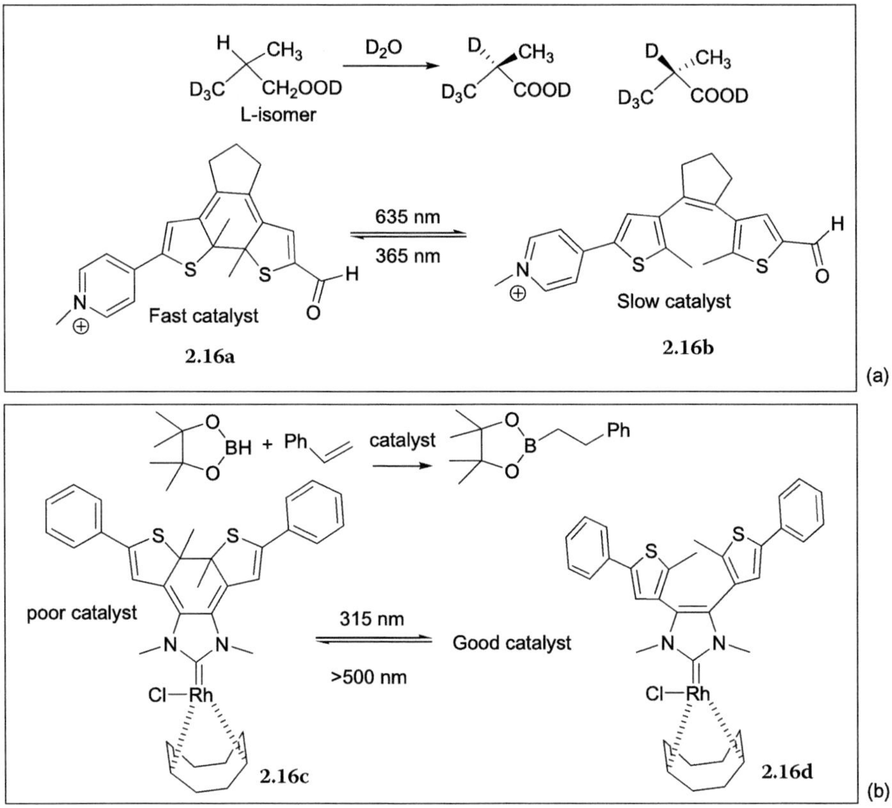

FIGURE 2.16 Catalytic (a) racemization (From D. Wilson and N. R. Branda, *Angew. Chem. Int. Ed.* 2012, 51, 5431–5434[16]); (b) hydroboration reaction controlled by irradiation. (From B. M. Neilson and C. W. Bielawski, *Organometallics* 2013, 32, 3121–3128.[17])

2.5 times faster over the similar crown ether derivative without the potassium ion **2.17c**. The oxidizing compound **2.17a** has the potassium ion bound to the crown ether part of the compound (Figure 2.17). It is more rigid than the corresponding crown without the potassium ion (**2.17c**), and it is also more hydrophobic. Due to the flexible part of the compound **2.17c**, it causes the dehydrogentive aromatization process at a comparably slower rate.

2.1.2 Intermediate Characterization

Basic informations and presumptions on the schemes of viable interactions between a catalyst and a substrate are of the highest priority to design a new catalytic reaction. In certain cases, products have an affinity to bind to another reacting species in the reaction medium, lead to stoichiometric reactions. Avoiding such interactions may turn the reaction into a catalytic reaction. This type of substrate or catalyst binding involving the product not only reduces the catalytic activity but in certain cases modifies the course of the reaction. There are other situations where catalyst

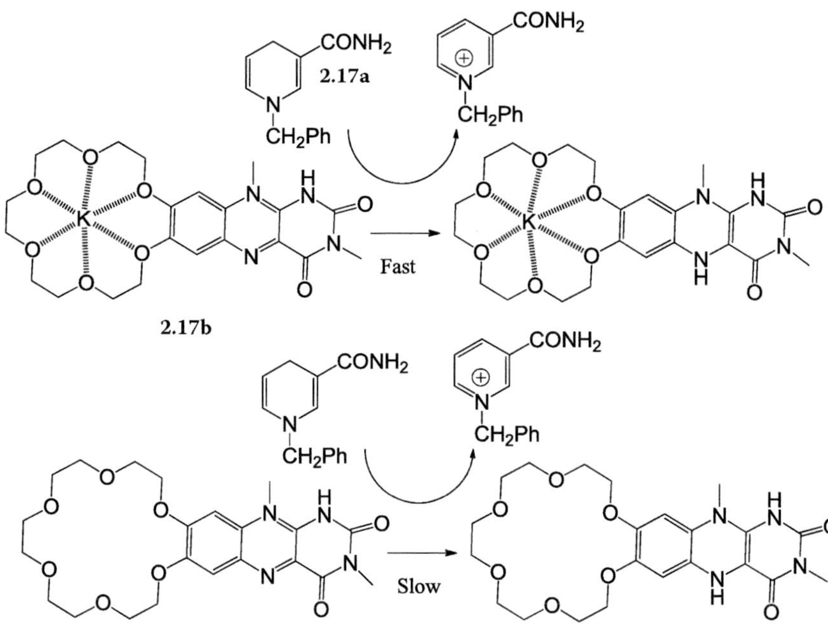

FIGURE 2.17 Influence of crown effect in catalytic dehydrogenative aromatization reaction. (From S. Shinkai et al. *J. Am. Chem. Soc.* 1987, 109, 923–924.[18])

modifications take place, but in such cases, the modified catalyst may react with substrates differently from the reactions of the reagent alone. On the other hand, a modified species or product formed during the course of a reaction acts as a catalyst to accelerate reactions. Such reactions are autocatalytic in nature. Thus, it is necessary to understand substrate—catalyst binding schemes. To enforce a selective path or control the reactivity of catalytic reactions passing through more than one intermediate, the characterization of each intermediate becomes essential. Many catalytic reactions pass through transient intermediates. The isolation of those intermediates may not be possible but it is possible to obtain evidence on the signatures of the transition states through *in-situ* study. In such a situation, the intermediate characterization is done by spectroscopic means when the lifetime of the intermediate matches with the detection limit of the spectroscopy under consideration. The isolation of the intermediate species in catalytic reactions is a challenge. In general, the variations of the reactants and modification of the catalyst helps to obtain model intermediate compound as replica intermediate of a fast reaction. The choice of spectroscopic tool to study a supramolecular catalytic reaction depends on the feasibility of the spectroscopic tool under consideration to monitor change in physical properties inherent to the catalyst, substrate—catalyst complex or product. UV-visible spectroscopy is conventionally used in ascertaining compositions, ligand environment and course of reactions. The observation on the maximum spectroscopic information about the course of the reaction, depends on the lifetime of a reactive species. On the other hand, various thermochemical, imaging and optical techniques apparently provide information on the suitability of a catalyst as well as on the thermochemical and morphological changes of a catalyst. Many heterogeneous catalytic reactions require scanning and transmission electron microscopy study to ascertain the morphology and sizes of the catalyst before and after catalysis. Infrared spectroscopy is routinely used in characterization of catalysts and intermediates of catalytic reactions. In many reactions, the course of reactions are monitored by following the changes in the intensity of one or more characteristic absorptions of reactants and products. Surface modifications are also monitored by infrared (IR) spectroscopy. When a heterogeneous and homogeneous catalysis occurs at conjunction, hot filtration and further analysis of the filtrate help one to know if it is homogeneous, heterogeneous or a combination of both during the course of the catalytic reactions.

Nuclear magnetic resonance (NMR) spectroscopy finds an important place in *in-situ* study of homogeneous catalytic reaction in solutions. NMR studies at variable temperatures helps to establish dynamic phenomena and reactive intermediates. NMR experiments have large versatility, and experiments using properties such as time of relaxation, spin-orbit coupling, the nuclear Overhauser effect, spin tickling and decoupling are routinely applied to understand the course of a reaction or transient species, including stereochemical consequences in catalytic reactions. However, for such NMR studies paramagnetic species are not suitable unless special experiments are devised. A particular experimental methodology fits to study a certain type of reactions under consideration may not be suitable to study another related reactions. The stereochemical consequences of a chemical reaction may be reflected in the nuclear overhauser effect (NOE) in NMR spectroscopy. For example, n-butyllithium is a highly reactive compound; it exists as a tetramer. In this compound, the nuclear Overhauser effect can differentiate the identification of NMR signals from methylene groups in close proximity to lithium and the methylene group located next to such a group depicting the overall environment of these methylene groups. The hetero-nuclear overhauser effect is used to arrive at the interactions between the lithium atom with the methylene groups. Any interaction occuring through space between lithium and the interacting $-CH_2$ group is graded by the distance and orientation of the interacting group with respect to the lithium ion. The polar nature of the lithium—carbon bond influences such interactions. Scalar coupling in the case of a covalent linkage is used to differentiate the coupling schemes of the methylene groups located at different locations away from lithium, this is possible as the geometrical arrangements are such that interactions through space with lithium ions are possible. Aggregate formed by butyl-lithium provides different J-couplings between multiple lithium ions as well as with alkyl groups. n-Butyl-lithium dissolved in tetrahydrofuran solvent has a ^{13}C NMR signal for a C-1 peak with a decet-splitting pattern. This splitting occurs due to the coupling of the carbon with three neighbouring lithium ions. This pattern supports the presence of the complex as a tetrameric n-buyllithium cluster $[Li_4(butyl)_4]$ (Figure 2.18). This observation also suggests that each lithium ion couples to the adjacent carbons and does not couple across the lithium ions located at the body diagonal positions of the tetrameric structure. Two-dimensional NOE study shows cross-peaks from the lithium ion that interact with the H_a and H_b protons of the butyl chain. This observation confirms their positions on the corners of the cube of the tetramer.

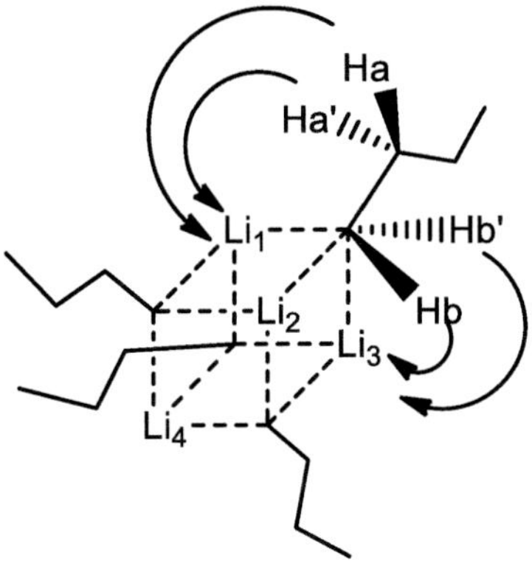

FIGURE 2.18 Structure of tetramer of n-butyllithium. (From M. O. Huaping and T. C. Pochapsky, *Prog. Nuclear Magnetic Resonance Spectroscopy*. 1997, 30, 1–38.[19])

To study a reaction involving paramagnetic species requires tools like electron paramagnetic resonance (EPR) or magnetic moment measurements. For example, in a redox couple involving a transformation between the paramagnetic and diamagnetic states of a transition metal controlled by a co-oxidant, the reaction can easily be monitored by EPR spectroscopy. Organic radicals catalyse many reactions under supramolecular conditions. Cucurbit[n]urils (**2.19a**) are urea derivatives with a cylindrical cavity, and the size of the cylindrical cavity depends on the number of monomeric subunits connected to form cyclic structures. The long alkyl chain attached to the dye possessing a radical cation **2.19c** is easily included into the cylindrical part of a cucurbit[7]uril molecule to activate the dye radical. The activation effect by cucurbit[7]uril in such a dye, shown in Figure 2.19, is reflected in the growth of concentration of the radical cation of the dye. The cation radicals formed in the absence and presence of the derivatives in **2.19a** are monitored by electron paramagnetic resonance spectroscopy. The formation of the radical cation from the dye in solution upon irradiation is reflected in the electron paramagnetic resonance spectroscopy. The radical has an EPR signal centred at $g = 2.0031$. In the presence of cucurbit[7]uril, the concentration of such a radical remains very low (<2%) during the oxidation process. Due to the large signal-to-noise ratio of such a low-intensity EPR peak, the EPR signals are too broad to be practically observed. The low concentration of the cation radical may be either due to a low rate of formation of the radical under supramolecular conditions or due to instant reaction under supramolecular conditions. The product formation in this example improves under supramolecular conditions. In this reaction, the radical cation is activated due to the supramolecular effect of the encapsulation of cucurbit[7]uril with carbonyl groups at the external periphery of the cylindrical hollow structure formed followed by its reaction. The latter step is fast; hence, the reaction is accelerated in the presence of cucurbit[7]uril. It is also possible to monitor the concentration of the cationic radical by UV-visible spectroscopy. UV-visible spectra of the dye during the course of oxidation shows a broad absorption band in the region of 550–700 nm due to the cation radical of the dye, and this absorbance is observed at the expense of the original absorption peak of the parent dye.

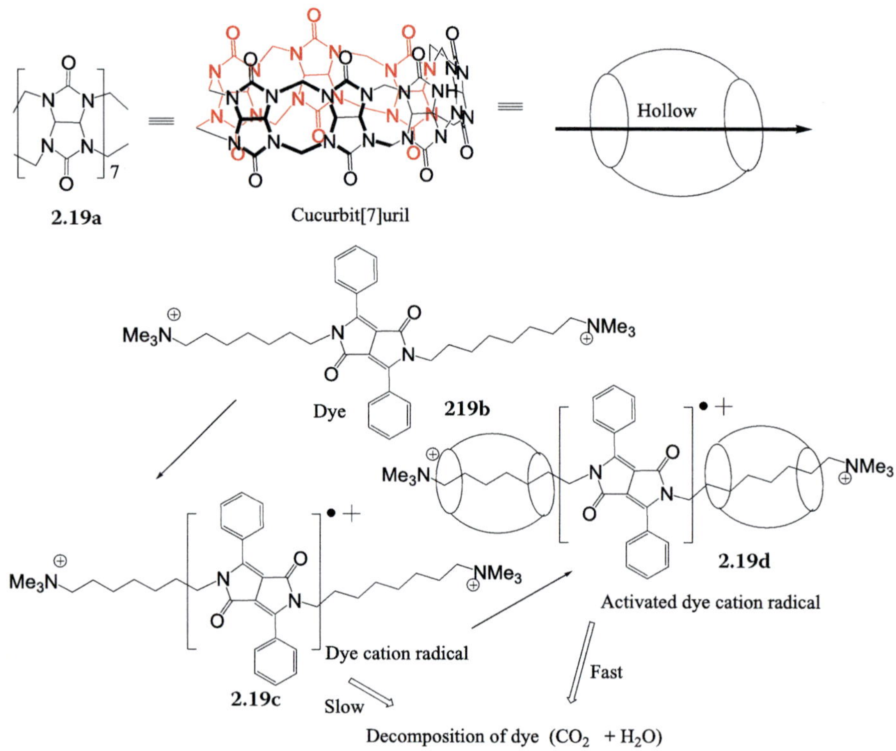

FIGURE 2.19 Supramolecular catalysis of a dye decomposition monitored by EPR. (From Y. Jiao et al. *Angew. Chem. Int. Ed.* 2016, 55, 8933–8937.[20])

The supramolecular features associated with substrate—catalyst and product—catalyst interactions have equal priority to address critical issues in the successful performance of a catalytic reaction. If the products formed in a reaction are strongly bound to the catalyst, one has to have a way to constantly recover the catalyst, rather to regenerate the activating species through product isolation. Initial interactions of a substrate with a catalyst and the composition of intermediate species are determined by spectroscopic titrations. Such titrations are generally done through fluorescence spectroscopy, UV-visible spectroscopy and microcalorimetric titrations. From such experiments, different thermodynamic parameters, including the stoichiometry and binding constant are determined.

The leaving group of a substrate also influences the supramolecular catalytic reaction. For example, dimethyl-arylphosphate triesters (**2.20a**) react with methanol under catalytic condition by lanthanum salt (Figure 2.20) and form trimethyl phosphate (**2.20b**). When the aryl groups contain an *o*-methoxycarbonyl substituent, these are very fast lanthanum (III) ion-catalysed methanolysis reactions. The catalytic reactions pass through the dimeric lanthanum species and follow second-order kinetics. The rates of these reactions are dependent on the leaving groups at different concentrations and pH values. The presence of an aryl-*o*-methoxycarbonyl unit increases the rate of cleavage of the phosphate triesters with poor leaving groups relative to that phosphate

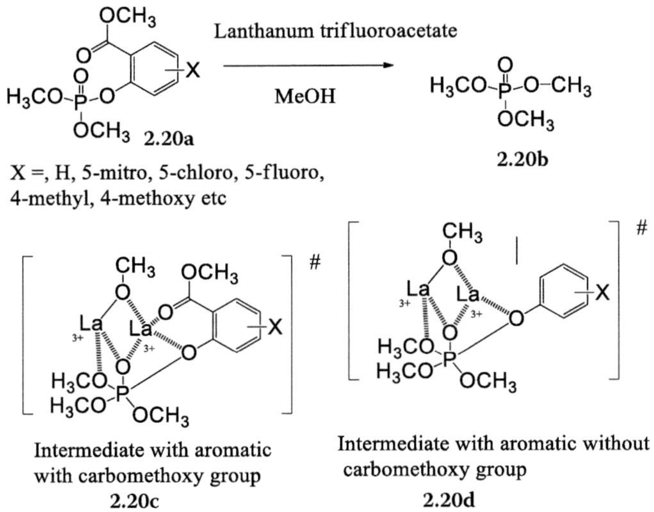

FIGURE 2.20 The effect of substituent on the leaving group. (From D. R. Edwards et al. *J. Am. Chem. Soc.* 2009, 131, 13738–13748.[21])

triester that possesses good leaving groups. This is due to the higher cleavage rate of the leaving group relative to the substrate that has no aryl-*o*-methoxycarbonyl group. This effect is due to the formation of the intermediate **2.20c**. The simple phenyl-substituted phosphate does not have a chelating site, as in the case of the one with an *o*-methoxycarbonyl group and hence forming an intermediate **2.20d**.

Dynamic assemblies mediated by guests are useful in supramolecular catalysis. Such dynamic processes involve multiple equilibriums of host—guest interactions and also interactions between different subassemblies. Some dynamic assembling processes are studied with the aid of high-pressure liquid chromatography by monitoring the reaction mixture at definite time intervals. For example, dynamic self-assembly is formed upon oxidations of the thiol groups of the mono-anion of 3,7-dimercapto-2-naphthoic acid by the oxygen present in air. The S—S bonded cyclic tetramer of 3,7-dimercapto-2-naphthoic acid **2.21b** is present as an equilibrium between a free tetramer and interlocked catenane type of self-assemblies. Such dynamic molecular networks are formed in response to the substrate, and each assembly acts as a transient catalyst. Hence, the catalytic process is governed by the overall dynamic assembling and disassembling process. The Aza-Cope reaction illustrated in Figure 2.21 is catalysed by the dynamic networks.

A redox active site on a catalyst with matching redox potential with respect to the reactant is the essence of many electrochemical catalytic reactions. To construct a redox catalytic cycle, the reversible redox processes associated with the catalyst and its ability to revert back and forth between oxidised and reduced states in the presence of other reactants are an utmost necessity. The environment created by the solvent and other supramolecular effects seriously affects the redox properties of organic or inorganic couples. Cyclic voltammetry helps to establish reaction paths of such redox reactions. Study using cyclic voltammetry also provides the overall impact of the supramolecular

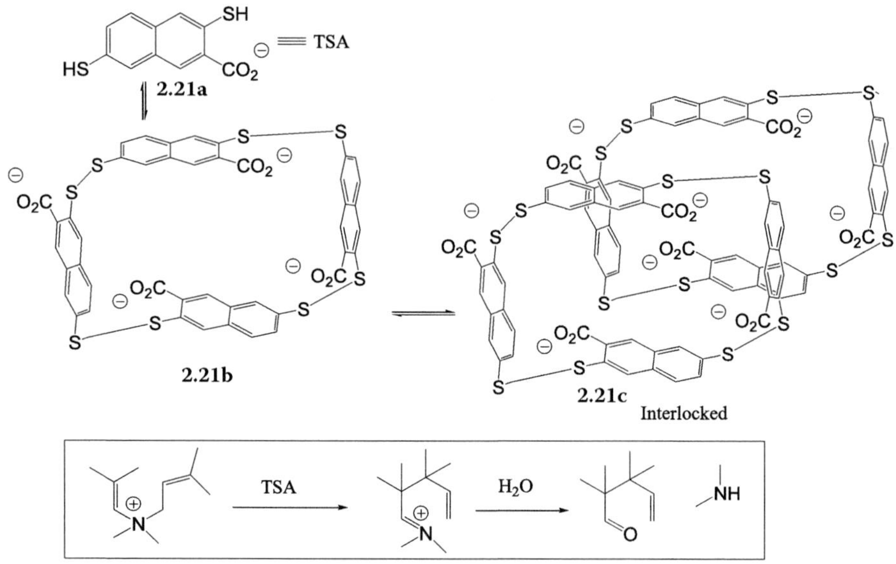

FIGURE 2.21 Dynamic assemblies assisting in catalysis. (From H, Fanlo-Virgs et al. *Angew. Chem. Int. Ed.* 2014, 53, 11346–11350.[22])

environment on a redox couple. Once information on the redox behaviour of a couple is established, a catalyst to function in specific conditions can be developed by applying the requisite electromotive force and necessary current. Based on such observations, numbers of supramolecular redox catalytic reactions are carried out in confined media.

The electroactive compounds with two C=O bonds with appropriate geometry, as in compounds like 9,10-phenanthrenequinone, acenaphthenequinone, form strong hydrogen-bonded cocrystals with urea or thiourea (Figure 2.22). The radical anions formed from these compounds are stabilised as hydrogen-bonded adducts, and their characteristic redox couples occur at different places from the free anionic radical. In cyclic voltammetry, the anodic peak of the radical anion formed in the presence of 1,3-diphenylurea of acenaphthenequinone is broader than the peak of the anion radical formed in the absence of 1,3-diphenylurea. The half-wave potential ($E_{1/2}$) of this peak shifts towards a higher value by 170 mV compared to the free anion radical peak. This shift is dependent on the solvent. Urea derivatives are well known to form self-assembly in solution by intermolecular hydrogen bonds between carbonyl and N—H bonds. Certain solvents also interact with urea to disfavour self-assembly of urea, but such an effect may favour the urea binding to the radical anion. In ^{1}HNMR of solution of urea in CD_2Cl_2, the addition of acenaphthenequinone causes a shift in the N—H signals of urea by 0.19 ppm towards upfield. When the urea solution is diluted, the same peak moves towards downfield and these changes occur to a certain concentration of urea, but beyond a point of dilution, the signal of the N—H again shifts to upfield. This suggests that there is competition of urea to self-aggregate in solution. The ^{1}H NMR spectra of 1,3-diphenylurea in DMF-d_7 is not affected by dilution, which suggests that polar solvents do not allow self-aggregation.

The assembling and disassembling processes are very often found in supramolecular catalysis. In the case of metal complexes, such a probability in solution exists through

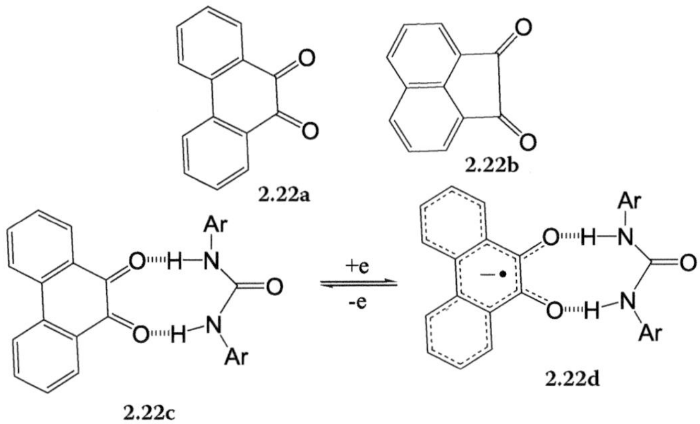

FIGURE 2.22 Example of two electroactive organic compounds that forms adduct with diarylurea derivatives. (From Y. Ge et al. *J. Org. Chem.* 2000, 65, 8831–8838.[23])

ligand reorganization or by increasing or decreasing nuclearity of metal clusters or assemblies (Figure 2.23). Hence, the stabilities of different possible forms are explored in solution. Self-assembly of a polynuclear complex may dissociate uniformly in solution to form multiple or smaller polynuclear complexes. For example, the complex $M_{12}L_{24}$ (where L is a terpyridine type ligand) is a complex with 12 metal ions in cuboctahedral dispositions. It decomposes to a less stable M_8L_{16} and relatively long-lived M_9L_{18} complexes. These complexes are characterised in solution by time-of-flight mass spectroscopy as well as by ^{1}HNMR spectroscopy recorded under different conditions. For example, mass spectrometry of the sample of the $Pd_{12}L_{24}$ species shows mass fragmentation for three species $Pd_{12}L_{24}$, Pd_9L_{18} and Pd_8L_{16}. The mass spectra also suggest

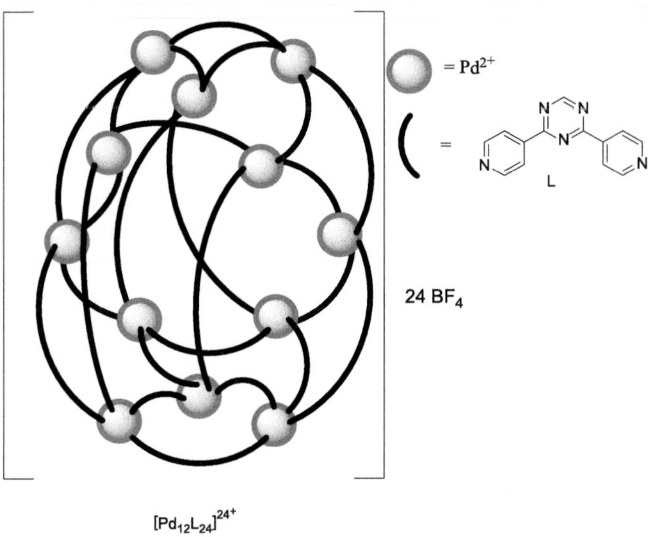

FIGURE 2.23 The structural representation of M_8L_{16}.

that the concentration of each species changes with temperature and time. For example, at 60°C, the mass peak for the Pd_9L_{18} species is observed, which is not observed after 1 hour, whereas the mass peak of the Pd_8L_{16}-species is observed after 1 hour under similar conditions, but the mass peaks for the Pd_8 species disappear after 6 hours. This shows that the stability of the Pd_8L_{16} species is higher than the Pd_9L_{18} species.

Framework structures are very important to provide spaces for the inclusion of substrate and catalyst to cause a reaction between them. They also provide active sites for a substrate for catalytic reaction. Metal organic frameworks belong to a class of macromolecules which often serve as heterogeneous catalysts. Some metal organic framework catalysts may lose their original structures in various ways; one process to do so is through desolvation of solvated complexes. Some of the disintegrated structures regain their original structures through post-treatments. For catalytic reactions, it is desirable to prepare highly pure frameworks with high thermal stability. The reaction conditions, including stirring, solvent, temperature, time, pH and pressure, contribute to the specific synthesis of metal organic frameworks. Various techniques relating to thermal stability, crystallinity and porosity are used conventionally. In such studies, thermogravimetric analysis helps to ascertain thermal stability, desolvation and so on. Particle size analysis and porosity studies are carried out to establish viable utility of those complexes as catalysts. Besides single crystal x-ray diffraction technique to determine the crystal structures, the powder x-ray diffraction technique is extensively used to ascertain crystallinity and phase of the supramolecular systems. In many cases, x-ray photoelectron spectroscopy is used to understand the electronic features and measure elemental composition. On the other hand, energy-dispersive x-ray spectroscopy helps to get evidences on the adsorption of catalysts when surface modification or surface-bound catalysts are considered. It becomes essential on many occasions to confirm whether a metal species is leached from a solid catalyst to the solution during the course of a catalytic reaction. In such cases, inductively coupled plasma spectroscopy is applicable.

2.1.3 Product Selectivity Including Stereo Specificity

The success of catalytic reactions lies not only in high atom economy and fast reactions under optimum conditions with high turnover but also in the high selectivity of the product. The fundamental approach of optical resolution from a racemic mixture is based on in the weak interactions of the racemic form of a compound with chiral substrate. Following competitive crystallizations, one optically active form specifically crystallises from a specific solution and provides a means to separate optical isomers. A chiral supramolecular catalyst guides a reaction for high entioselectivity due to catalyst—substrate or catalyst—product interactions. Such interactions enforce the formation of specific optical isomer. Chiral induction using chiral ligands is very common. Many reactions carried out in a chiral-confined environment result in stereo-selective products. Reaction conditions and reagent variations are two ways to modulate the selectivity to achieve optical induction. The reaction between a primary thiol and α-β-unsaturated carbonyl compounds is catalysed by *o*-fluorobenzoic acid or phosphorous-based acids. Both the catalytic reactions take place in the presence of a quinidine base (Figure 2.24), giving an enantioselective product. The use of a phosphorous-based acid along with the quinidine derivative (**2.24a**) in acetone provides

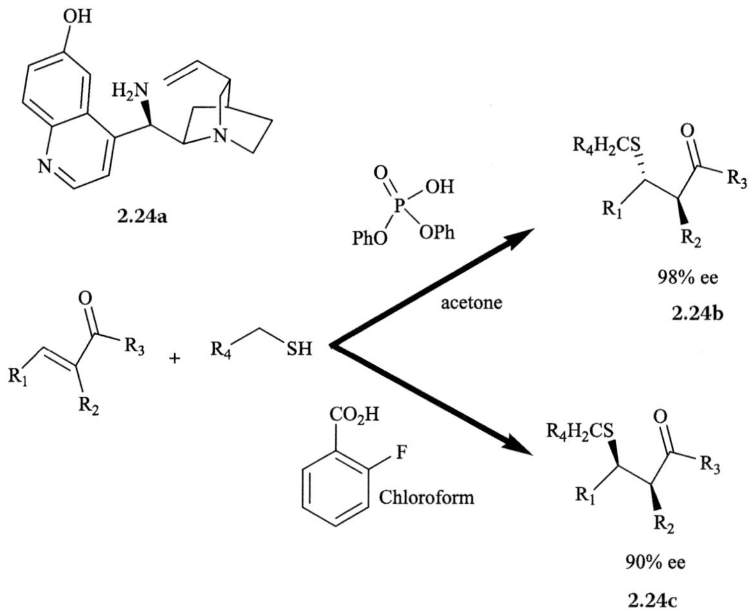

FIGURE 2.24 Enantioselectivity in presence of quinidine. (From T. Govender et al. *Chem. Rev.* 2016, 116, 9375–9437.[24])

higher enantiomeric excess over the catalytic reaction done by *o*-fluorobenzoic acid with quinidine derivatives in dichloromethane.

When more than one reactive site undergoes chemical reactions, the use of a catalyst leads to different isomers as products from the same reagent. A catalyst binds differently to the substrate, changing the course of the reaction to provide chemoselectivity, or the catalyst causes specific site-selective reaction. For example, epoxidation of terpenoid **2.25a** takes place at the double bonds connecting C11-C10, C7-C6, C3-C2, as shown in Figure 2.23, upon reaction with hydroperoxides. The catalyst **2.25b** causes selective epoxidation with high enantiomeric excess at the C3-C2 position, while a catalyst **2.25c** is used, changes the reactivity and the epoxidation reaction takes place more selectively at the C7-C6 position. These observations are attributed to the hydrogen-bonded intermediate involved in this reaction. The interaction of the terpenoid **2.25a** with the catalyst **2.25c** provides an adduct, which uses the ethereal oxygen and a peptide N—H bond to hold the O-H group. Hence, the catalyst adopt a locked geometry through intramolecular hydrogen bonds between a carbonyl of a proline unit and an NH of an amide unit. This geometry **2.25d** prevents the attack of the peroxy compounds at the C3-C2 position, and the reaction takes place at the second favourable position at C6-C7 (Figure 2.25).

Supramoulecular catalysis provides avenues to link multiple active sites of the same or different types in a host to act as a catalyst. In such reactions of a catalyst with multifunctional compounds, the interplay of weak interactions plays a major role in deciding the faith of the product formed. One such example the compound **2.26a** shown in the Figure 2.26 has two thiourea groups as active sites; this compound has interesting

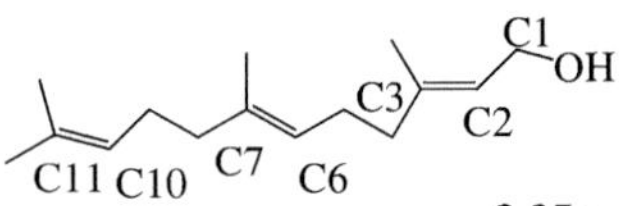

2.25a

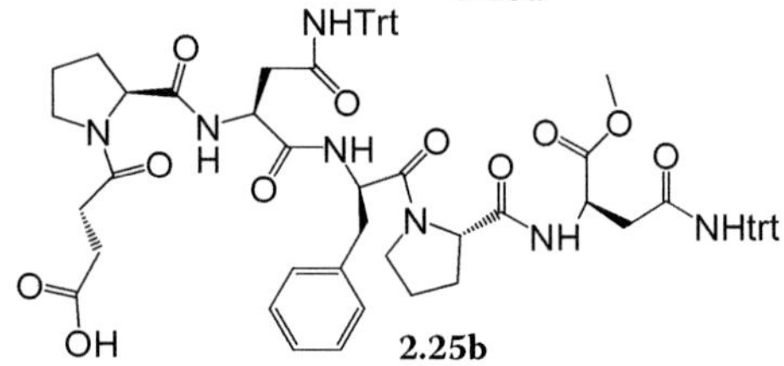

2.25b

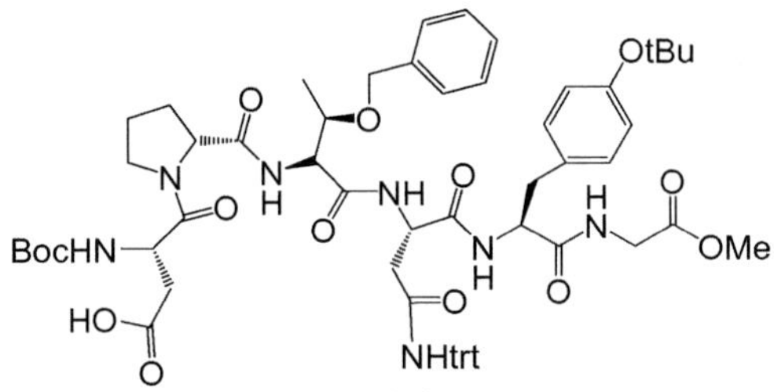

2.25c

With **2.25b** expoxidation ratio at C11-C10 : C7-C6 : C3-C2 is 1 : 1 : 9.8 with 86% ee)
With **2.25c** expoxidation ratio at C11-C10 : C7-C6 : C3-C2 is 1.2 : 8.2 : 1.0
(selectivity in C7-C6 attack)

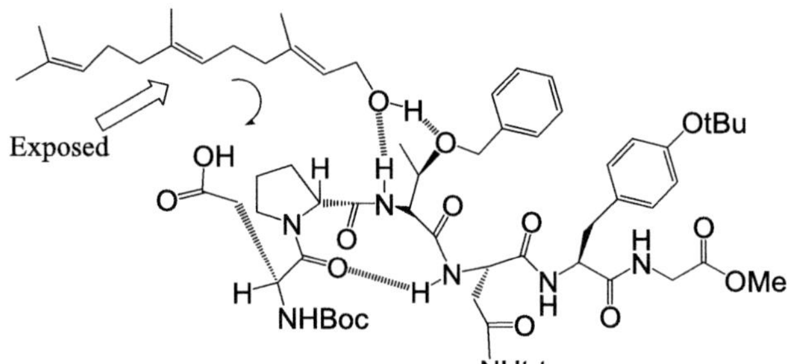

Transition state
2.25d

FIGURE 2.25 Selectivity in epoxidation by using different peptides. (From P. A. Lichtor and S. J. Miller, *J. Am. Chem. Soc.* 2014, 136, 5301–5308.[25])

solvent-dependent catalytic properties. This compound acts as a catalyst for the coupling reaction between imine and dimethylmalonate catalysed by **2.26a** yields the same product **2.26b** in independent reactions carried out in m-xylene as well as in acetonitrile. But the difference is that the enantioselectivity observed from the two solvents is different. The different enantioselectivity is attributed to the interplay of weak interactions of the solvents, substrate and catalyst while forming the reactive intermediate.

It is not a necessary condition that a confining molecule have a fixed barrier through covalent linkages. A molecule with a base with functional groups can be organised to provide the shape of an enclosure to recognise guest molecules. Such

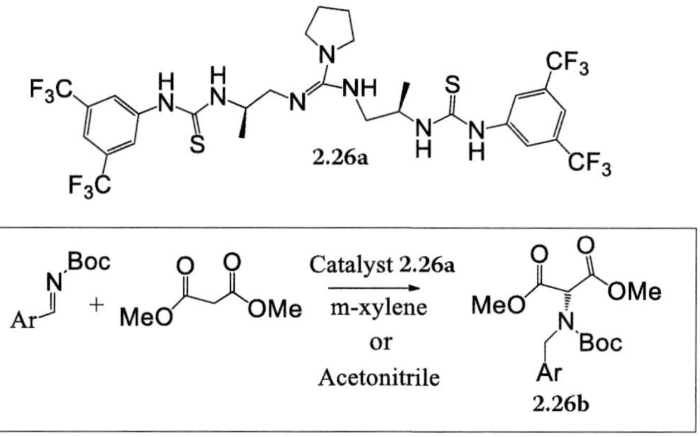

FIGURE 2.26 Change in enantioselectivity by solvent. (From Y. Sohtome et al. *Angew. Chem. Int. Ed.* 2010, 49, 9254–9257.[26])

molecules find utility in supramolecular catalysis. For example, the compound **2.27a** is a catalyst for cycloaddition reactions; it has a chiral enclosure created by using three functional units anchored to the binol. This compound has functional groups spread and positioned across the naphthalene ring, where the naphthalene ring acts like a base to provide enclosure of cylindrical shape. Such enclosures accommodate the diene and dienophile for cycloaddition. The cavity within the enclosure has chiral space due to the chirality associated with the substituents. The compound **2.27a** acts as catalyst for the 4+2 Diel-Alder cycloaddition reaction in a very selective manner with high endoselectivity. The intermediate **2.27c** formed in the reaction results in high optical induction (Figure 2.27). On the other hand, the compound **2.27b** is not a chiral molecule, does not have enough substituents to form an enclosure to form inclusion complex and is unable to cause optical induction.

Selective transformations with high enantioselectivity are also achieved through reactions in confined media or by using container molecules as catalysts. A thermochemical 4+2 cyclo-addition reaction between a diene and substituted benzoquinone derivative passes through a high amount of selectivity when catalysed by cyclodextrin (Figure 2.28). No significant reaction between **2.28a** and **2.28b** takes place without a cyclodextrin catalyst, but in the presence of a catalyst, the reaction takes place with a good yield. It provides two isomers **2.28c** and **2.28d** in a 12:88 ratio. Formation of a higher amounts of the isomer **2.28d** indicates that the quinone is encapsulated into the cavity so that the less substituted end goes to the interior of the rim of cyclodextrin; hence, the product with more methyl groups on the other exposed side predominates.

Weak interactions such as the C—H···O interaction are prominent to guide a catalytic reaction to form a selective product. The C—C coupling reactions between naphthalen-2-yl methanesulfonate (**2.29b**) with ((4,5-dihydrofuran-2-yl)oxy)trimethylsilane (**2.29c**) are catalysed by palladium acetate in the presence of a ligand **2.29a** (Figure 2.29). The reaction is carried out in tert-butylmethyl ether solvent in the presence of lithium acetate at room temperature (20°C). A high selectivity in the formation of steteroselective product 3-(naphthalen-2-yl)dihydrofuran-2(3H)-one (**2.29d**) with 84% enantiomer excess is observed. Stereochemical consequence is attributed to the formation of intermediate

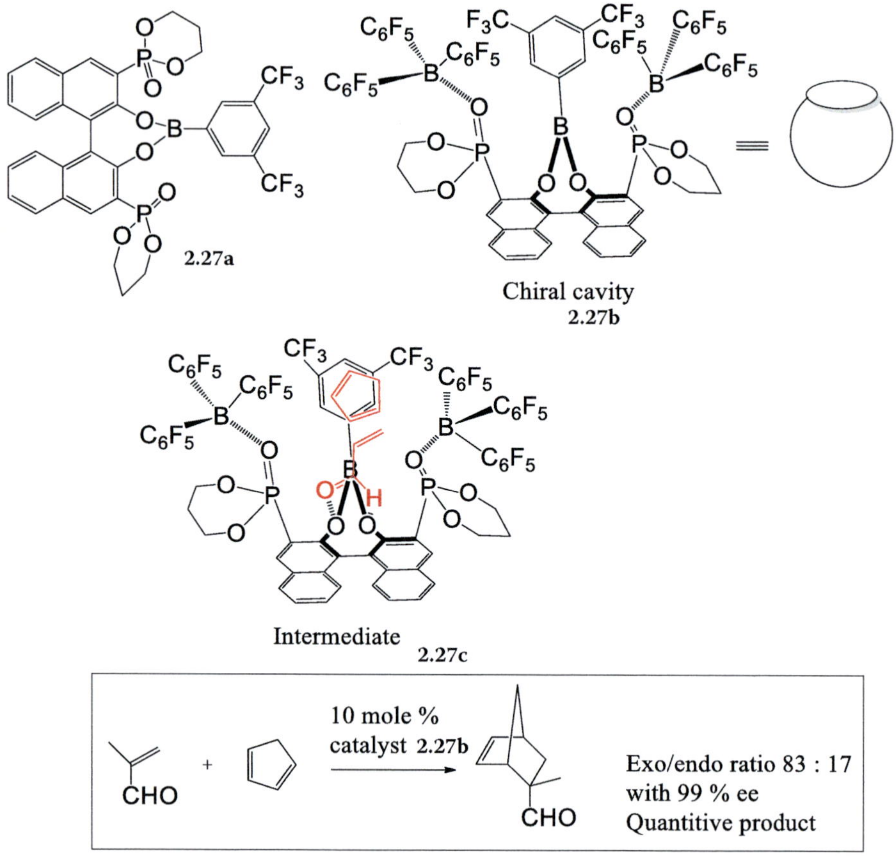

FIGURE 2.27 Encapsulation of reactants by extending peripheral embrace. (From M. Hatano and K. Ishihara, *Chem. Commun.* 2012, 48, 4273–4283.[27])

species **2.29e**. The intermediate has C—H···O interactions as a stabilizing factor. The intermediate **2.29e** has two hydrogen atoms at the *peri* positions of the naphthalene ring that are involved in hydrogen bonding with the keto form of the reactant connected to palladium by a Pd—C bond. The naphthyl group acts as a nucleophile and is bonded to the palladium by a palladium—carbon bond. The high selectivity in the product is due

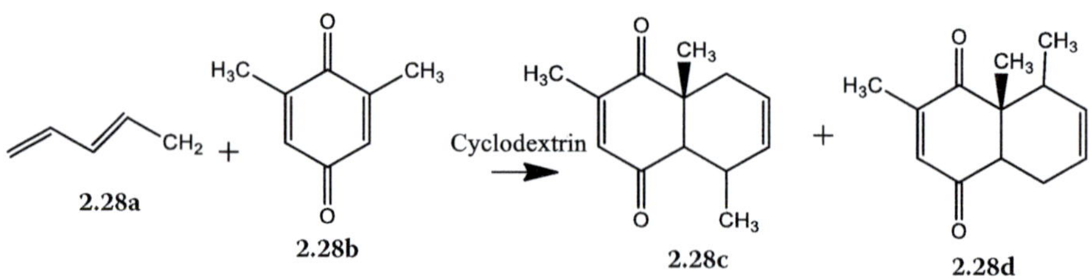

FIGURE 2.28 Selectivity observed in product formation in the presence of cyclodextrin. (From W.-S. Chung and J.-Y. Wang, *J.C.S. Chem. Commun.* 1995, 971–972.[28])

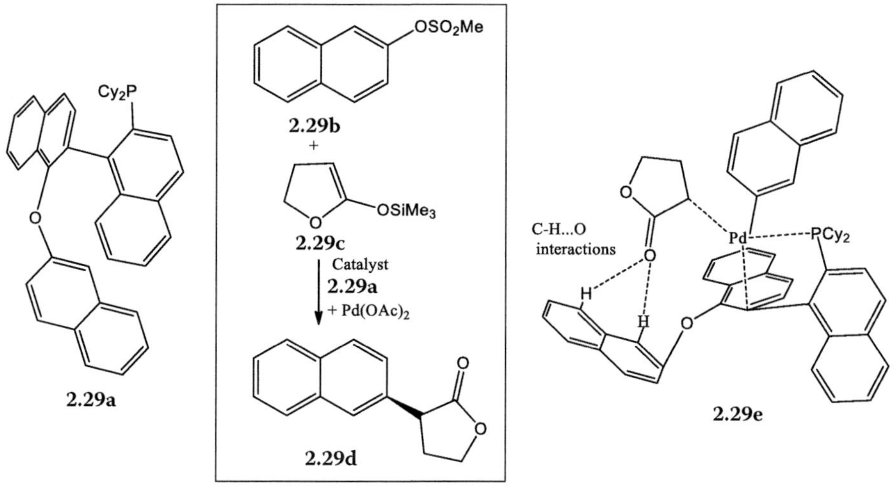

FIGURE 2.29 High selectivity in product due to C—H···O interactions. (From Z. Huang et al. *Angew. Chem. Int. Ed.* 2013, 52, 5807–5812.[29])

to the preferential attack of the nucleophile namely the naphthyl anion from one direction of the intermediate. The other sides are not easily accessible due to steric reasons.

Hydration of alkynes is conventionally done by mercury salts which provide ketones from terminal alkyne compounds. However, the use of designed pendent groups in a bifunctional catalyst provides highly selective products and also shows enzymelike activities. This is observed in the case of ruthenium complex **2.30a**. The complex has a piano-stool-like structure with a cyclopentadienyl anion as the basal plane of the stool. Three other ligands are like the legs of the stool. Among them, one leg of the stool is an acetonitrile molecule coordinating to a ruthenium (II) ion and the other two are from phosphorus-based ligands coordinating to the ruthenium ion. The phosphorus ligand has N-methylamine pyridine as a pendent group. The complex **2.30a** is a catalyst for the reaction between water and acetylene. The aldehyde is formed as an anti-Markovnikov product (Figure 2.30). In this reaction, the intermediate is a ruthenium—acyl-bonded hydride complex. The intermediate undergoes reductive elimination to form aldehyde. The reaction occurs under neutral conditions, has activity comparable to enzymes and forms an insignificant amount of ketone.

Metal complexes are immobilised by a “self-supporting” approach by interacting among themselves through hydrogen bonds. Such an approach of self-assembling inorganic complexes leads to heterogeneous chiral catalysts by the combination of multitopic chiral ligands with metallic ions. The catalysts are without a solid support or a support that makes the catalyst insoluble. This makes them different from conventional heterogeneous catalysts. Design of such a catalyst stems from the steric effects observed in the chiral (1R,2R)-1,2-diphenylethane-1,2-diamine ligand (**2.31a**) containing ruthenium complex [$RuCl_2(PPh_3)_2$(**2.31a**)] (**2.31b**). The ruthenium complex **2.31b** is a catalyst for asymmetric hydrogenation of 1-acetophenone to form 1-(1-naphthyl)ethanol with 75% enantiomeric excess. A similar complex with tri(3,5-xylyl)phosphine ligand instead of triphenylphosphine is also a catalyst for the same reaction but shows better

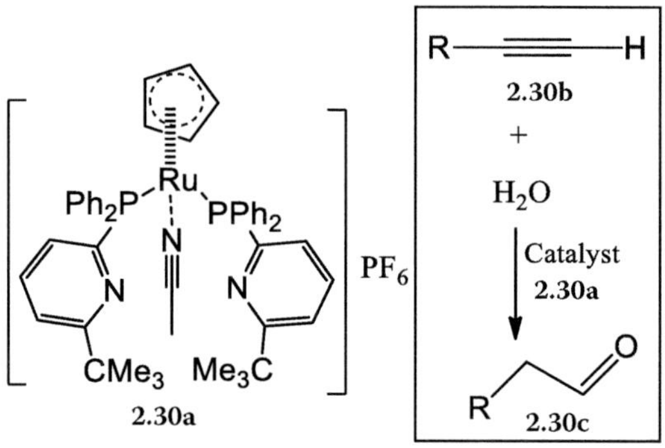

FIGURE 2.30 Anti-Markovnikov addition of water to acetylene. (From D. B. Grotjahn et al. *J. Am Chem. Soc.* 2004, 126, 12232–12233.[30])

enantioselectivity. There are higher steric repulsions between the two monophosphine ligands upon introduction of the methyl groups. The methyl-phenyl groups on the aromatic ring bring the aryl rings of the phosphine ligands into a closer vicinity to 1,2-diphenylethane-1,2-diamine ligand to make a better candidate for chiral induction. Based on this observation, the ligands **2.31c** and **2.31d** are developed (Figure 2.31).

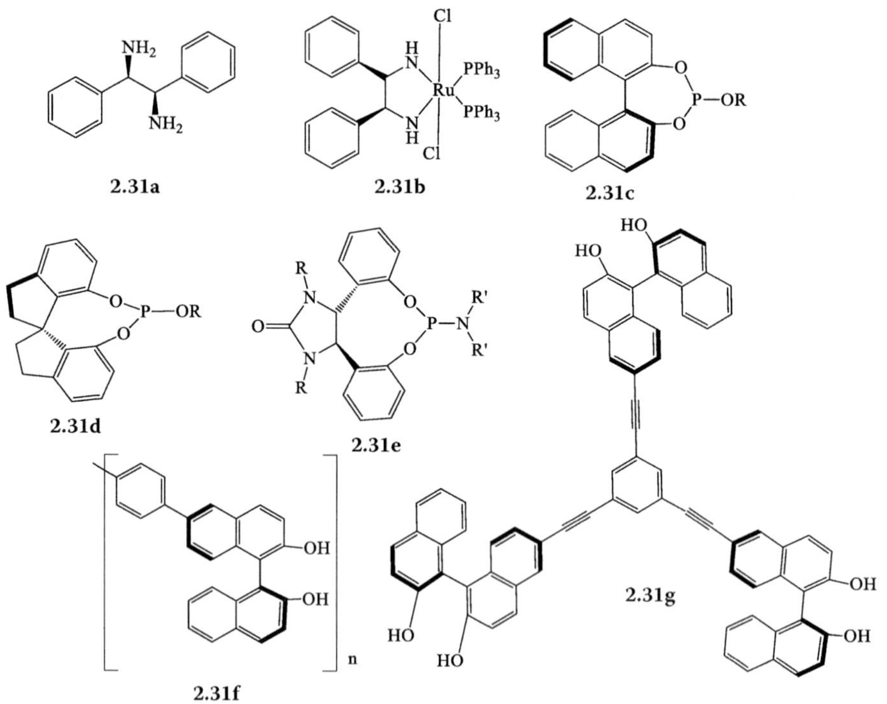

FIGURE 2.31 Design of ligands for high asymmetric induction. (From K. Ding, *Pure Appl. Chem.* 2006, 78, 293–301.[31])

These two ligands respectively show 88.8% and 89.0% ee of the 1-phenylethanol from the reduction of acetophenone. Similar ruthenium complex—possessing monodentate phosphoramidite ligands **2.31d** catalyse asymmetric hydrogenations and enhance asymmetric inductions. The ligand of this complex, there is a steric effect from both R and R′ groups to enhance asymmetric induction of the hydrogenated products.

Assemblies of chiral titanium complexes with multitopic ligands **2.31e** and **2.31f** are generated *in situ* with $Ti(OiPr)_4$. Such catalysts provide excellent enantioselectivity in both the carbonyl-ene reaction of α-methylstyrene with ethyl glyoxylate (up to 98% ee) to form C—C bonds between them and the oxidation of sulphides (up to >99% ee) to sulphoxide. The multitopic ligands **2.31g** have bridging spacers, and the ruthenium complex of this ligand is a useful catalyst for enantioselective catalysis for epoxidation of α-β-unsaturated carbonyl compounds. Due to the larger extension angles and longer linkage spacers, the complex containing this ligand shows good activity to provide enantiomers.

These examples have shown the roles of weak interactions, geometrical aspects, choice of catalyst and choice of ligands on a catalyst, and the reactive intermediates controlling the selectivity in the formation of product. It is imperative that use of the correct conditions and fine-tuning of supramolecular interactions pave the way to synthesise enantioselective products through catalytic reactions.

2.2 INORGANIC SUPRAMOLECULAR CATALYSIS

The central point of inorganic supramolecular catalysis relies on oxidation states, coordination numbers and affinity as well as the ability to change the coordination numbers of the metal ions. The supramolecular aspects are largely controlled by the type of ligand attached to the metal ion. Metal ions play roles the guide the directional nature of self-assembled structures by providing stability. Furthermore, the solvent plays a role in modifying supramolecular interactions. As a matter of fact, in many such reactions, the second coordination sphere built up through weak interactions on the outer periphery of the ligands provides sites for activation. The second coordination sphere recognises particular isomers to provide specific products and brings together multiple components to a common platform for a catalytic reaction. Certain organic compounds with supramolecular features alone are not necessarily the factors to qualify a compound for a certain catalysis. For example, the ligand in **2.32a** has an intermolecular hydrogen-bonded structure and quinoline as a binding site for a hydrogen bond. It also has a cyclic intramolecularly hydrogen-bonded structure that resembles its structure as a ligand in a metal complex. This ligand alone is not a catalyst for the hydrolysis of an imine. However, when this ligand forms a zinc complex **2.32b**, the complex acts as catalyst. The corresponding zinc complex **2.32b** in acidic conditions is a catalyst for the hydrolysis of (Z)-N-((anthracen-10-yl) methylene)-4-nitrobenzenamine (**2.32c**). The hydrolysis yields 4-nitroaniline and anthracene aldehyde (**2.32d**). The complex **2.32b** has a ligand that binds to a zinc ion as a tridentate ligand occupying three of five coordination sites, and the other two coordination sites of the zinc ion are occupied by two acetonitrile molecules (Figure 2.32). In solution, the acetonitrile dissociates from the complex, causing coordination unsaturation and makes way for coordination of imine. The imine thus is activated and hydrolyses under mild acidic conditions.

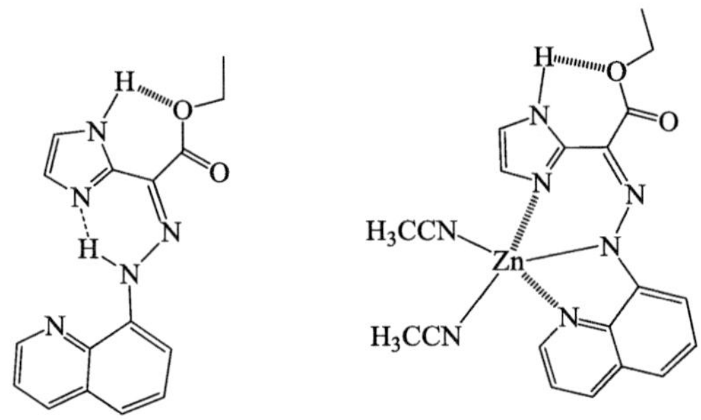

2.32a

2.32b

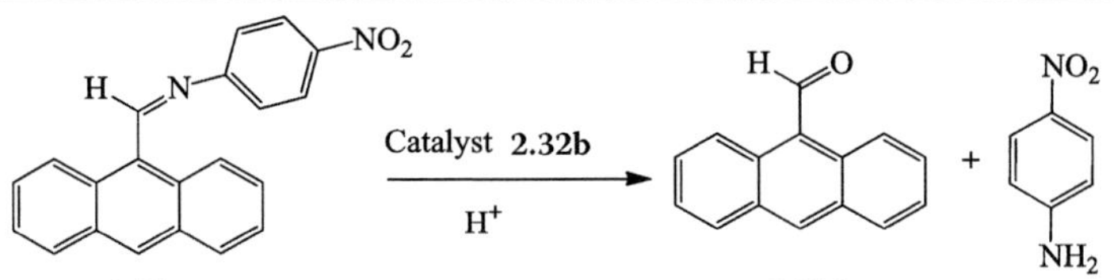

2.32c

2.32d

FIGURE 2.32 A zinc complex as catalyst for hydrolysis of imine. (From J. T. Foy et al. *Chem. Sci.* 2015, 6, 209–213.[32])

The metal ions as well as the ligands of certain metal complexes are responsible to independently activate reactants. The nitro-aldol condensation reaction is catalysed by the complex **2.33a**. The complex has cobalt(III) ion in five coordination geometry. The Salene part of the compound holds the zinc ions. The two sides of the Salene ligand have two thiourea units anchored at two sides, flanking the Schiff base portion of the ligand. The complex expands the coordination number to form a hexa-coordinated complex **2.33b** with an aldehyde by coordination through the oxygen of the C=O group of an aldehyde. The urea part forms a complementary hydrogen bond with the ene-form of the anion of the nitro-ethane derivative. This anion is stabilised by complementary hydrogen bonds. The metal centre activates the aldehyde, whereas the urea site of the ligand stabilises the anion. This facilitates the formation of the C—C bond between the two reactants (Figure 2.33).

Though conventionally, the central metal ion of a complex plays a crucial role in a catalytic reaction, the catalytic activity of certain complexes gets passivised due to the coordination of particular ligands to the metal ion. Such effects occur due to the poor lability of those ligands, as strong binding to the metal ion inhibits coordination unsaturation. The inhibition of the catalytic effect also occurs as a result of steric effect of ligands to prohibit the approach of another ligand. The electronic environment is also modified for a ligand site. For example, in many complexes, the *trans*-effect or back-bonding of a ligand changes electronic factors at another coordination site. This depends on alignment of the incorporated ligand into the coordination sphere of a metal ion in a complex. The complex **2.34a** is a carbene-based ruthenium complex; it catalyses the ring-opening reaction of norbornene derivative to polymerise (Figure 2.34). This complex, when reacted with pyridine, replaces one of the chlorides and forms a complex **2.34b**. The complex **2.34b** is not the catalyst for the above reaction. This is due to changes in

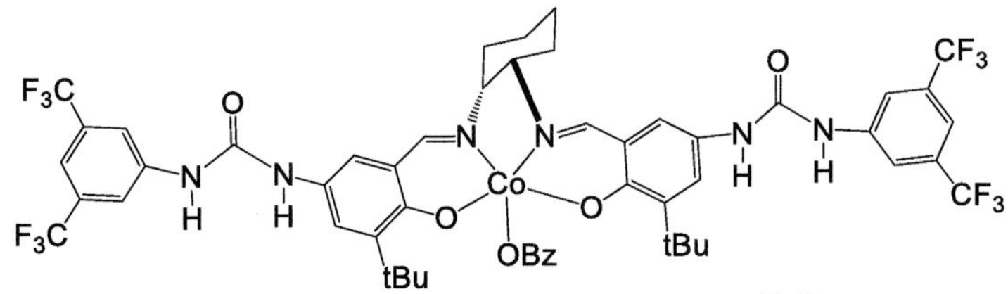

2.33a

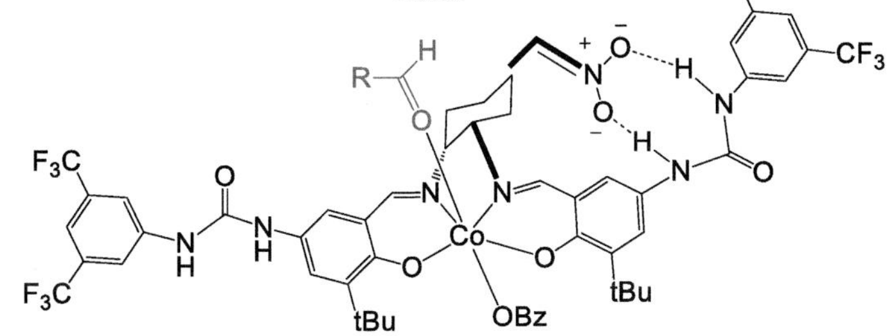

2.33b

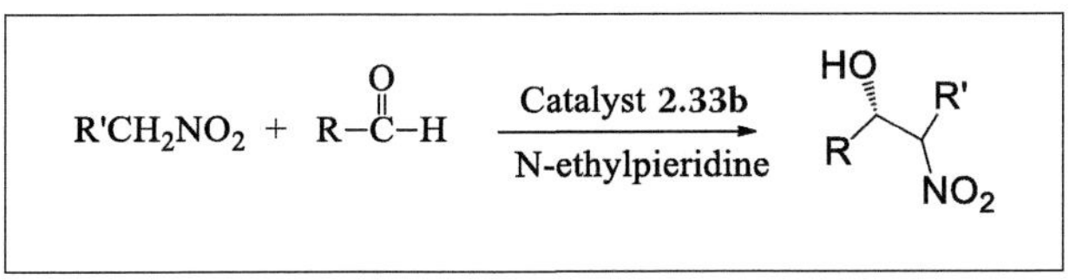

FIGURE 2.33 Nitro-aldol reaction catalysed by a cobalt complex. (From K. Lang et al. *Angew. Chem. Int. Ed.* 2012, 51, 1620–1624.[33])

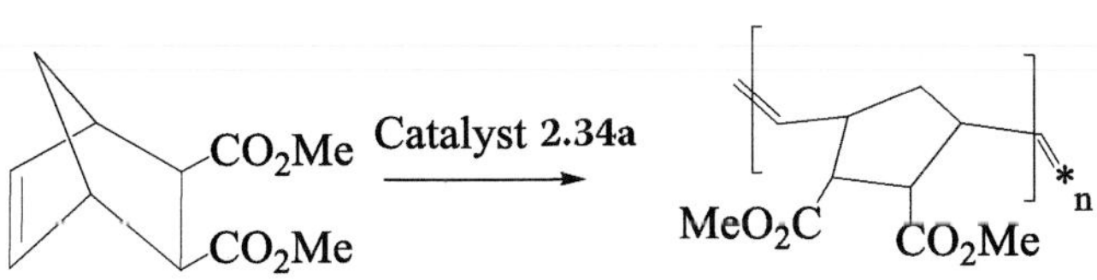

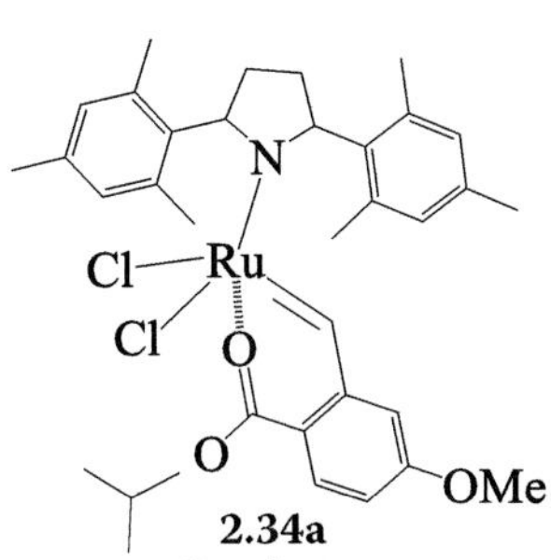

2.34a
Catalyst

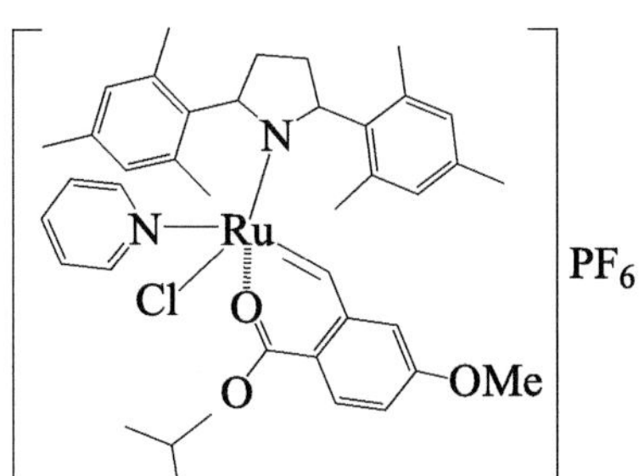

2.34b
Not catalyst

FIGURE 2.34 Catalytic control by changing coordinating ligand. (From L. Kovbasyuk and R. Kramer, *Chem. Rev.* 2004, 104, 3161–3187.[35])

TABLE 2.1
Changes of the Electronic Environment Causing Differences in the Reactivity

Reaction	Catalyst	Not catalyst
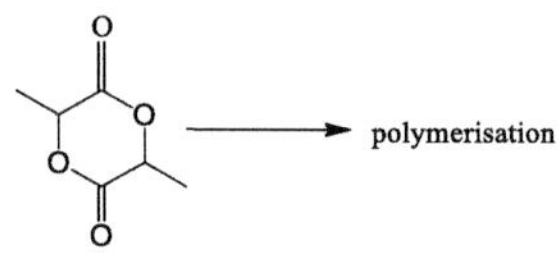	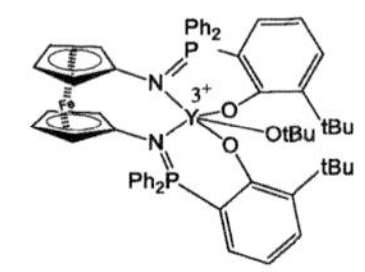	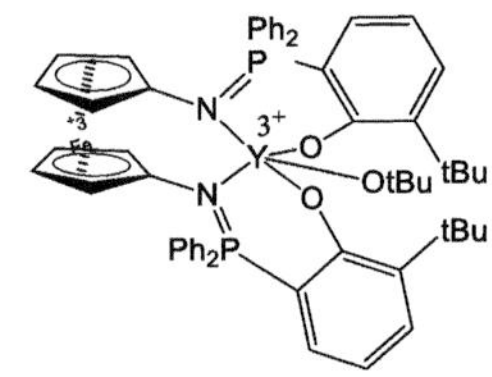
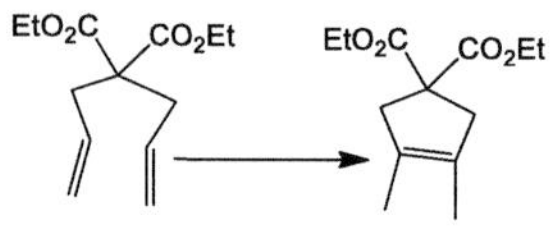	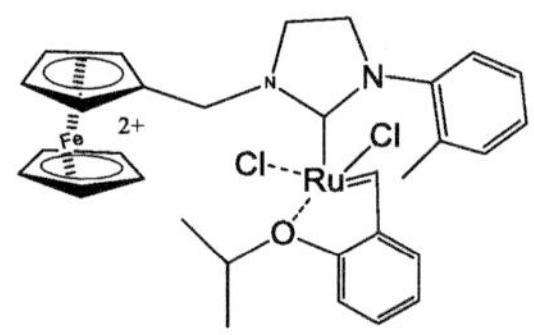	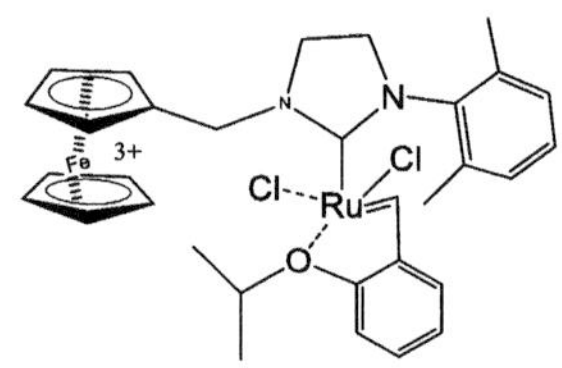

the coordination environment around the ruthenium causing changes in the electronic properties of the ruthenium centre. This example is on the catalytic activity changes due to changes in the local environment, but changes in the redox property at a site away from the reactive site also influence catalytic activity. The examples given in Table 2.1 have ferrocene moiety attached to a catalytically active centre used for ring open reactions. On oxidation of the ferrocene unit to the ferricenium cation, the catalytic activity of these complexes are lost.

Many catalysts are developed based on specific designed schemes. For example, a self-assembly between two metal complexes may be formed by hydrogen bonds at two remote sites of the complexes. One of those complexes may serve as a source to provide a continuous reagent by reacting with another substrate, whereas another component provides a template effect to hold the substrate to do the catalysis. For example, in Figure 2.35, the complex **2.35a** has an N-oxide-containing part that reacts with the substrate and is subsequently transformed continuously to form N-oxide by reaction with oxygen molecules. This portion of the molecule acts as a sink for the oxy-transfer reaction. The catalyst is a combination of two complexes that are assembled together through supramolecular interactions of the exterior portions of the ligands. Supramolecular assembly formed by amide—amide interactions brings the two metal complexes to encapsulate the reactant to cause a catalytic reaction (Figure 2.35). The self-assembled complex **2.35a** has N-oxide in the ligand part of one complex, and the other has a planar porphyrin ring that coordinates to the nitrogen-containing substrate. The complex shows distinguishable catalytic reactivity to cause an epoxidation reaction between styrene and vinyl-pyridine. The vinyl-pyridine binds to the zinc ion of the complex, whereas styrene does not. Thus, the epoxidation reaction is faster with vinyl-pyridine than styrene.

The projection of reactive sites matters in a supramolecular catalyst to act in a concerted manner to perform catalytic reactions. Active sites may be appropriately

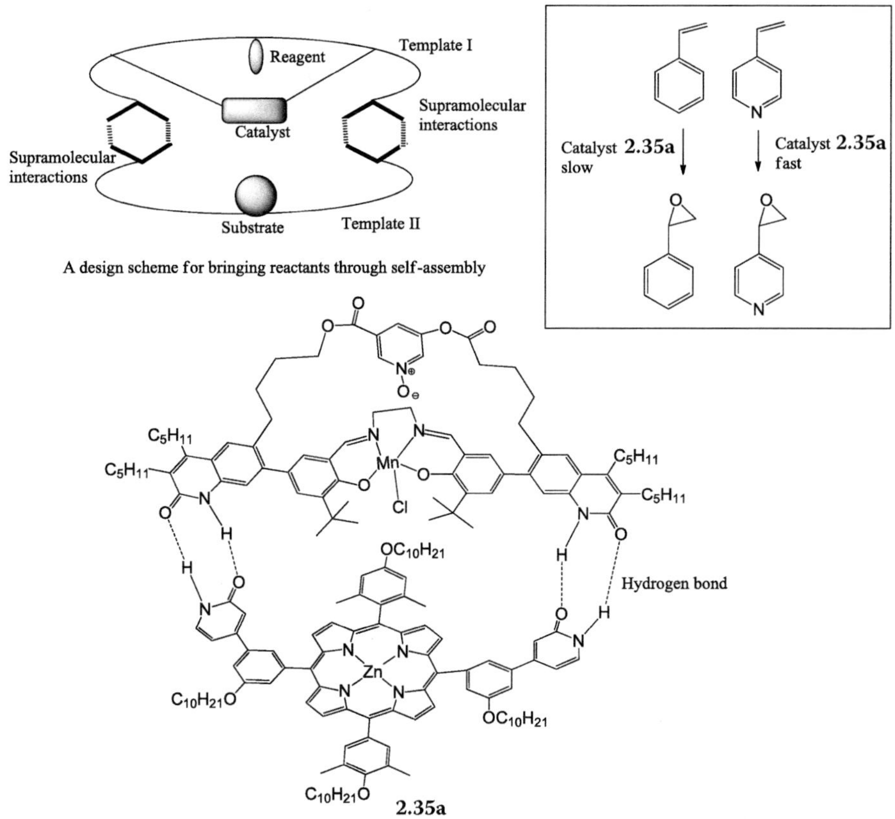

FIGURE 2.35 Specific activation of a substrate through supramolecular approach. (From S. Jonsson et al. *Chem. Commun.* 2005, 549–551.[34])

controlled from a structural modification at a distant site or at a site near a catalytic centre. This is done routinely in flexible systems, where geometry is controlled by a coordination effect at a suitable binding site. For example, the complex **2.36a** adopts a bend structure where the amine cation is brought close to a reactive metal site. In this case, the scaffold with polyether bonds folds upon interaction with a bivalent barium ion. The complex **2.36b**, on the other hand, brings the two copper sites into close proximity by complexing with palladium (Figure 2.36).

2.2.1 Design of Mononuclear Inorganic Complexes for Supramolecular Catalysis

Due to many possibilities for varying reaction parameters, it is difficult to differentiate and list a series of inorganic complexes in a definite category of catalyst, but it is relatively easy to classify complexes based on the nuclearity withholding their applications. However, classification may be based on the number of metal ions in the complex that act as catalysts. But it is of utmost importance to systematically

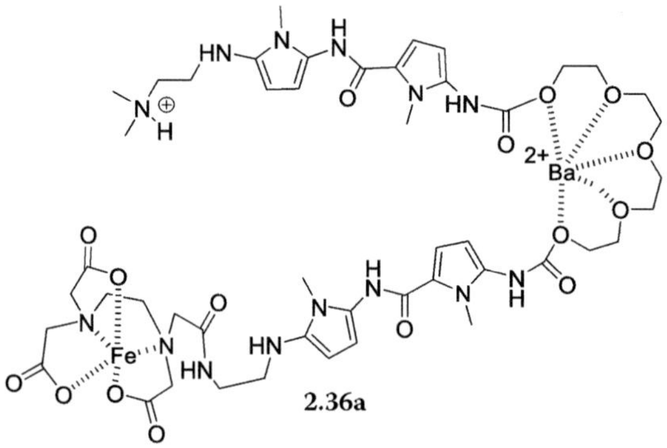

2.36a

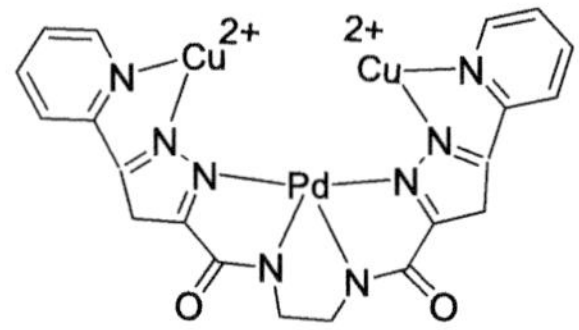

2.36b

FIGURE 2.36 Allosteric effect bringing reactive sites to close proximity. (From L. Kovbasyuk, R. Kramer, *Chem. Rev.* 2004, 104, 3161–3187.[35])

study the principles to develop catalysts based on nuclearity of the metal complexes. The metal ions of a supramolecular inorganic catalyst, together with the ligands, directly or indirectly cause the activation of a substrate. In a conventional inorganic mononuclear catalyst, the activation of a reagent takes place by direct metal substrate interactions. The supramolecular interactions of ligands with one or more substrate are very useful to bring together reactants. The metal ion in a complex provides the appropriate geometry for interaction to cause the activation of the reactants. There are several ways to activate a reactant by a mononuclear complex for supramolecular catalytic reaction. Two such possibilities are listed in Equations A and B in Figure 2.37. One of the ligands of the hexa-coordinated metal complex of equation A of Figure 2.37 has supramolecular sites to interact with a substrate. Upon interaction of the complex with a substrate, the binding strength of the metal ion-bound ligand relatively weakens. In such an event, the ligand—substrate interactions activate the substrate and help in release of the ligand to accommodate another substrate or ligand in the place of this particular ligand. In the other instance, shown in equation B of Figure 2.37, a chelating ligand binds to a substrate through supramolecular interactions. This modifies one binding site of the chelating ligand and eventually breaks a metal—ligand bond at this particular site. This results in coordination unsaturation and modifies the coordination behaviour on the end of this chelating ligand. The weakened metal—ligand bond eventually facilitates the binding of a substrate to the metal ion at this particular site. On the other hand, the situation is different in the case of electro- or photocatalytic reactions. In electrochemical reactions, the oxidation state or formal redox properties of the metal ions change with the environment and change the course of a catalytic reaction. In some photochemical reactions, the excitation of the metal complex occurs as depicted in Equation C of Figure 2.37. The lability of a ligand in an excited state of the metal complex differs from the lability of the same ligand at ground. Supramolecular features further provide guidance to know whether an exchange in energy from an excited state will take place. In particular, the inner-sphere or outer-sphere electron transfer process also has a role in ligand exchange reactions. In the event of changes in oxidation

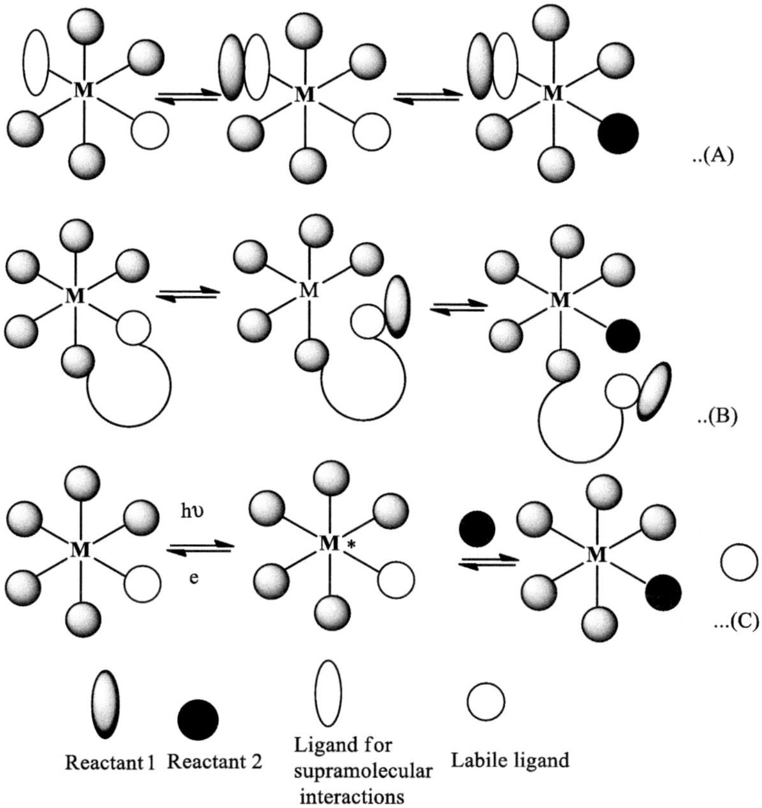

FIGURE 2.37 Schematic representations of some prominent ways of activation of ligands and substrates in mononuclear complexes.

numbers, the coordination behaviour and lability of the metal complexes change. For example, an octahedral cobalt (III) complex is inert, whereas a cobalt (II) complex is labile. Furthermore, ligands also influence the electronic properties of the metal ion. A ligand with high crystal field stabilization energy results in a diamagnetic complex, whereas a weak one changes it to paramagnetic. Apparently, many reactions have supramolecular features, though hydrogen bonds or other related weak interactions are used to explain an exception, but to control the reactivity, the internal electronic features are as important as those. The role of solvents, ability to expand or contract coordination numbers of a metal ion and ability to undergo oxidative addition and reductive eliminations are controlled by supramolecular features which dictate the overall activity and selectivity. Based on these properties, supramolecular catalysts from various inorganic complexes are designed. One such design principle for a mononuclear inorganic complex catalyst is shown in Figure 2.38. A four-coordinate metal complex **2.38a** has a pincer-type ligand tethered to crown-ether functionality. The crown ether part is amenable to complex with the metal ion and also to bind to the alkali metal cation. The binding of the crown ether to the metal ion is modulated by adding a monodentate ligand or coordinating solvent molecules. The ligand/solvent

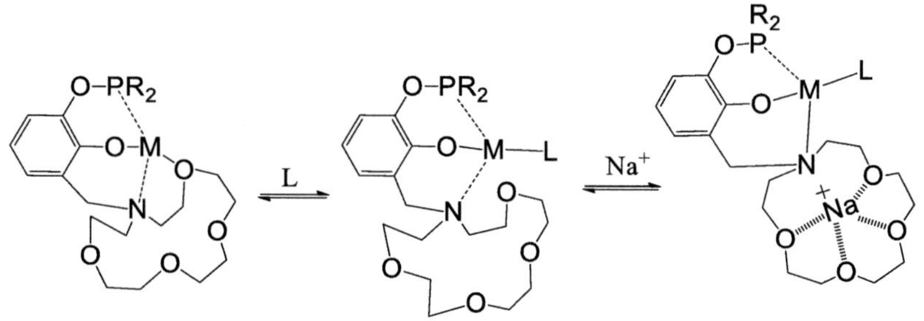

FIGURE 2.38 Design of supramolecular catalytic site. (From A. J. M. Miller, *Dalton Trans.* 2017, 46, 11987–12000.[36])

molecules go on to form new metal—ligand bond; this replaces the M—O bond of the ether **2.38b**. This is due to different bond strengths of the metal ligand bonds. The weak metal—ligand interaction of the metal ion between the metal ion with the ethereal oxygen of the crown ether part is dislodged in a strong ligating solvent or by a competing ligand. The geometry of the crown ether may be further modulated by interactions of the crown ether part with alkali metal ion to form a rigid structure **2.38c**. Hence, the reactivity of the metal centre is modulated by the solvent or ligand or through cation binding.

The above principle is applied in iridium-catalysed isomerization of terminal to internal olefin (Figure 2.39). Depending on the amount of acetonitrile, the complex **2.39a** adopts the structure **2.39b** or **2.39c**. The flexible polyether part of the ligand has the preference to bind an alkali cation and forms a complex **2.39d** by reacting with the $NaBPh_4$ in acetonitrile. The complex **2.39a** in acetonitrile and $NaBPh_4$ is a good catalyst. It isomerises olefin through multiple paths. The key feature of the reaction is the control of the coordination sites and binding of the incoming ligand to the iridium catalyst to change the geometry of the flexible crown ether part.

The conformational dynamics of hydrogen exchange and different intermediates are established through NMR studies. The equilibriums between the intermediates **2.40a–d** (Figure 2.40) suggest that in the lithium-containing complex, the oxygen atom of the crown ether part is not coordinating to the iridium. In this case, there is fast hydrogen to deuterium exchange, whereas in the absence of a lithium ion, one oxygen atom of the polyether coordinates to the iridium ion. This slows down the exchange process between externally supplied deuterium with the iridium—hydride bond. Hence, coordination unsaturation caused by alkali-metal ion binding to the polyether part of the ligand provides crown-ether—type geometry which facilitates intake of hydrogen molecules. This activation causes an increase in the reaction rate.

Ion pairs formed between the ligands in a metal complex become a guiding factor for selectivity in catalytic reactions. Hydroboration reaction by pinacolate borane (HBpin) on benzyl ammonium tostylate salt **2.41b** is facilitated in a stoichiometric manner by the bipyridine salt **2.41a** to form the salt **2.41e**. This reaction results in low *meta*-isomer selectivity and forms the *meta* to *para* isomer of the product **2.41d** in a 2:1 ratio. The iridium complex **2.41c** is a good catalyst for this reaction.

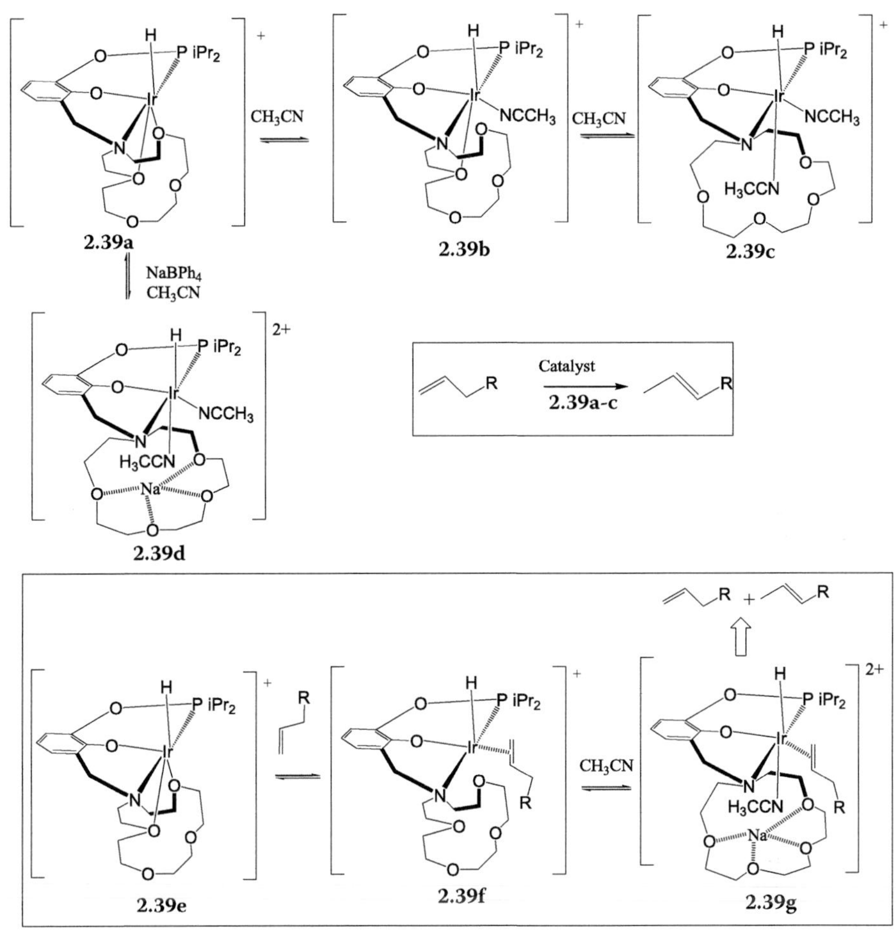

FIGURE 2.39 Catalytic isomerization reaction of olefin by a supramolecular iridium catalyst. (From A. J. M. Miller, *Dalton Trans.* 2017, 46, 11987–12000.[36])

In the hydroboration reaction catalysed by the iridium complex, *meta*-isomer is formed three times more than *ortho*-isomer. The reaction is carried out with 1.5 mole % of catalyst. In this example, the iridium ion of the complex **2.41c** activates the borane by forming an iridium—boron bonded intermediate complex. The benzyl ammonium cation forms an ion pair with the sulphonate group of the bipyridine-derived ligand anchored in the complex and provides the intermediate complex **2.41d**. The C—H located at the *meta*-position of the reactant, namely the benzyl ammonium cation, is in close proximity to the iridium ion, which is coordinately unsaturated. Thus, the *meta* C—H is also in the close proximity to the iridium—boron bonds. Hence, high *meta*-selectivity in the product is observed from this catalytic reaction (Figure 2.41).

π-Interactions between substrate and reactant improve enantiomer selectivity. The 4 + 2 cycloaddition reaction between 2-methyl-1,3-butadiene reacts with the compound 3-((E)-but-2-enoyl)-4-alkyl-oxazolidin-2-one (**2.42a**) to yield isomers

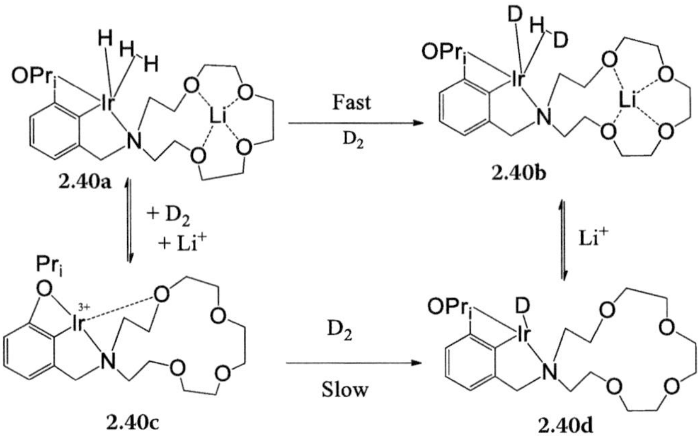

FIGURE 2.40 Conformation dynamics of deuterium and hydrogen exchange. (From A. J. M. Miller, *Dalton Trans.* 2017, 46, 11987–12000.[36])

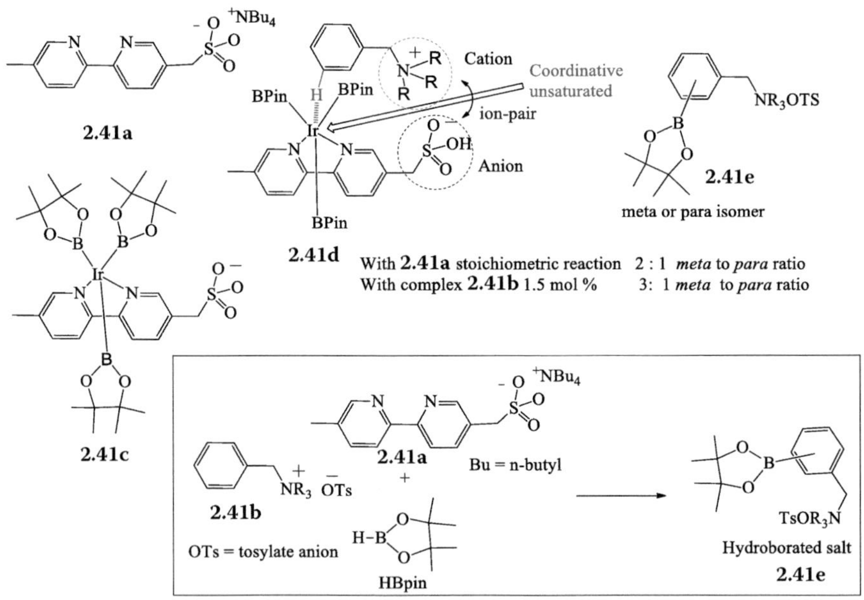

FIGURE 2.41 Ion pair in iridium catalysed hydroboration reaction. (From H. J. Davis et al. *J. Am. Chem. Soc.* 2016, 138, 12759–12762.[37])

(**2.42b** and **2.42c**). This reaction is catalysed by diethyl aluminium chloride; the ratio between the two isomers is highly dependent on the R-group on the reactant. When the R-group is a benzyl group, the ratio between the isomers (**2.42b** and **2.42c**) is 21:1, whereas in the similar reaction with R = cyclohexyl methyl, the ratio is 9:1. The large difference in the selectivity arises as consequence of the π-interactions between aromatic rings with the double bond of the dienophile. Due to such interactions

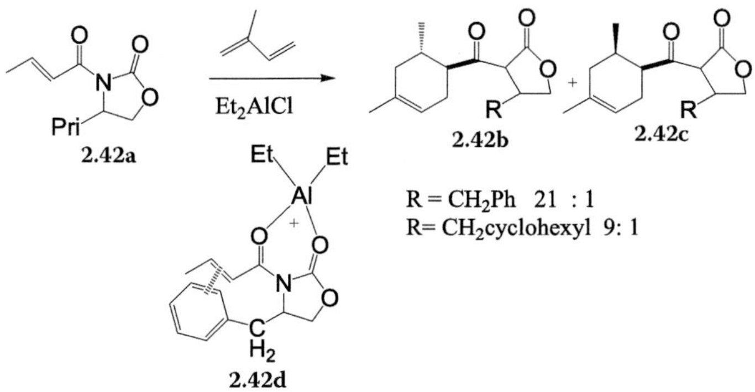

FIGURE 2.42 π-interactions contributing to selectivity. (From D. A. Evans et al. *J. Am. Chem. Soc.* 1988, 110, 1238–1256.[38])

and to the coordination of the substrate to the aluminium ion-forming chelate, chemoselectivity is observed in the case of the substrates with benzyl substituent. The intermediate species **2.42d** causes the specific reaction to give a particular stereoisomer in higher amounts over the other isomer (Figure 2.42).

The role of C—H···π interactions in stabilization of intermediate species is revealed in ruthenium-arene complex-catalysed dehydro-aromatisation reactions of heterocyclic compounds (Figure 2.43). The reaction passes through a hydrogen-bonded intermediate (**2.43h**) that possesses a C—H···π interaction. An arene-ruthenium complex **2.43b** with

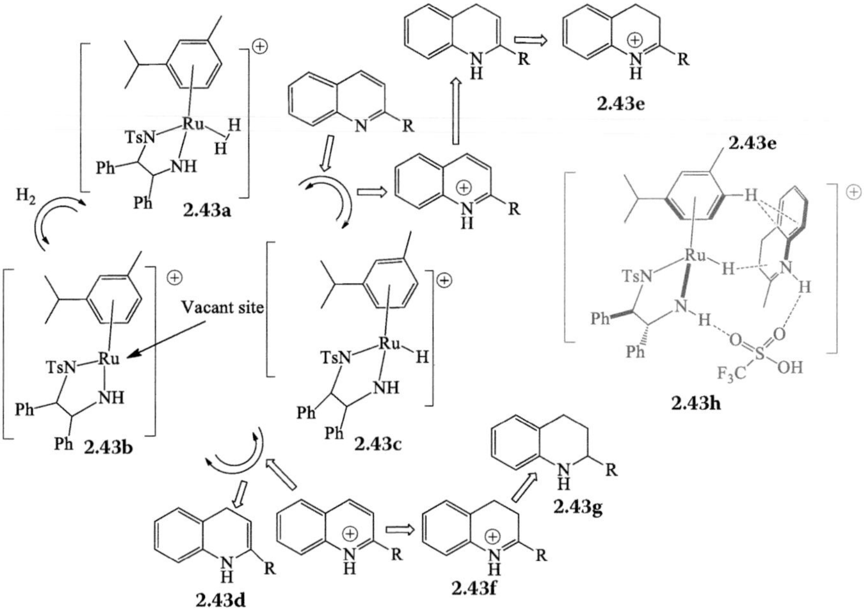

FIGURE 2.43 C—H···π interaction contributing to a catalytic dehydrogenative aromatisation reaction. (From T. Wang et al. *J. Am. Chem. Soc.* 2011, 133, 9878–9891.[39])

NN-*bis*-chelating ligands forms a dihydrogen complex **2.43a** with molecular hydrogen. The dihydrogen complex utilises the activated hydrogen molecules to hydrogenate the quinoline derivative. The dihydrogen ruthenium complex transforms to a hydride complex **2.43c**. At this step, the quinoline is reduced through a sequence of reactions to yield the product **2.43e**. The hydride complex **2.43c** hydrogenates **2.43e** to form the hydrogenated product 2-alkyl-1,2,3,4-tetrahydroquinoline **2.43g**. The transfer of a hydride **2.43c** generates another ruthenium complex **2.43b**. This complex has a vacant coordination site, so it is coordinatively unsaturated. This complex is ready to take up hydrogen molecules to form the complex **2.43a** to restart another catalytic hydrogenating cycle. In the reduction, an intermediate **2.43h** is formed in which tetrahedral trifluoromethane sulphonate anion acts a bridge anchoring the N—H of the heterocycle and the N—H bonds of the ancillary ligand. In the intermediate **2.43h**, there are substantial contributions of the C—H···π interaction between a hydrogen atom of the ligating arene ring with the π-cloud on the aryl ring of the partially reduced alkyl-quinoline. The Ru—H bond also interacts with the double bond of the partially reduced quinoline moiety (Figure 2.43). These interactions make the system an efficient catalyst for hydrogenation reaction. Such reactions are highly specific when trifluoromethane sulphonic acid is used as a reagent in the reaction.

Pre-organised ligands are important to achieve selectivity in catalytic reactions. *Bis*-indole-based pincer-type ligands provide scaffolds to hold carboxylic acid, and when such units are part of a ligand, it helps in tuning reactivity. For example, the hydroformylation of unsaturated carboxylic acid by [Rh(CO)acac] is nonselective, and such reactions result in terminal and internal hydrogenated products of the olefin (Figure 2.44), but the use of a ligand **2.44a** with the complex [Rh(CO)acac] catalyses the hydroformylation reaction and yields selective product. The highest selectivity is observed with unsaturated carboxylic acid when $n = 3$. In this case, the terminal attack is prominent, it is in close proximity to the rhodium ion, and the carboxylic end of the substrate is tightly held to the ligand in the intermediate **2.44b**.

The allosteric effect of potassium ion in the palladium complex (**2.45**) provides a suitable geometry to construct a hydrophobic pocket. This hydrophobic pocket of the complex (**2.45**) retains the η^3-π-allyl palladium part of the complex. This complex is highly catalytically active for C—C bond formation through allylic activation of allylic acetates with dimethylmalonate (Figure 2.45). This reaction has higher selectivity in forming an enantioselective product than a reaction where a potassium ion is not used. The conformation rigidity imparted to the intermediate of the reaction by the potassium ion is the key to the enhanced selectivity.

The structure of the complex (**2.46a**) has a porphyrin ring exposed to bind a substrate for catalytic reactions. It is a good catalyst for C—C bond formation between *p*-nitro benzaldehyde and acetylacetone (Figure 2.46), whereas the complex segregates to form a mononcear zinc complex (Figure 2.46b), as well as a copper bipyridine complex. The mononuclear zinc complex **2.46b** is converted to the complex **2.46a** reversibly by controlling the copper(I) ion in solution. This complex has one axial coordination site is coordinated to a nitrogen of the ligand and has a rigid structure, with one side blocked for the approach of a reactant. This inhibits catalytic activity, and the complex **2.46b**, it is not a catalyst for the condensation reaction between aldehyde and acetylacetone.

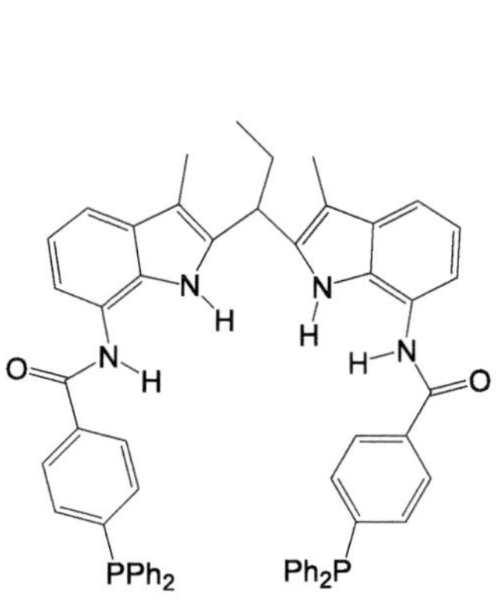

2.44a

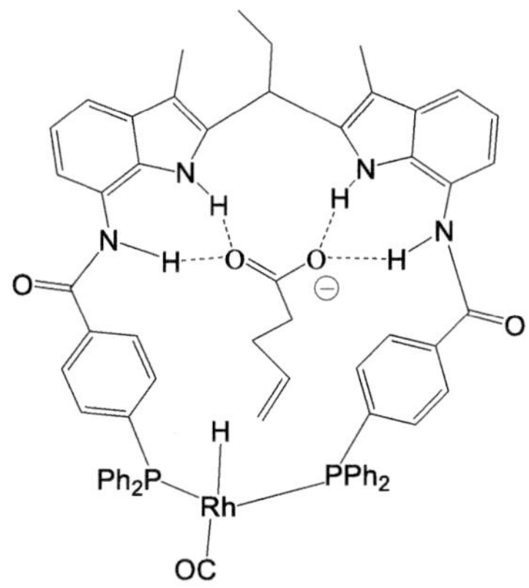

2.44b

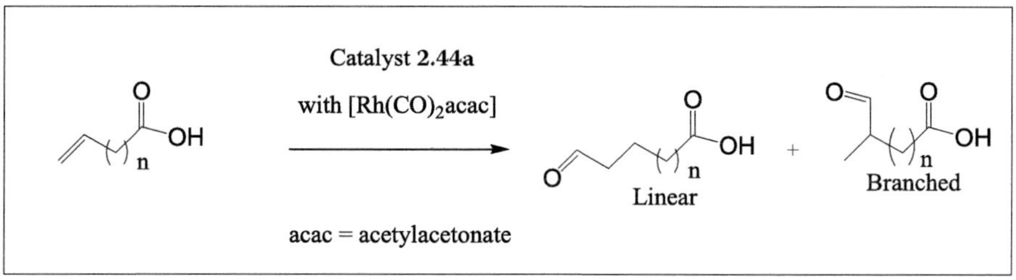

FIGURE 2.44 Selectivity in hydrogenation due to complementing hydrogen bonding through functional group at a remote site. (From P. Dydio et al. *Angew. Chem. Int. Ed.* 2011, 50, 396–400.[40])

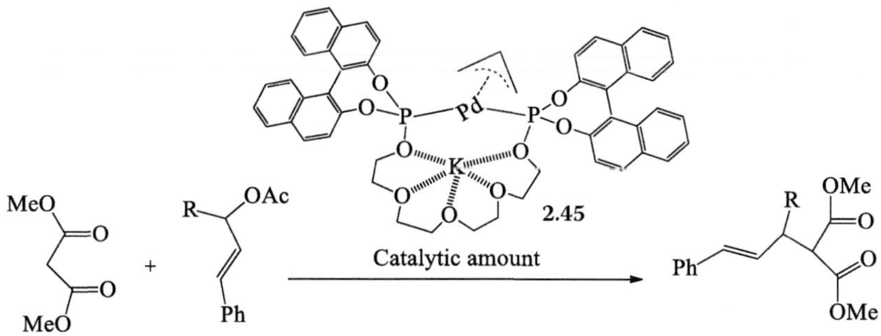

FIGURE 2.45 Allosteric effect of potassium cation assisting a catalytic C—C bond formation reaction. (From L. Rovira et al. *Organometallics* 2016, 35, 528–533.[41])

Conversion of water to oxygen is one of the highly sought-after reactions. The mononuclear complex $[Ru(H_2O)(NPM)(pic)_2]^{2+}$ (NPM = 4-*t*-butyl-2,6-di-(1′,8′-naphthyrid-2′-yl)-pyridine, pic = 4-methylpyridine) (**2.47a**) is a catalyst for water splitting. The water ligand has a role in the activation process. Similar to such an observation, the complex $[Ru(tda)(py)_2]$ (tda = 2,2′:6′,2″-terpyridine- 6,6″-dicarboxylate, py = pyridine) **2.47b** shows high catalytic activity with turnover frequency 5×10^5 sec^{-1} at pH 10. This complex has a free carboxylate group on the ligand. The free carboxylate

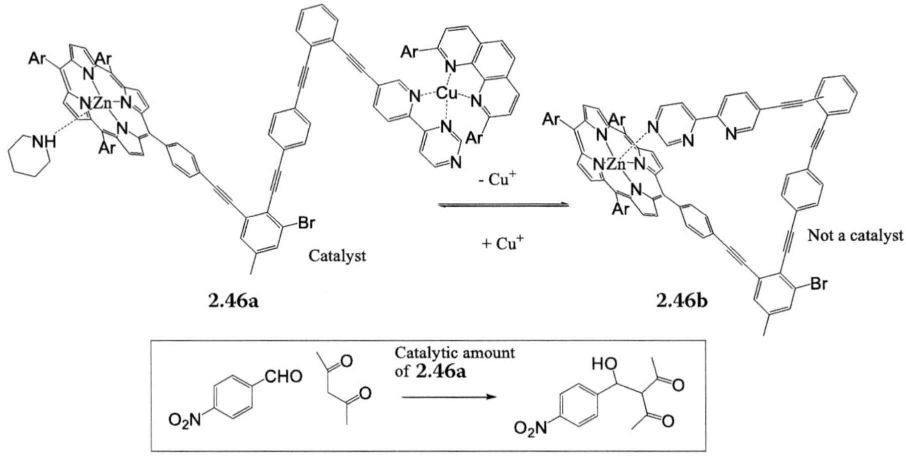

FIGURE 2.46 Lock of orientation through coordination influencing catalytic activity. (From M. Schmittel et al. *Angew. Chem. Int. Ed.* 2012, 51, 3832–3836.[42])

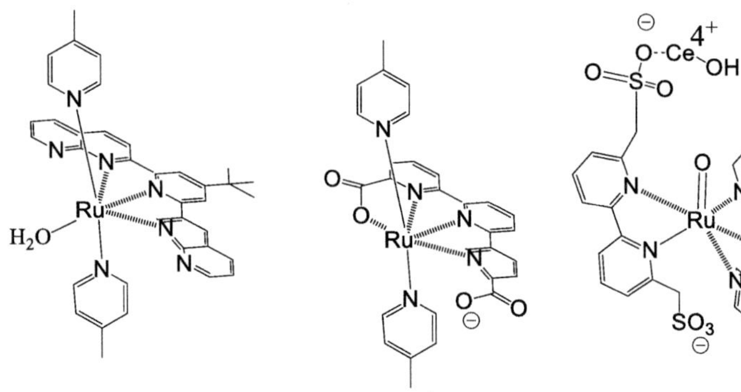

2.47a **2.47b** **2.47c**

FIGURE 2.47 Examples of mononuclear ruthenium catalysts for water splitting. (From V. Kunz et al. *Adv. Energy Mater.* 2017, 7, 1602939.[43])

group serves as activating site; it hydrogen bonds with the water molecule and brings it close to the active site. On the other hand, the sulphonate groups on the ligand of [Ru(tpy)(bpyms)(OH_2)] (bpyms = 2,2′-bipyridine-5,5′-bis(methanesulfonate) **2.47c** are at the appropriate location in the coordination sphere to anchor cerium (IV) hydroxide. The complex shows higher water-splitting ability in the presence of cerium (IV) ions than a complex without the cerium ion. These examples are explicit demonstrations of ligands controlling catalytic activity through supramolecular interactions (Figure 2.47).

2.2.2 Polynuclear Inorganic Complexes as Supramolecular Catalyst

Polynuclear metal complexes have the advantage of contributing in several more ways than mononuclear metal complexes by providing multiple functional and active sites.

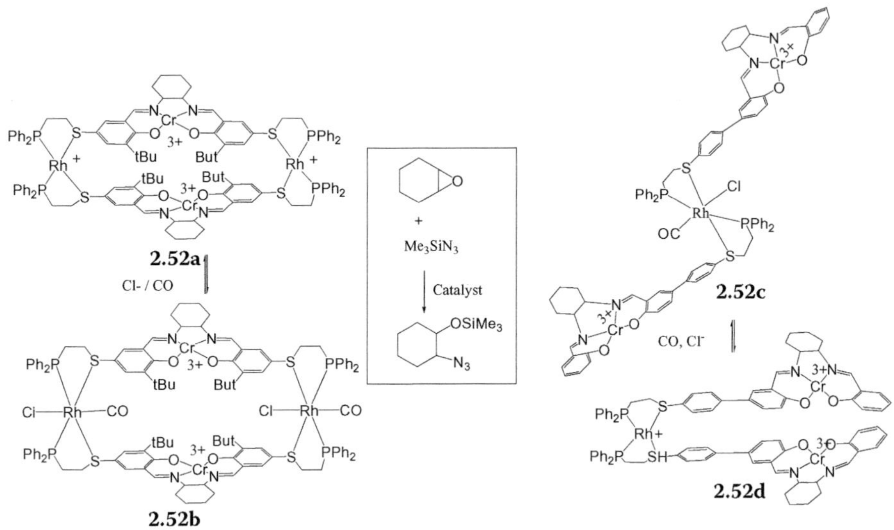

FIGURE 2.52 Control of reactivity of a chromium-Salene complex for cyano-silylation reaction. (From E. N. Jacobsen, *Acc. Chem. Res.* 2000, 33, 421–431.[48])

part of the ligand ligating to copper to form a cyclic structure. The coordination of the ligand provides a five-coordinated environment around a copper (I) ion. The reaction between cycloheptene and ethyl 2-azidoacetate to form bicyclo[5.1.0]oct-3-ene **2.53d** and diethyl fumarate **2.53e** is catalysed by the mononuclear copper (I) complex **2.53b** only when iron (II) ions are added. Without the iron ion complex **2.53b**, it is not a catalyst for this reaction. Upon addition of iron(II) ions, it transforms to a trinuclear complex **2.53c**. The transformation takes place by the reorganization of the ligating sites to accommodate an iron (II) ion. The complex **2.53c** has terpyridine parts located at the two ends of the two independent ligands coordinated to the iron ion. The other ends of the ligands bind to two independent copper(I) ions through the phenanthroline (Figure 2.53). Hence, this copper(I) complex has two copper(I) centres located at remote sites. These sites are available for the catalytic cyclopropanation reaction, whereas the complex **2.53b** has coordinatively saturated copper; thus, the active site is concealed and unable to catalyse the cyclopropanation reaction alone.

The reactivity of complexes of polyfunctional ligands is modulated with externally added ligands. A simple scheme may be constructed by taking a ligand with three different kinds of ligating sites at three arms. In Figure 2.54, the two arms of the ligand **2.54a** are extended to bind to a common metal ion, whereas the other arm independently binds to another metal ion to form a bimetallic complex **2.54b**. On addition of another chelating ligand, one site, which is coordinatively unsaturated, is ligated by the ligand to form **2.54c**. This construction provides two different metal sites for carrying out catalytic reactions. Depending on the type of ligating sites on each end, reactivity modulation at two sites is achieved by adding ligands from external sources.

For example, competitive bindings between ligands, namely porphyrin, terpyridine, phenanthroline and cyclohexylamine, are utilised in the complex **2.54d–f** to design

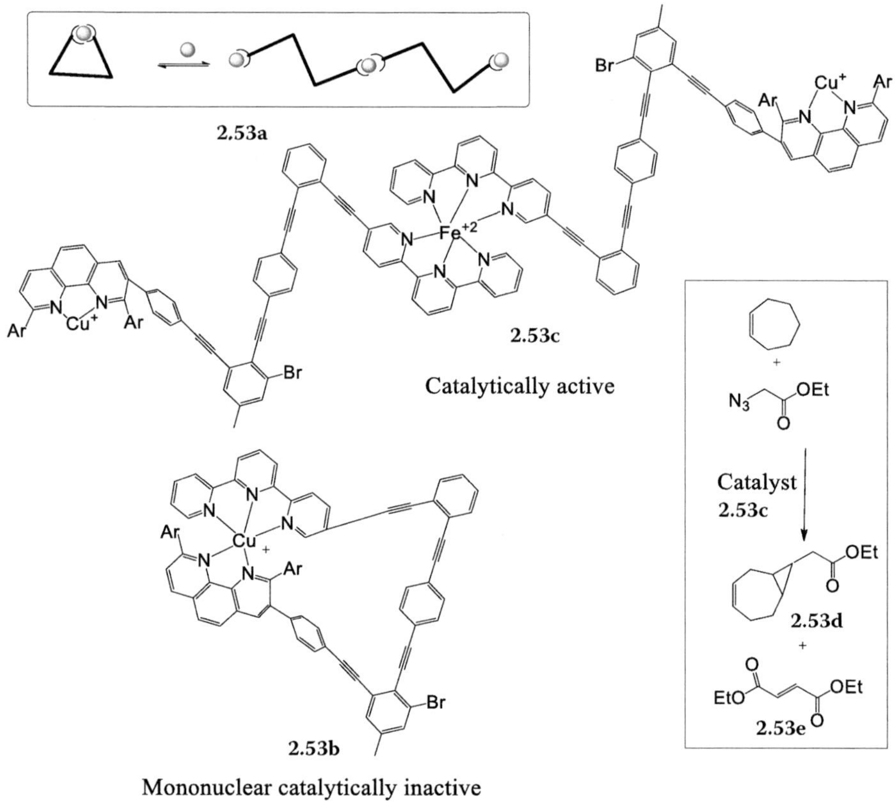

FIGURE 2.53 Expanding nuclearity of a complex to provide active sites. (From S. De et al. *Dalton Trans.* 2014, 43, 10977–10982.[49])

a catalyst for catalytic C—C bond formation between diethylmalonate and p-nitro benzaldehyde. The complex **2.54e** has coordination unsaturation around the porphyrin ring. This complex, upon reaction with cyclohexyl amine, forms a coordinatively saturated amine complex **2.54d**. Besides four nitrogen of porphyrin rings coordinating to the zinc ion, there is an amine ligating at the axial coordination site. The other axial site has nitrogen coordinating to zinc from the arm of the porphyrin ligand. Thus, all the coordination sites of the zinc ion are blocked. This complex is not a catalyst for the reaction between diethylmalonate and p-nitro benzaldehyde, but the complex **2.54e** is an active catalyst. This complex has one side of zinc-porphyrin left open for catalytic reactions. The activity of the complex **2.54e** is enhanced by transforming it to a complex **2.54f** by reaction with 3-ferrocenyl-1,10-phenanthroline. The ligand 3-ferrocenyl-1,10-phenanthroline reacts at the copper site; hence, this site is not available for complexation with the reactants (Figure 2.54). This process allows the zinc-porphyrin part to catalyse the reaction without its interference.

The complexes of supraphos ligand (**2.55a**) with rhodium complexes such as Rh(cyclooctadiene)tetrafluoroborate are catalysts for asymmetric hydrogenation of *N*-(3,4-dihydro-2-naphthalenyl) acetamide (Figure 2.55). The catalytic activity of

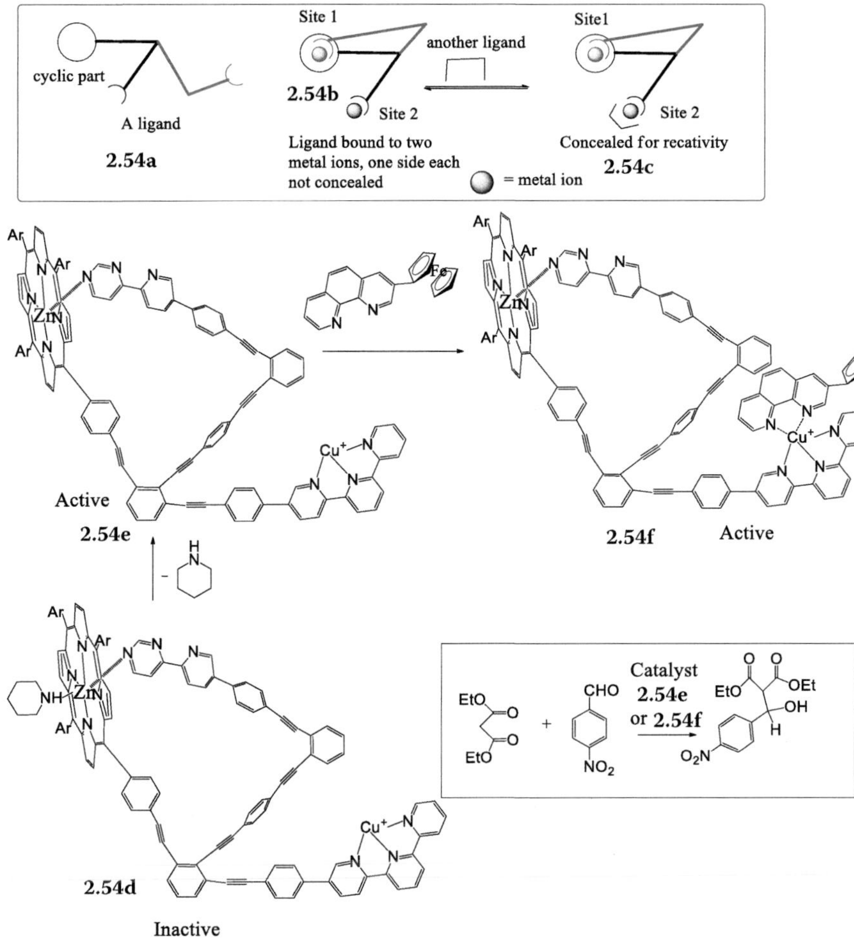

FIGURE 2.54 Self-assembling by added ligand facilitating catalysis. (From S. De et al. *Angew. Chem. Int. Ed.* 2014, 53, 14255–14259.[50])

the supraphos complex increases in the presence of piperazine. The piperazine ligand locks the structure by connecting two porphyrin units by providing an allosteric effect to form a complex **2.55b**. This complex provides the requisite spaces for the reactants on another site that is the rhodium site of the complex. The complex **2.55b** has binol units to provide control on the incoming reactant. This enhances catalytic efficiency for the asymmetric hydrogenation reaction of N-(1,2-dihydronaphthalen-3-yl)acetamide to form N-(1,2,3,4-tetrahydronaphthalen-3-yl)acetamide.

As the number of metal complexes of cyclic structure increases, the reactivity may increase. The macrocyclic *bis*-porphyrin zinc complex involving an intermediate **2.56a** is a good catalytic intermediate for phosphate ester hydrolysis. In a similar tetra-porphyrin complex, more active sites of a similar kind can be introduced. In the intermediate **2.56b**, the tetra-porphyrin complex simultaneously activates more ester derivatives by providing appropriate spaces to perform catalytic reactions. Due to the

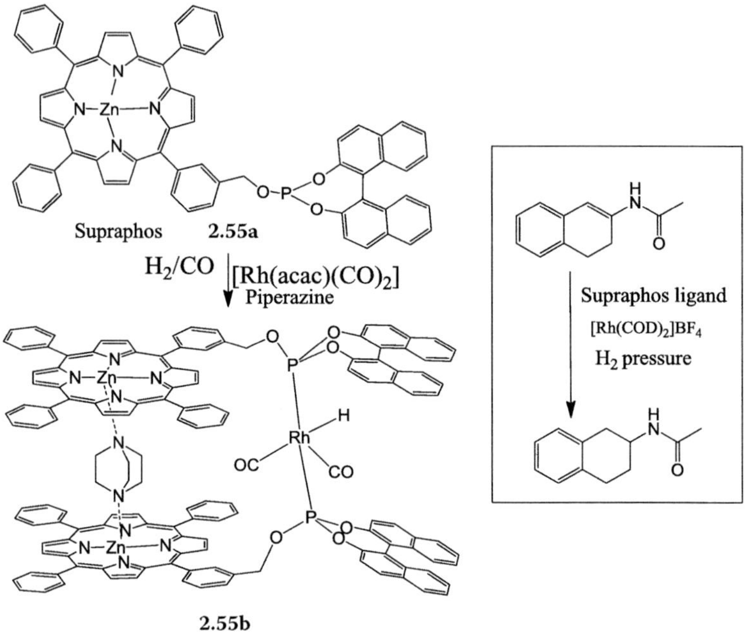

FIGURE 2.55 Allosteric effect and supramolecular ligand for hydrogenation reaction. (From V. F. Slagt et al. *Angew. Chem. Int. Ed.* 2003, 42, 5619–5623.[51])

formation of the intermediate **2.56b**, the tetranuclear complex has higher catalytic activity to hydrolise phosphate esters (Figure 2.56).

In general, the mononuclear complexes of ruthenium with additional free ligating sites conducible for hydrogen bonds are useful in water splitting. However, there are also alternative approaches to perform the activation process. One such method is based on catalytic cycles using a multinuclear complex of ruthenium (Figure 2.57). Persulphate is used as an oxidant for the oxidizing agent in such reactions. A dye-sensitised photochemical process is used in the reaction to maintain the cyclic catalytic redox reaction of Ru(2,2′-bipyridine)$_3$]$^{2+}$ (**2.57a**). The redox cycle of the Ru(2,2′-bipyridine)$_3$]$^{2+}$ in conjunction with the reactivity of a ruthenium metallacycle **2.57b** is highly catalytic in the water-splitting reaction. The advantages of using a metallacycle in this reaction are twofold. In mononuclear complexes, the axial ligand dissociates; that does not take place in a metallaocycle **2.57b**. Secondly, the rigid structure of the metallacycle prevents the dimerization of catalytic species, which is a cause of lowered activity in mononuclear complexes.

2.2.3 Metal Organic Frameworks and Coordination Polymers

Metal organic frameworks are self-assemblies formed by complex or free cationic systems as nodes with polytopic organic ligands acting as linkers to form three-dimensional network structures. From a structural point of view, the structures of metal organic frameworks are analogous to zeolite frameworks (Figure 2.58). The

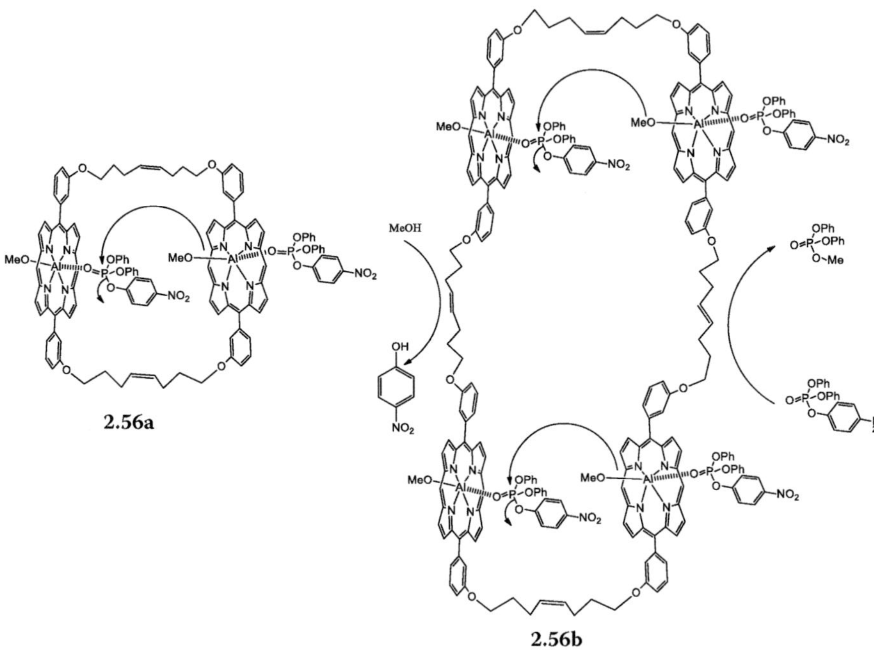

FIGURE 2.56 Multiple sites for substrates and reactants through multinuclear complex. (From R. K. Totten et al. *Chem. Commun.* 2012, 48, 4178–4180.[52])

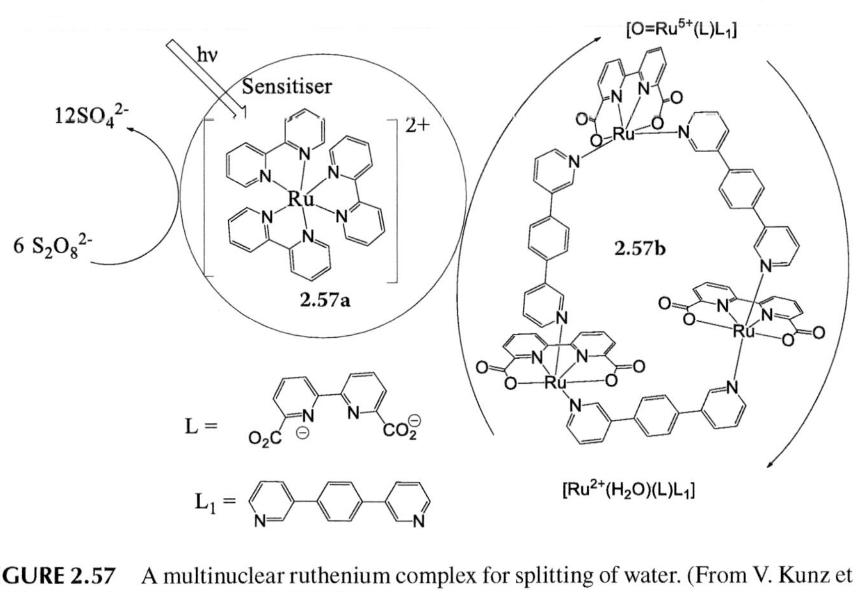

FIGURE 2.57 A multinuclear ruthenium complex for splitting of water. (From V. Kunz et al. *ACS Energy Lett.* 2017, 2, 288–293.[75])

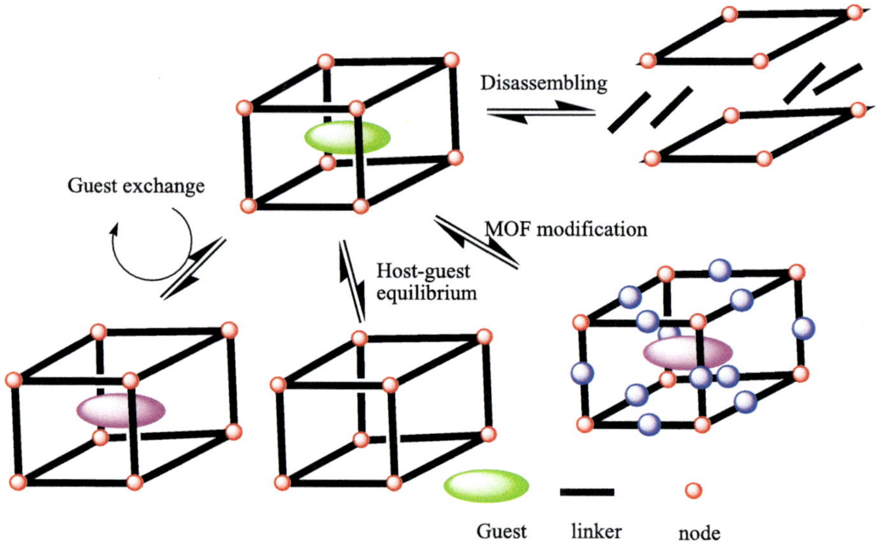

FIGURE 2.58 Different methods of modification of metal organic frameworks.

secondary building unit (SBU) approach is applied in the construction of metal organic frameworks. Due to the combinations of SBUs by linkers, different porous structures at nanodimensions are achieved. One of the advantages of utilizing metal organic frameworks over zeolites originates from the fact that large varieties in structures are brought about by changing the nodes and spacers. According to dimensionality, MOF-based materials are broadly classified into two classes. One class is composed of inorganic moieties organised into either 1D chains or 2D layers with intervening organic molecules as pillars, and another class contains open-framework coordination polymers possessing zero-dimensional "inorganic" clusters or isolated metal ions that are connected by bridging organic polytopic ligands. To a certain extent, the classification also establishes distinctions in reactivity and serves as a referral point to reactivity. For example, photocatalytic and Lewis-type catalytic reactions are often caused by zero-dimensional inorganic clusters, whereas MOFs with one-dimensional structure units are useful for Brönsted-type catalysis.

The voids and channels in metal organic frameworks are uniform, continuous and permeable. These serve as active sites by providing surfaces and assisting in transport of reactants/products. Alternatively, these empty spaces in porous metal organic frameworks are analogous to reaction vessels with the reactive sites to perform catalytic reactions. The structures of metal organic frameworks, together with the reactivity of the metal ions in them, control supramolecular activation of substrates. Redox active sites, coordinative unsaturated metal sites or function groups of the ligands guide and control the chemical reactivates of different reactants. Such artificially created microenvironments are influenced by charge, polarity, chirality, redox potential and photo-activity. Due to inherent affinity for polar, nonpolar or ionic molecules and specific choices to bind with specific stereoisomers, the reactions are very specific. The structural tenability of MOFs as supramolecular catalysts is appropriate for nanoscale confinements to perform multicomponent reactions. The poor solubility of metal organic frameworks means catalytic reactions pass through heterogeneous conditions with several advantages over conventional catalysts. The foremost advantage of metal organic framework catalysts is that they provide a means for continuous processing. Due to multiple active sites in the frameworks, they generally have enhanced catalyst reactivity and stability. Second, the permeable channels and organised voids allow recognition and selective passes of substrates to reach catalytically active sites. There are many examples of structurally nonrigid metal organic frameworks that respond to stimuli. It is possible to tune reactivity of MOFs to obtain selective products. It is also pertinent that metal organic frameworks are structurally modified by microenvironments created by the interplay of weak interactions. MOFs provide suitable transition states with a low activation barrier. Finally, the ease of separating and reusing them as heterogeneous catalysts and the possibility of designing catalysts from biodegradable molecules as enzyme mimics give MOF catalysts a definite edge over other catalysts.

Some common MOFs are listed in Figure 2.59. The MOF $[Cu_3(btc)_2]_n$ is derived from copper(II) ions with 1,3,5-benzene tricarboxylate (btc) and is commonly known as HKUST-1 (HKUST for Hong Kong University of Science and Technology); it has a three-dimensional structure. The pores have three different internal diameters: 9.0, 5.0 and 3.5 Å, respectively. The copper (II) ions of the cubic network in the MOF have Cu2 clusters formed as a paddle-wheel type of structure that is similar to the structure of copper(II) acetate dimers. The axial positions of the paddle wheels are occupied by water molecules. These water molecules are weakly bound and project inwards to the centres of the voids.

MIL-100 (Fe), named based on the Materials of Institut Lavoisier where it was discovered, has the composition $[Fe_3F(H_2O)_2O(btc)_2]_n$. The three-dimensional network of this MOF has two types of mesoporous cages. The free apertures are approximately 25 and 29 Å in diameter; these pores are accessible through voids of 5.5 and 8.6 Å. MOF MIL-101(Cr) chromium 1,4-benzenedicarboxylate has the formula $[Cr_3X(H_2O)_2O(bdc)_3]_n$ (where X = F, OH; bdc = 1,4-benzenedicarboxylate), and has trimeric chromium(III) octahedral secondary building blocks. Each SBU has removable water molecules at terminal positions. It has two types of quasispherical

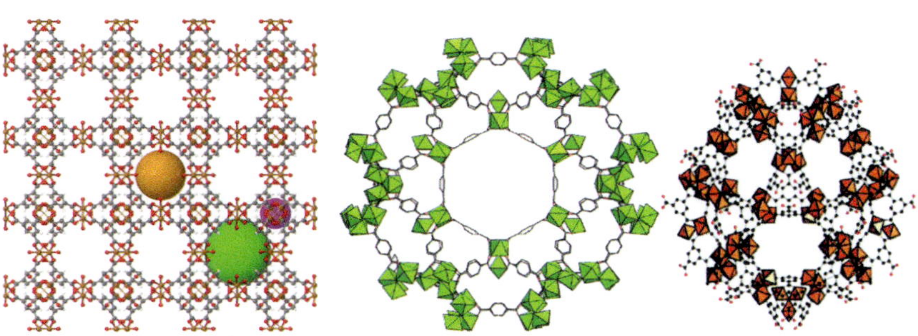

HKUST-1 MIL100Fe MIL101Cr

FIGURE 2.59 Pores in metal organic frameworks.

cages. These cages have diameters of 29 and 34 Å. These voids are accessible through microporous windows with dimensions of 12 and 16 Å. As the terminal water molecules can be reversibly removed, the removal of the water molecules makes large pore areas to serve as spaces for binding reactants within the cavities.

MIL-47(V) is a vanadium (IV)-based metal organic framework with $[\{V(IV){=}O(bdc)\}_n$. It has a framework with benzene dicarboxylate (bdc) linkers. The structure of the MOF is composed of infinite $V(IV)O_6$ octahedrons in chainlike arrangements. The pores in the MOF have 10.5 Å × 11.0 Å dimensions and are distributed along the *a*-crystallographic axis. The MIL-47(V) catalyst with tetrabutyl hydroperoxide in water is a very good oxidation catalyst of hydrocarbons. Decane is catalytically oxidised in aqueous medium or in mixed solvents containing water to cyclohexene oxide, cyclohexane-1,2-diol, tert-butyl-2-cyclohexenyl-1-peroxide and 2-cyclohexen-1-one as isolable products. In this reaction, a portion of vanadium ions leaches out from the MOF; thus, the reaction is believed to occur in a combination of heterogeneous and homogeneous processes.

The aluminium MOF with 1,4-benzene dicarboxylic acid MIL-53(Al) is composed of $Al_2(bdc)_3$. The structure of this MOF is composed of corner-sharing aluminium (III) ions, where each aluminium ion is in an octahedral environment. Each aluminium ion is linked to the others by two carboxylate groups of bdc ligands. The pore dimension of the MOF is 8.5 Å, and the surface area is 1140 $m^2\ g^{-1}$. This MOF acts as a heterogeneous catalyst for hydrogenation of olefins in the presence of hydrizine (Figure 2.60). Hydrazine acts as a source for hydrogen for hydrogenation reactions. The catalyst retains its crystallinity after four cycles of catalytic hydrogenation reaction.

Upon dehydration, many MOFs transform to coordinatively unsaturated MOFs. The vacant sites created around the metal ions by the removal of the water ligands are utilised to introduce new ligands. When ligands with functional groups are introduced, newly functionalised MOFs are obtained. For example, MIL-101,

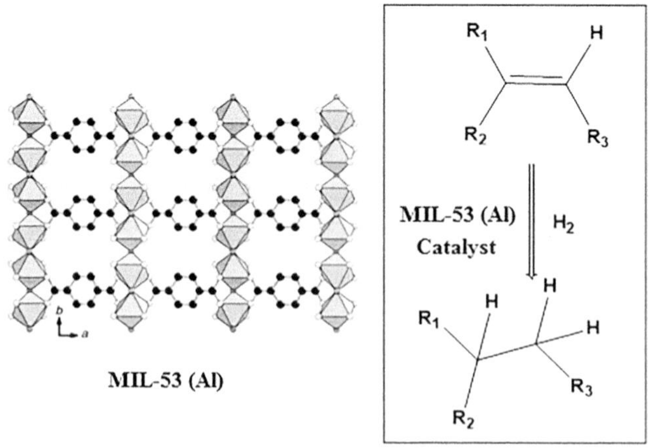

FIGURE 2.60 MIL-53(Al)-catalysed hydrogenation reaction. (From A. Dhakshinamoorthy et al. *Adv. Synth. Catal.* 2009, 351, 2271–2276.[53])

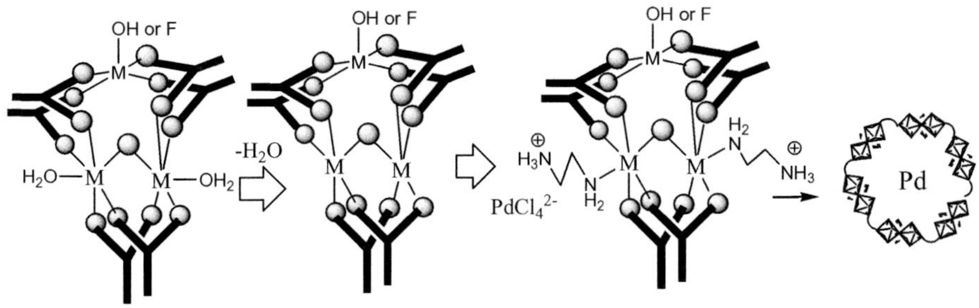

FIGURE 2.61 Selective encapsulation of noble metal through functionalisation of coordinatively unsaturated metal sites of MIL-101. (From Y. K. Hwang et al. *Angew. Chem. Int. Ed.* 2008, 47, 4144–4148.[54])

upon dehydration, generates five coordinated metal sites which are used to anchor ethylenediamine to transform it to an ethylenediamine-derived palladium complex, as illustrated in Figure 2.61. This functionalised MOF, on reduction by reducing agents such as hydrazine hydrate, converts palladium (II) to palladium (0) and encapsulates the noble metal ion. The encapsulated nanodimensional metal particles in MOFs are highly active catalysts for the C—C and C—heteroatom bond formation reactions that are conventionally carried out with palladium at a zero oxidation state. Metal organic frameworks from many ligands with different metal ions are used to carry out catalytic reactions. Some of the nodes and connections used for specific reactions are listed in Table 2.2.

The zinc-based metal organic framework IRMOF-3 (isoreticular MOF-3) has composition $[Zn_4O(bdc\text{-}NH_2)_3]_n$. It is an amino-benzenedicarboxylate MOF; it has linkers with free amine groups. This amine containing MOF has voids with diameter of 9.6 Å. The amine moieties react with salicylaldehyde molecules to form the Salene type of units. The Salene units serve as ligands to form different complexes, as illustrated in Figure 2.62. The new MOF (**2.62b**) formed by the Salene-derived MOF with $V(O)(acac)_2.H_2O$ (acac = acetylacetonate) has vandium ions as active sites. These active sites make the MOF suitable for epoxidation of cyclohexene by tertiarybutyl hydroperoxide. Similarly, a manganese complex containing MOF (**2.62c**) is a catalyst for a three-component reaction between 2-ethynyl-N-tosylbenzenamine, aldehyde and secondary amine to yield heterocycle, as illustrated in Figure 2.62.

The derivatisation of aromatic units in an MOF results in a new MOF bearing specific characteristic properties. The MIL-53 (Al) metal organic framework undergoes sulphonation to form SMIL-53(Al) (Figure 2.63). The MOF has significant difference in acidity and solubility from the parent MIL-53(Al). The MOF is a catalyst for the esterification reaction; for example, it catalytically esterifies acetic acid with n-pronanol to form propylacetate.

Tapelike hydrogen-bonded subassemblies are integral parts of self-assemblies formed by urea derivatives. Such subassemblies are barriers for catalytic activation of substrates. This is due to substrate binding needing to overcome the competition of self-assembly among urea molecules to utilise N—H sites, but by linking the urea units in MOFs, the N—H bonds are exposed outwards and placed in appropriate positions

TABLE 2.2
MOFs and Their Catalytic Reactions

MOF Components	Specific Name	Catalytic Reaction
Iron(III) ion + fluoride + water + 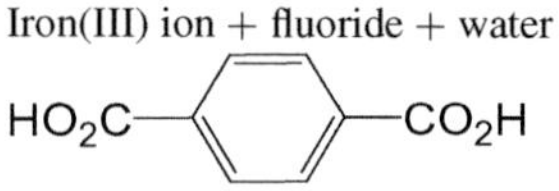	MIL 100 Fe	Friedel-Craft benzylation Oxidation of hydrocarbon Claisen Schmidt condensation Thiol oxidation to disulphide
Chromium(III) ion + fluoride + 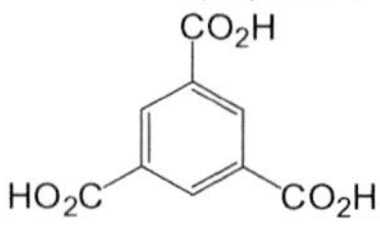	MIL 101Cr	Oxidation of tetralene Cyano-silylation reaction Oxidation of sulphide Cycloaddition of carbon dioxide
Copper(II) ion + 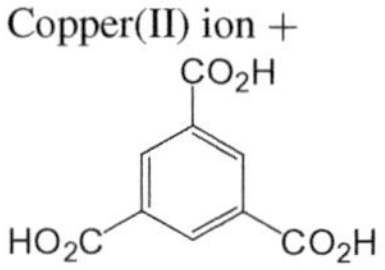 = btc	$Cu_3(btc)_2$	Cyclopropanation Cyanosilylation of aldehyde Friedlander reaction Pechmann reaction
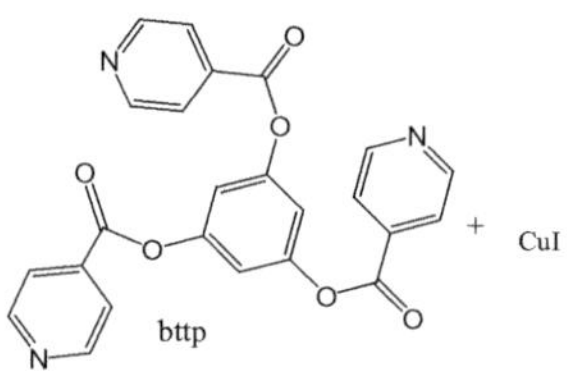	$Cu_2I_2(bttp)_4$	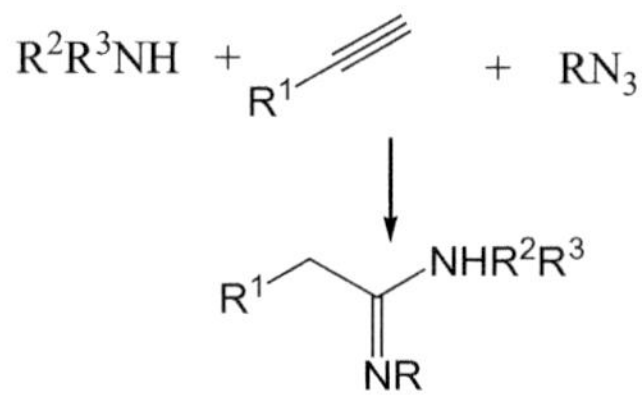
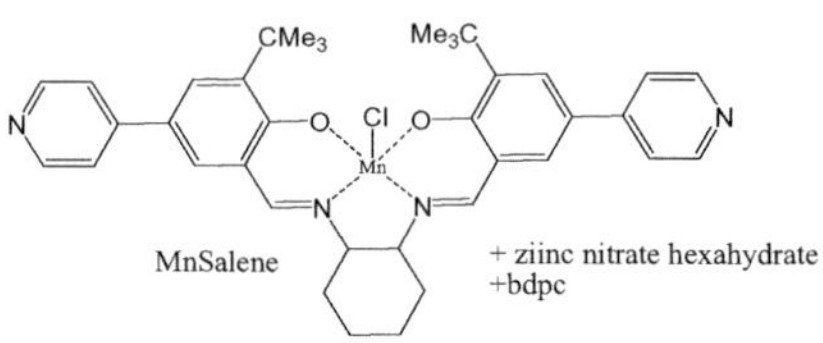	Zn_2(bpdc) (MnSalene)	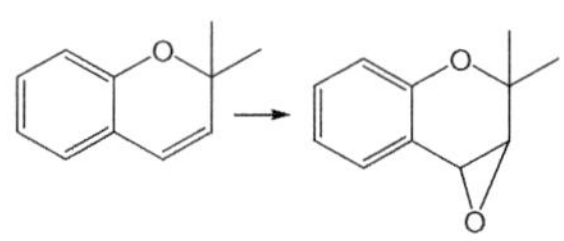
V(IV)=O + 	MIL-47(V)	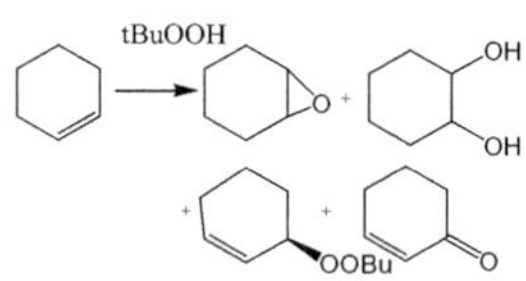
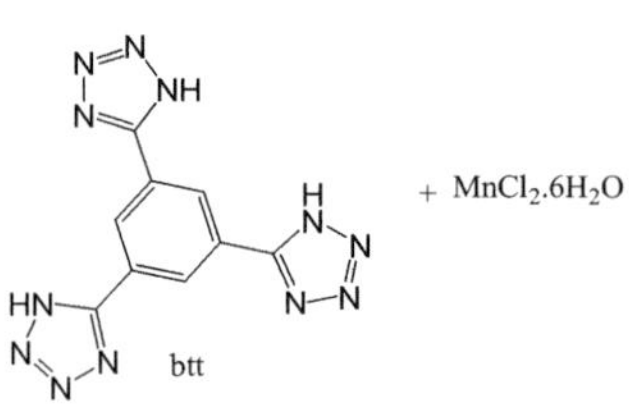	Mn-btt	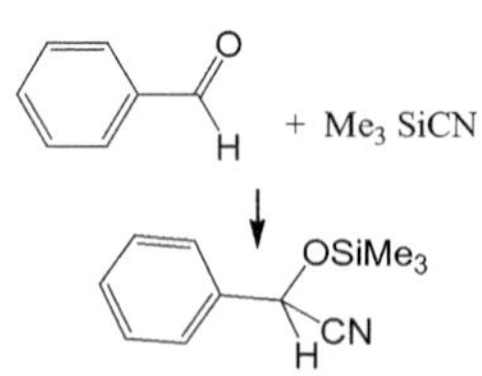

Source: From I. Nath et al. *Chem. Soc. Rev.* 2016, 45, 4127–4170.[55]

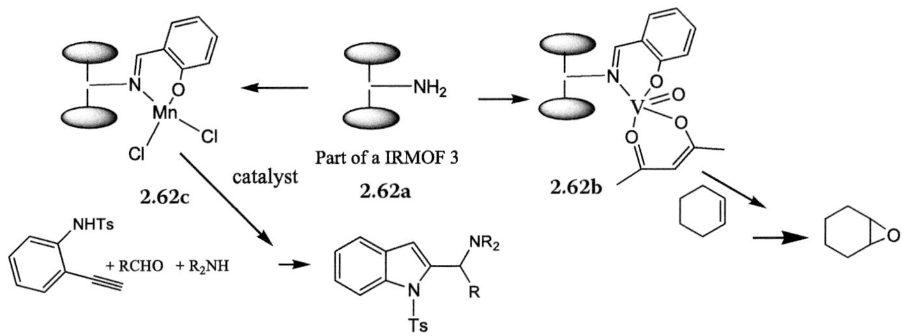

FIGURE 2.62 Functionalised MOF anchoring different metal complexes as catalyst. (From J. L. C. Roswell, O. M. Yaghi, *J. Am. Chem. Soc.* 2006, 128, 1304–1315;[56] M. J. Ingleson et al. *Chem. Commun.* 2008, 2680–2682.[57])

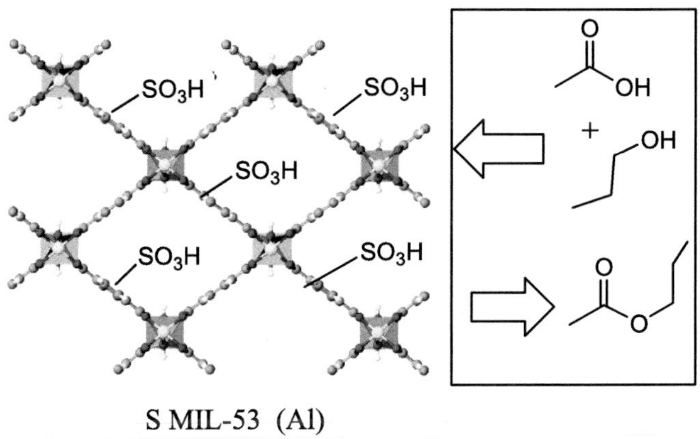

FIGURE 2.63 Acid-functionalised MOF as catalyst. (From A. Dhakshinamoorthy et al. *Catal. Sci. Technol.*, 2013, 3, 2509–2540.[77])

so that each urea unit acts like a discrete unit to bind to the substrate. Urea units connected to two 3,5-benzenedicarboxylic acid units are used as building ligands for MOF with 4,4′-bipyridine ligands. Such MOFs have urea units inside the voids for substrate binding and activation. An example of an MOF where the urea units remain segregated is depicted in Figure 2.64 to bind substrates. Such metal organic frameworks provide pre-organised systems for substrate binding. The urea containing tetracarboxylic acid **2.64a** forms an MOF with a zinc ion with 4,4′,-bipyridine linkers. It acts as a catalyst for C—C bond formation between pyrrole derivatives with nitro-olefins. The MOF in this case not only serves as a container but also activates the nitro-olefin by forming the complementary hydrogen bonds between urea and the nitro group (Figure 2.64).

Certain functional groups of MOFs easily transform to other functional groups. Thus, the use of such functional groups as precursors for construction of MOFs requires protection and deprotection techniques. The concept is executed in dopamine containing MOF MIL 101(Cr). Dopamine has a catechol unit, and this particular

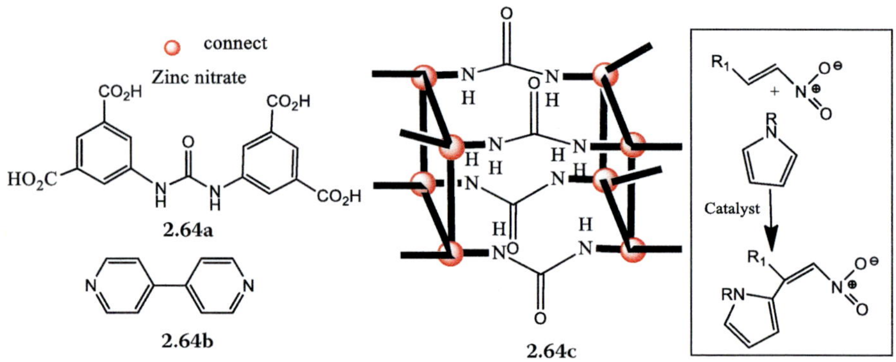

FIGURE 2.64 Different components of a MOF where urea tethers as activating sites are present and one catalytic reaction by such MOF. (From J. Y. Lee et al. *Chem. Soc. Rev.* 2009, 38, 1450–1459.[58])

dihydroxyaromatic unit is protected tert-butyldimethylsilyl (TBDMS) groups to yield corresponding silylether. The chromium (III) sites at the MOF external surface are grafted to MIL101(Cr), yielding the product **2.65b**. When 2-(methylthio)ethylamine is used as a secondary modifier to this grafted MOF, the 2-(methylthio)ethylamine molecules get inside the voids to form the product **2.65c**. This modifies the internal surface of the MOF. The grafted MOF reacts with $VO(acac)_2$ to form vanadium complex-anchored MOF **2.65c**. The vanadium-containing MOF catalytically converts thioanisole in the presence of tetrabutylhydroperoxide to the corresponding sulfoxide (57%) and sulfone (26%) (Figure 2.65). This vanadium-containing MOF serves as a heterogeneous catalyst. After the first cycle of reaction, it is filtered and used for several catalytic cycles.

2.3 ARTIFICIAL ENZYMES

Enzymes are supramolecular assemblies with one or more polymeric peptide sequence components. Active sites of enzymes possess diverse functional groups and show wide ranges of chemical biotransformations under physiological conditions. Enzymes have nanometer-sized three-dimensional structures created by the folding and self-assembly of polymeric chainlike components through supramolecular interactions. The molecular understanding of their catalytic activities becomes complicated due to a high amount of conformational flexibilities, diversities and complexities in their structures. Based on the understanding of natural enzymes, artificial enzymes are designed. One of the primary points in such design involves the choice of supramolecular scaffolds such as macrocycles, self-assembled nanometer-sized templates or covalently linked units that are pre-organised or reorganised with multiple functional units that are suitable to bind substrates for activation. Many enzymes exist as ensembles of multiple conformational states, and such a process replicates in the mimics. Those ensembles change conformations upon substrate binding. The flexible intermediate states of enzymes increase stability to the transition state and hence a lower-energy path occurs than in nonenzymatic reactions. For

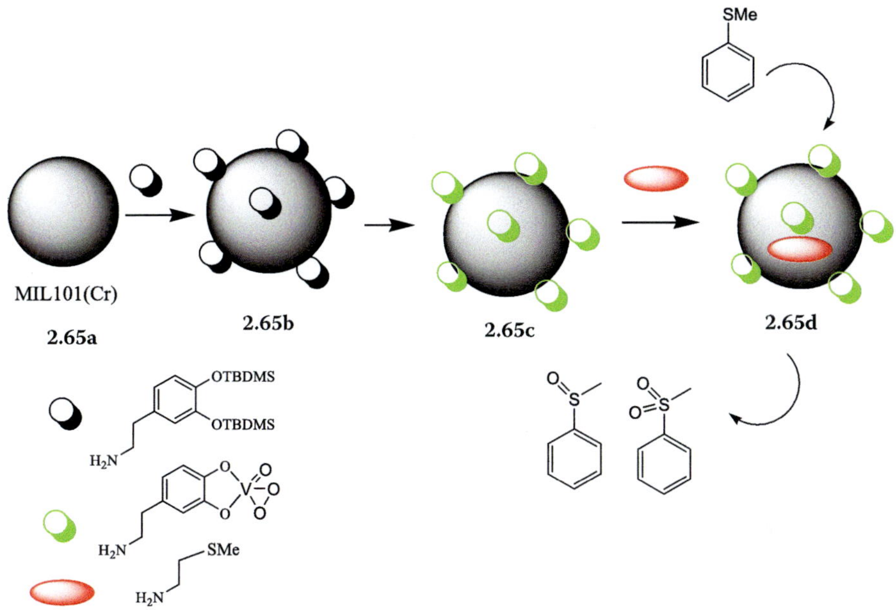

FIGURE 2.65 Protection and deprotection to introduce catecholate complex into MIL 101 (Cr). (From J. Liu et al. *Chem. Soc. Rev.* 2014, 43, 6011–6061.[59])

successful mimic development, it is necessary to decrease product inhibition. This is done by using templates to which the product has lower affinity than the substrate.

Functionalised container molecules such as cyclodextrins, calixarenes, cyclophanes, cavitands, capsules and molecular cages are extensively used in catalysis. Cyclodextrin molecules are easily Functionalised common hosts to achieve enzyme mimics. The ability of cyclodextrin to accommodate small guest molecules in its hydrophobic cavities forms inclusion complexes. Some of these inclusion complexes are equivalent to the intermediate of selective catalytic reactions. The reaction of hypohalous acid with anisole in the presence of a catalytic amount of β-cyclodextrin is very specific. In this reaction, para-halogenated anisole is formed (Figure 2.66).

Enzymelike activities are observed in the catalytic acetylation reactions of cyclodextrin. The catalytic acetylation reactions of cycodextrins are dependent on the substrates. For example, the compounds **2.67a** and **2.67b** acylate cyclodextrins, but the reactivity of the two esters differs. This is due to the difference in the transition states in these two cases. In such reactions, one of the peripheral hydroxyl groups of cyclodextrin forms a tetrahedral transition state upon interaction with an ester-containing compound. An aromatic ester is generally activated by inclusion of the aromatic part in the cavity of cyclodextrin. Thus, the included aromatic ring should be at a position such that it places the ester group in the closest reach of the hydroxyl group of the cyclodextrin. This effect causes a large difference in the relative rate of functionalisation of a cyclodextrin by esters **2.67a** and **2.67b** (Figure 2.67). In the case of **2.67a**, the aromatic ring has to be partly pulled up from the cavity to form the tetrahedral intermediate species. In the case of the compound **2.67b**, the

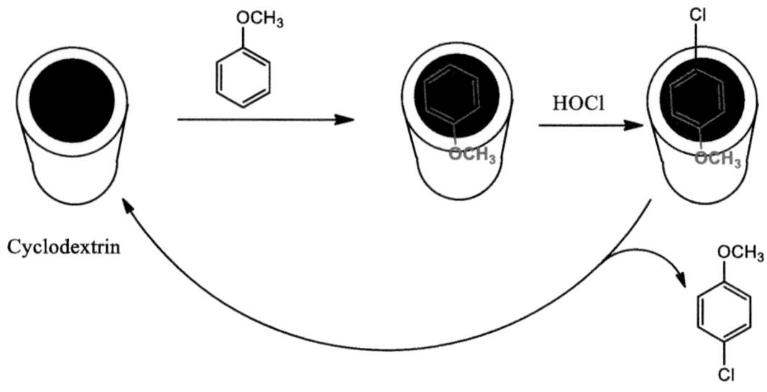

FIGURE 2.66 Cyclodextrin-catalysed selective catalytic halogenation. (From R. Breslow, *Acc. Chem. Res.* 1995, 28, 146–153.[60])

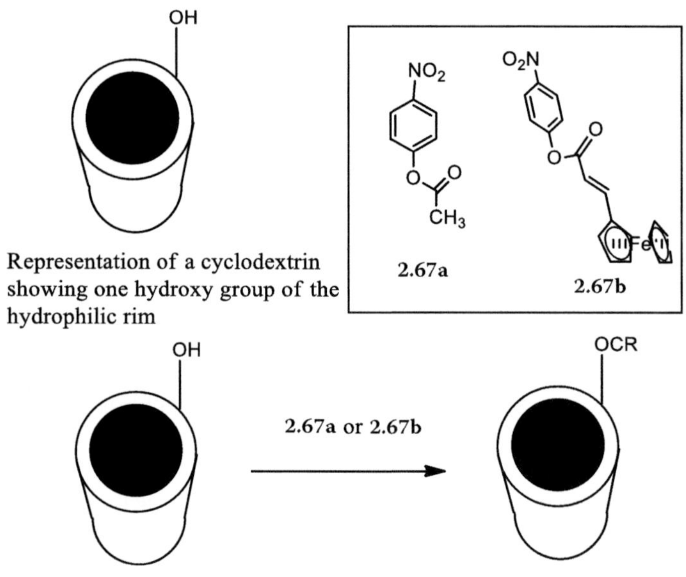

FIGURE 2.67 Difference in rate of acylation of cyclodextrin. (From R. Breslow, *Acc. Chem. Res.* 1995, 28, 146–153.[60])

ferrocene unit is included inside the cyclodextrin cavity and remains as support to facilitate formation of tetrahedral intermediate for acylation. This effect causes a large difference between the two esters in acylation of cyclodextrin. Though the nitro-aromatics are encapsulated, the ring has the ability to stabilise the tetrahedral transient species; it is not as conducive as in the case of the ferrocene unit—containing compound **2.67b** (Figure 2.67). These acetylation reactions mimic the first step involved the hydrolysis of esters and peptides by serine proteases. The transition states in such a case during acylation reactions also involve tetrahedral intermediates.

Phosphates of pyridoxal and pyridoxamine derivatives are coenzymes for transamination reactions. Based on this, the cyclodextrin derivative (**2.68**) with

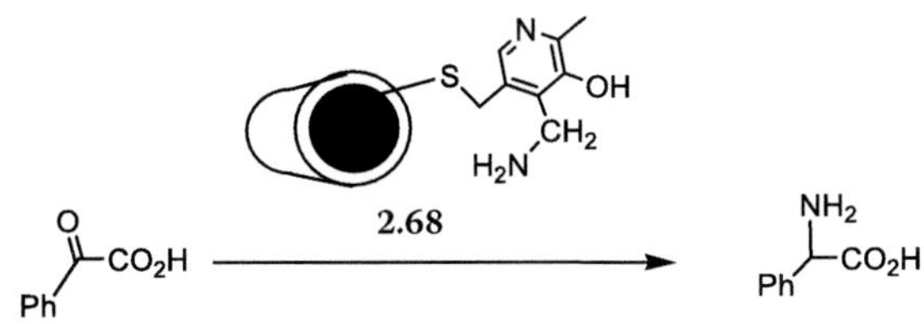

FIGURE 2.68 Transamination reaction catalysed by cyclodextrin derivative. (From R. Breslow, *Acc. Chem. Res.* 1995, 28, 146–153.[60])

pyridoxamine is designed for catalytic transamination reactions (Figure 2.68). The catalytic transamination caused by the cyclodextrin derivative **2.68** shows a preference for phenyl or indole pyruvic acid. The difference in the relative rate is due to the encapsulation of the phenyl group of phenyl pyruvic acid to the cavity of the cyclodextrin molecule. The cyclodextrin **2.68** undergoes a condensation reaction with the keto-carboxylic acid, such as phenyl or indole pyruvic acid, to form the corresponding imine. The imine yields the transamination product through a shift of the imine group followed by hydrolysis. The reaction is very specific and is a mimic of a transaminase enzyme, which also contains a pyridoxamine unit. The cyclodextrin derivative—catalysed transamination reaction of phenyl pyruvic acid is 15,000 times faster than the reaction without the cyclodextrin derivative.

The three isomeric β-cyclodextrin *bis*-imidazoles **2.69a–c** differ in structures. The imidazole units are located at different positions of the cyclodextrins. One isomer has two imidazoles attached to two neighbouring glucose units; in another isomer, these groups are anchored to two glucose units located at positions which have an intervening glucose, whereas in the third isomer, these groups are anchored to two glucose located at positions with two intervening glucose units. At a suitable pH where only one of the imidazole units is protonated and the other imidazole remains free, the isomers catalyse the hydrolysis of phosphate anhydride **2.69d**. Among the isomers, the first isomer **2.69a** is the most efficient catalyst (Figure 2.69). A greater amount of product **2.69c** is formed as a result of encapsulation of the phosphate anhydride. This happens as a result of encapsulation of the anhydride in the cyclodextrin ring during the reaction. The encapsulation controlling the approach of the nucleophile is directed by the oxygen atom located at the *ortho*-position.

Bis-imidazole containing cyclodextrins facilitates cyclisation reactions of aromatic appended compound 6-(4-(tert-butyl)phenyl)-6-oxohexanal (**2.70d**) to form (4-(tert-butyl)phenyl)((2R)-2-hydroxycyclopentyl)methanone (**2.70e**). The compound **2.70e**, on subsequent dehydration, forms (4-(tert-butyl)phenyl)(cyclopent-1-en-1-yl)methanone (**2.70f**). The *bis*-imidazole isomer of cycodextrin that has imidazole units anchored at the first and third sugar moieties is suitable to meet stereoelectronic requirements of this reaction (Figure 2.70). This cyclisation reaction facilitated by cyclodextrin **2.70a** involves monoprotonated imidazole; the protonated imidazole forms a hydrogen bond with the carbonyl group next to the aromatic unit trapped in the cyclodextrin ring, as illustrated as **2.70b**. The adduct **2.70b** stabilises the enol form **2.70c**, which undergoes a cyclisation reaction to yield (4-(tert-butyl)phenyl)((2R)-2-hydroxycyclopentyl) methanone. This transformation follows the dehydration reaction to for **2.70f**.

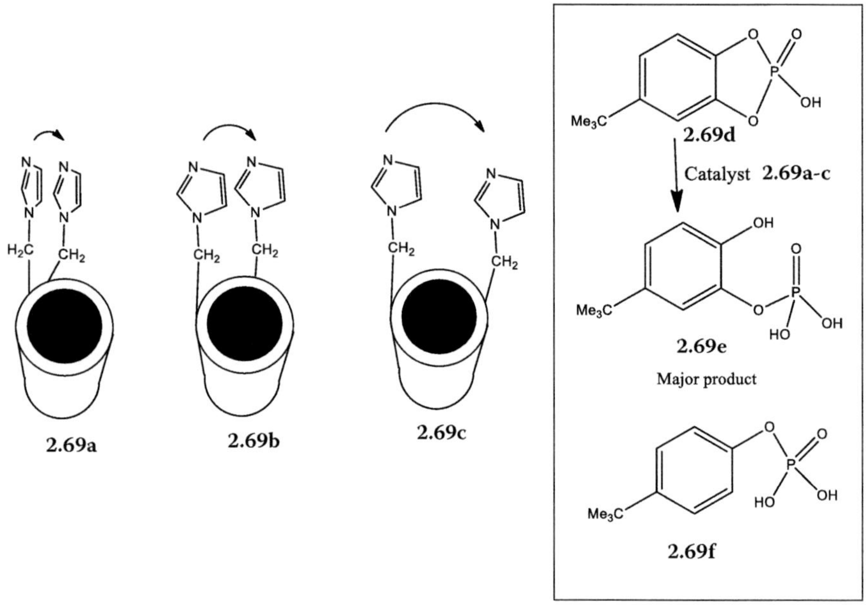

FIGURE 2.69 Rate differences due to distance between the imidazole tethers of cyclodextrin. (From R. Breslow, *Acc. Chem. Res.* 1995, 28, 146–153.[60])

Adequately anchored pairs of cyclodextrin molecules in lateral orientations by a bridging unit brings them into the vicinity to provide space to a reactant by partially encapsulating two ends. Such bridging units having ligating sites to form an inclusion complex as well as to bind to metal ions. Thus, such complexes provide sites for inclusion through two cyclodextrin rings and also provide an additional catalytic site on the bridge. The β-cyclodextrin derivative (**2.71**) is such a complex. It is a copper complex linked by 2,2′-bipyridine units that bridge to two cyclodextrin units at two sides (Figure 2.71). This copper complex is a good catalyst for the hydrolysis of 4-nitrophenyl ester derivative (**2.71b**). The two cyclodeextrin units encapsulate two ends of the derivative **2.71b** in their cavities. Such encapsulation keeps the ester unit disposed at an appropriate geometry for activation by the copper ion (**2.71c**). The cyclodextrin-copper complex facilitates very fast hydrolysis comparable to the enzyme.

In biological systems, the rate of a reaction changes through conformational changes or denaturation. The relative rates of cyclodextrin-catalysed hydrolysis reactions are controlled by adding an external substrate that preferentially goes into the interior of the cyclodextrin rim and stops the reactivity. This is done by adding *trans* stilbene derivatives to inhibit the cyclodextrin-catalysed hydrolysis of esters. The catalytic hydroloysis reaction of **2.72c** by cyclodextrin is inhibited by a compound **2.72b**, whereas the same reaction proceeds on irradiation. The *trans*-stilbene derivative **2.72b** is encapsulated into the cavity and inhibits the catalytic hydrolysis reaction of 4-nitrophenoxyacetate (Figure 2.72), whereas the guest-included complex, upon UV irradiation, releases the guest as it transforms to a *cis*-form **2.72a** to make empty space for substrate activation. Hence, the catalytic hydrolysis occurs efficiently,

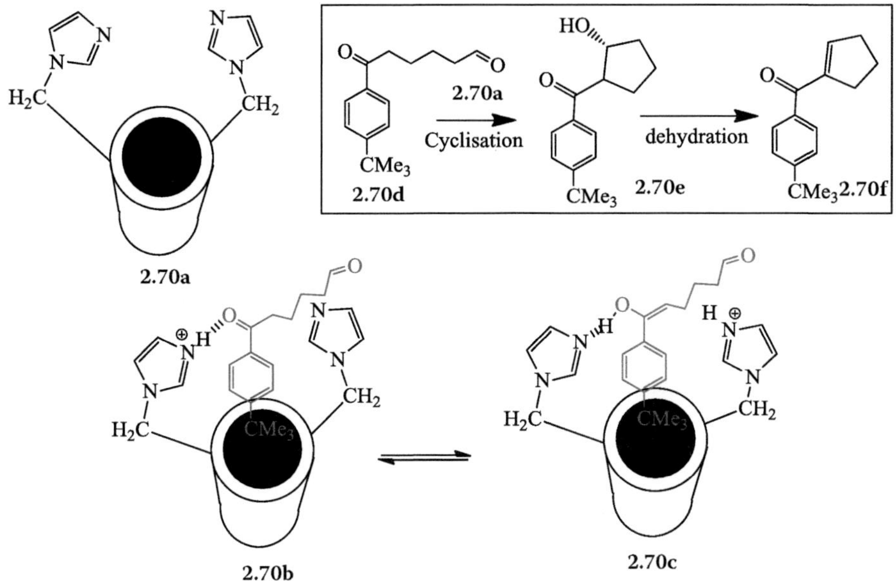

FIGURE 2.70 Proton transfer in catalytic cyclisation to for carbocycle by cyclodextrin. (From R. Breslow, *Acc. Chem. Res.* 1995, 28, 146–153.[60])

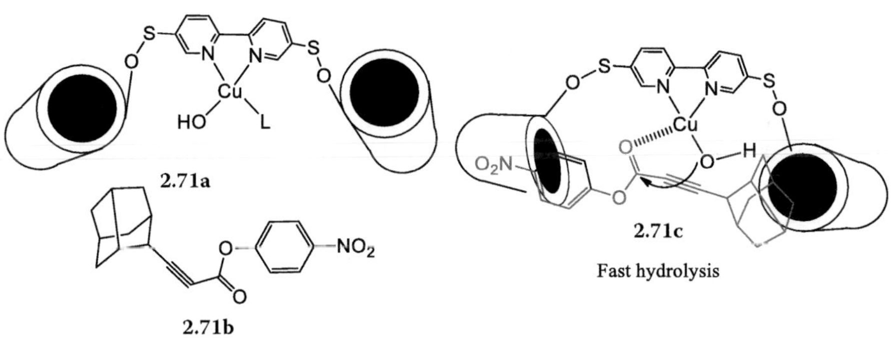

FIGURE 2.71 Cyclodextrin metal complex assisting in hydrolysis. (From O. Bistri, O. Reinaud, *Org. Biomol. Chem.* 2015, 13, 2849–2865.[61])

but the same reaction is inhibited upon heating. Heating converts the *cis*-form of the guest to the *trans*-form. The latter isomer again goes inside the cavity of cyclodextrin to inhibit the catalytic hydrolysis reaction. Based on this sequence of reactions, the relative rates of reactions are controlled by changing stimuli.

Certain enzymes have pyridoxamine or pyridoxal phosphate as a cofactor to catalyse proton transfer reactions. In certain biological reactions, base groups bind to metal ions, where the metal ion acts as a Lewis acid. One such mimicking reaction is the transamination reaction performed by pyridoxamine containing zinc complex (**2.73c**), which has an internally linked base group. Upon reaction of the complex

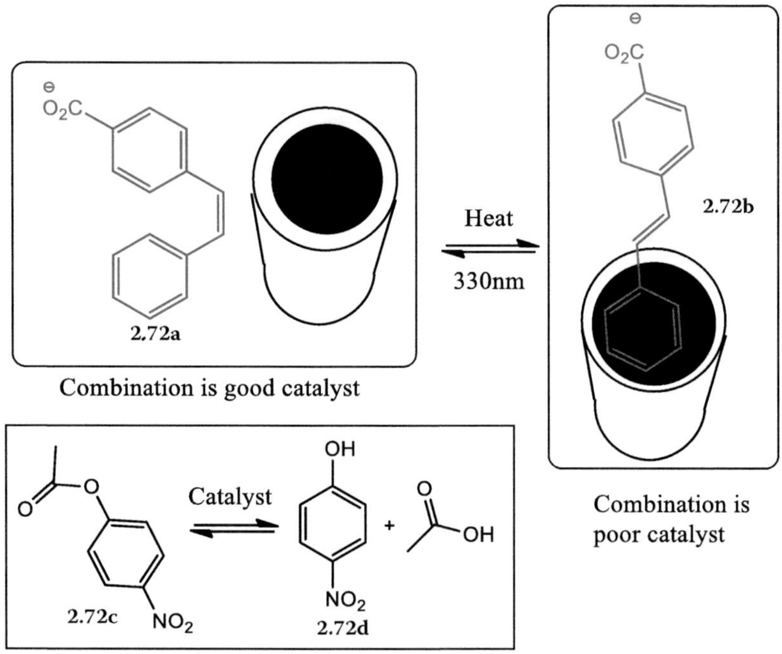

FIGURE 2.72 Control of cyclodextrin catalysis by stimuli guided guest encapsulation. (From R. Breslow, *Acc. Chem. Res.* 1995, 28, 146–153.[60])

with keto-carboxylic acid, a transamination reaction occurs. The complex is chiral; hence, a stereoselective amino acid is formed (Figure 2.73). A proton from the CH_2 group is extracted by the dimethylamino group and the proton is transferred stereospecifically to form an intermediate (**2.73d**). The intermediate hydrolyses to form chiral amino acid (**2.73f**). The reaction is catalytic, as further reaction of amino acid with the **2.73e** initial complex is regained and a catalytic cycle is maintained.

The active site of phosphotriesterase (PTE) is an enzyme that hydrolyses the phosphate ester and has one hydroxyl ligand bridging two zinc (II) metal ions. Of the two zinc ions in the enzyme, one binds to the oxygen of the phosphate ester and activates it for hydrolysis. The other transfers the —OH group for nucleophilic displacement of the —OR group of the substrate. There are aspartate (Asp) and histidine (His) amino acid residues in the near proximity to the active site. They play the role of the base to transfer the protons to effectively cause the catalytic cycle, as illustrated in Figure 2.74. Many enzyme mimics are based on different design principles are available as PTE mimic. For some of them the calixarene ring acts as the anchoring unit to hold the reactant, and the zinc ion activates the ester, as shown in Figure 2.75.

The dinuclear complex **2.75b** has a locked conformer of calix-arene. It has a pre-organised structure suitable to perform hydrolysis of ester in an efficient manner. As the loading of the metal ions increases, the complex hydrolyses with high degree of activity. The trinuclear complex has a high degree of efficiency for selective cyclic ester formation from the compound **2.75e**. The compound has free OH groups which result in a cyclic structure by replacing the 4-nitrophenolate group. To do so, the

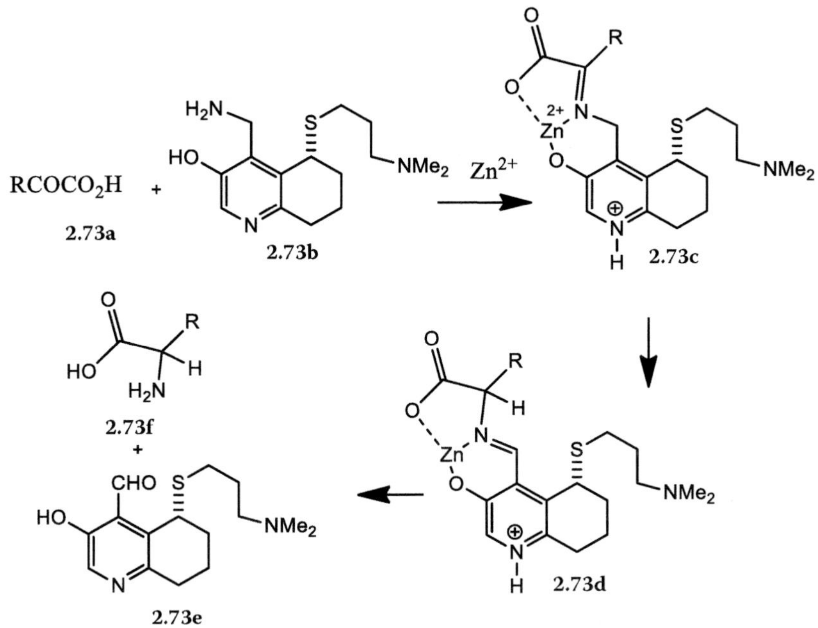

FIGURE 2.73 Transamination reaction by zinc complex. (From S. C. Zimmerman, R. Breslow, *J. Am. Chem. Soc.* 1984, 106, 1490–1491.[62])

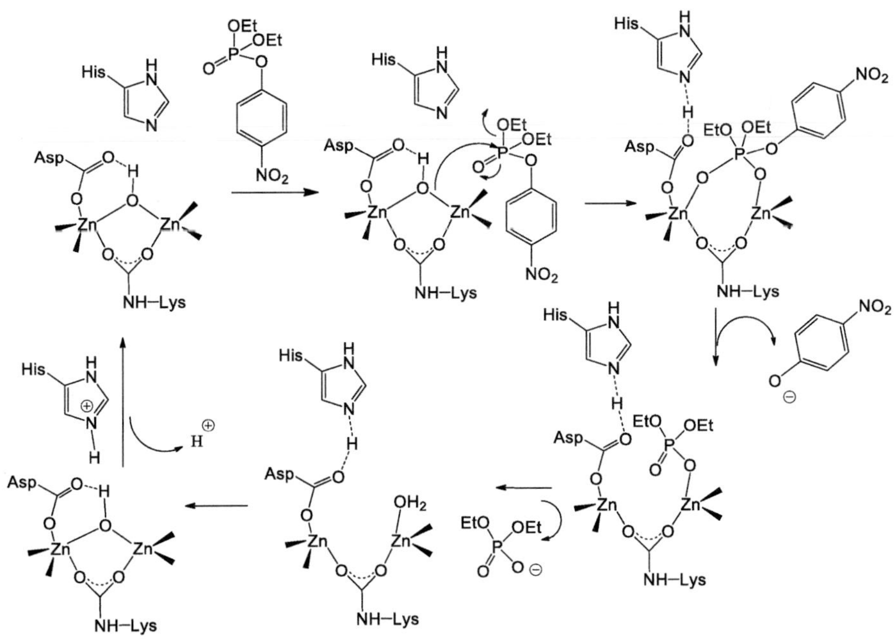

FIGURE 2.74 Active site and hydrolysis of ester by a zinc enzyme. (From S. D. Aubert et al. *Biochemistry* 2004, 43, 19, 5707–5715.[63])

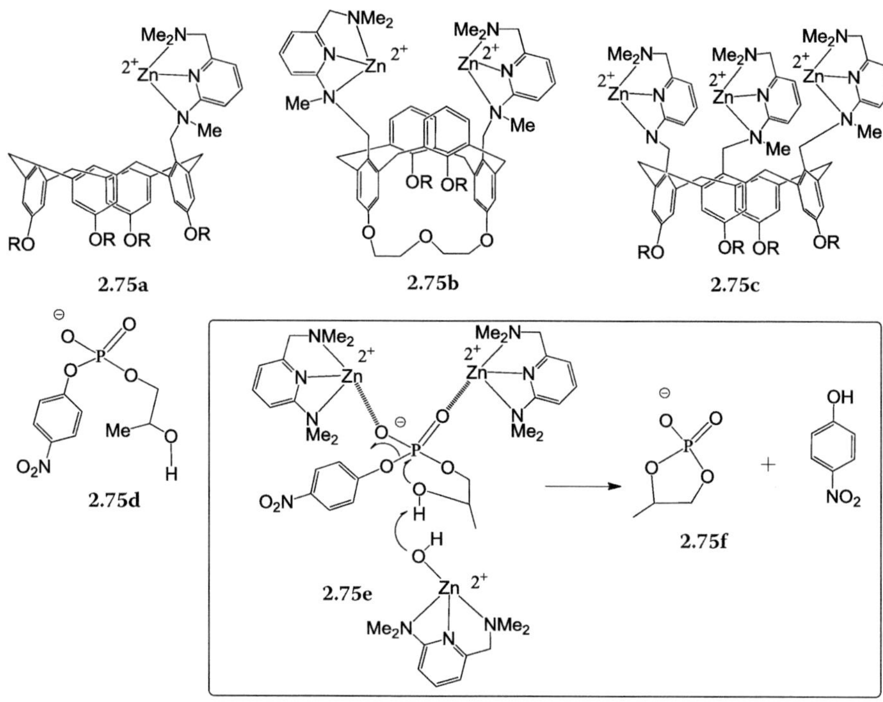

FIGURE 2.75 Some zinc complexes as model for phosphate ester hydrolysis and a selective cyclisation reaction of a phosphate ester derivative. (From P. Molenveld et al. *J Org Chem.* 1999, 64, 3896–3906.[64])

phosphate ester initially forms a bridged structure involving two zinc ions. The third zinc ion picks up a proton with the help of the hydroxyl ligand. This results in the formation of the cyclic ester (**2.75f**).

Various ligating scaffolds with chiral centres are designed by holding two or more ligating sites at distant ends. Some of these are used to induce chirality in products. In the example shown in **2.76a**, there are two quinoline units that bind to substrates or metal ions. This compound has a chiral heterocyclic unit to imbibe chirality. Design of a bifunctional catalyst through phosphine coordinating to the rhodium ion and the indoles holding carboxylic acid tethered to thiourea is shown in Figure 2.76. The *bis*-phosphine ligand **2.76b** coordinates with a rhodium ion to form acyclic enclosure. This cyclic enclosure has two indole rings that provide anchoring sites to substrates like carboxylic acid for molecular recognition, forming an adduct like that in **2.76c**. This adduct provides urea as the active sites for activation of the substrate as well as at the rhodium centre and it acts as a bifunctional catalyst.

2.3.1 Biomimetic Metal Organic Frameworks

The active sites of metalloenzymes are in supramolecular environments, and specific reactions of enzymes are associated with the microenvironments created by them. Metal organic frameworks perform many reactions as enzyme mimics. The models

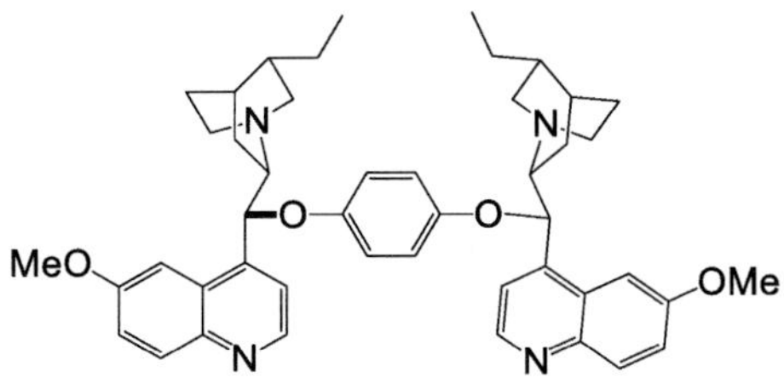

2.76a

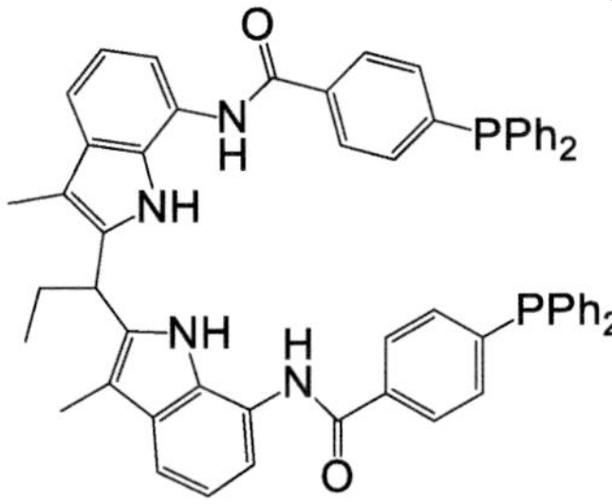

Achiral precatalyst

2.76b

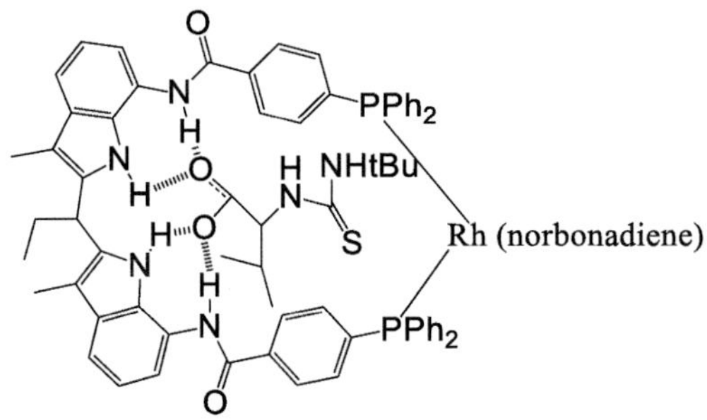

Chiral precatalyst 98 % ee for hydrogenation

2.76c

FIGURE 2.76 Cleft-type ligands and example of a noncovalent substrate binding. (From P. Dydio et al. *Org. Lett.* 2010, 12, 1076–1078.[65])

prepared from MOFs to mimic phosphodiesterase, carbonic anhydrase, cytochrome P450, chymotrypsin and hydrogenase are well known.

The zirconium-based metal organic framework UiO-66 (UiO = Universitetet i Oslo) is made up of benzene 1,4-dicarboxylate. It has Zr6-octahedron, each capped by four μ^3-O, which is represented by $Zr_6O_6(CO_2)_{12}$. The adjacent zirconium ions are bridged benzene-1,4-dicarboxylate by carboxylate groups forming an eight-coordinated zirconium coordination environment. The framework structure of this MOF remains structurally unchanged upon removal or addition of hydroxy bridges. The MOF UiO-66 serves as a mimic of phosphodiesterase reactivity.[78] It causes catalytic hydrolysis of methyl paraoxon (dimethyl-4-nitrophenylphosphate). The half-life of methanolysis of dimethyl-4-nitrophenylphosphate is about 40 to 50 minutes at 333 K. Since the UiO-66 has apertures of 6 Å, the inclusion of the dimethyl-4-nitrophenylphosphate molecule at the interior of the MOF is not possible. This particular catalysis occurs at the outer surface of the catalyst.

Cyt P450 enzymes cause oxidation of different organic substrates; they have active prosthetic groups of iron porphyrin complexes. In these enzymes, the porphyrin holds a Fe(III) metal ion whose fifth coordination site is occupied by a cysteine residue (Cys 400) of the peptide chain. The iron sites are coordinatively unsaturated; the sixth coordination site in each case is vacant. Oxygen is activated at this vacant site, and the catalytic oxidation processes involve the tetravalent intermediate Fe(IV)=O radical cation. The radical cation binds the organic substrate. Two strategies are adopted to prepare the metal organic framework as a Cyt P450 enzyme mimic. Functionalised porphyrins such as **2.76a** and **2.76b** form different MOFs in the reaction with cobalt (II) and iron

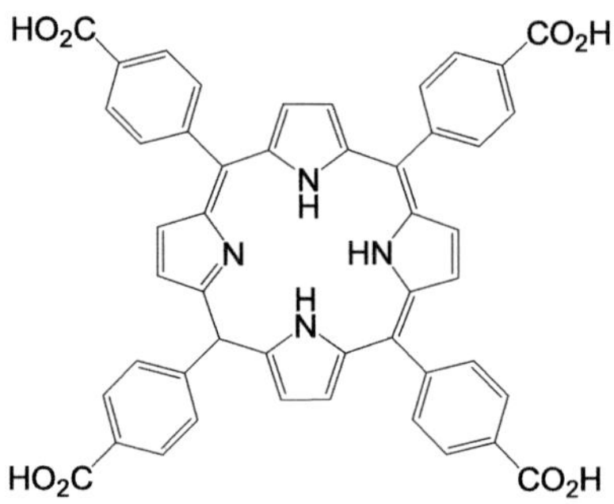

2.77a

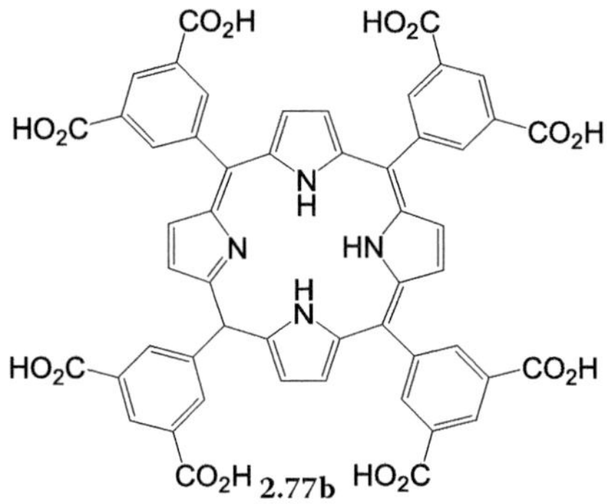

2.77b

FIGURE 2.77 Porphyrin-based linkers for metal organic frameworks. (From C. Zou et al. *Inorg. Chem.* 2013, 52, 3620–3626.[66])

(II) complexes. These MOFs are biological mimics of CytP450 (Figure 2.77). They catalytically convert (E)-1,2-diphenylethene to 2,3-diphenyloxirane in the presence of oxygen. The rates are much slower than those of the natural enzymes; the constraints are on the necessity of diffusing a reactant into the interiors of the pores and the discharge of the product. In principle, in an MOF, the number of active sites increases due to the surface of the voids as catalytic sites and multiple metal ions, but the synergic constraints of diffusion reduce the extent of activation one would have anticipated.

Carbonic anhydrase enzymes have mononuclear zinc as an active site; the active site has three metal—nitrogen bonds originating from three different imidazoles of histidine residues. A mononuclear cationic cobalt complex of 2,6-bis(2-benzimidazolyl) pyridine trapped in the MOF shows carbonic anhydrase activity. This cobalt complex has 2,6-bis(2-benzimidazolyl)pyridine, along with a water molecule, occupying the coordination sites (Figure 2.78). The complex is encapsulated into the voids of MOF formed by 4,4′,4″-(1,3,5-triazine-2,4,6-triyl)tribenzoic acid with terbium (III) nitrate pentahydrate. The cobalt complex that has a 11.7 Å lateral dimension; gets into the voids of the terbium MOF that have void diameters of 39 Å × 47 Å. The cobalt complex, upon reaction of CO_2, forms a carbonate complex which has lateral size 27.4 Å and can remain encapsulated. The aperture of the MOF is about 13 Å × 17 Å. It is also possible to immobilise the cobalt complex *in-situ* in basic conditions; hence, encapsulation provides similar reactivity to the MOF as that of carbonic anhydrase. These reactions occur at slightly basic conditions (Figure 2.79).

UiO-66 is a robust MOF; it is formed by the reaction of zirconium tetrachloride with 1,4-benzenedicarboxylic acid. This MOF is modified through exchange of the benzene dicarboxylate by reacting with $[FeFe](bdt)(CO)_6$ (bdt = benzenedithiolate) (**2.79c**). The MOF **2.79c** is a good catalyst for reduction of protons. The linker-modified Fe-UiO-66 catalyses transformation of H^+ to H_2 under heterogeneous conditions (Figure 2.79).

A catalytic cycle using ruthenium *tris*-bipyridine complex for electron transfer is shown in Figure 2.80. In this catalytic cycle, the ascorbic acid or light is used to reduce the ruthenium complex. The reduced form of ruthenium supplies the electron to the Fe-UiO-66 MOF for production of hydrogen.

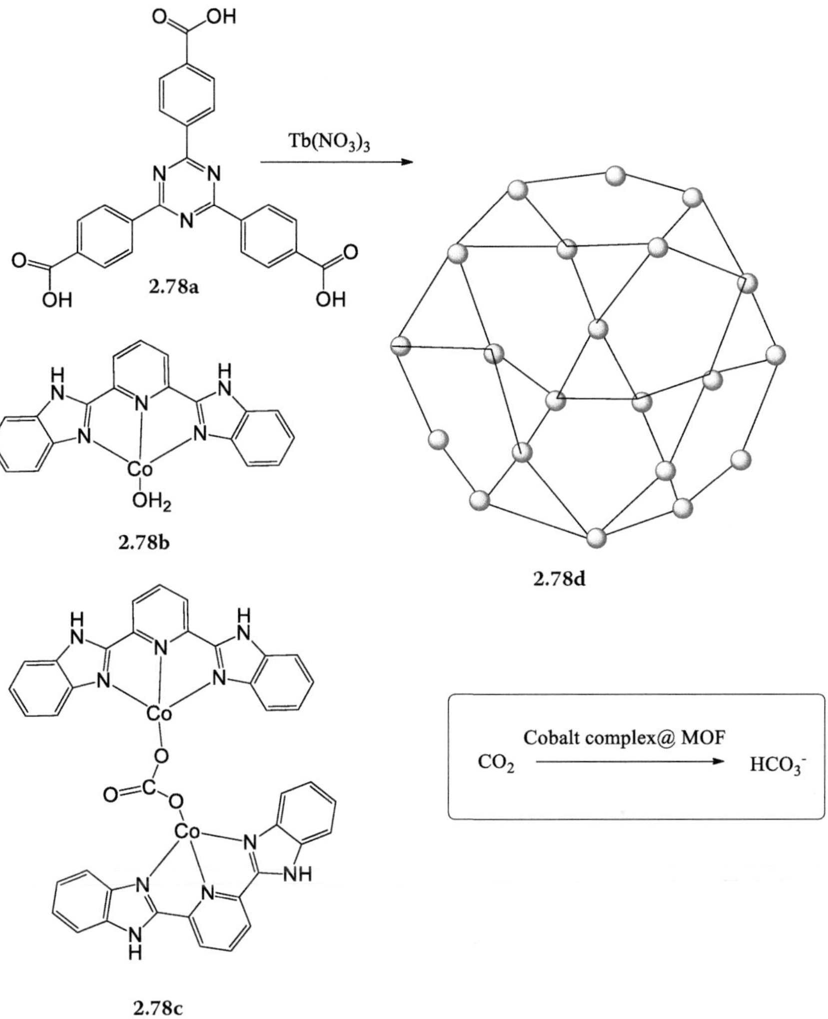

FIGURE 2.78 Carbonic anhydrase by cobalt complex trapped in MOF. (From P. C. Sahoo et al. *J. Cryst. Growth.* 2013, 373, 96–101.[67])

Composites of enzymes with MOF are utilised for activation of hydroperoxides to alcohol and oxygen. Some such reactions are useful to sense organic coloured dye substances through oxidation. Cytochrome c—embedded MOF ZIF-8 is formed *in situ* in the presence of the enzyme, 2-methylimidazole and zinc nitrate. The morphology of the parent MOF remains intact in the enzyme-immobilised MOF. The composite material decomposes hydrogen peroxide to form oxygen and water 10 times faster than the free enzyme. The enhanced rate is due to a conformational change that increases the exposure of the heme group to act as a catalyst.

The enzyme chymotrypsin acts as hydrogen bond—donating (HBD) catalyst with similar catalytic activity to a Lewis acid catalyst. Zinc MOFs possessing urea

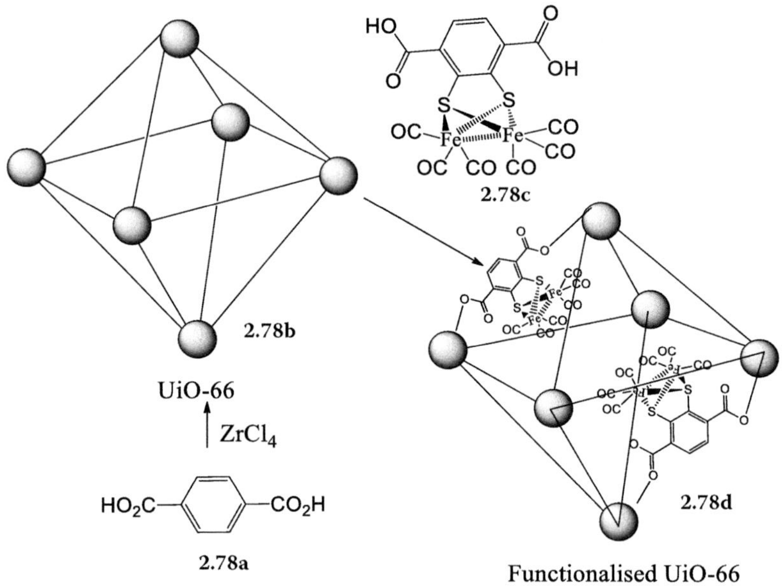

FIGURE 2.79 Synthesis of MOF for mimic of hydrogenase activity. (From S. Pullen et al. *J. Am. Chem. Soc.* 2013, 135, 16997–17003.[68])

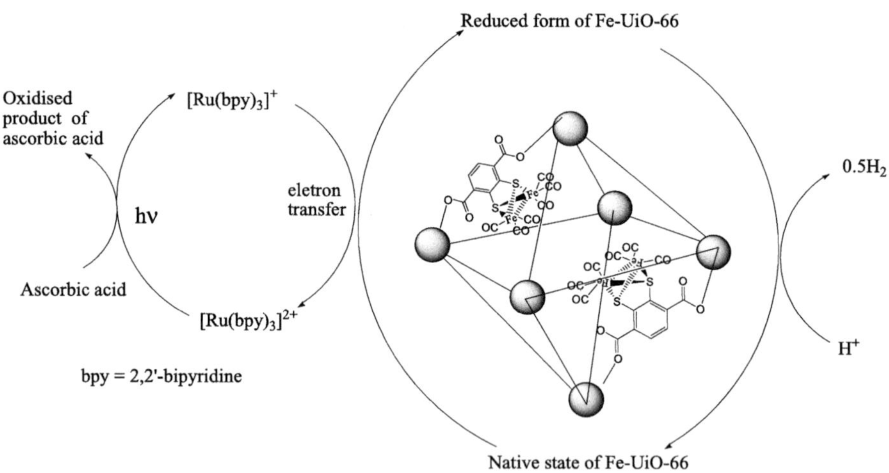

FIGURE 2.80 A catalytic cycle for water splitting using MOF with enzymelike activity. (From S. Pullen et al. *J. Am. Chem. Soc.* 2013, 135, 16997–17003.[68])

and thiourea as part of the linkers are good hydrogen bond—donating catalysts for activation of various carbonyl compounds. Such activation causes a consequent nucleophilic attack that mimics the protease enzyme and follows a mechanistic path similar to chymotrypsin.

2.4 APPLIED ORGANOCATALYTIC REACTIONS TAKING ADVANTAGE OF SUPRAMOLECULAR FEATURES

The next century is looking forward to environmentally benign processes for product development. Control of reactivity with selectivity is the prime concern to increase the holistic aspects of industrial needs. Supramolecular catalysis has not made clear industrial headway other than in certain enzymatic reactions. It may be mentioned that enzyme catalysis was utilised by humans before the scientific aspects were developed. One of the needs of industrial methods is to use environmentally friendly catalysts with high atom economy. Composites, encapsulated catalysts and metal organic frameworks with supramolecular features are the future materials for catalytic reactions; many of them have great promise as heterogeneous catalysts. Sugar-based catalysts are of great promise as biodegradable catalysts for highly enantio-selective reactions. For example, biodegradable catalysts based on sugar-based acid catalysts are easily prepared by functionalisation of sugar molecules to sulphonic acid derivatives.

Oligo-sugar molecules are biodegradable and nontoxic, and when they are transformed to sulphonic acid derivatives, they are water soluble and useful catalysts for multicomponent reactions. Due to the sulphonic acid group in a sugar, the sulphonated sugar acts as an acid. Such oligomers have several sites for weak interactions and are optically active. Some of them are suitable catalysts for three-component reactions between benzaldehyde, aniline and dimethyl phosphonate to form dimethyl (phenyl(phenylamino)methyl)phosphonate (Figure 2.81). The catalytic cycle shown in Figure 2.81 involves activation of the aldehyde by the sulphonic acid groups; the activation is followed by imine formation. The aldimine hydrogen-bonded to sulphonate reacts with dimethyl phosphonate to form the hydrolised product.

The synthesis of chiral natural products is of interest, and such reactions with high selectivity are achieved through supramolecular catalysis. The ligand modification to introduce supramolecular effects is achieved in many chiral-phosphine ligands. One such reaction of phosphorous-based chiral ligand (S)-2-(2-(diphenylphosphino)phenyl)-4-methyl-4,5-dihydrooxazole is shown in Figure 2.82. In this reaction, allylic activation controls the phosphine ligand either to incorporate the allyl group to form a C—C bond or to form a product by simple loss of the allyl-ester part. The ligand, in combination with palladium dibenzoylacetonate complex, causes chiral induction of **2.82a**, resulting in allyl 9-benzyl-3-(2-cyanoethyl)-4-oxo-2,3,4,9-tetrahydro-1H-carbazole-3-carboxylate by decarboxylative C—C bond formation through allylic activation (Figure 2.82). In this reaction, the compound (R)-3-(3-allyl-9-benzyl-4-oxo-2,3,4,9-tetrahydro-1H-carbazol-3-yl)propanenitrile (**2.82b**) is formed as a major product with a 19-fold difference with respect to the decarboxylated compound 3-(9-benzyl-4-oxo-2,3,4,9-tetrahydro-1H-carbazol-3-yl)propanenitrile (**2.82c**) formed without allyl incorporation.

Aromatic ligands provide steric and π-interactions as well as electronic effects that help to selectively cyclise the (E)-10-(3,5-dimethoxyphenyl)-7-methyldeca-1,7-dien-3-ol through allylic activation to form (1S,4aS,10aS)-5,7-dimethoxy-4a-methyl-1-vinyl-1,2,3,4,4a,9,10,10a-octahydrophenanthrene (Figure 2.83). For this purpose, a dimeric iridium cyclooctadiene complex with a binaphthalenediol (binol)-based ligand **2.83c** is used. The ligand used for the reaction is a phosphorus-nitrogen-phosphorus (PNP) ligand with binol directly connected to the phosphorus centre.

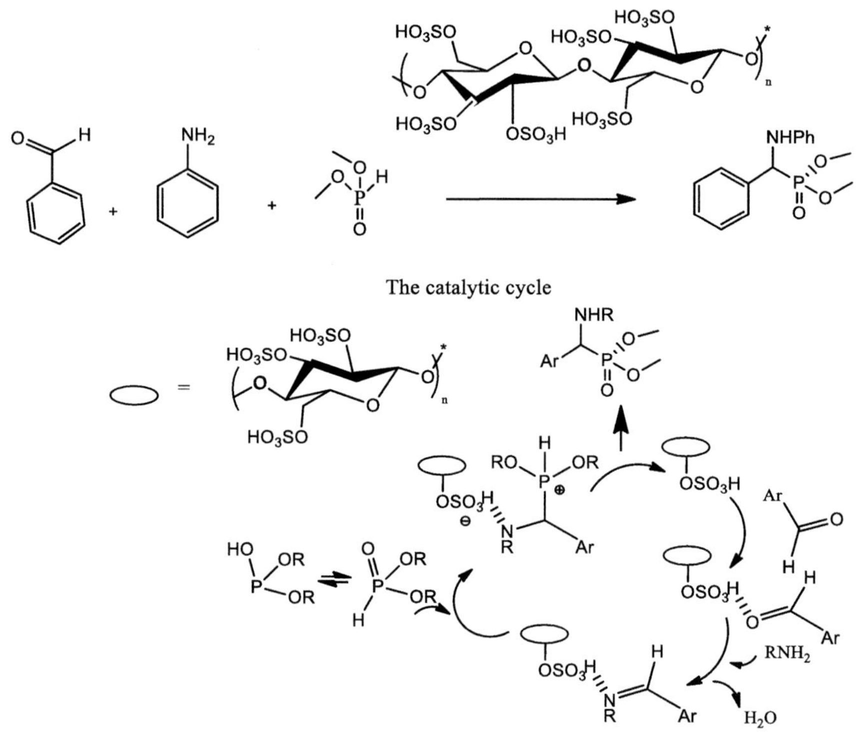

FIGURE 2.81 A Functionalised sugar catalysed three component reaction. (From S. Rostamnia, A. Hassankhani, *Supramolecular Chem.* 2014, 26, 736–739.[69])

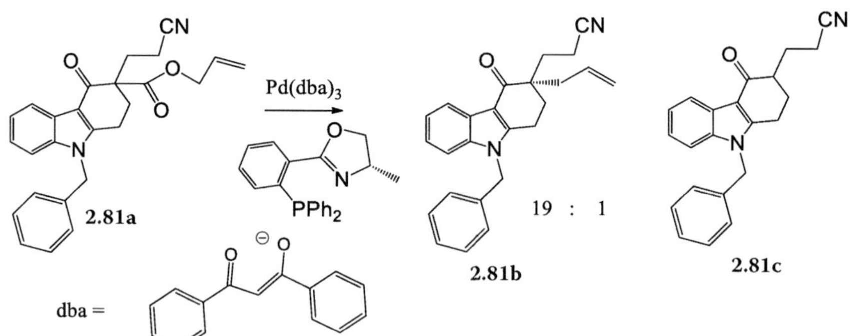

FIGURE 2.82 Enhancing selectivity in product. (From Z. Li et al. *Angew. Chem. Int. Ed.* 2013, 52, 4117–4121.[70])

The catalytic reaction occurs in the presence of zinc(II) triflate. The PNP ligand coordinates to the intermediate iridium complex and forms a mononuclear reactive species which causes the allylic activation of **2.83a** for the cyclisation reaction to form the product **2.83b**. The role of the zinc ions is to activate the alcohol so that it forms an allylic complex with iridium to facilitate the reaction.

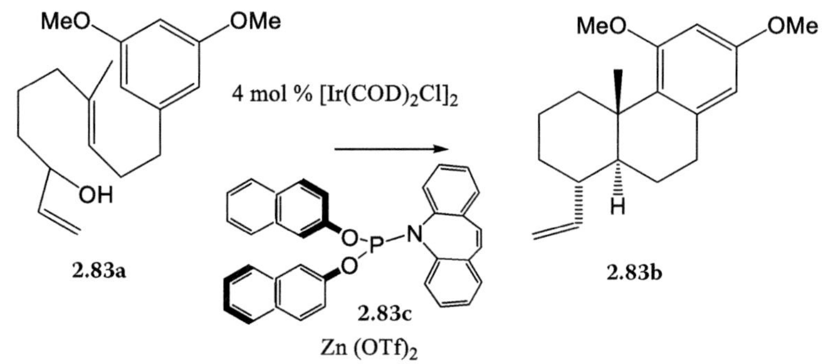

FIGURE 2.83 Selective cyclisation through choice of ligand. (From J. Y. Hamilton et al. *Angew. Chem. Int. Ed.* 2013, 52, 7532–7535.[71])

Choice and design of ligands help in using them in selective product formation. 4-methoxy-2-((2E,6E)-3,7,11-trimethyldodeca-2,6,10-trien-1-yl)phenol cyclises with high stereo-specificity to form (6aS,12aS,12bR)-10-methoxy-4,4,6a,12b-tetramethyl-2,3,4,4a,5,6,6a,12,12a,12b-decahydro-1H-benzo[a]xanthene through consecutive cyclisation by C—C and C—O bond formations in the presence of tin complex **2.84a**. A ligand developed from the 3-ethoxy-2′-ethyl-[1,1′-biphenyl]-2-ol compound with anisole, cyclohexa-1,3-diene, fluorobenzene and methane (1:1:2:1:2) **2.84b** has the ability to form a complex **2.84a** with stannic chloride, which is a catalyst in the presence of trifluoroacetic acid for the cyclisation reaction (Figure 2.84) of **2.84c** to form **2.84d**. High specificity in chiral induction is achieved through the tin complex as a ligand modification by providing appropriate geometry to the transition state, which is otherwise not possible.

Chiral salts are used for selective photochemical catalytic reactions. One such example is shown in Figure 2.85. This photochemical reaction of 1-(pent-4-enoyl) piperidin-4-one by irradiation at 366 nm is catalysed by the salt **2.85a** to form (41R,7aR,8aS)-hexahydrocyclobuta[ij]quinolizine-1,5(41H,6H)-dione. In this catalytic reaction, the aluminium ion activates the reactant **2.85b** and the chiral salt provides the stereo control to selectively form the compound **2.85c**.

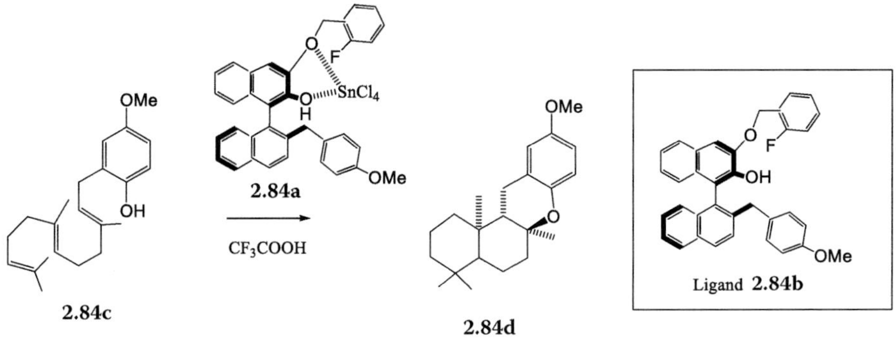

FIGURE 2.84 Lewis metal complex catalysed cyclisation reaction. (From H. Ishibashi et al. *J. Am. Chem. Soc.* 2004, 126, 11122–11123.[72])

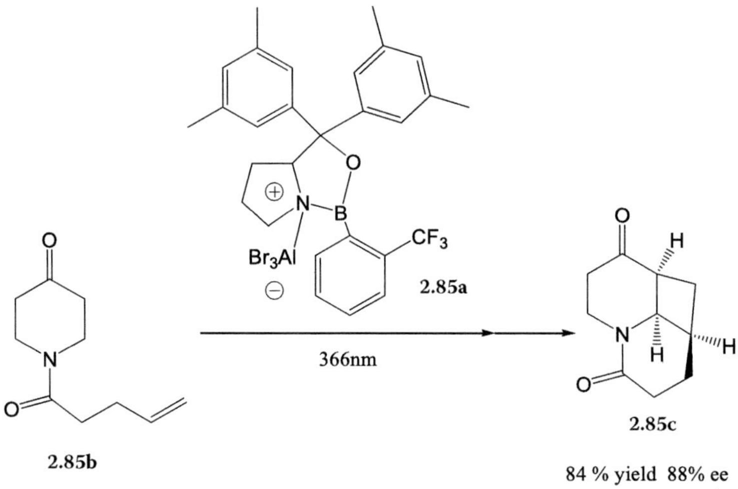

FIGURE 2.85 Ion-pair catalyst for photo-catalytic cyclisation. (From R. Brimioulle, T. Bach, *Science* 2013, 342, 840–843.[73])

Enantioselective reaction as a key step to prepare the natural product baclopfen (**2.86e**) is achieved following a reaction path where a designed bifunctional urea-amine supramolecular catalyst **2.86a** is used. The reaction between diethylmalonate with unsaturated nitro compound (E)-1-nitro-prop-1-ene is catalysed by a bifunctional urea-based catalyst **2.86a**; it forms selective C–C—bonded product

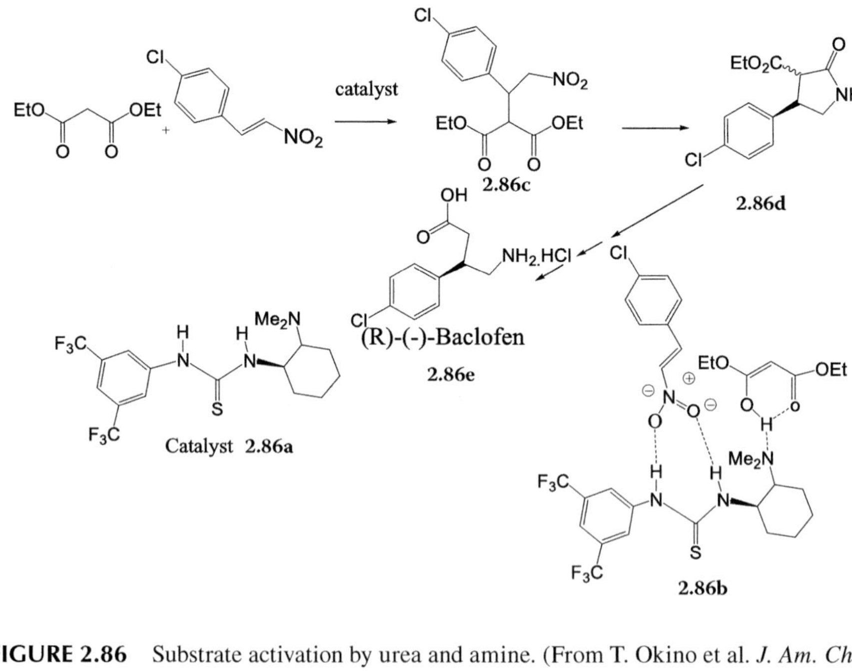

FIGURE 2.86 Substrate activation by urea and amine. (From T. Okino et al. *J. Am. Chem. Soc.* 2005, 127, 119–125.[74])

diethyl 2-(1-(4-chlorophenyl)-2-nitroethyl)malonate **2.86c**. The reactants interact with thiourea and amine to form an intermediate **2.86b** (Figure 2.86) which yields a compound **2.86c**. In the hydrogen-bonded intermediate, urea activates the nitro group. The enol form of the diethylmalonate is stabilised by the NMe_2 group. The compound **2.86c** undergoes cyclisation to form a compound **2.86d**. Subsequent transformations through several steps give the natural product baclofen **2.86e**.

REFERENCES

1. M. Raynal, P. Ballester, A. Vidal-Ferran, P. W. N. M. van Leeuwen, Supramolecular catalysis. Part 2: Artificial enzyme mimics. *Chem. Soc. Rev.* 2014, 43, 1734–1787.
2. E. M. Broderick, N. Guo, T. Wu, C. S. Vogel, C. Xu, J. Sutter, J. T. Miller, K. Meyer, T. Cantat, P. L. Diaconescu, Redox control of a polymerization catalyst by changing the oxidation state of the metal centre. *Chem. Commun.* 2011, 47, 9897–9899.
3. T. G. Ribelli, M. Fantin, J. C. Daran, K. F. Augustine, R. Poli, K. Matyjaszewski, Synthesis and characterization of the most active copper ATRP catalyst based on tris[(4-dimethylaminopyridyl)methyl]amine. *J. Am. Chem. Soc.* 2018, 140, 1525–1534.
4. L. R. MacGillivray, G. S. Papaefstathiou, T. Friscic, D. B. Varshney, T. D. Hamilton, Template-controlled synthesis in the solid-state. *Topics in Current Chem.* 2004, 248, 201–221.
5. H. C. Kiefer, W. I. Congdon, I. S. Scarpa, I. M. Klotz, Synthetic derivatives of polyethyleneimine with enzyme-like catalytic activity (synzymes). *Proc. Natl. Acad. Sci. U.S.A.* 1972, 69, 2155–2159.
6. L. Liu, W. Zhou, J. Chruma, R. Breslow, Transamination reactions with multiple turnovers catalyzed by hydrophobic pyridoxamine cofactors in the presence of polyethylenimine polymers. *J. Am. Chem. Soc.* 2004, 126, 8136–8137.
7. C. C. Bai, B. R. Tian, T. Zhao, Q. Huang, Z. Z. Wang, Cyclodextrin-catalyzed organic synthesis: Reactions, mechanisms, and applications. *Molecules*. 2017, 22, 1475.
8. Z. Zhang, P. R. Schreiner, (Thio)urea organocatalysis—What can be learnt from anion recognition? *Chem. Soc. Rev.* 2009, 38, 1187–1198.
9. M. Mahlau, B. List, Asymmetric counter anion-directed catalysis: Concept, definition, and applications. *Angew. Chem. Int. Ed.* 2013, 52, 518–533.
10. J.-R. Chen, Y.-Q. Zou, L. F. Fan, R. Fen, T. W.-J, Xiao, Highly enantioselective Michael addition of aldehydes to nitro-olefins catalyzed by primary amine thiourea organocatalysts. *Tetrahedron* 2010, 66, 5367–5372.
11. S. L. Balof, S. J. P'Pool, N. J. Berger, E. J. Valente, A. M. Shiller, H.-J. Schanz, Olefin metathesis catalysts bearing a pH-responsive NHC ligand: A feasible approach to catalyst separation from RCM products. *Dalton Trans.* 2008, 5791–5799.
12. R. S. Stoll, M. V. Peters, A. Kuhn, S. Heiles, R. Goddard, M. Buhl, C. M. Thiele, S. Hecht, Photoswitchable catalysts: Correlating structure and conformational dynamics with reactivity by a combined experimental and computational approach. *J. Am. Chem. Soc.* 2009, 131, 357–367.
13. L. Osorio-Planes, C. Rodrıguez-Escrich, M. A. Pericas, Photoswitchable thioureas for the external manipulation of catalytic activity. *Org. Lett.* 2014, 16, 1704–1707.
14. I. M. Lorkovic, R. R. Duff, M. S. Wrighton, Use of the redox-active ligand 1,1′-bis(diphenylphosphino)cobaltocene to reversibly alter the rate of the rhodium(I)-catalyzed reduction and isomerization of ketones and alkenes. *J. Am. Chem. Soc.* 1995, 117, 3617–3618.
15. O. Coulembier, S. Moins, R. Todd, P. Dubois, External and reversible CO_2 regulation of ring-opening polymerizations based on a primary alcohol propagating species. *Macromolecules* 2014, 47, 486–491.

16. D. Wilson, N. R. Branda, Turning "on" and "off" a pyridoxal 5′-phosphate mimic using light. *Angew. Chem. Int. Ed.* 2012, 51, 5431–5434.
17. B. M. Neilson, C. W. Bielawski, Photoswitchable metal-mediated catalysis: Remotely tuned alkene and alkyne hydroborations. *Organometallics* 2013, 32, 3121–3128.
18. S. Shinkai, K. Kameoka, K. Ueda, O. Manabe, Remote control of flavin reactivities by an intramolecular crown ring serving as a metal-binding site. *J. Am. Chem. Soc.* 1987, 109, 923–924.
19. M. O. Huaping, T. C. Pochapsky, Intermolecular interactions characterized by nuclear Overhauser effects. *Prog. Nuclear Magnetic Resonance Spectroscopy.* 1997, 30, 1–38.
20. Y. Jiao, W. L. Li, J.-F. Xu, G, Wang, J. Li, Z. Wang, X. Zhang, A supramolecularly activated radical cation for accelerated catalytic oxidation. *Angew. Chem. Int. Ed.* 2016, 55, 8933–8937.
21. D. R. Edwards, C. T. Liu, G. E. Garrett, A. A. Neverov, R. S. Brown, Leaving group assistance in the La^{3+} $^{-}$catalyzed cleavage of dimethyl (*o*-methoxycarbonyl)aryl phosphate triesters in methanol. *J. Am. Chem. Soc.* 2009, 131, 13738–13748.
22. H, Fanlo-Virgs, A.-N. R. Alba, S. Hamieh, M. Colomb-Delsuc, S. Otto, Transient substrate-induced catalyst formation in a dynamic molecular network. *Angew. Chem. Int. Ed.* 2014, 53, 11346–11350.
23. Y. Ge, L. Miller, T. Ouimet, D. K. Smith, Electrochemically controlled hydrogen bonding. *o*-quinones as simple redox-dependent receptors for arylureas. *J. Org. Chem.* 2000, 65, 8831–8838.
24. T. Govender, P. I. Arvidsson, G. E. M. Maguire, H. G. Kruger, T. Naicker, Enantioselective organocatalyzed transformations of β-ketoesters. *Chem. Rev.* 2016, 116, 9375–9437.
25. P. A. Lichtor, S. J. Miller, Experimental lineage and functional analysis of a remotely directed peptide epoxidation catalyst. *J. Am. Chem. Soc.* 2014, 136, 5301–5308.
26. Y. Sohtome, S. Tanaka, K. Takada, T. Yamaguchi, K. Nagasawa, Solvent dependent enantiodivergent Mannich-type reaction: Utilizing a conformationally flexible guanidine/bisthiourea organocatalyst. *Angew. Chem. Int. Ed.* 2010, 49, 9254–9257.
27. M. Hatano, K. Ishihara, Conformationally flexible chiral supramolecular catalysts for enantioselective Diels–Alder reactions with anomalous *endo*/*exo* selectivities. *Chem. Commun.* 2012, 48, 4273–4283.
28. W.-S. Chung, J.-Y. Wang, Control of regioselectivity in the Diels-Alder reactions of alkyl-substituted 1,4-benzoquinones by β-cyclodextrin and its derivatives. *J.C.S. Chem. Commun.* 1995, 971–972.
29. Z. Huang, Z. Chen, L. Hui, L. Gia, C. P. Quang, H. Hirao, J. (Steve) Zhou, Weak arene C-H···O hydrogen bonding in palladium-catalyzed aylation and vinylation of lactones. *Angew. Chem. Int. Ed.* 2013, 52, 5807–5812.
30. D. B. Grotjahn, D. A. Lev, M. Tokunaga, General bifunctional catalyst for the anti-Markovnikov hydration of terminal alkynes to aldehydes gives enzyme-like rate and selectivity enhancements. *J. Am Chem. Soc.* 2004, 126, 12232–12233.
31. K. Ding, Development of homogeneous and heterogeneous asymmetric catalysts for practical enantioselective reactions. *Pure Appl. Chem.* 2006, 78, 293–301.
32. J. T. Foy, D. Ray, I. Aprahamian, Regulating signal enhancement with coordination-coupled deprotonation of a hydrazone switch. *Chem. Sci.* 2015, 6, 209–213.
33. K. Lang, J. Park, S. Hong, Urea/transition-metal cooperative catalyst for anti-selective asymmetric nitroaldol reactions. *Angew. Chem. Int. Ed.* 2012, 51, 1620–1624.
34. S. Jonsson, G. J. O. Fabrice, P.-O. Norrby, K. Warnmark, A dynamic supramolecular system exhibiting substrate selectivity in the catalytic epoxidation of olefins. *Chem. Commun.* 2005, 549–551.
35. L. Kovbasyuk, R. Kramer, Allosteric supramolecular receptors and catalysts. *Chem. Rev.* 2004, 104, 3161–3187.

36. A. J. M. Miller, Controlling ligand binding for tunable and switchable catalysis: Cation-modulated hemilability in pincer-crown ether ligands. *Dalton Trans.* 2017, 46, 11987–12000.
37. H. J. Davis, M. T. Mihai, R. J. Phipps, Ion pair-directed regiocontrol in transition-metal catalysis: A meta-Selective C–H borylation of aromatic quaternary ammonium salts. *J. Am. Chem. Soc.* 2016, 138, 12759–12762.
38. D. A. Evans, K. T. Chapman, J. Bisaha, Asymmetric Diels-Alder cycloaddition reactions with chiral α-β-unsaturated acyloxazolidinones. *J. Am. Chem. Soc.* 1988, 110, 1238–1256.
39. T. Wang, L.-G. Zhuo, Z. Li, F. Chen, Z. Ding, Y. He, Q.-H. Fan, J. Xiang, Z.-X. Yu, A. S. C. Chan, Highly enantioselective hydrogenation of quinolines using phosphine-free chiral cationic ruthenium catalysts: Scope, mechanism, and origin of enantioselectivity. *J. Am. Chem. Soc.* 2011, 133, 9878–9891.
40. P. Dydio, W. I. Dzik, M. Lutz, B. de Bruin, J. N. H. Reek, Remote supramolecular control of catalyst selectivity in the hydroformylation of alkenes. *Angew. Chem. Int. Ed.* 2011, 50, 396–400.
41. L. Rovira, H. Fernández-Pérez, A. Vidal-Ferran, Palladium-based supramolecularly regulated catalysts for asymmetric allylic substitutions. *Organometallics* 2016, 35, 528–533.
42. M. Schmittel, S. De, S. Pramanik, Reversible ON/OFF nanoswitch for organocatalysis: Mimicking the locking and unlocking operation of CaMKII. *Angew. Chem. Int. Ed.* 2012, 51, 3832–3836.
43. V. Kunz, D. Schmidt, M. I. S. Röhr, R. Mitric, F. Würthner, Supramolecular approaches to improve the performance of ruthenium-based water oxidation catalysts. *Adv. Energy Mater.* 2017, 7, 1602939.
44. V. Martinez-Agramunt, S. Ruiz-Botella, E. Peris, Nickel-cornered molecular rectangles as polycyclic aromatic hydrocarbon receptor. *Chem. Eur. J.* 2017, 23, 6675–6681.
45. S. M. Rummelt, K. Radkowski, D.-A. Rosca, A. Furstner, Interligand interactions dictate the regioselectivity of *trans*-hydrometalations and related reactions catalyzed by [Cp*RuCl]. hydrogen bonding to a chloride ligand as a steering principle in catalysis. *J. Am. Chem. Soc.* 2015, 137, 5506–5519.
46. H. J. Yoon, J. Kuwabara, J.-H. Kim, C. A. Mirkin, Allosteric supramolecular triple-layer catalysts. *Science* 2010, 330, 66–69.
47. C. G. Oliveri, N. C. Gianneschi, S. T. Nguyen, C. A. Mirkin, C. L. Stern, Z. Wawrzak, M. Pink, Supramolecular allosteric cofacial porphyrin complexes. *J. Am. Chem. Soc.* 2006, 128, 16286–16296.
48. E. N. Jacobsen, Asymmetric catalysis of epoxide ring opening reactions. *Acc. Chem. Res.* 2000, 33, 421–431.
49. S. De, S. Pramanik, M. Schmittel, A monomer–dimer nanoswitch that mimics the working principle of the SARS-CoV 3CLpro enzyme controls copper-catalysed cyclopropanation. *Dalton Trans.* 2014, 43, 10977–10982.
50. S. De, S. Pramanik, M. Schmittel, A toggle nanoswitch alternately controlling two catalytic reactions. *Angew. Chem. Int.Ed.* 2014, 53, 14255–14259.
51. V. F. Slagt, P.W. N. M. van Leeuwen, J. N. H. Reek, Multicomponent porphyrin assemblies as functional bidentate phosphite ligands for regioselective rhodium-catalyzed hydroformylation. *Angew. Chem. Int. Ed.* 2003, 42, 5619–5623.
52. R. K. Totten, P. Ryan, B. Kang, S. Joong, L. Linda, J. Broadbelt, R. Q. Snurr, J. T. Hupp, S. B. T. Nguyen, Enhanced catalytic decomposition of a phosphate triester by modularly accessible bimetallic porphyrin dyads and dimers. *Chem. Commun.* 2012, 48, 4178–4180.
53. A. Dhakshinamoorthy, M. Alvaro, H. Garcia, Metal-organic frameworks (MOFs) as heterogeneous catalysts for the chemoselective reduction of carbon-carbon multiple bonds with hydrazine. *Adv. Synth. Catal.* 2009, 351, 2271–2276.

54. Y. K. Hwang, D.-Y. Hong, J.-S. Chang, S. H. Jhung, Y.-K. Seo, J. Kim, A. Vimont, M. Daturi, C. Serre, G. Frey, Amine grafting on coordinatively unsaturated metal centers of MOFs: Consequences for catalysis and metal encapsulation. *Angew. Chem. Int. Ed.* 2008, 47, 4144–4148.
55. I. Nath, J. Chakraborty, F. Verpoort, Metal organic frameworks mimicking natural enzymes: A structural and functional analogy. *Chem. Soc. Rev.* 2016, 45, 4127–4170.
56. J. L. C. Roswell, O. M. Yaghi, Effects of functionalization, catenation, and variation of the metal oxide and organic linking units on the low-pressure hydrogen adsorption properties of metal–organic frameworks. *J. Am. Chem. Soc.* 2006, 128, 1304–1315.
57. M. J. Ingleson, J. P. Barrio, J. B. Guilbaud, Y. Z. Khimyak, M. J. Rosseinsky, Framework functionalisation triggers metal complex binding. *Chem. Commun.* 2008, 2680–2682.
58. J. Y. Lee, O. K. Farha, J. Roberts, K.A. Scheidt, S. B. T. Nguyen, J. T. Hupp, Metal–organic framework materials as catalysts. *Chem. Soc. Rev.* 2009, 38, 1450–1459.
59. J. Liu, L. Chen, H. Cui, J. Zhang, L. Zhang, C.-Y. Su, Applications of metal–organic frameworks in heterogeneous supramolecular catalysis. *Chem. Soc. Rev.* 2014, 43, 6011–6061.
60. R. Breslow, Biomimetic chemistry and artificial enzymes: Catalysis by design. *Acc. Chem. Res.* 1995, 28, 146–153.
61. O. Bistri, O. Reinaud, Supramolecular control of transition metal complexes in water by a hydrophobic cavity: A bio-inspired strategy. *Org. Biomol. Chem.* 2015, 13, 2849–2865.
62. S. C. Zimmerman, R. Breslow, Asymmetric synthesis of amino acids by pyridoxamine enzyme analogs utilizing general base-acid catalysis. *J. Am. Chem. Soc.* 1984, 106, 1490–1491.
63. S. D. Aubert, Y. Li, F. M. Raushel, Mechanism for the hydrolysis of organophosphates by the bacterial phosphotriesterase. *Biochemistry* 2004, 43, 19, 5707–5715.
64. P. Molenveld, W. M. G. Stikvoort, H. Kooijman, A. L. Spek, J. F. J. Engbersen, D. N. Reinhoudt, Dinuclear and trinuclear Zn(II) calix[4]arene complexes as models for hydrolytic metalloenzymes: Synthesis and catalytic activity in phosphate diester transesterification. *J Org Chem.* 1999, 64, 3896–3906.
65. P. Dydio, T. Zieliński, J. Jurczak, 7,7′-Diureido-2,2′-diindolylmethanes: Anion receptors effective in a highly competitive solvent, methanol. *Org. Lett.* 2010, 12, 1076–1078.
66. C. Zou, T. Zhang, M.-H. Xie, L. Yan, G.-Q. Kong, X.-L. Yang, A. Ma, C.-D. Wu, Four metalloporphyrinic frameworks as heterogeneous catalysts for selective oxidation and aldol reaction. *Inorg. Chem.* 2013, 52, 3620–3626.
67. P. C. Sahoo, Y. N. Jang, S. W. Lee, Enhanced biomimetic CO_2 sequestration and $CaCO_3$ crystallization using complex encapsulated metal organic framework. *J. Cryst. Growth.* 2013, 373, 96–101.
68. S. Pullen, H. Fei, A. Orthaber, S. M. Cohen, S. Ott, Enhanced photochemical hydrogen production by a molecular diiron catalyst incorporated into a metal–organic framework. *J. Am. Chem. Soc.* 2013, 135, 16997–17003.
69. S. Rostamnia, A. Hassankhani, Application of biodegradable supramolecular polymer-supported catalyst for multicomponent synthesis of α-aminophosphonates Kabachnik–Fields reaction. *Supramolecular Chem.* 2014, 26, 736–739.
70. Z. Li, S. Zhang, S. Wu, X. Shen, L. Zou, F. Wang, X. Li, F. Peng, H. Zhang, Z. Shao, Enantioselective palladium catalyzed decarboxylative allylation of carbazolones: Total synthesis of (–)-aspidospermidine and (+)-kopsihainanine. *Angew. Chem. Int. Ed.* 2013, 52, 4117–4121.
71. J. Y. Hamilton, D. Sarlah, E. M. Carreira, Iridium-catalyzed enantioselective allylic alkynylation. *Angew. Chem. Int. Ed.* 2013, 52, 7532–7535.
72. H. Ishibashi, K. Ishihara, H. Yamamoto, A new artificial cyclase for polyprenoids: Enantioselective total Synthesis of (–)-chromazonarol, (+)-8-*epi*-puupehedione, and (–)-11′-deoxytaondiol methyl ether. *J. Am. Chem. Soc.* 2004, 126, 11122–11123.

73. R. Brimioulle, T. Bach, Enantioselective Lewis acid catalysis of intramolecular enone [2+2] photocycloaddition reactions. *Science* 2013, 342, 840–843.
74. T. Okino, Y. Hoashi, T. Furukawa, X. Xu, Y. Takemoto, Enantio- and diastereoselective Michael reaction of 1,3-dicarbonyl compounds to nitroolefins catalyzed by a bifunctional thiourea. *J. Am. Chem. Soc.* 2005, 127, 119–125.
75. V. Kunz, M. Schulze, D. Schmidt, F. Würthner, Trinuclear ruthenium macrocycles: Toward supramolecular water oxidation catalysis in pure water. *ACS Energy Lett.* 2017, 2, 288–293.
76. D.-Y. Ma, K. Warnmark, Mechanoassisted supramolecular catalysis in solid state synthesis. *Chem. Cat. Chem.* 2010, 2, 1059–1060.
77. A. Dhakshinamoorthy, M. Opanasenko, J. Cejka, H. Garcia, Metal organic frameworks as heterogeneous catalysts for the production of fine chemicals. *Catal. Sci. Technol.*, 2013, 3, 2509–2540.
78. M. J. Katz, S.-Y. Moon, J. E. Mondloch, M. H. Beyzavi, C. J. Stephenson, J. T. Hupp, O. K. Farha, Exploiting parameter space in MOFs: A 20-fold enhancement of phosphate-ester hydrolysis with UiO-66-NH_2. *Chem. Sci.* 2015, 6, 2286–2291.

3 Containers and Vessels for Supramolecular Catalysis

3.1 CONTAINER MOLECULES

The principal inspiration for container molecules in catalysis stems from reactions occurring in nature. Whereas for synthetic chemists dealing with host—guest chemistry, the catalytic ability of cyclodextrins is a system among different hosts that has generated interest to develop many other related systems with cavities to promote and perform reactions inside their cavities. The rate and selectivity of a chemical reaction are greatly influenced by environment; analogous effects are observed from shielded inner phases of synthetic host systems. In general, the confinement of a guest inside a host lowers the symmetry of the guest, which provides scope for asymmetric reactions that mimic biological systems. Host systems may act as containers to hold reactants and reagents to carry out reactions. The container may be composed of a single covalently linked system or may be self-assembled coordination cages and organic capsules. Container molecules have voids to include the guest within or a concave space, as in the case of cavitands. The functional modification of the surroundings of a container molecule creates different possibilities to bind guests, thereby increasing complexity but also, in certain cases, ease of performing selective reactions. The acid—base properties of container molecules are used in carrying out certain unconventional reactions. A minute change in the structure of a host molecule influences the binding of the substrate significantly; these container molecules have a complete hold on the specific reaction rate of reactants under special conditions.

The ability to confine a substrate in a cavity or bowl-like container molecule may be understood from the everyday utilities of bowl-like vessels or containers. We use the open spaces of bowls, for example, cups, plates, baskets and so on, to store household materials. For storage purposes, one can maximise the utility of the spaces of an independent bowl, which may or may not be possible with multiple bowls kept over each other. Two baskets combined may provide an altogether different space for materials which may not be suitable to accommodate the amount of substrate that would have been accommodated in two independent baskets. But the arrangement of two baskets may provide a specific space and support for a particular geometry to make it conducive to accommodating a particular item of a suitable shape. Baskets arranged in multiple numbers in independent places (Figure 3.1) would also be suitable for specific purposes. On the other hand, the phase of matter matters in storage; for example, a cup or glass may store liquids, where the liquid will maximise to utilise the shape, but to accommodate appropriate solid shapes or pretreatment of

3.1a 3.1b 3.1c

FIGURE 3.1 Bowls commonly used for storage of household items.

the solid for, proper fitting into a container is necessary. The strength and stability of containers also matter in storage. Some of these processes are like pre-organised and reorganised concepts in the chemical and biological sciences. A similar analogy may be used for molecules that can adopt bowl-like shapes. Conformational flexible bowl-shaped molecules would change shape with interaction with a guest or by stimuli, whereas two or more bowl-shaped molecules may interact together and provide no space or a space that is completely different from the parent shape or space. Such molecules will recognise guests much differently than the parent molecule.

As an example, the cyclic molecule **3.2a** organises in various ways under different pH values (Figure 3.1). The compound **3.2a** has a cyclic structure with rigid sp^2 and sp-carbon centres in the cyclic frame. The ring has a carboxylic acid at one side and a pyridine unit connected at an exactly opposite end of the acid group. Both these groups are located in the interior of the cyclic structure. The pyridine and carboxylic acid form a hydrogen bond and have a planar structure (Figure 3.2). In the event of deprotonation of the carboxylic acid by external reagent or changing pH of the medium, the hydrogen bond between the carboxylic acid and pyridine is absent, as it forms an anionic species **3.2b**. Due to the loss of the hydrogen bond by generating carboxylate anion, the conformation of the parent molecule is changed. The conformation of the salt changes to a boat-type conformation from the planar arrangements originally found in the neutral compound. Besides losing planarity, the compound becomes

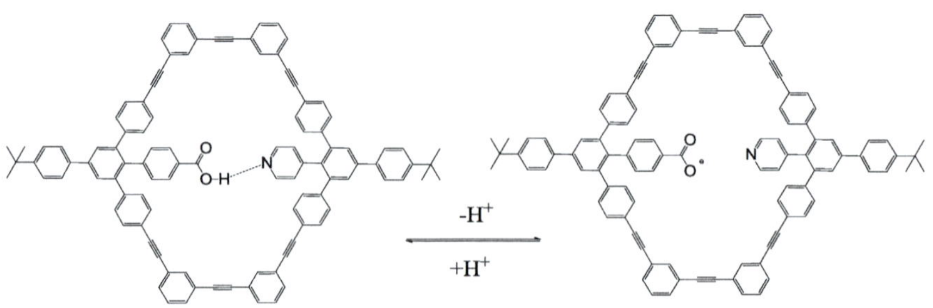

3.2a 3.2b

FIGURE 3.2 Neutral and anion of a cyclic host affecting intramolecular hydrogen bonds. (From A. B. Vliegenthart et al. *Org. Lett.* 2015, 17, 4172–4175.[1])

more soluble in polar solvents. The lack of planarity upon deprotonation provides self-assemblies of the compound with relatively fewer π-interactions among the aromatic rings. The respective conformation in neutral conditions or of the deprotonated form reversibly change upon adjustment of the pH. This example suggests that cooperation between functional groups within a macrocycle decides the nature of voids.

The zero-dimensional cage-like enclosures of molecular architectures are usually built by connecting two or more templates by connecting each other through bridging units. The flexibility of the linking units of the templates, interactions among the rings and interaction of the linkers with other molecules or ions are factors to provide better control to maintain a definite shape. The flexible *bis*-zinc-porphyrin cage **3.3a** has four covalent linkers binding the two zinc-porphyrin units. Each linker is composed of two N-phenyl-1,2,3-triazolyl linked to each other by a -CH_2-$O(CH_2)_2O$-CH_2- unit. The cage encloses 1,4-diazabicyclo[2.2.2]octane as a guest. The cage undergoes conformational changes upon interaction of different solvents and ions. For example, eight triazole rings of the assembly are ligands those coordinate to four silver (I) ions. The coordinations provides a rigid structure of the silver complex **3.3b**. Due to the complex formation through the coordination of nitrogen atoms to silver ions, a conformational locked geometry of the complex is observed (Figure 3.3). In this conformation, two porphyrin rings are placed face to face in a parallel manner. The cage changes size; it expands upon protonation. The expansion occurs to repel the positively charged species created by protonation on the uncoordinted nitrogen sites in the triazole rings. The complexation and protonation are reversible and provide structural changes which are generally referred to as chemically induced breathing of flexible structures.

Binols with skeletal features as shown in **3.4a** are generally used in chiral recognitions. They are in demand due to commercial availability and ease of transformations to various derivatives. Two hydroxyl groups on binols are appropriately linked to one or more polyfunctional connecting units or ions/atoms acting as a template, or a pivot generates different topologies in the form of cages or containers. Thus, a molecule derived from naphthalenediol may be functionalised in different ways, as shown in Figure 3.4. One of the examples shown in this figure has three naphthalenenediol units linked covalently to a tri-methylenebenzene template **3.4b**. In another case,

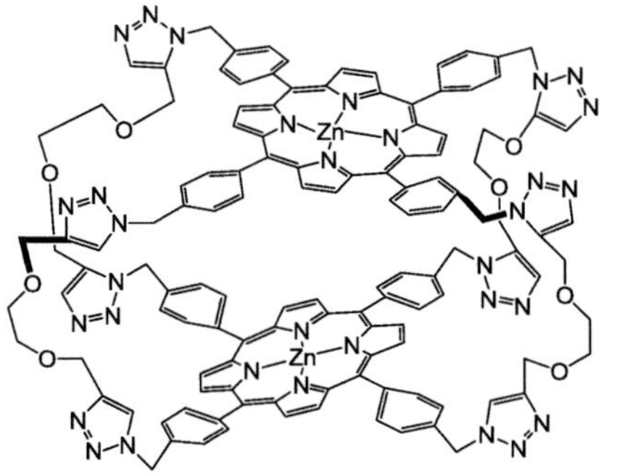

3.3a

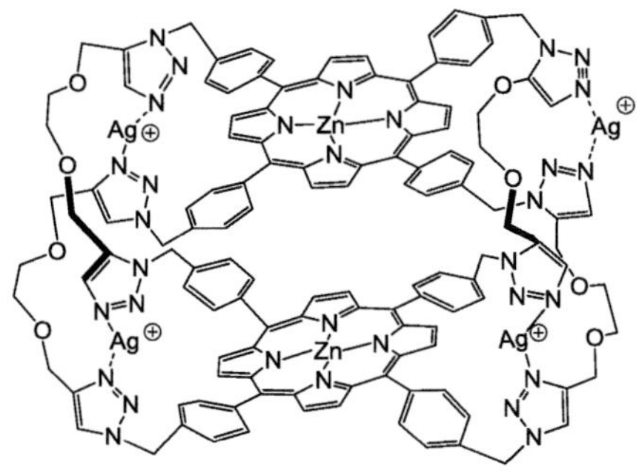

3.3b

FIGURE 3.3 Changes in conformation of cages upon formation of silver complex. (From L. Schoepff et al. *J. Org. Chem.* 2017, 82, 5845–5851.[2])

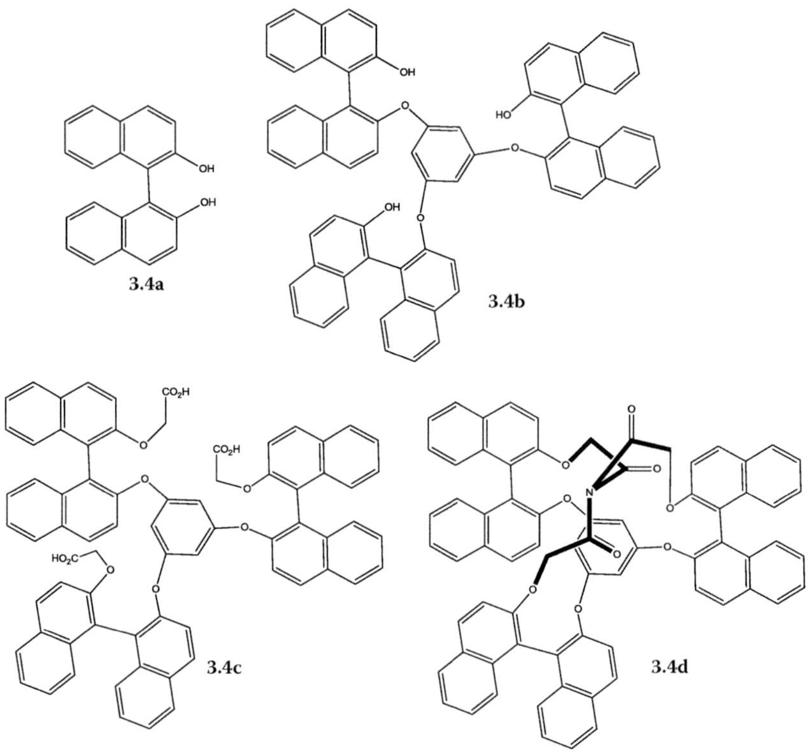

FIGURE 3.4 Designing a container or cage-like structure from binol. (From D. Zhang et al. *Chem. Rev.* 2017, 117, 4900–4942.[3])

phenolic OH groups of binol are linked to $-CH_2COOH$ to form the product **3.4c**, and, finally, each carboxylic group is converted to an amide functional group to form a cage. The three carboxylic groups once attached to a central nitrogen atom pivot, which converts each carboxylic group to amide, providing the cage-like structure **3.4d**. This cage-like molecule has flexible arms, and it shows interesting properties as a ligand while coordinating to metal ions. For example, it adopts a bowl-like structure when the central nitrogen is connected to oxido-vanadium ion **3.5a**. This bowl-shaped complex is a very good catalyst for the oxidation reaction of sulphide to sulphoxides with high enantioselectivity. The catalyst is very efficient and performs at a rate more than 300 times higher than that of a similar complex that does not have the enclosure part. The observed enantioselectivity is due to the presence of a chiral pocket above the vanadium centre. There are various other ways to construct bowl-like shapes of vanadium complexes such that inclusion of V=O is in the deep interior of a confined arrangement. One such example with an interconnected external rim is the complex **3.5b**. Another bowl-shaped complex **3.5c** has the binol components at the exterior part of the bowl that are not connected to one another. In this case, the binol units are the barriers to provide cavitylike spatial arrangements (Figure 3.5).

Cyclophane terms refer to compounds that have one or more aromatic groups linked together to form a cyclic basal unit and such basal units connected to

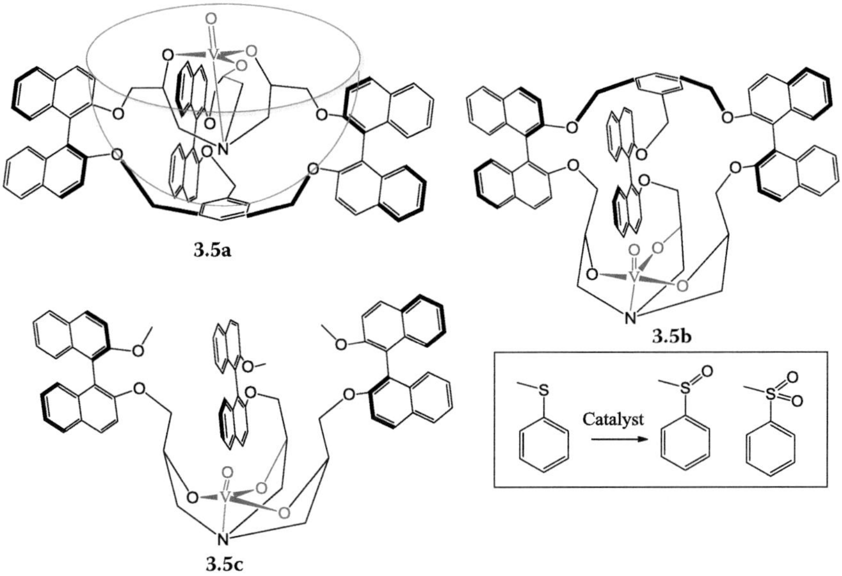

FIGURE 3.5 Vanadium complexes adopting (**3.5a**) bowl-shaped structure with closed outside periphery; (**3.5b**) inclusion of V=O in the interior of the cavity; (**3.5c**) a bowl-shaped complex formed by barriers. (From D. Zhang et al. *Chem. Sci.* 2017, 8, 789–794.[4])

another unit bridging at two nonadjacent positions. Cyclophane molecules provide confinement to a substrate or ions. Hence, they are able to cause many catalytic reactions. Thiamine diphosphate is a coenzyme that assists keto-transfer reactions, nonoxidative decarboxylations of β-keto acids and oxidative decarboxylations of α-keto acids. The active site of the coenzyme **3.6b** has a six-member heterocyclic ring with an amine as well as imine unit; the amine group is stabilised by hydrogen bonding with a glutamate residue of the protein. This heterocyclic ring has the propensity to aromatise by losing a proton (Figure 3.6).

Aromatic esters are formed from aldehydes under mild conditions by the enzyme in a combined approach of encapsulation and chemical activation followed by electrochemical reaction. Covalently linked flavin causes enhancement of the reaction rate, and it is not capable of causing benzoin condensation alone and further electrochemical process is required for the catalytic reaction. The reaction mechanism involves encapsulation of aldehyde followed by C—C bond formation. The C—C-bonded compound is oxidised and followed by an attack of methanol that provides the ether. During the course of the reaction, the flavin is reduced. It is oxidised electrochemically for functioning of the catalytic cycles. A similar catalysis is also performed by a cyclophane derivative combined with flavone **3.7a**. The catalytic reaction sequences are shown in Figure 3.7. The cyclophane part of thiazolio-cyclophane **3.7a** binds to aromatic aldehydes in basic medium to form the product **3.7b**. The enol form of the intermediate is oxidised with the aid of the flavin part to form the product. The product **3.7c** is formed by a hydride transfer. The adduct **3.7c** reacts with methanol to form methyl 2-naphthoate and the compound

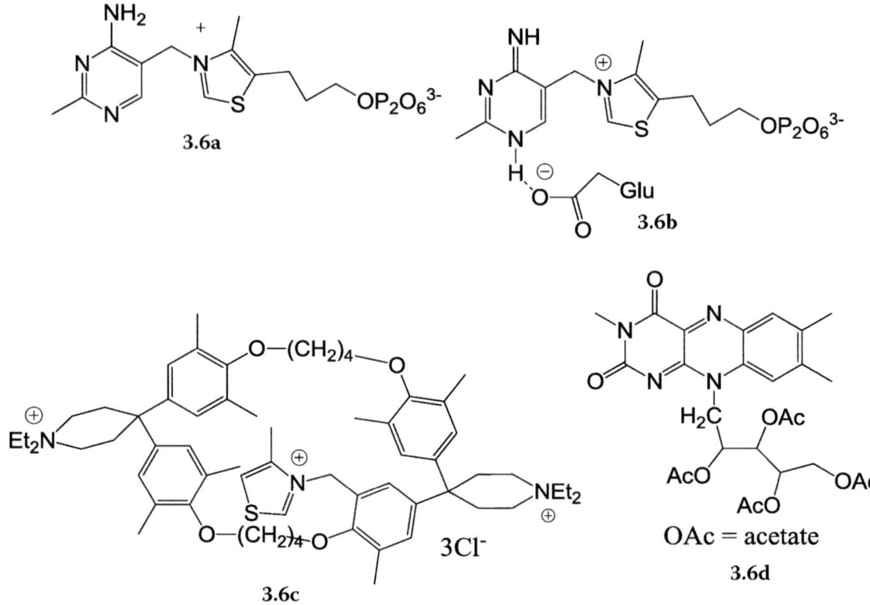

FIGURE 3.6 The biological active site of thiamine diphosphate (**3.6a**); environment of imine form involved in the redox process of the enzyme (**3.6b**), and a cyclophane to encapsulate aromatic aldehyde (**3.6c**) and a flavin for electron transfer. (From S.-W. Tam-Chang et al. *Hel. Chim. Acta* 1993, 76, 2616–2639.[5])

3.7d. This compound is the reduced form of the cyclophane derivative. The reduced form, on subsequent oxidation by electrochemical means, reverts to the original cyclophane **3.7a**. This oxidation involves the transfer of a hydride equivalent to the added flavin derivative. During the course of the recation an acylthiazolium species is formed, which reacts with methanol to form a 2-naphthyl methyl ester. The original cyclophane is regenerated electrochemically when the dihydroflavin is reoxidised. A combination of cyclophane without covalently linked flavin has poor catalytic performance. This is due to naphthaldehyde competing with flavin to get encapsulated inside the cavity of the cyclophane. In such a situation, reactants are not readily accessible for intermolecular oxidation by an external flavin molecule. To overcome this, flavin moiety is covalently linked in a compound **3.7a**.

The folded structures found in nature are very important as applied materials. Among them, proteins are well known to adopt folded structures. The helical structures of DNA are of general interest in gene-expressions and various other activities as biomaterials. Folded compounds provide avenues to expose some active sites outwards to bind substrates. Certain ligands with inherent helicity bind to metal ions to adopt folded structures. Among those compounds, certain ones also possess active sites for asymmetric catalysis. On the other hand, a catalyst is modified to react with higher efficiency by using foldamer host. One such example is the oxidation of methylphenyl-methanol to acetophenone by an N-oxide radical (Figure 3.8) in the presence of a foldamer. The reaction is catalysed by the foldamer in conjunction

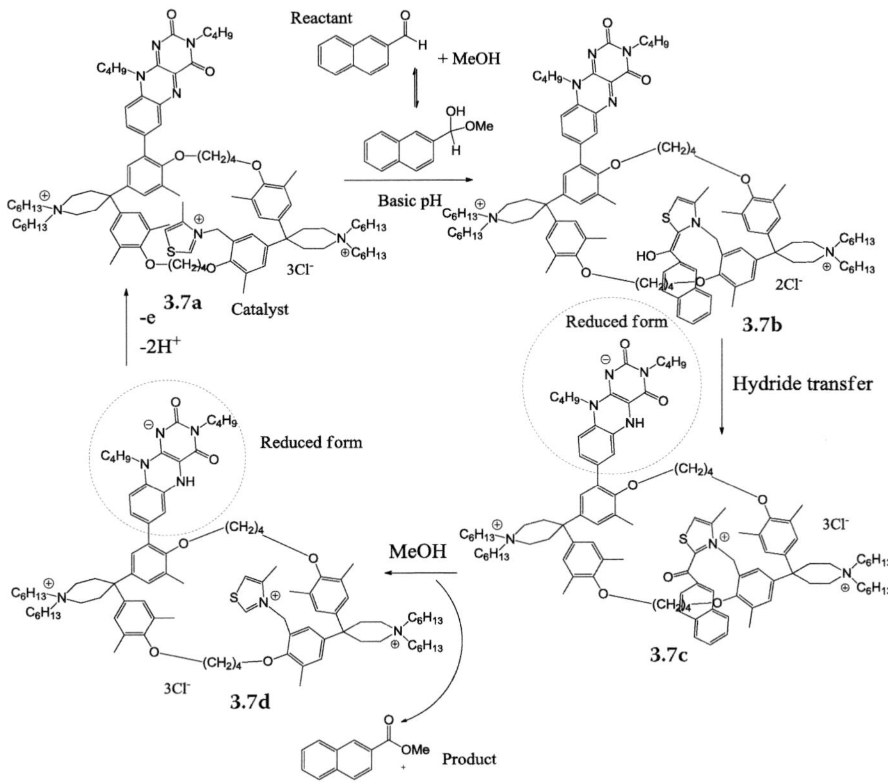

FIGURE 3.7 Mechanism of catalytic ester formation from aldehyde by cyclophane derivative. (From S.-W. Tam-Chang et al. *Hel. Chim. Acta* 1993, 76, 2616–2639.[5])

with externally added N-oxide or a covalently linked N-oxide on a foldamer. In these reactions, the parts of the foldamer linked to the reagent through covalent links provide higher efficiency and sensitivity. The catalytic activity of the compound to oxidize methylphenyl-methanol to acetophenone is enhanced by using a catalyst **3.8c** over a catalyst **3.8b** in combination with a compound **3.8a**. This happens due to the availability of the N-oxide radical **3.8c** on the host that binds the substrate and causes the reaction on the same template. But the use of a combination of the compounds **3.8a** and **3.8b** has a relatively lower probability of the reactants coming together to the host.

3.1.1 Cages and Containers for Supramolecular Catalysis

The basic utility of containers in supramolecular catalysis is to accommodate reactants with or without an additional catalyst to cause a catalytic reaction. Thus, the prerequisites are the specific ability to bind to substrate, sizes and shapes and activation mechanisms to cause catalytic reactions.

Noncovalently linked self-assembled coordination cages and organic capsules (Figure 3.9) have advantages in catalysis over conventional organic catalysts. They are conventionally prepared by a bottom-up approach and are organised in different

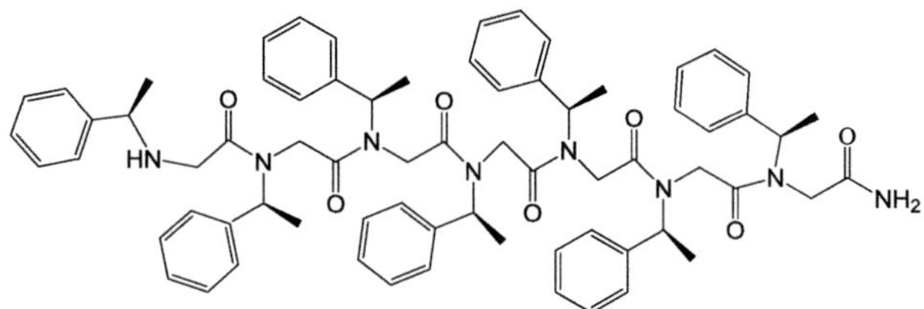

3.8a

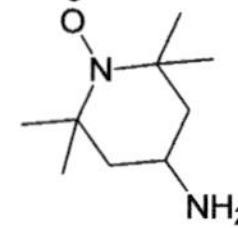

3.8b

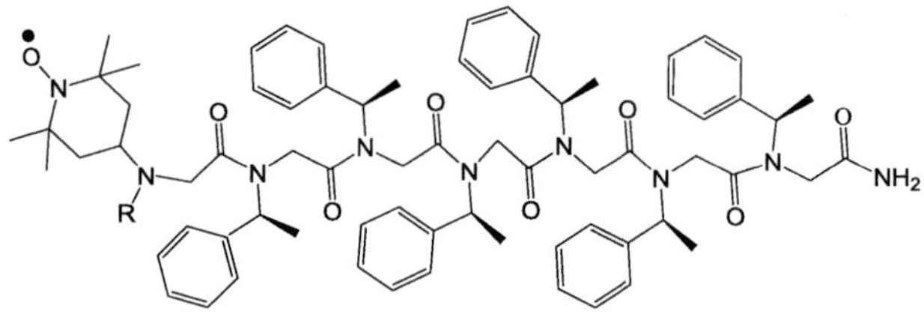

3.8c

FIGURE 3.8 A foldamer of oligopeptide, a nitroxide radical and a foldamer attached to nitroxide radical. (From G. Maayan et al. *Proc. Nat. Acad. Sci. (USA)* 2009, 106, 13679–13684.[6])

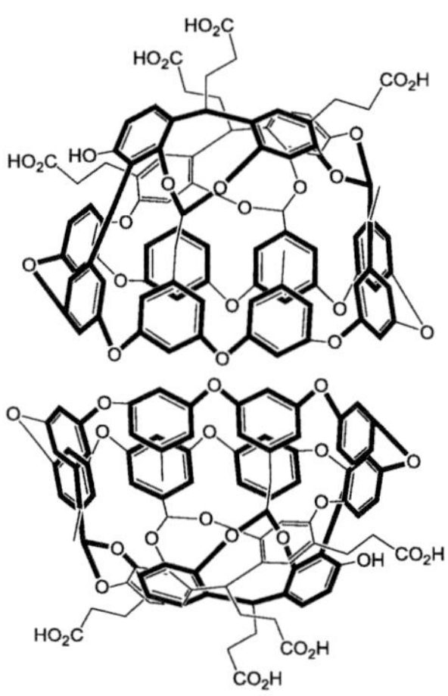

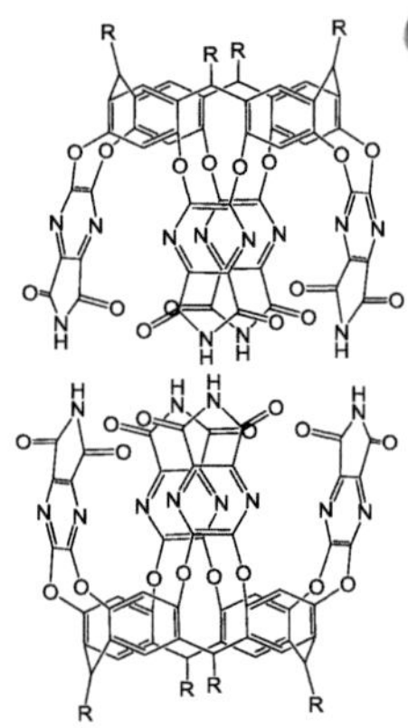

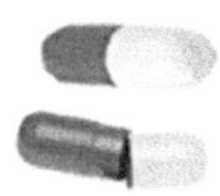

3.9d

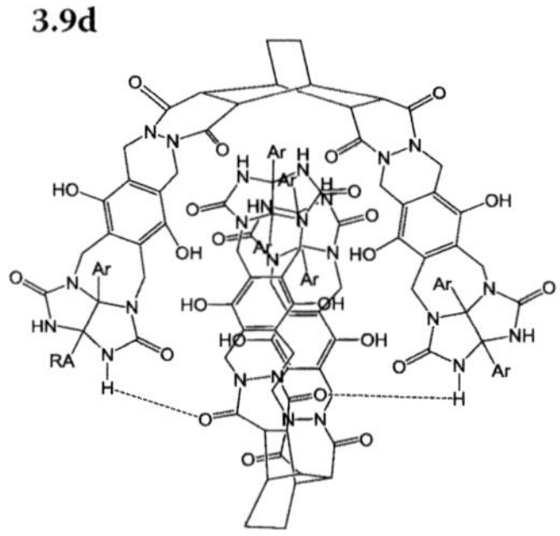

3.9a **3.9b** **3.9c**

FIGURE 3.9 Some capsulelike assemblies.

ways but in modular fashions. Some of them are developed from combinations of multiple units of smaller building blocks. There are large numbers of host molecules with substantial amounts of flexibility in their structures. Among them, some show dynamic properties due to conformational changes. Due to the flexibility of those molecules, they have the ability to intake guest molecules in a reversible manner. While forming self-assemblies, they also stabilise different conformations with or without guests. Many catalysts are developed from container molecules with well-defined hydrophobic cavities that resemble the pockets of enzymes.

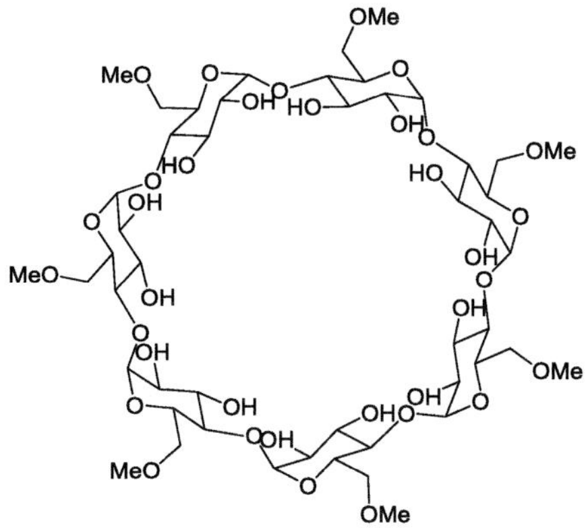

3.10a

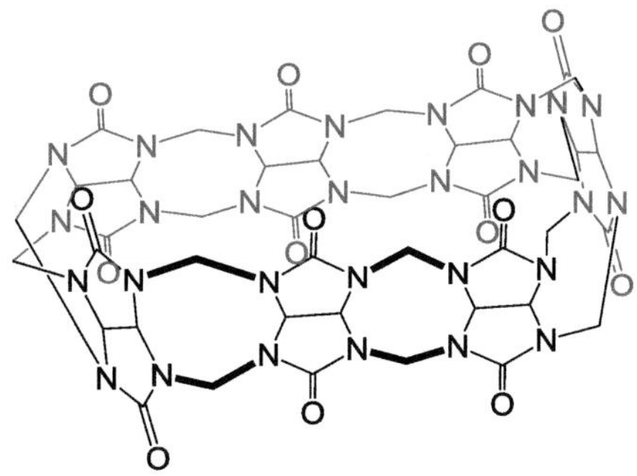

3.10b

FIGURE 3.10 Dimethyl-β-cyclodextrin and cucurbit[7]uril.

As the hydrophilicity or hydrophobicity of the cavities varies, so the catalytic abilities of different container molecules vary. The ability of container molecules to stabilise transition states of catalytic reactions differs. Cucurbit[7]uril (CB7) and β-cyclodextrin (β-CD) have cyclic structures with wide differences in the building units, but each has a cavity. β-cyclodextrin has one side decorated with primary hydroxyl groups and the other with secondary hydroxyl groups. Methylated cyclodextrin (DM-β-CD) **3.10a** is cyclodextrin derivative with the primary hydroxyl groups methylated (Figure 3.10). DM-β-CD has 20 times higher solubility in water than β-CD. The dissolution of this derivative does not change the properties of the cavity. On the other hand, the two sides of the cucurbit[7]uril **3.10b** are symmetric with two identical openings. Both these molecules have similar cavities and abilities to include aromatic guest molecules. The solvolysis reactions of different substituted benzoyl chloride to the corresponding ester or acid are different in the presence of cucurbit[7]uril (CB7) and methylated cyclodextrin (DM-β-CD). The difference in reactivity is due to the mechanism of the solvolytic reactions. The reactions involving an associative mechanism are catalysed by DM-β-CD, but such reactions are inhibited by CB7. The CB7 cavity stabilises the transition state involving acylium ion by electrostatic interactions; hence, the substituted benzoyl chlorides that pass through a dissociative mechanism are catalysed by CB7. The solvolysis reactions of benzoyl chloride derivatives with electron-donating substituent pass through a dissociative mechanism. Those reactions following the dissociative path are catalysed by CB7 but inhibited by DM-β-CD. In the case of cyclodextrin, the hydroxyl group participates as a nucleophile, so the catalysis of cyclodextrin to cause solvolysis involves an associative mechanism. Due to this, those substrates undergoing solvolysis through dissociative processes are inhibited by methylated cyclodextrin.

3.1.2 Noncovalently Linked Cages for Supramolecular Catalysis

Many reactions are catalysed by cages formed by self-assembly of supramolecular hosts serving as containers. For example, six molecules of resorcinarene **3.11a**,

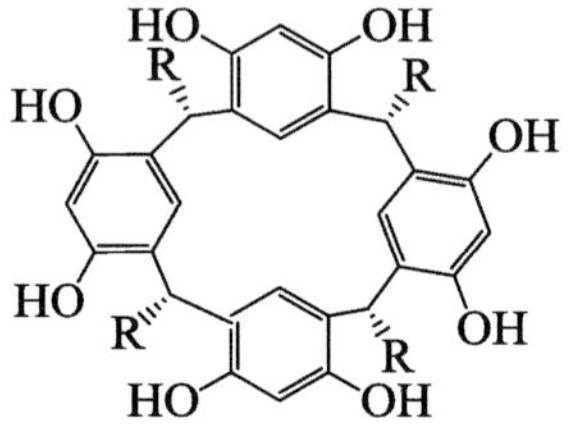

3.11a

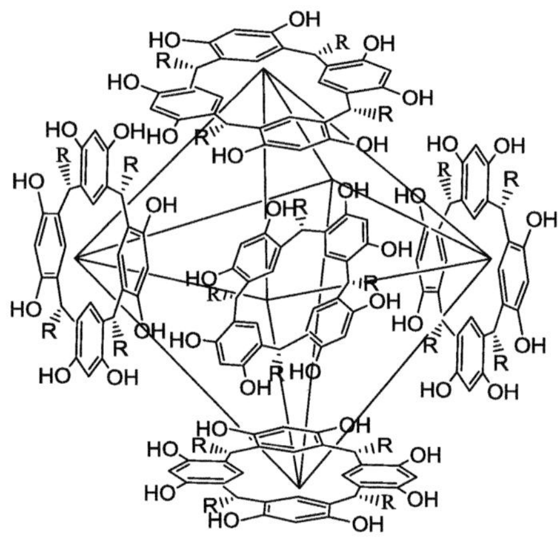

3.11b

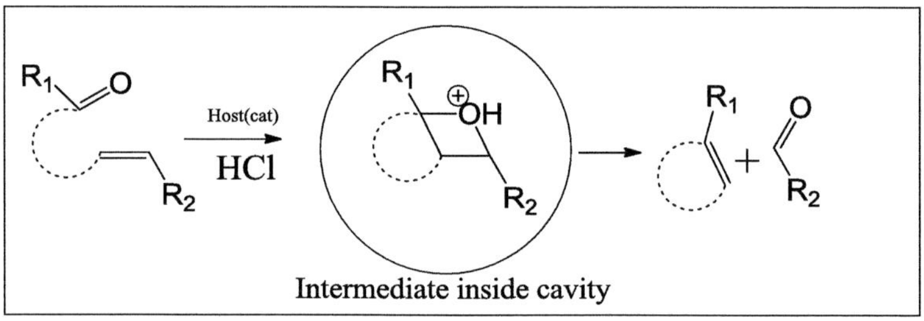

FIGURE 3.11 Resorcinarene-catalysed carbonyl-olefin metathesis reaction. (From L. Catti, K. Tiefenbacher, *Angew. Chem. Int. Ed.* 2018, 57, 1–5.[7])

together with eight molecules of water, self-assemble to form a cage-like structure **3.11b**. The self-assembly catalyses the intramolecular carbonyl—olefin metathesis reaction illustrated in Figure 3.11. Generally, the carbonyl-olefin metathesis reaction proceeds with an organometallic catalyst and these reactions do not take place under the Brønsted acid condition. However, a catalytic amount of the self-assembled cage of resorcinarene in the presence of hydrochloric acid acts as a catalyst. The self-assembly catalyses the reaction by encapsulating acid and reactants. The reactants inside the cavity of the host form a protonated cyclic ether as an intermediate. This intermediate breaks down as an aldehyde and an olefin.

3.1.3 Cavitands for Supramolecular Catalysis

Cavitands are bowl-shaped molecules with concave surfaces, and they are widely used in supramolecular catalysis. Among concave molecules, resorcinarene provides significant avenues for functionalisation of the concave shape to prepared decorated units on the periphery. Such modification increases spaces to accommodate guest molecules. The groups that provide the requisite folds to form bowl-like structures with adequate rigidity provide the correct geometry. For this purpose, resorcinarene functionalised with amide groups provides cavitands with a deeper cavity (Figure 3.12). Such structures are stabilised by intramolecular hydrogen bonds.

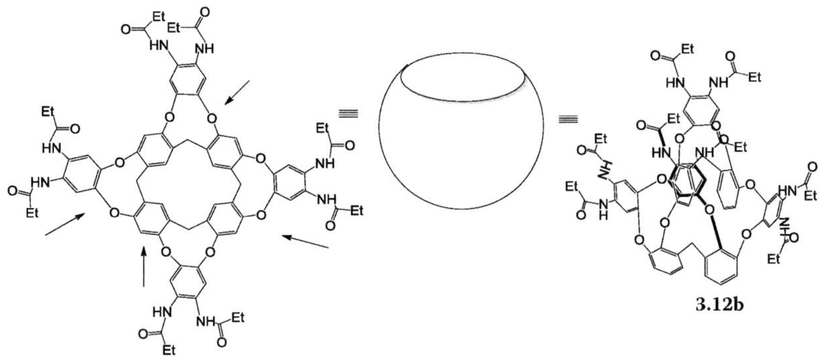

3.12a

3.12b

FIGURE 3.12 A functionalised resorcinarene as an example of a cavitand. (From K. Kobayashi, M. Yamanaka, *Chem. Soc. Rev.* 2015, 44, 449–466.[8])

Bowl-like structures are easily modified as desired and different functional groups are incorporated. For example, the cavitand **3.13a** is converted to the form **3.13b** or **3.13c** by functionalising the diamine units with appropriate functional groups. Certain functional groups connected at the outer periphery of cavitands are useful to hold reactants as guests. For example, an anthracene-carboxylic acid group anchored to a cavitand derived from resorcinarene **3.13b** acts as a site to react with isopropyl-isocyanide. The isopropyl cyanide is protonated and encapsulated into the cavitand to form an intermediate **3.13d**. The protonation caused by the carboxylic acid generates a carboxylate anion which reacts with the protonated isopropyl isocyanide to form an imine **3.13e**. This imine reacts with water to form an amide derivative **3.13f**. Similarly, a cyclohexyl carboxylic acid unit connected to a similar cavitand **3.13c** is effective to cause a rearrangement reaction of cyclohexyldiimide-condensed product with carboxylic acid. In this reaction, the carboxylic acid of **3.13c** reacts with diisopropyl-carbodiimides to form an intermediate **3.13g**. This intermediate rearranges inside the cavity to form the amide derivative **3.13h** (Figure 3.13).

Catalytic reactions such as the deuterium exchange in methyl acrylate and ring expansion of epoxide derivative 4-(3,3-dimethyloxiran-2-yl)-2-methylbutan-2-ol are performed using these cavitands. The compound 4-(3,3-dimethyloxiran-2-yl)-2-methylbutan-2-ol (**3.14a**) possess an epoxide group and a hydroxyl group at suitable positions to form six- or five-membered cyclic ethers (Figure 3.14). These cyclisation reactions are less specific, but when they are catalysed by the cavitand **3.14c**, there is higher selectively to form the five-membered ring containing the compound 2-(5,5-dimethyltetrahydrofuran-2-yl)propan-2-ol as the major product. The 2,2,6,6-tetramethyltetrahydro-2H-pyran-3-ol (**3.14b**) is formed from the hydrogen-bonded epoxide intermediate **3.14d**. This five-membered compound is encapsulated into the void of the cavitand. The transition state **3.14d** shown in Figure 3.14 has preferential selectivity to include the five-membered ring. The carboxylic acid of the host **3.13c** forms a hydrogen bond with the oxygen atom of the epoxide, facilitates the attack of the free hydroxy group at the other end of the compound. The reaction is 50 times faster than a similar reaction without using the cavitand catalyst.

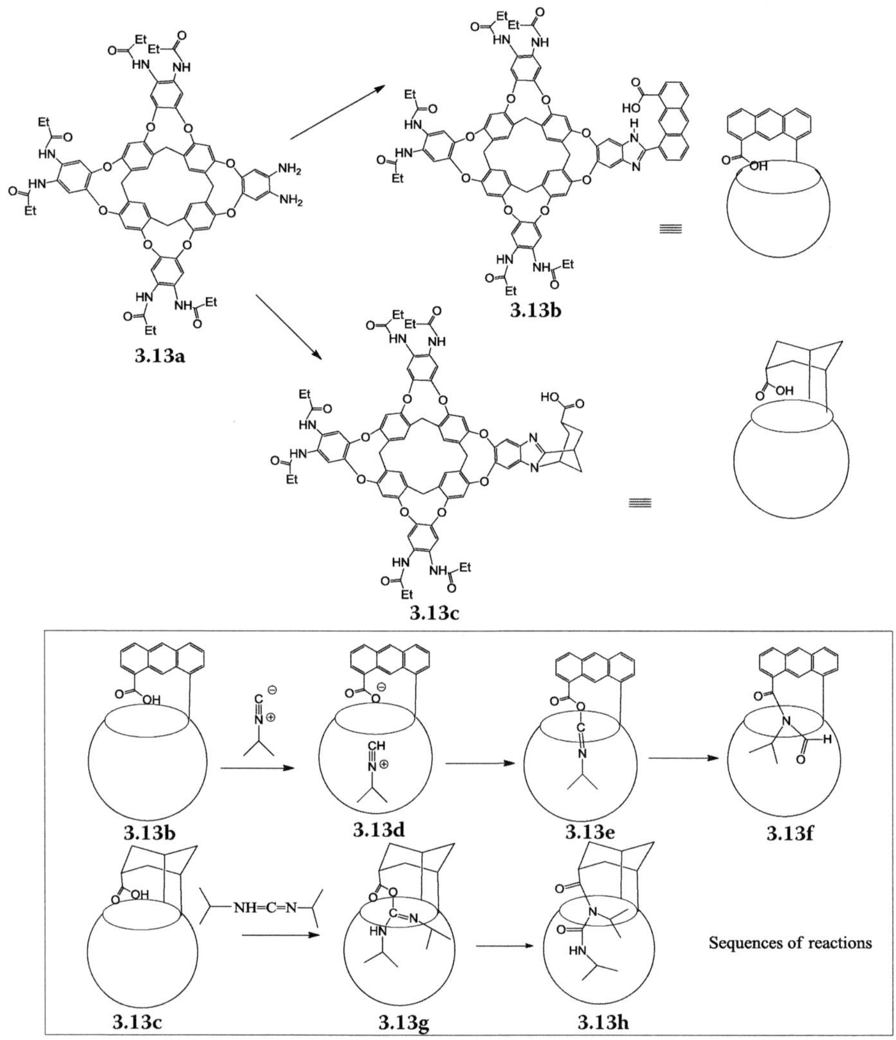

FIGURE 3.13 Cavitand functionalisation through encapsulation of reactant. (From B. W. Purse, J. Rebek Jr., *Proc. Nat. Acad. Sci. (USA)* 2005, 102, 10777–10782.[9])

In another method, a facilitating component, 1,4-diazabicyclo[2.2.2]octane (DABCO), is encapsulated into the cavitand. The encapsulated DABCO interacts with the substrate methylacrylate **3.14e** to form a transient carbon nitrogen bond in the intermediate species **3.14f**. The intermediate is an enol form and exchanges labile deuterium from deuterated solvent, which is acetone-d_6 in the present case. Once the intermediate takes up a deuterium from the deuterated solvent, it reverts back to the keto form. Finally, the loss of a hydrogen atom and charge neutralisation provide the deuterated methylacrylate and DABCO-encapsulated cavitand **3.14h**). The recovered cavitand **3.14h** once again participates in the reaction to act as a catalyst.

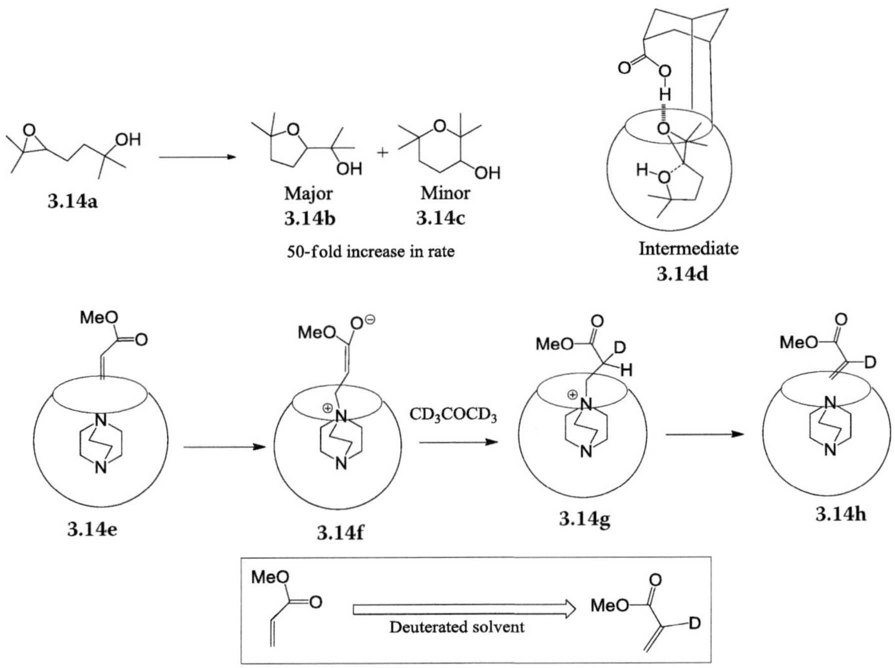

FIGURE 3.14 Catalytic reactions of cavitand based on resorcinarene. (From S. R. Shenoy et al. *J. Am. Chem. Soc.* 2008, 130, 5658–5659.[10])

Certain metal complexes also adopt a bowl shape, and among them, specific ones help in catalysis by accommodating reactants. The palladium complex (**3.15a**) with composition Pd_6L_4 (**L** = triazine cored terpyridine complex) has a bowl-like architecture. The bowl-like complex has *cis* end-capped ethylenediamine ligands bound to palladium (II) ions and triazine-cored tridentate ligands. The triazine-cored terpyridine ligand has a bend angle of 112° which makes it possible to connect three palladium ions on one face of the polyhedron. The dimension of the bowl is about 3 nm. The aromatic rings surrounding the cavity provide hydrophobicity to the cavity, and the charge of the palladium ion present on the outside surface has hydrophilicity. The bowl assembles as a dimeric capsule in the solid state. This complex is a catalyst for 1,3-dipolar cycloaddition reactions between anthracene derivatives with unsaturated imides (Figure 3.15) to form a cycloadded product **3.15h**.

3.1.4 Foldamers for Supramolecular Catalysis

Foldamers are folded structures of oligomers with specific compact conformations. The structure of foldamers varies with conditions and is guided by weak intramolecular interactions. The folded and stretched forms of such compounds may be compared to a millipede. While in motion, a millipede is in the open form; a millipede goes to a protective mode and adopts a folded form (Figure 3.16). Folded molecules with active sites in reorganised and pre-organised states are useful for catalytic reactions.

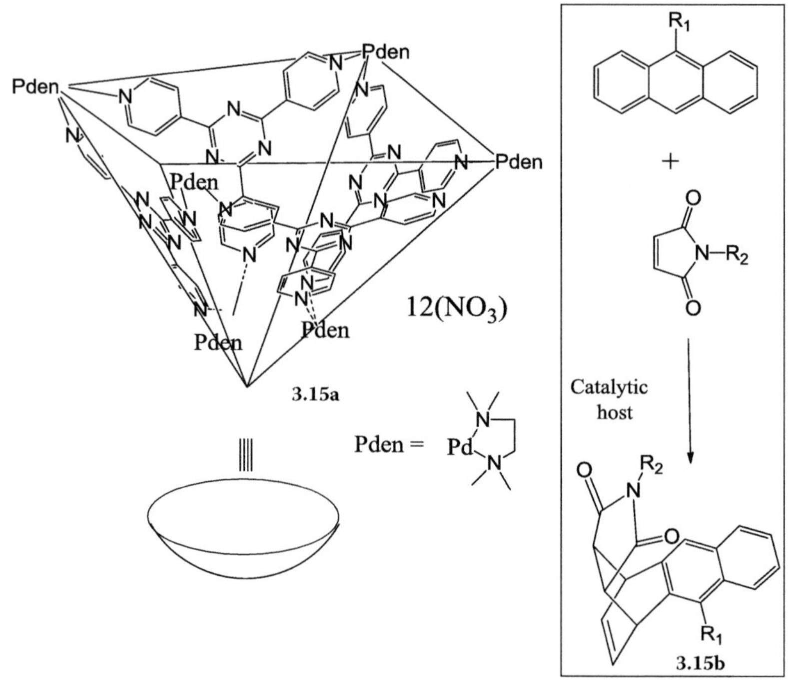

FIGURE 3.15 Cavitand of palladium complex catalysing 1,3-dipolar addition reaction. (From S.-Y. Yu et al. *J. Am. Chem. Soc.* 2000, 122, 2665–2666.[11])

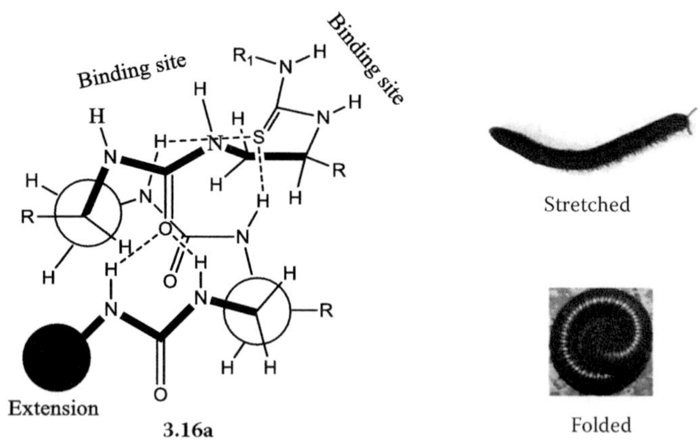

FIGURE 3.16 An urea derivative as foldamer (left) and a millipede while moving and in protective form (right). (From D. Bécart et al. *J. Am. Chem. Soc.* 2017, 139, 12524–12532.[12])

For example, the poly-urea derivative **3.17a** has a folded structure. This urea derivative with flexible tethers has a thiourea unit attached at one end. The folded structure has two urea sites exposed outwards from the folds, which are readily available to bind substrates to activate for catalytic reactions. Based on such a model, catalysts are

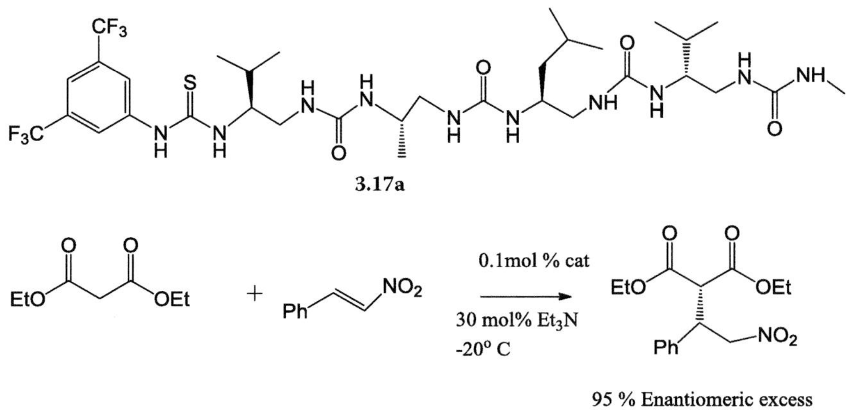

FIGURE 3.17 A thiourea derivative as a catalyst for asymmetric C—C bond forming reaction. (From D. Bécart et al. *J. Am. Chem. Soc.* 2017, 139, 12524–12532.[12])

developed. The foldamer **3.17a** is used for enantioselective carbon—carbon bond formation. The reaction between 1-((E)-2-nitrovinyl)benzene and diethylmalonate is catalysed by a peptide to form diethyl 2-((R)-2-nitro-1-phenylethyl)malonate in 95% enantionmeric excess (Figure 3.17).

The foldamer **3.18a** is a bifunctional catalyst. The reaction between 1-((E)-2-nitrovinyl)benzene and dimethymalonate is catalysed by the foldamer **3.18a** (Figure 3.18). The thiourea unit of the foldamer **3.18a** hydrogen-bonds to the nitro

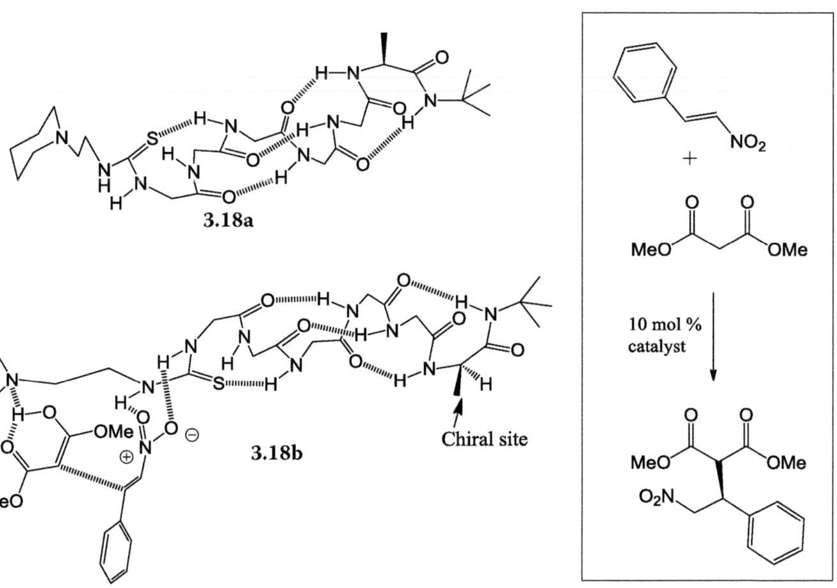

FIGURE 3.18 A bifunctional catalyst of foldamer. (From D. Bécart et al. *J. Am. Chem. Soc.* 2017, 139, 12524–12532.[12])

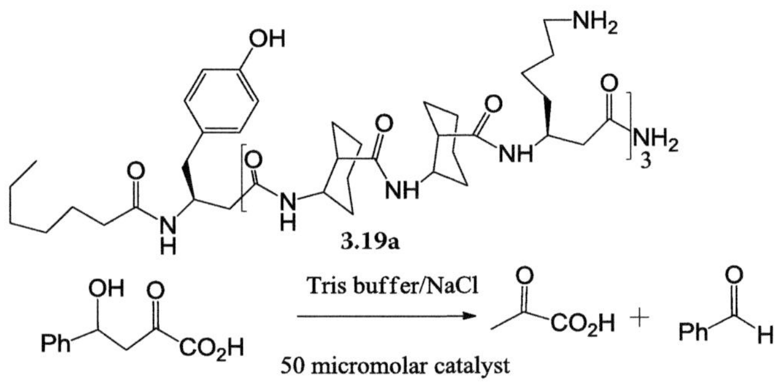

FIGURE 3.19 A foldamer catalysed retro-aldol reaction. (From M. M. Miller et al., *Angew. Chem. Int. Ed.* 2009, 48, 922–925.[13])

group of the nitro olefin, and the enol form of the dimethylmalonate hydrogen bonds to the amine unit of the foldamer; hence, it acts as a bifunctional catalyst. The transition state responsible for the reaction is **3.18b**. In this particular example, beside the catalytic site, there is a chiral centre at a remote site from the active sites of the catalyst. This site has a directing effect on the nucleophile for chiral induction. The catalytic C—C bond formation reaction is highly enantioselective and yields 85% products with 82% of S isomer and 18% of R isomer.

The oligopeptide **3.19a** has cyclohexyl units bridging the peptide bonds. The oligopeptide has a tyrosine residue and there are (-CH_2)$_4$-NH_2 as side chain functionalities. Due to intramolecular hydrogen bonds, it adopts a folded structure and the side-chain functionalities help in catalyzing the retro-aldol reaction (Figure 3.19). The retro-aldol reaction of 4-hydroxy-2-oxo-4-phenylbutanoic acid-catalysed by the oligopeptide in **3.19a** is 3,000 times faster than the uncatalysed reaction. The catalytic reaction yields ketocarboxylic acid and benzaldehyde.

The optically active tetra-peptide **3.20c** has a β-hairpinlike structure. This compound is an excellent catalyst to convert 3-arylquinazolin-4(3H)-ones to tribrominated compound. The brominated compound, on methylation, yields 3-(2,4,6-tribromo-3-methoxyphenyl)-2-methylquinazolin-4(3H)-one (**3.20b**) at a 99% yield (Figure 3.20). From a series of similar derivatives of the peptide constructed with functional variations, the chosen catalyst shows higher efficiency. The hairpinlike structure stabilises the atropisomer of the reactant with a suitable geometry; this provides the extra efficiency to the catalyst.

3.2 INORGANIC CONTAINER MOLECULES

The coordination ability of a metal ion depends on the nature of the central metal ion. On the other hand, stability of a metal complex is guided by the lability of ligands. The lability depends on various factors. Various types of closed or open architectures of metal complexes are possible with different metal ions. Basically, metal ions, when placed in the vertices of different polyhedrons with the coordination sites occupied by suitable ligands, generate cages. Certain ligands with metal ions forms secondary building blocks called "connectors," acting as spacers to link the metal ions and as

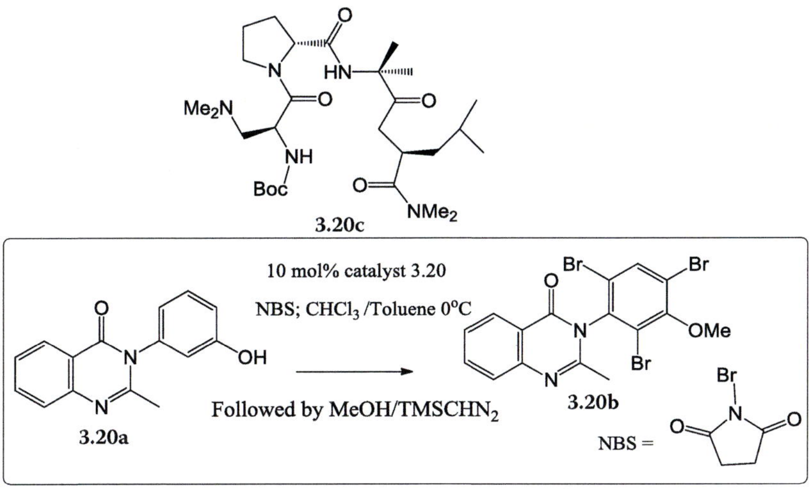

FIGURE 3.20 Atrop selective bromination by a foldamer catalyst. (From A. J. Metrano et al. *Chem. Commun.* 2016, 52, 4816–4819.[14])

bridging units to connect the metal ions, often called "linkers." There are definite geometrical approaches to design closed polyhedral structures of metal complexes. Some complexes adopt cage-like structures with voids occupied by cationic or anionic species. For example, the dianion of the ligand **3.21a** forms a self-assembly with the neutral ligand in the complex **3.21b**. The anionic self-assembly encapsulates *tris*-phenanthroline iron (II) cation; the complex has the structure shown in Figure 3.21. The complex has a reversible redox Fe(II)/Fe(III) couple at $E_{1/2} = 1.08V$ ($i_p/i_c = 1.0$ and $\Delta Ep = 0.062V$). The complex, in conjunction with tertiary butyl hydroperoxide, is a good catalyst for epoxidation reactions of olefin compounds.

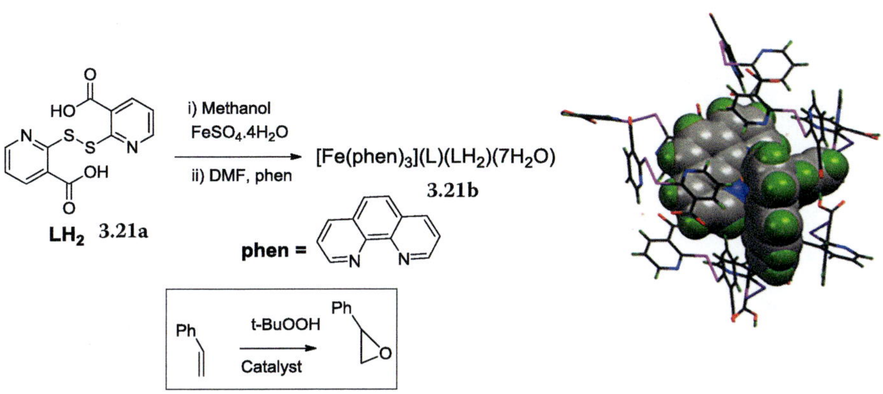

FIGURE 3.21 Encapsulation of tris-1,10-phenanthroline in the assembly of anions. (From W. M. Singh et al. *J. Chem. Cryst.* 2012, 42, 775–782.[15])

3.2.1 Polydentate Ligands and Metal Coordination

Ligands containing two or more ligating sites are called polydentate ligands. Some polydentate ligands with nitrogen and oxygen as ligating atoms are shown in Figure 3.22. The denticity, flexibility and stereochemical aspects and the length of the intervening groups between the two connecting atoms provide the characteristic geometrical features of the ligands to bind to form a metal complex. However, the coordination behaviours of the central metal have the decisive role in providing the overall structural

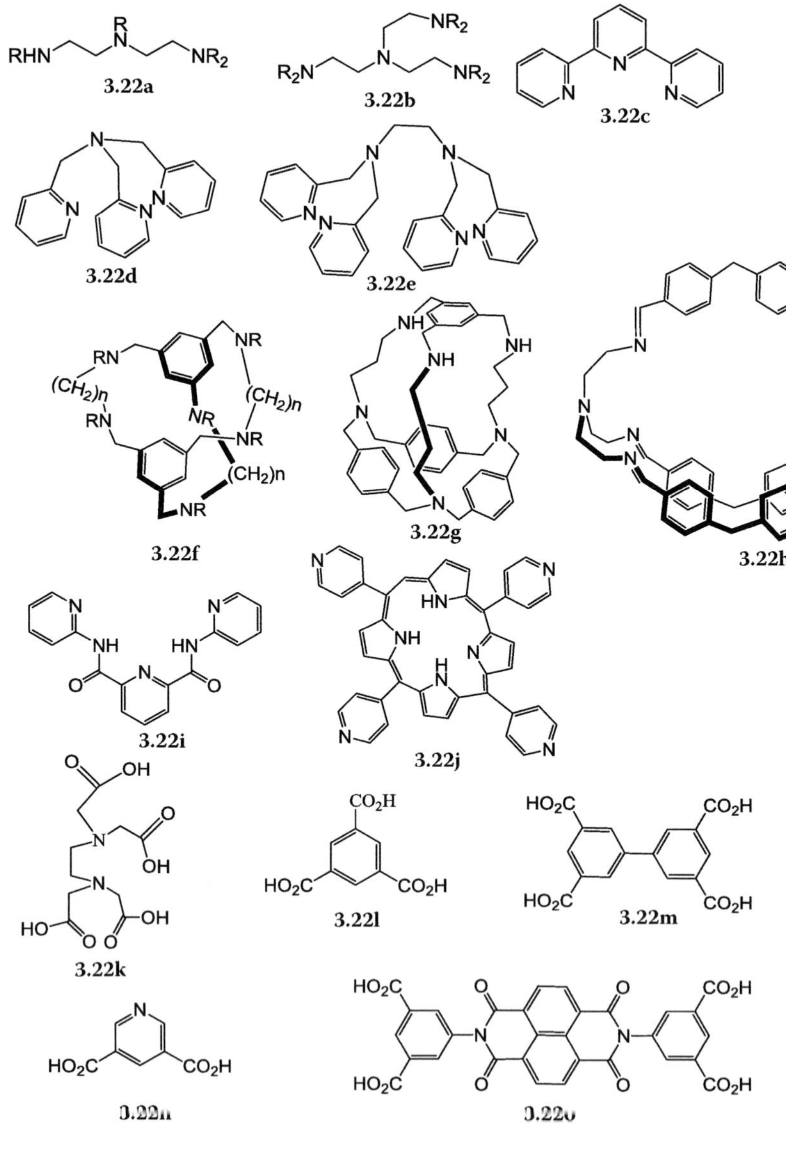

FIGURE 3.22 Some common examples of polydentate N and O donor ligands.

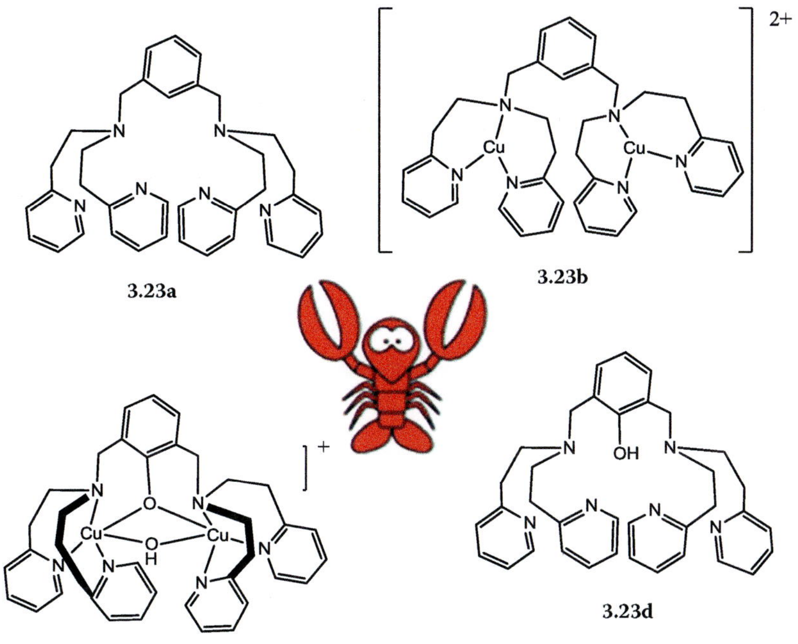

FIGURE 3.23 Functionalisation of a polydentate ligand. (From K. D. Karlin et al. *J. Am. Chem. Soc.* 1984, 106, 2121–2128.[16])

frame of the metal complex. In the context of supramolecular catalysis, the ligands have a major role to activate a substrate. Certain ligands have supramolecular sites available to noncovalently anchor guest molecules. Cyclic or bicyclic ligands are extensively used in supramolecular chemistry. Macrocycles, corands, podands and cryptands are different classes of supramolecular ligands that find extensive uses in supramolecular chemistry. Polydentate ligands increase the complexity of ligating, but there are large numbers of mononuclear complexes of polydentate ligands. Ligands such as those **3.22f** and **3.22g** adopt a cage-like structure; those **3.22h** and **3.22j** adopt a bowl-like structure and are very selective in binding a substrate or forming a complex.

Certain complexes of polydentate ligands undergo ligand centred-reactions, or the ligand is functionalised to another polydentate ligand by remaining as a part of a new complex. The poly-dentate ligand **3.23a** forms a dinuclear copper (I) complex **3.23b**. This complex reacts with oxygen and modifies the ligand; accordingly, a new hydroxide and phenoxy bridged complex **3.23c** is formed. Upon removal of the copper ion from the complex, a new ligand **3.23d** with a phenolic unit is obtained (Figure 3.23).

Ethylene diamine tetracaetic acid (EDTA) is a common ligand known for its binding to calcium and magnesium ions. It is often used for estimation of hardness in water. The polydentate ligand EDTA has multiple sites for deprotonation and remains in different forms with different pKa values. The complex with ruthenium has a free carboxylic acid group, and the two nitrogen and four carboxylate groups coordinate to the ruthenium ion along with a water molecule. The complex has $pK_{a1} = 2.4$ and

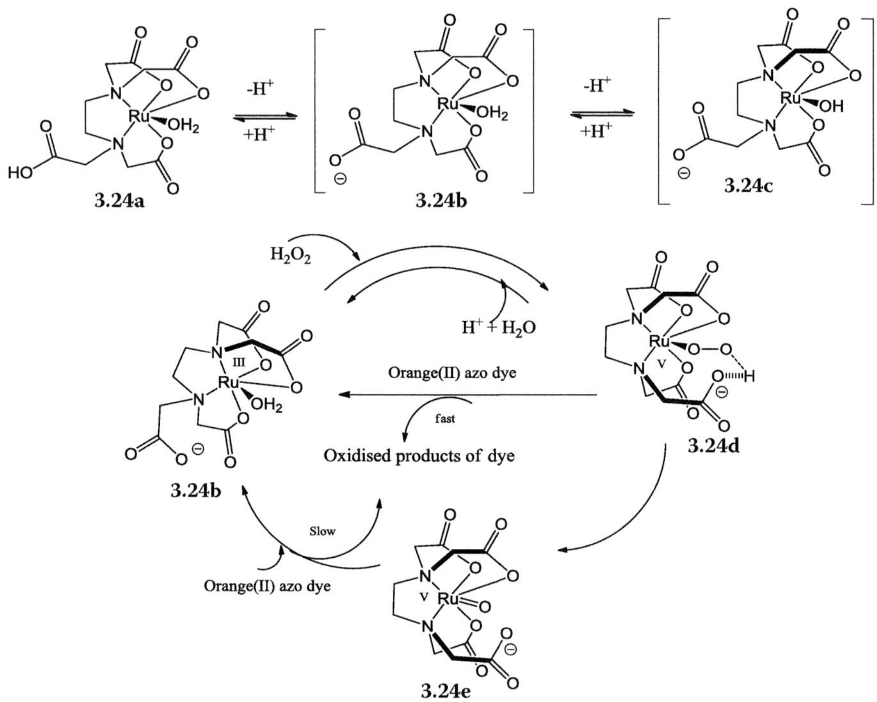

FIGURE 3.24 Ruthenium-ethylenediaminetetracaetic acid complex in catalytic dye decomposition. (From D. Chatterjee et al. *Dalton Trans.* 2011, 40, 10473–10480.[17])

$pK_{a2} = 7.6$; the first pKa is associated with deprotonation of the carboxylic acid, and the second is for deprotonation of the coordinated water molecule.

The ruthenium (III) EDTA complex **3.24a** is a useful catalyst for decomposition of orange(II) azo-dye (Figure 3.24). The complex **3.24a** reacts with hydrogen peroxide to form hydroperoxo complex $[Ru(edta)(OOH)]^{2-}$ complex **3.24d** at pH 1–5. This complex acts as the catalytic species for the degradation of the orange(II) azo dye in the presence of hydrogen peroxide. The hydroperoxo complex **3.24d** has intramolecular hydrogen bonds between the hydroperoxy ligand with the free carboxylate anion of the ethylenediaminetetraacetate ligand. The hydroperoxo ruthenium complex further transforms to the ruthenium (V) oxo-complex **3.24c**. The complex **3.24c** decomposes the dye, but the reactivity is three times slower than the ruthenium-hydroperoxo complex **3.24d**.

The polypeptide-functionalised polydentate ligand **3.25a** is an interesting example where a folded structure of long-chain polypeptide holds the ligating unit. The compound **3.25a** has a folded structure and a terminal end with polydentate ligating sites. The ligating site binds to a zinc(II) ion to form a hexa-coordinated complex **3.25b**. The complex has flexible geometry, and the structure reversibly changes upon deprotonation and subsequent protonation (Figure 3.25).

π-Interactions among aromatic rings provide avenues to design metal complexes with the aromatic part interacting with another π-aromatic part. The palladium

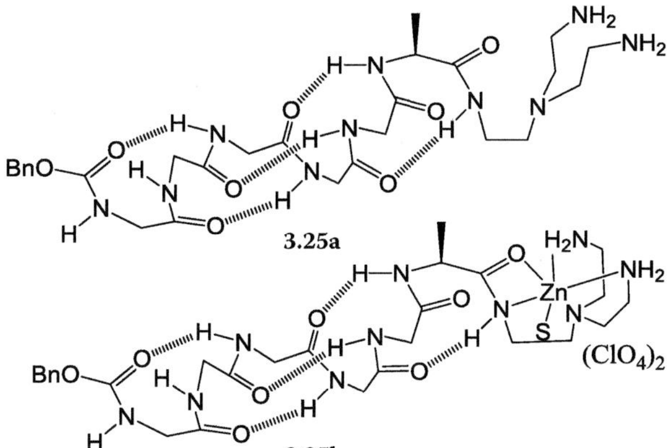

3.25a

3.25b

OBz = benzoate; S = solvent

FIGURE 3.25 A polydentate ligand anchored to a polypeptide and its zinc complex. (From B. A. F. Le Bailly et al. *Angew. Chem. Int. Ed.* 2016, 55, 2132–2136.[18])

complexes **3.26b** and **3.26c** are two complexes with a pyrene unit attached to a ligand. The pyrene ring of the complexes interacts with the π-cloud of graphene oxide. The complexes use graphene oxide as a base through π-interactions to form a supported metal complex. Each complex has individual characteristic catalytic properties. The graphene oxide template complex **3.26b** is a hydrogenating catalyst; it converts stilbene to saturated analogue. The complex **3.26c** is a catalyst for the oxidation of secondary benzylic alcohols (Figure 3.26). There are also examples in which graphene oxide acts as a π-template to hold multiple complexes with π-aromatic rings to form π-π interactions. The two complexes anchored on the graphene oxide surface behave as a combined unit. For example, the complexes **3.27a** and **3.27b** are anchored to graphene oxide, forming the graphene oxide-supported complex **3.27c**. The supported complex has very specific catalytic activity. It causes defluorination of fluoro-aromatic compounds (Figure 3.27).

3.2.2 Inorganic Container Molecules from Mono- and Dinuclear Complexes

The confinement provided by ligands changes the hydrophobic or hydrophilic microenvironment. It has a large influence on the performance of a catalyst. The labiality associated with a metal—ligand bond and the coordination unsaturation provide access to exchange ligands. The exchanges of ligands in solution contribute to catalytic reactivity. However, it is important that the ligand fields of different ligands decide the electronic configuration of central metal ions and control reactivity. The *trans*-effect and back-bonding properties of a ligand control the reactivity of a complex by influencing the strength of the metal—ligand bond located at another coordination site within the same metal complex. Hence, the performance of a metal complex in catalytic activity depends on the surrounding environment of the ligand. If a metalcentric reaction is focused, the metal site should have access to a reactant.

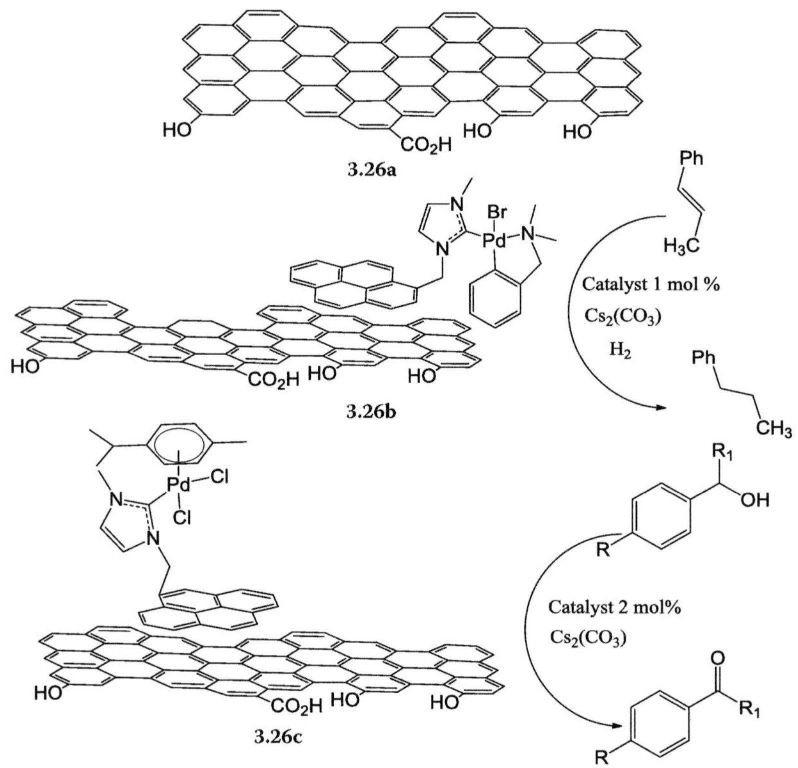

FIGURE 3.26 Graphene-oxide surface with π-interactions with π-cloud of ligands and their catalytic reactivity.

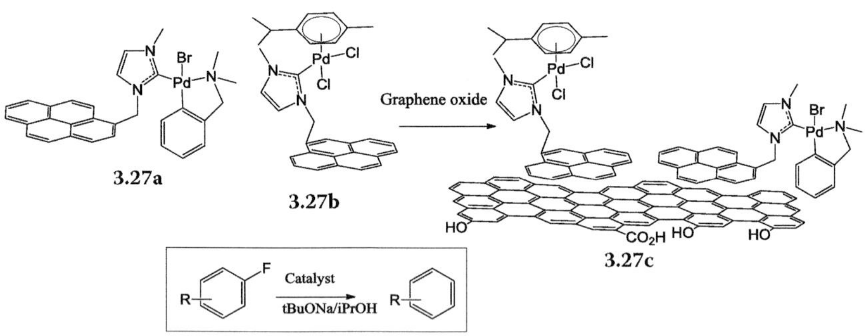

FIGURE 3.27 π-interactions of two complexes with graphene-oxide and catalytic defluorination reaction. (From S. Sabater et al. *ACS Catal.* 2014, 4, 2038–2047;[19] J. Mata et al., Organometallics 2015, 34, 1186–1190.[20])

However, there are exceptions of metal complexes where the reactions may not require direct interaction of substrate with the metal ion. Energy and electron transfer occurring through mediators, space or a part of the ligands perform redox processes. Most importantly, there are ligands that do not directly participate in a catalytic reaction but dictate the course of the redox reaction by changing the rate and selectivity. Those ligands are known as spectator ligands. In the chemistry of mononuclear complexes, spectator ligands contribute significantly to product selectivity.

Cyclophane compounds form cage-like structures. Various ligands are designed by putting appropriate ligating groups in a cyclophane. Some metal complexes with those ligands find utility in supramolecular catalysis. For example, the complex **3.28c** is a palladium complex of a cyclophane ligand **3.28a**. It is a catalyst for polymerization of olefins. The complex **3.28b** is without a cyclic enclosure. This complex has less efficiency than one with an enclosure. The ligand **3.28a** has two nitrogen atoms that coordinate to the palladium form hydride-complex, along with an olefin. The complex has the ability to release olefin more easily than the complex **3.28b**. The higher labiality of olefin in the complex **3.28c** is due to the agostic interactions of hydrogen atoms of a phenyl ring of cyclophane, as illustrated in Figure 3.28. As compared to acyclic catalysts, the binding ability of ethylene is reduced in the cyclophane complex. H-agostic interactions stabilise a hydride intermediate and facilitate the release of the olefin component attached to the palladium. Due to this, the dissociation of the olefin-terminated polymer is favourable, and from these reactions, low-molecular-weight oligomers are formed rather than high-molecular-weight polymers.

The binuclear palladium complex **3.29a** has a cage-like structure. It is composed of two palladium linked to rigid V-type ligands with the nitrogen atoms as donors

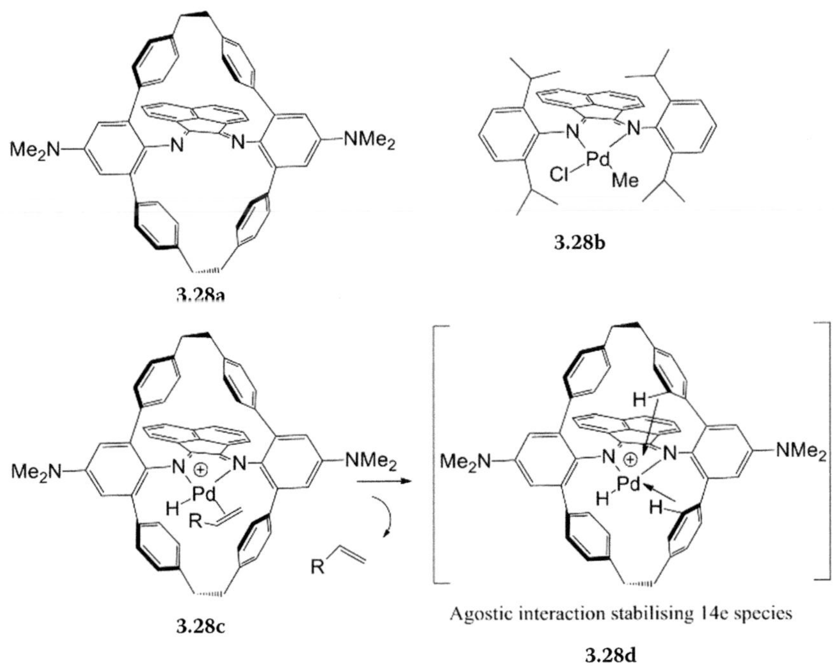

FIGURE 3.28 Confinement provided by cyclophane to palladium complex and agostic interactions (shown by arrows in d). (From C. S. Popeney et al. *Organometallics* 2011, 30, 2432–2452.[21])

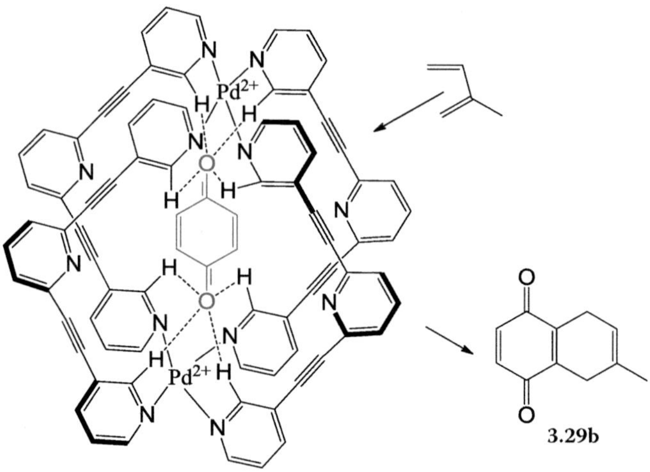

3.29b

3.29a

FIGURE 3.29 A dinuclear palladium complex with cage-like structure encapsulating benzoquinone catalyses a cyclo-addition reaction. (From V. Martí-Centelles et al. *J. Am. Chem. Soc.* 2018, 140, 2862–2868.[22])

at the distal ends to coordinate to the palladium ions. It accommodates a molecule of 1,4-benzoquinone inside the cavity. The guest is anchored by weak C—H···O interactions. The benzoquinone with 2-methyl-1,3-propanediene-included complex undergoes catalytic 4 + 2 cycloaddition reaction (Figure 3.29) to form a cycloadded product **3.29b**. The turnover number of this catalytic reaction is 1,000.

Metal complexes are used as building blocks to prepare gridlike structures to provide a barrier through coordinating to a connecting unit. Such complexes provide definite shapes and sizes to carry out catalytic reactions. For example, the complex **3.30a** is a polyporphyrin complex. The porphyrin rings of this complex bear pyridine rings. Thus, in addition to metal coordination at the central position of porphyrin; each porphyrin unit behaves as a ligand for further modifications. Accordingly, they are connected to $Re(CO)_3Cl$ units to make square grids acting as a surrounding barrier for a central porphyrin unit that has a manganese ion (Figure 3.30). From a catalytic activity point of view, this buildup is analogous to biological systems as a biomimic. The assembly is unique in having multiple metal sites to modulate reactivity. This molecular assembly is used in selective epoxidation reaction. For example, an equimolar mixture of two olefins, (Z)-1,2-diphenylethene and (Z)-1,2-bis(4-(tert-butyl)phenyl)ethane, undergoes a catalytic epoxidation reaction yielding 2,3-diphenyloxirane in higher selectivity over 2,3-bis(4-(tert-butyl)phenyl)oxirane. Formation of 2,3-diphenyloxirane is three and half times higher than that of the other olefin (Figure 3.30).

3.2.3 Molecular Flasks

Molecular flasks are compounds that mediate or catalyse chemical transformations inside their cavities. From an inorganic complex point of view, many polyhedral coordination cages act as hollow containers for guest encapsulation. These molecules

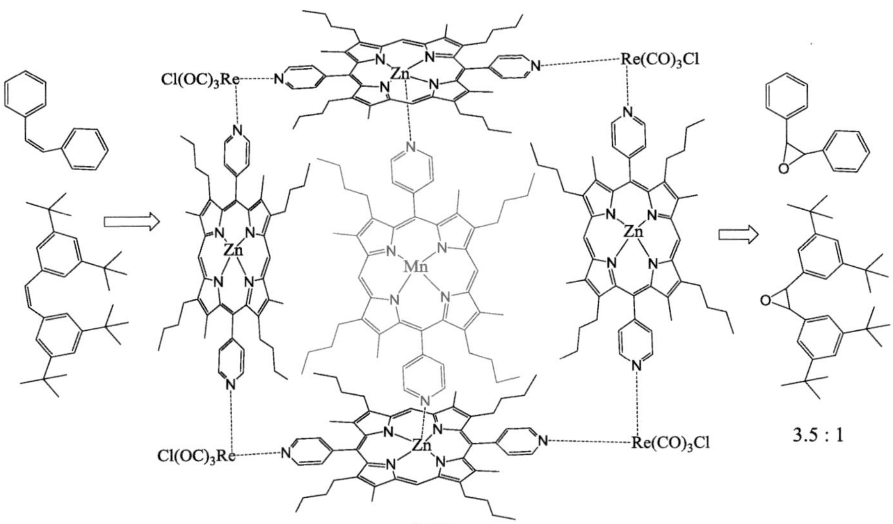

3.30a

FIGURE 3.30 Assembling of porphyrins for selective catalysis. (From M. L. Merlau et al. *Angew. Chem. Int. Ed.* 2001, 40, 4239–4232.[23])

activate, modulate and guide reactivity of substrates for efficient catalytic reactions to form selective products. Since the sizes and shapes of containers are important, there is great interest among synthetic chemists to generate hollow containers from inorganic complexes. In one approach, synthetic chemists modify container molecules, such as in the case of cages in external surfaces, edges or corners, or in the interiors to perform various functional activities and selective guest binding. Alternatively, as per the choices and requirements, ligands are also modified and reacted with metal salts or complexes to form container molecules. Though such syntheses are apparently simple, in actual practice, selective synthesis of container molecules is a challenge. This is because of the necessity of seeking one among several options to form container capable of performing the desired catalytic reaction. The environment in the interiors of the cages, or rather the surroundings of the cavities and their binding selectivity of cage-like complexes, provides an advantage to using them in catalytic reactions. There may be single or multiple types of cavities in cage-like structures, but the beauty is that they are uniformly distributed. Hence, single or multiple types of cavities can differentiate molecules of different sizes and shapes. The process of differentiation enables one to perform size-selective catalysis. Furthermore, certain cavities' shape and size change by solvation, desolvation or application of external stimuli. The changes in cavity sizes and the intrinsic properties of the complex cause differences in binding abilities to substrates in each state. Due to the difference, the reaction rate of a catalytic reaction can be controlled in addition to obtaining a specific product. For example, crystalline molecular flasks are built with trigonal linker 2,4,6-tris(4-pyridyl)-1,3,5-triazine (tpt), coupled with zinc iodide, which shows dynamic behaviour. The framework structure formed by including nitrobenzene has the composition $[(ZnI_2)_3(tpt)_2]\cdot 6C_6H_5NO_2$. The complex loses nitrobenzene molecules easily on heating, but the parent framework is retained. After losing nitrobenzene, the structure shrinks and the original structure is brought back after reuptake of nitrobenzene. The complex also shows flexibility of structure upon exchange of included solvent molecules without crystal degradation. Metal organic frameworks provide ideal examples of crystalline molecular flasks. The catalytic aspects of MOFs are discussed in Chapter 2. In the next section, container molecules or molecular flasks of polyhedral discrete units are discussed.

The circle, triangle, rectangle and so on constitute two-dimensional voids, but in three-dimensional structures, different polyhedral geometries are needed as descriptors of voids. In the rational design of complexes with two- or three-dimensional cavities, the overall shape of the resulting polyhedrons forming cages is important. Only the proper choice of ligands with matching shape and symmetry will provide the desired cage. Hence, cage-like structures concern understanding different polyhedrons. When the focus is on inorganic cages, polyhedral structures are to be constructed by connecting ligand and metal ions in the correct geometry to provide a requisite polyhedron. In general, metal ions or parts of metal complexes are placed on the corners of polyhedrons, which are called connects, and the ligands, called linkers, bridge two or more sites depending on the ligands or geometric constraints of a ligand to form cages. However, one may also imagine many polyhedrons to build by using two-dimensional architectures connected through bridging ligands. The two-dimensional structures act as the faces of such polyhedrons. For example, square faces are present in a cube; triangular faces are present in a prism. For formation of cages, ligands are connected to metal ions in a definite manner, and the corners occupied by metal sites should utilise their coordination to such an extent as to avoid peripheral growth. Peripheral growth leads to extension of the polyhedron, leading to three-dimensional structures to build metal organic framework structures (Figure 3.31).

As mentioned, polyhedrons have one or more type of faces; these may be of triangular, square, pentagonal or hexagonal shape. On the other hand, each polygon has a characteristic turning angle within its angular components. For example, three linear units combined together with 60° turns make a molecular triangle. Similarly, 90°, 108° and 120° are the prerequisite turns to form a square, pentagonal and hexagonal structure, respectively. To design a three-dimensional polyhedral, at least one type of linking component must have more than two coordination sites. The combination of tridentate angular components with linear linking units generates a triangular prism (Figure 3.31). Combining tridentate angular components with bidentate components generates a cube or octahedron **3.31c**. Accordingly, two different kinds of rigid connecting ligands are used as connectors to generate different polyhedral structures in an easily understandable and predictable manner.

3.2.4 Inorganic Cages for Supramolecular Catalysis

Confinement, surface area and active sites are certain points that are enhanced by using multiple active sites. Multinuclear complexes provide a high amount of versatility in guest inclusion and have an extraordinary ability to guide specific chemical reactions. The reactions that are conventionally catalysed by metal catalysts are modified by using cage-like complexes. Such an example of a cage is built by four triazine-based

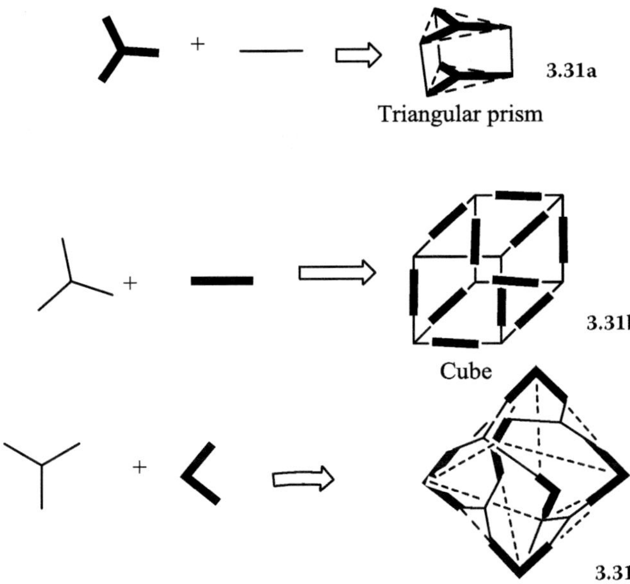

FIGURE 3.31 Design of polyhedral structures by directional properties of ligands. (From P. J. Stang, B. Olenyuk, *Acc. Chem. Res.* 1997, 30, 502–518.[24])

terpyridine ligands (**3.32a**) linked to six palladium ions; in addition to these, each palladium ion has one teteramethyleethylenediamine ligand as chelate. The cage **3.32b** has tetrahedral symmetry and has a large central void (Figure 3.32). Cage-like structures of palladium terpyridine complexes have voids ~2–5 nm diameter to accommodate guest molecules. It catalyses 1,3-dipolar additions of N-cyclohexylmaleimide with anthracene. The interior of the void has the geometrical requirements to provide a favourable space to the reactants to form a selective product. The catalytic reaction by the palladium cage of (anthracen-10-yl)methanol (**3.32c**) with 1-cyclohexyl-1H-pyrrole-2,5-dione (**3.32d**) at 80°C gives 98% of the compound **3.32e**. This yield is double the amount formed with an ordinary palladium catalyst under identical conditions.

The internal structure of a platinum cage formed by the dipyridine ligand **3.33b** has suitable supramolecular features to bind sulphonate and carboxylate group by guanidine units. The palladium cage prepared from the ligand **3.33b** with $[Pt(CH_3CN)_4(OTf)_2$ encapsulates gold complex $Au(PPh_2C_6H_4SO_3)Cl$ as well as pent-4-enoic acid (Figure 3.33). A representative internal structure of a portion of the cage with hydrogen bonding units holding the gold complex is depicted as **3.32c**. The gold complex—encapsulated host further includes the pent-4-enoic acid to form the product **3.32d**. Two guanidine N—H···O hydrogen bonds form with a sulphonate group. On the other hand, a pent-4-enoic acid molecule also forms hydrogen bonds with guanidine within the cavity. The triple bond of the pent-4-enoic acid is coordinated to the gold ion of the included complex. This activation causes facile conversion of the pent-4-enoic acid to 5-methylenedihydrofuran-2(3H)-one (Figure 3.32). Thus, the platinum cage remaining intact within the cage acts as a catalyst for the cyclisation reaction.

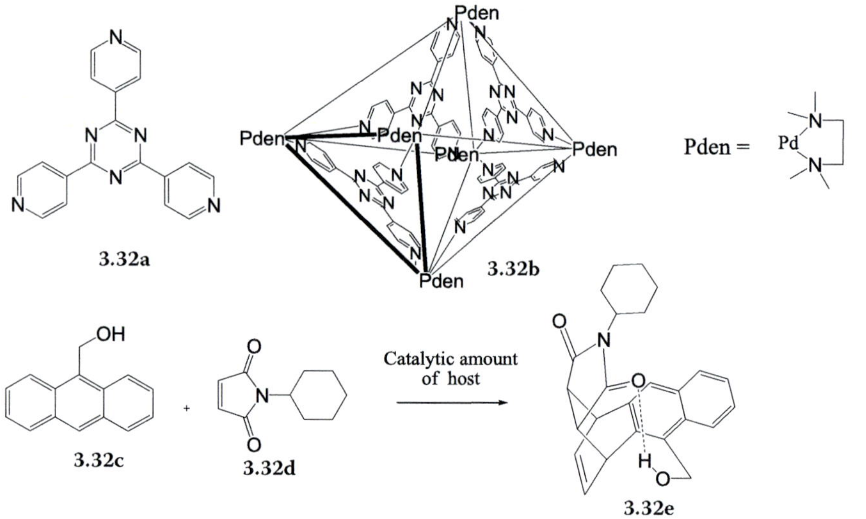

FIGURE 3.32 A palladium cage catalyst for 1,3-dipolar addition reaction. (From M. Fujita et al. *Nature* 1995, 378, 469–471.[25])

There are many examples of tetrahedral cages of polydentate ligands with metal ions. Such cages with appropriate internal spaces are excellent catalysts for specific reactions. For example, a tetranuclear cage is formed by a ligand **3.34a** with gallium ions. The ligand **3.34a** has a naphthalene group connected to two catecoholate units through amide units as tethers. This ligand is suitable for tetrahedral clusters (Figure 3.34). Each ligand connects two gallium (III) ions at two ends by coordinating through catecoholate. Each gallium ion located at the vertices of the tetrahedron is connected by three ligands to provide a composition $[Ga_4L_6]^{12-}$. The interior cavity has a dimension of 450 Å^3. The complex is soluble as well as stable in water.

This gallium cluster in acidic condition reversibly binds guest molecules through several equilibrium steps illustrated in Figure 3.35. At first, the protonated host acidifies a guest, and inclusion of the protonated guest follows. The protonated guest included in the cavity of the host loses the proton to form a neutral guest-included complex. Finally, the released guest is in equilibrium with the encapsulated and free form. The intrinsic acidity associated with the host and guest affects the equilibrium to decide the coexistence of host and guest together or independently in solution.

Acetals, hemiacetals and orthoformates are hydrolysed by the gallium complex **3.34b**. The catalytic process follows the sequence shown in Figure 3.36. In the process, an acetal group is protonated and encapsulated. Inside the host assembly, alcohols are lost to provide the carbonyl compound back. The gallium cage catalyses selective cyclisation reactions of unsaturated carbonyl compound 3,3,7-trimethyloct-6-enal in acidic conditions (Figure 3.37). The 3,3,7-trimethyloct-6-enal is in equilibrium with 3,3,7-trimethyloct-6-ene-1,1-diol. While forming the guest-encapsulated cage **3.37a**, the solvent molecules of the gallium complex **3.37b** are replaced by 3,3,7-trimethyloct-6-enal. In the cage, the alcohol and aldehyde groups of the 3,3,7-trimethyloct-6-enal are close to each other due to a hairpinlike geometry of the **3.37c**. The hairpinlike

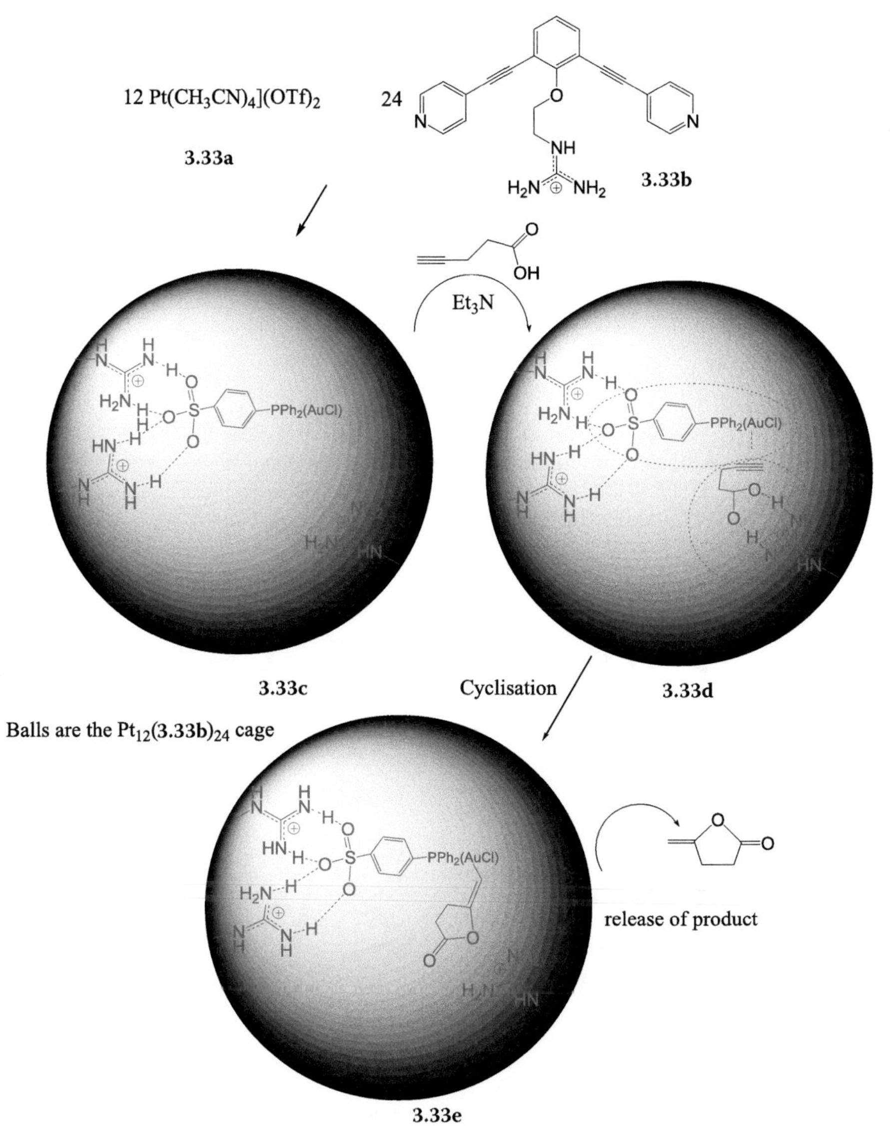

FIGURE 3.33 A platinum cage containing guanidine units encapsulating gold complex as supramolecular catalyst. (From Q.-Q. Wang et al. *Nat. Chem.* 2016, 8, 225–230.[26])

geometry is suitable for a cyclisation reaction. The carbonyl group of the reactant encapsulated in the complex **3.37b** is protonated to form an intermediate **3.37d**. The complex **3.37d** directly forms a product-encapsulated cage **3.37a**, or it transforms to an intermediate species **3.37e** with a carbocation species. The **3.37d** finally transforms to the solvent included form and replaces the cyclised product to form the catalyst **3.37a**.

The gallium cage is a versatile catalyst due to its ability to encapsulate different metal catalysts. It is a host for the iridium catalyst $[Cp^*(PMe_3)Ir(Me)(C_2H_4)]^+$. The

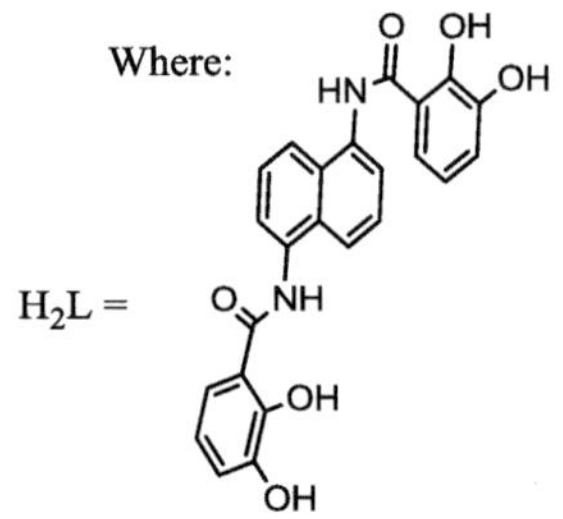

3.34a

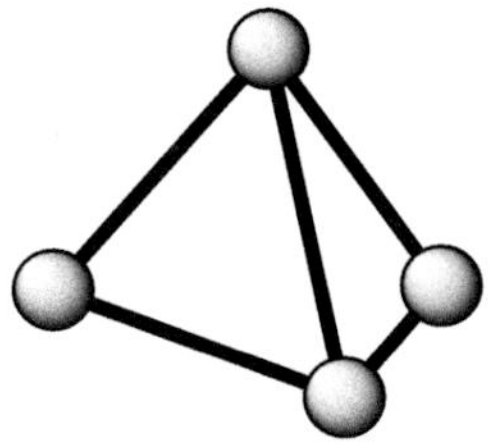

Arrangement of metal ions in $[Ga_4L_6]^{12-}$
3.34b

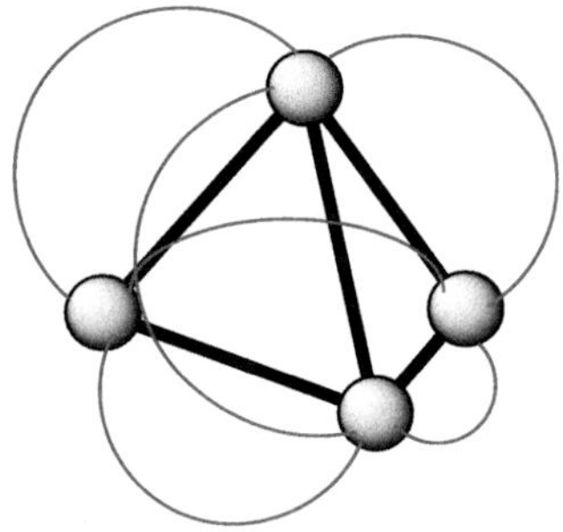

Structural skeleton of $[Ga_4L_6]^{12-}$
3.34c

FIGURE 3.34 A ligand for tetrahedral cage-like complex and the skeleton of $[Ga_4L_6]^{12-}$ cluster. (From M. D. Pluth et al. *Science* 2007, 316, 85–88;[27] M. D. Pluth et al., *J. Am. Chem. Soc.* 2008, 130, 11423–11429.[28])

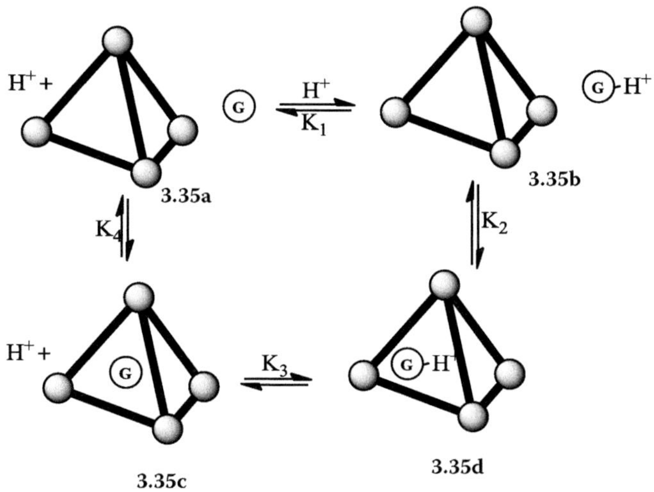

FIGURE 3.35 The acid base equilibrium of tetra nuclear cage with change of pH.

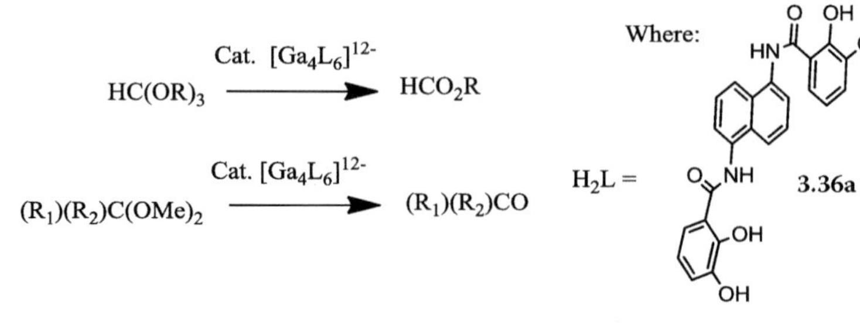

FIGURE 3.36 A gallium-cage–catalysed deprotection reaction.

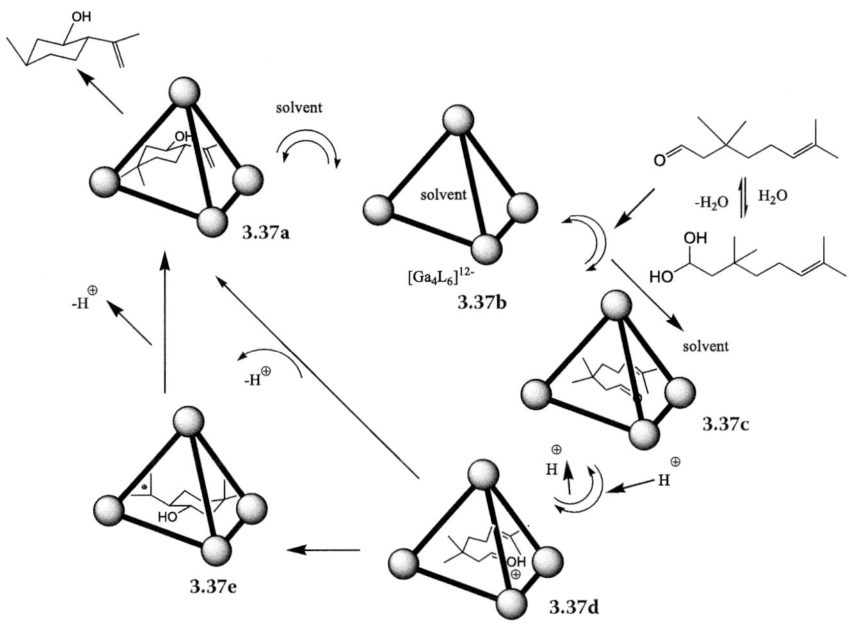

FIGURE 3.37 A gallium-cage catalyzing cyclisation reactions. (From C. J. Hastings et al. *Chem. Eur. J.* 2014, 20, 3966–3973.[29])

$[Cp^*(PMe_3)Ir(Me)(CO)]^+$ complex encapsulated inside the gallium cage activates the aldehyde C—H bond of acetaldehyde and decomposes it to liberate ethylene and methane (Figure 3.38). From a mixture of acetaldehyde and benzaldehyde, the catalyst **3.38a** reacts only with acetaldehyde, leaving aside benzaldehyde, whereas the catalyst

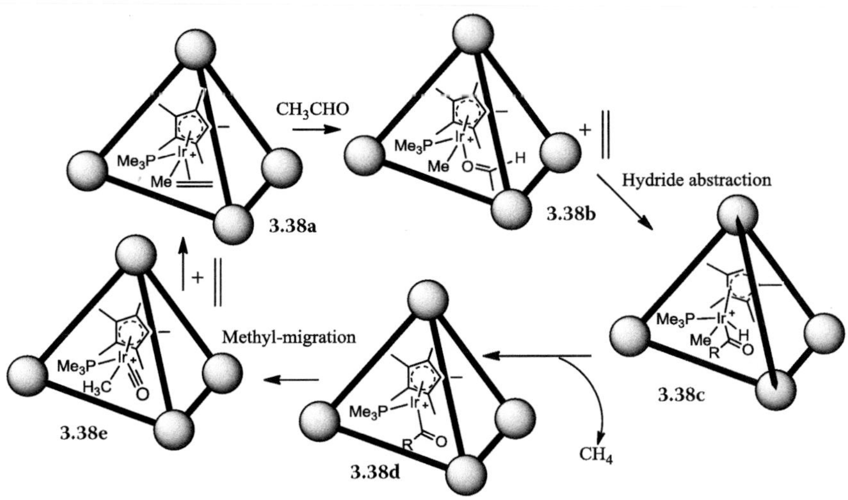

FIGURE 3.38 Selective reaction of acetaldehyde to form methane in a gallium cage. (From C. J. Brown et al. *Chem. Rev.* 2015, 115, 3012–3035.[30])

$[Cp^*(PMe_3)Ir(Me)(C_2H_4)]^+$, in the absence of the gallium cage, transforms both these aldehydes without showing a discriminating reactivity for them. Control experiments suggest that the difference in reactivity arises from differences in the sizes of the aldehydes. The reaction mechanism for catalytic decomposition of acetaldehyde involves an encapsulated catalyst **3.38a** undergoing displacement of olefin by aldehyde to form the intermediate **3.38b**. The intermediate **3.38b** formed by ligation of acetaldehyde transforms to another intermediate **3.38c**. The intermediate has methyl, carbonyl, acyl and hydride ligands. It loses methane to form the intermediate **3.38d**. The migration of the methyl group from the acyl unit to the iridium ion results in the carbonyl complex **3.38e**. The catalytic reactions of the gallium cluster are not only applicable to size-selective catalysis but also they differentiate the diastereoselective products. For example, an independent reaction of butyraldehyde or isobutyraldehyde gives a different diasteriomeric ratio of products. The reaction of butyraldehyde shows better diastereoselectivity than the corresponding reaction of isobutyraldehyde.

The ligand **3.39a** designated as L is suitable to form a coordination cage $[CuI_4L_4]^{4+}$ (Figure 3.39), and the cage is capable of encapsulating anions. The encapsulated anions modulate the redox properties of the copper ions. Due to this change in the redox property by anions, the modulation of reactivity in the catalytic oxidation and hydroxylation reactions of 1,2,3,4-tetrahydronaphthalene in the presence of tertiary butylhydroperoxide are affected by changing anions.

This cage has copper (I) ions as the active sites. The catalysis in this case occurs at the surface of the cage. The cage has an anion inside the cage, which provides an example to modulate reactivity through host—guest interactions affecting the coordination environment, which influences the redox activity. The catalytic reaction primarily yields to carbonyl compound 3,4-dihydronaphthalen-1(2H)-one, whereas 1,2,3,4-tetrahydronaphthalen-1-ol is formed as a minor product. There is high catalytic activity to form 3,4-dihydronaphthalen-1(2H)-one when the nitrate is used as the counter ion of the catalyst. The other anions such as trifluoroacetate, perchlorate and tosylate provide lower yields of the product.

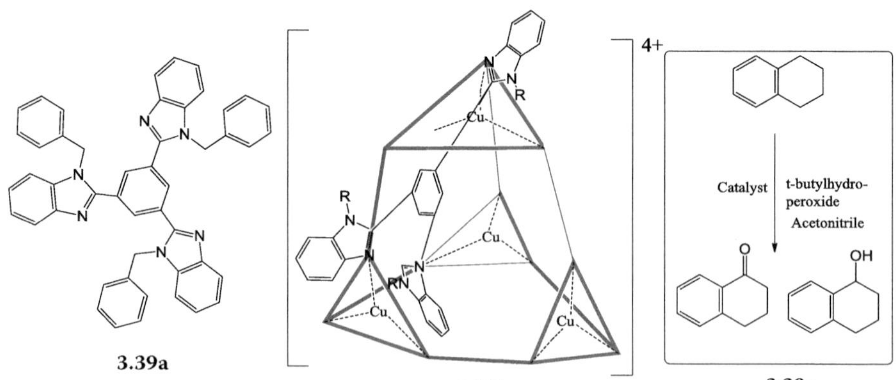

3.39a 3.39b 3.39c

FIGURE 3.39 Anion-dependent catalysis by a tetranuclear copper (I) cage. (From Q.-T. He et al. *ACS Catal.* 2013, 3, 1–9.[31])

3.2.5 Confining Inorganic Active Sites

The stabilisations of reactive sites by concealing within hydrophobic pockets eventually lead to isolate reactive species. Importantly, the catalytic sites of metalloenzymes causing dioxygen activation involve various steps. The binding of O_2 at a reduced metal centre, formation of metal-superoxo or peroxo species and oxygen—oxygen bond cleavage reactions are some such reactions. Studies on model compounds have shown that mononuclear complexes have the ability to form metal-dioxygen ($M\text{-}O_2$) complexes which are used for various oxidation reactions. Dioxygen binds to metal ions in different ways, as shown in Figure 3.40. The macrocyclic N-tetramethylated cyclam ligands listed in Figure 3.40 stabilise different metal-oxygen complexes. The ability to stabilise depends on the size and type of metal ion used. Mononuclear complexes with different first-row transition metal ions possess end-on metal-superoxo and side-on metal-peroxo species. The complexes have different binding modes depending on the metal ions and the size of the macrocyclic ring. The macrocyclic ligands stabilise each such species through the control provided by the geometric and electronic requirements available due to the ligands to stabilise a particular species.

Generally, peroxo and superoxo complexes are formed by reactions of dioxygen with transition metal complexes (Figure 3.40). These complexes have applications in olefin polymerization and C—H activation reactions. Many such complexes are stable

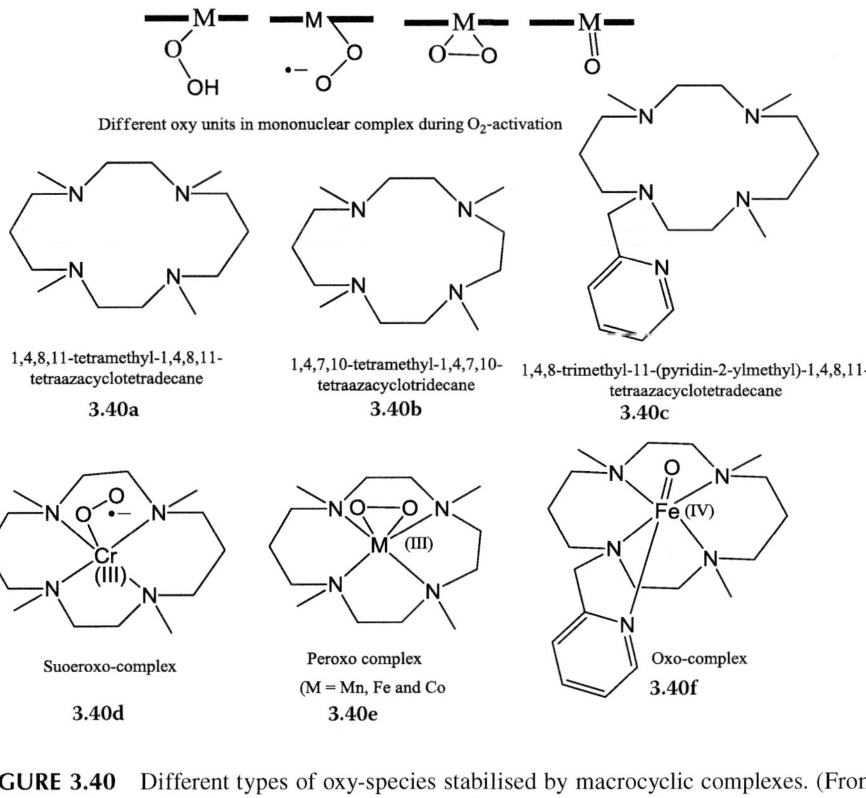

FIGURE 3.40 Different types of oxy-species stabilised by macrocyclic complexes. (From J. Cho et al. *Acc. Chem. Res.* 2012, 45, 1321–1330.[32])

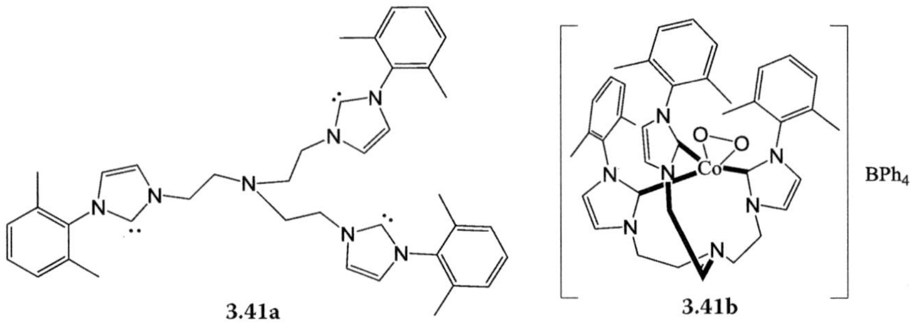

3.41a **3.41b**

FIGURE 3.41 A *tris*-carbene ligand and its cobalt complex with peroxy-group. (From X. Hu et al. *J. Am. Chem. Soc.* 2004, 126, 13464–13473.[33])

due to the hydrophobic, electronic and steric effect provided by ancillary ligands. Metal carbene complexes are also stabilised in hydrophobic environments. Tripodal ligands not only stabilise such cobalt complexes but also stabilise reactive oxygen molecules together in such complexes. For example, *tris*[2-(3-alkylimidazol-2-ylidene)ethyl (**3.41a**) is a polydentate N-heterocyclic carbene ligand. This *tris*-carbene ligand binds to the cobalt (III) ion and activates molecular oxygen by forming a carbene complex **3.41b** with a peroxo ligand (Figure 3.41). The peroxo ligand is formed from molecular oxygen. Peroxo complexes show IR stretching around 930–740 cm^{-1}, whereas a superoxo complexes show IR stretch in the range of 1200–1070 cm^{-1}. In the infrared spectra of the *tris*-carbene, the cobalt peroxo-complex shows O—O stretching at 890 cm^{-1}. The assignment of this stretching frequency is confirmed by isotopic labelling experiments. The ^{18}O-labelled compound has a stretching absorption of 840 cm^{-1}. The observed shift is in agreement with the theoretically anticipated 51 cm^{-1} shift upon incorporation of the ^{18}O isotope in lieu of the naturally abundant ^{16}O isotope (Figure 3.42).

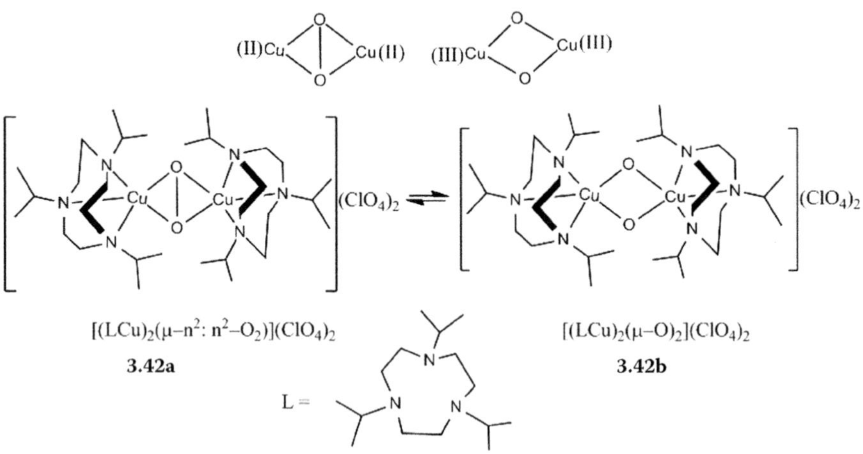

FIGURE 3.42 Two copper dioxygen species and equilibrium between [(LCu)$_2$(μ-η^2:η^2-O$_2$)](ClO$_4$)$_2$ and [(LCu)$_2$(μ-O)$_2$](ClO$_4$)$_2$ complex. (From P. L. Holland, W. B. Tolman, *Coord. Chem. Rev.* 1999, 190–192, 855–869.[34])

Aggregation of a metal complex stabilises certain forms of reactive species more than another reactive species of similar kind, but with slightly different structures. For example, many metal dioxygen model complexes of enzymes have a planar (μ-η^2:η^2-peroxo)dicopper(II) unit **3.42a** as an active site. The complex **3.42b** has a core composed of bis(μ-oxo)dicopper(III). These two complexes derived from the same macrocyclic ligand 1,4,7-triisopropyl-1,4,7-triazonane (L) have different Cu_2O_2 cores in the copper complexes $[(LCu)_2(\mu\text{-}\eta^2{:}\eta^2\text{-}O_2)](ClO_4)_2$ and $[(LCu)_2(\mu\text{-}O)_2](ClO_4)_2$ exist in equilibrium. The macrocycle ligand has nitrogen atoms functionalised by isopropyl groups. The complexes have similar energies. The isomer $[(LCu)_2(\mu\text{-}\eta^2{:}\eta^2\text{-}O_2)](ClO_4)_2$, dissolved in dichloromethane upon dilution with tetrahydrofuran, transforms to the complex $[(LCu)_2(\mu\text{-}O)_2](ClO_4)_2$. The reverse process happens on dilution of a tetrahydrofuran solution of $[(LCu)_2(\mu\text{-}O)_2](ClO_4)_2$ with dichloromethane. The solvent dependency, dilution and conductance data suggest selective aggregation of the $[(LCu)_2(\mu\text{-}O)_2](ClO_4)_2$ isomer.

Planar carbocations are selectively stabilised in the voids of cavitands. For example, the host **3.43a** has a resocinarene frame with flexible substituents of long-chain hydrocarbons. It also has semirigid substituted thiophene carboxylates (Figure 3.43). The thiophene carboxylates, together with the ring, provide a cavitylike structure the overall structure adopt a C_{2v}-symmetry. This compound is self-complementary to reversibly form dimer mediated by a planar tropylium cation. A tropylium cation is encapsulated between the spaces provided by the two faces of the cavitands. The tetrafluoroborate anion is encapsulated within the hydrophobic n-pentyl substituents, as depicted in the diagram showing the structural representation. The extra stability of this 2:1 host—guest complex is due to charge transfer and has an orange colour due to absorbance at $\lambda_{max} = 463$ nm in dichloromethane.

In addition to these, structurally rigid container molecules derived from zeolites are used as hosts for various organic compounds. Zeolites are aluminosilicate

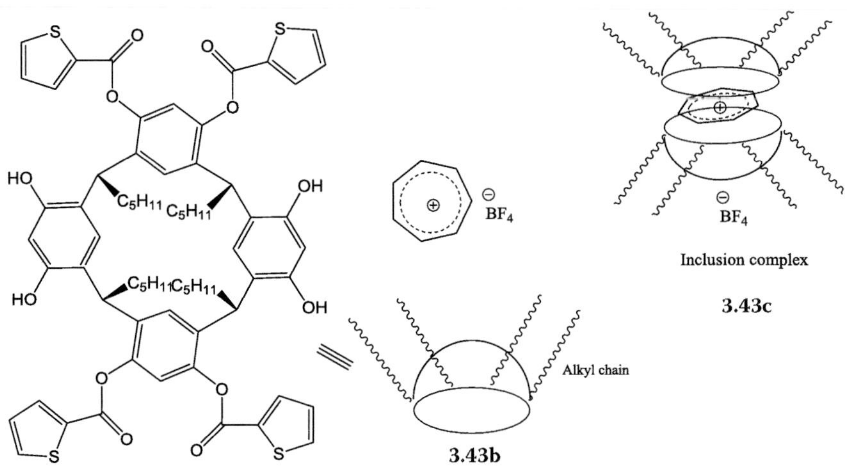

3.43a

3.43b

3.43c

FIGURE 3.43 Host–guest complex stabilizing tropylium cation. (From A. Shivanyuk et al. *Angew. Chem. Int. Ed.* 1999, 38, 2906–2909.[35])

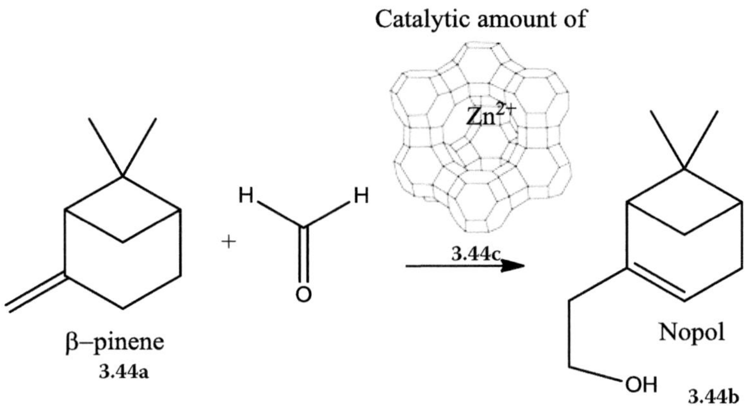

FIGURE 3.44 Conversion of β-pinene to nopol. (From V. S. Marakatti et al. *Catal. Sci. Technol.* 2014, 4, 4065–4074.[36])

and they have porous structures. Due to intrinsic acid-base properties associated with aluminosilicate, they are good catalysts for selective organic transformations in heterogeneous conditions. Some of the metal ion-exchanged zeolites are active catalysts. The acidity of zeolites changes upon incorporation of zinc ions and acts as a catalyst for the Prin reaction. The Prin reaction refers to transformation of activated aldehyde; it is an electrophilic addition reaction and is generally catalysed by an acid. Such reactions in homogeneous conditions are nonspecific. Zinc-incorporated β-zeolite has high selectivity in the conversion of β-pinene upon reaction with formaldehyde to form nopol (Figure 3.44). The reaction proceeds when 0.31 mmol/g of zinc ion is loaded onto the H-beta zeolite with a turnover number of 138. In Brönsted acid-catalysed reactions of ß-pinene, it rearranges to α-pinene, limonene, camphene and terpinolenes, which is not the case in the confined media of zeolite. The incorporated zinc ions in the zeolite decrease the Brönsted acidity and enhance the reaction rate to form a specific product. This reaction is presumably through the stabilization of tertiary carbocation, which is held inside the cavity through the oxygen atom of the condensation product formed between the reactants.

Fishing equipment to trap fish in confined space; opening on the upper side is to facilitate hold and also to insert hand to catch once fish is trapped.

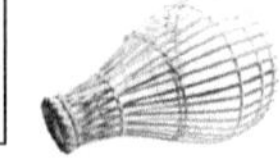

REFERENCES

1. A. B. Vliegenthart, F. A. L. Welling, M. Roemelt, R. J. M. K. Gebbink, M. Otte, Synthesis and characterization of a Brønsted pair functionalized shape-persistent macrocycle. *Org. Lett.* 2015, 17, 4172–4175.
2. L. Schoepff, L. Kocher, S. Durot, V. Heitz, Chemically induced breathing of flexible porphyrinic covalent cages. *J. Org. Chem.* 2017, 82, 5845–5851.

3. D. Zhang, A. Martinez, J.-P. Dutasta, Emergence of hemicryptophanes: From synthesis to applications for recognition, molecular machines, and supramolecular catalysis. *Chem. Rev.* 2017, 117, 4900–4942.
4. D. Zhang, K. Jamieson, L. Guy, G. Gao, J.-P. Dutasta, A. Martinez, Tailored oxido-vanadium(V) cage complexes for selective sulfoxidation in confined space. *Chem. Sci.* 2017, 8, 789–794.
5. S.-W. Tam-Chang, L. Jimenez, F. Diederich, A thiazolio-cyclophane as model for pyruvate oxidase and one-pot synthesis of aromatic esters by electrochemical oxidation of aldehydes mediated by bis(coenzyme) catalysis. *Hel. Chim. Acta* 1993, 76, 2616–2639.
6. G. Maayan, M. D. Ward, K. Kirshenbaum, Folded biomimetic oligomers for enantioselective catalysis. *Proc. Nat. Acad. Sci. (USA)* 2009, 106, 13679–13684.
7. L. Catti, K. Tiefenbacher, Brønsted acid-catalyzed carbonyl-olefin metathesis inside a self-assembled supramolecular host. *Angew. Chem. Int. Ed.* 2018, 57, 1–5.
8. K. Kobayashi, M. Yamanaka, Self-assembled capsules based on tetrafunctionalized calix[4]resorcinarene cavitands. *Chem. Soc. Rev.* 2015, 44, 449–466.
9. B. W. Purse, J. Rebek Jr., Functional cavitands: Chemical reactivity in structured environments. *Proc. Nat. Acad. Sci. (USA)* 2005, 102, 10777–10782.
10. S. R. Shenoy, F. R. P. Crisostomo, T. Iwasawa and J. Rebek Jr., Organocatalysis in a synthetic receptor with an inwardly directed carboxylic acid. *J. Am. Chem. Soc.* 2008, 130, 5658–5659.
11. S.-Y. Yu, T. Kusukawa, K. Biradha, M. Fujita, Hydrophobic assembling of a coordination nano bowl into a dimeric capsule which can accommodate up to six large organic molecules. *J. Am. Chem. Soc.* 2000, 122, 2665–2666.
12. D. Bécart, V. Diemer, A. Salaün, M. Oiarbide, Y. R. Nelli, B. Kauffmann, L. Fischer, C. Palomo, G. Guichard, Helical oligourea foldamers as powerful hydrogen bonding catalysts for enantioselective C−C bond-forming reactions. *J. Am. Chem. Soc.* 2017, 139, 12524–12532.
13. M. M. Miller, M. A. Windsor, W. C. Pomerantz, S. H. Gellman, D. Hilvert, A rationally designed aldolase foldamer. *Angew. Chem. Int. Ed.* 2009, 48, 922–925.
14. A. J. Metrano, N. C. Abascal, B. Q. Mercado, E. K. Paulson, S. J. Miller, Structural studies of β-turn-containing peptide catalysts for atroposelective quinazolinone bromination. *Chem. Commun.* 2016, 52, 4816–4819.
15. W. M. Singh, B. R. Jali, J. B. Baruah, Iron(II) and manganese(II) complexes of 2-{2-(3-carboxypyridin-2-yl)disulfanyl} pyridine-3-carboxylic acid through C—S bond cleavage. *J. Chem. Cryst.* 2012, 42, 775–782.
16. K. D. Karlin, J. C. Hayes, Y. Gultneh, R. W. Cruse, J. W. McKown, J. P. Hutchinson, J. Zubieta, Copper-mediated hydroxylation of an arene: Model system for the action of copper monooxygenases. Structures of a binuclear Cu(I) complex and its oxygenated product. *J. Am. Chem. Soc.* 1984, 106, 2121–2128.
17. D. Chatterjee, E. Ember, U. Pal, S. Ghosh, R. van Eldik, Remarkably high catalytic activity of the RuIII(edta)/H_2O_2 system towards degradation of the azo-dye orange. *Dalton Trans.* 2011, 40, 10473–10480.
18. B. A. F. Le Bailly, L. Byrne, J. Clayden, Refoldable foldamers: Global conformational switching by deletion or insertion of a single hydrogen bond. *Angew. Chem. Int. Ed.* 2016, 55, 2132–2136.
19. S. Sabater, J. A. Mata, E. Peris, Catalyst enhancement and recyclability by immobilization of metal complexes onto grapheme surface by noncovalent interactions. *ACS Catal.* 2014, 4, 2038–2047.
20. J. Mata, E. Peris, S. Sabater, Immobilization of pyrene-tagged palladium and ruthenium complexes onto reduced graphene oxide: An efficient and highly recyclable catalyst for hydrodefluorination. *Organometallics* 2015, 34, 1186–1190.

21. C. S. Popeney, C. M. Levins, Z. Guan, Systematic investigation of ligand substitution effects in cyclophane-based nickel(II) and palladium(II) olefin polymerization catalysts. *Organometallics* 2011, 30, 2432–2452.
22. V. Martí-Centelles, A. L. Lawrence, P. J. Lusby, High activity and efficient turnover by a simple, self-assembled artificial Diels—Alderase. *J. Am. Chem. Soc.* 2018, 140, 2862–2868.
23. M. L. Merlau, M. del, P. Mejia, S. B. T. Nguyen, J. T. Hupp, Artificial enzymes formed through directed assembly of molecular square encapsulated epoxidation catalysts. *Angew. Chem. Int. Ed.* 2001, 40, 4239–4242.
24. P. J. Stang, B. Olenyuk, Self-assembly, symmetry, and molecular architecture: Coordination as the motif in the rational design of supramolecular metallacyclic polygons and polyhedra. *Acc. Chem. Res.* 1997, 30, 502–518.
25. M. Fujita, D. Oguro, M. Miyazawa, H. Oka, K. Yamaguchi, K. Ogura, Self-assembly of 10 molecules into nanometre-sized organic host frameworks. *Nature* 1995, 378, 469–471.
26. Q.-Q. Wang, S. Gonell, S. H. A. M. Leenders, M. Dürr, I. Ivanovic-Burmazovic, J. N. H. Reek, Self-assembled nanospheres with multiple endohedral binding sites pre-organize catalysts and substrates for highly efficient reactions. *Nat. Chem.* 2016, 8, 225–230.
27. M. D. Pluth, R. G. Bergman, K. N. Raymond, Acid catalysis in basic solution: A supramolecular host promotes orthoformate hydrolysis. *Science* 2007, 316, 85–88.
28. M. D. Pluth, R. G. Bergman, K. N. Raymond, Supramolecular catalysis of orthoformate hydrolysis in basic solution: An enzyme-like mechanism. *J. Am. Chem. Soc.* 2008, 130, 11423–11429.
29. C. J. Hastings, R. G. Bergman, K. N. Raymond, Origins of large rate enhancements in the Nazarov cyclization catalyzed by supramolecular encapsulation. *Chem. Eur. J.* 2014, 20, 3966–3973.
30. C. J. Brown, F. D. Toste, R. G. Bergman, K. N. Raymond, Supramolecular catalysis in metal—ligand cluster hosts. *Chem. Rev.* 2015, 115, 3012–3035.
31. Q.-T. He, X.-P. Li, L.-F. Chen, L. Zhang, W. Wang, C.-Y. Su, Nano sized coordination cages incorporating multiple Cu(I) reactive sites: Host—guest modulated catalytic activity. *ACS Catal.* 2013, 3, 1–9.
32. J. Cho, R. Sarangi, W. Nam, Mononuclear metal-O_2 complexes bearing macrocyclic N-tetra methylated cyclam ligands. *Acc. Chem. Res.* 2012, 45, 1321–1330.
33. X. Hu, I. Castro-Rodriguez, K. Meyer, Dioxygen activation by a low-valent cobalt complex employing a flexible tripodal N-heterocyclic carbene ligand. *J. Am. Chem. Soc.* 2004, 126, 13464–13473.
34. P. L. Holland, W. B. Tolman, Dioxygen activation by copper sites: Relative stability and reactivity of (μ-η^2:η^2-peroxo)- and bis(μ-oxo)dicopper cores. *Coord. Chem. Rev.* 1999, 190-192, 855–869.
35. A. Shivanyuk, E. F. Paulus, V. Böhmer, Guest-controlled formation of a hydrogen-bonded molecular capsule. *Angew. Chem. Int. Ed.* 1999, 38, 2906–2909.
36. V. S. Marakatti, A. B. Halgeri, G. V. Shanbhag, Metal ion-exchanged zeolites as solid acid catalysts for the green synthesis of nopol from Prins reaction. *Catal. Sci. Technol.* 2014, 4, 4065–4074.

4 Interlocked Systems in Catalysis and Switching

4.1 PRINCIPLES OF DYNAMIC INTERLOCKED SYSTEMS IN CATALYSIS

The catalysis based on dynamic systems stems from three fundamental aspects.[1–2] These aspects are: (a) to utilise conformational flexible or fluxional systems for catalysis, (b) to take advantage of building an interlocked system by controlled reactivity and (c) to limit or enhance reactivity of a particular site by organizing the interlocked system differently. In the general practice of catalysis, there is a preference for a particular conformation to enhance activity; hence, the first aspect is routine. This aspect is evident in functional features of transient species imparting selectivity in catalytic reactions. The second and third aspects are on the utilization of interlocked systems. In supramolecular chemistry, interlocked systems are formed between covalently linked multicomponent systems. They may arise from the insertion of a dumbbell-shaped axle or an axlelike structure accommodating one or more molecules with ringlike structures. These types of interlocked systems are known as rotaxanes. A rotaxanelike structure may be created from a single-component system where a long chain is attached to a ring and the chain adopts adequate geometry to insert the free end into the ring. The interlocked structures formed between two or more rings or closed structures interlocked to each other are known as catenanes. One or more components of interlocked molecules have a definite trajectory and path for movement within the interlocked system. In the case of rotaxanes, the paths of movement of the ring/s over the different identified positions with functional groups located on the axle are of major concern in catalysis. Similarly, for a catenane catalyst, the circular path traced by a component is a point of focus. A halt of a ring over a catalyst would conceal a catalytic site, whereas moving the ring away would expose the catalytic site. The situation may be described by the schematic diagrams shown in Figure 4.1. These motions are characteristic of interlocked systems and are guided by stimuli, the solvent, the reactant, light, heat, pressure, temperature and so on. As the positions of the rings are easily changed, such changes affect and control the reactivity of catalytic sites by providing suitable space and orientation to the catalytic site to react or not with a substrate. The circular and translational motion of the ring component is represented in drawing marked as **4.1a**. In the case of the rotation of two interlocked systems, it may be unidirectional or in the opposite directions. These motions are modulated by the hierarchy of binding units located at different points on the rings or axle.

The construction of a rotaxane through catalytic reactions based on an active site on a cyclic macrocycle is illustrated in Figure 4.2. A cyclic derivative of 1,2-cyclohexyldiamine **4.2a** forms a copper (I) complex **4.2b** upon reaction with

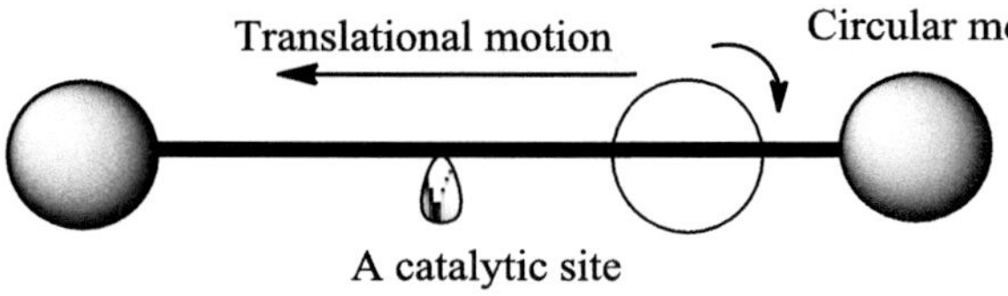

4.1a

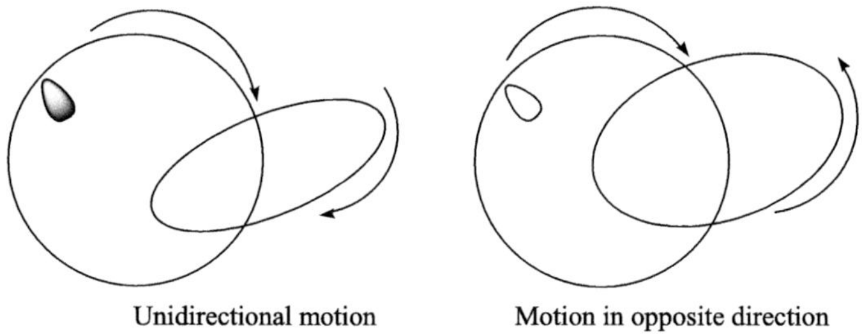

4.1b

FIGURE 4.1 Motion of rings in rotaxanes and catenanes over a catalytic site.

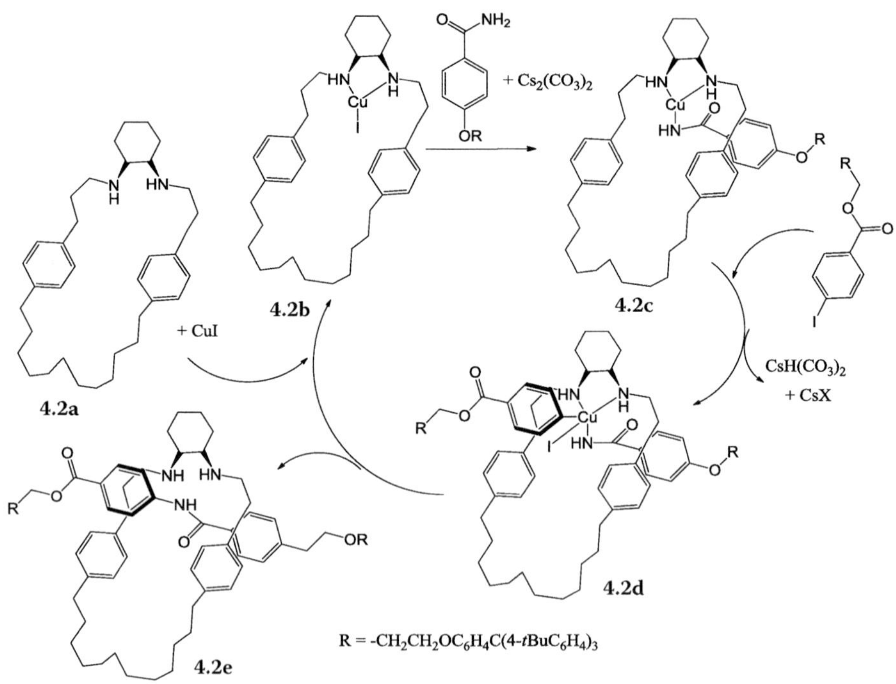

FIGURE 4.2 Rotaxane formation through catalytic reaction. (From S. Hoekman et al. *J. Am. Chem. Soc.* 2015, 137, 7656–7659.[3])

copper iodide. This complex, with a catalytic amount of cesium carbonate, reacts with an amide derivative, thereby forms the complex **4.2c**. This copper(I) complex **4.2c** has a deprotonated amide coordinated to the copper ion. During the course of this reaction, cesium bicarbonate is formed from cesium carbonate. The amido-copper (I) complex **4.2c** undergoes an oxidative addition reaction with iodo-derivative forming a reaction intermediate **4.2d**. This intermediate is ready for the C—N bond formation reaction, and the reductive elimination forms the desired rotaxane **4.2e** and releases copper (I) iodide for the next catalytic cycle. Copper (I) acetate can also be used instead of copper (I) iodide.

Ammonium cations are generally recognised by crown ethers. Due to the high binding affinity of ammonium cations, rotaxane molecules are formed, as the insertion of an axle to a ringlike geometry is favoured. Crown ether—connected chiral phosphate acts as a ligand and recognises a cation (Figure 4.3). The complex **4.3a**, a rhodium complex, has a crown ether—functionalised chiral phosphate ligand coordinating to the rhodium ion through a phosphorus atom.

The complex has two other ligands; one is cyclo-octadiene and other is a phosphine-based ligand bearing long chain with an amine group at a distal side. The complex has suitable geometry for adopting an interlocked structure. A portion of the long-chain ligand is inserted through the macrocyclic ring. This happens as the primary ammonium cation is recognised by the crown ether part of the other ligand. The rhodium complex **4.3a** is a catalyst for an asymmetric hydrogenation reaction. Asymmetric induction is due to the orientation of the binol part of the ligand guided by cation recognition caused by the crown ether forming the interlocked structure.

Coordination of a rotaxane at an appropriate site on the axle and ring of a rotaxane with a metal ion locks conformation. This causes a catalytic reaction of the axle to transform to another form. Such an example is improvised in the palladium-assisted magnesium ethoxide—catalysed cyclisation reaction of propargyl amide derivative.

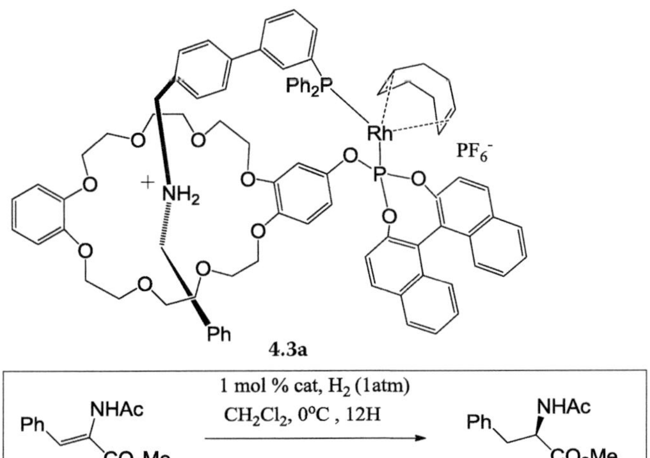

FIGURE 4.3 A catalytic reaction due to a rotaxane complex. (From G. Hattori et al. *J. Am. Chem. Soc.* 2007, 129, 12930–12931.[4])

The fundamental cyclisation reaction involved in the changes in geometry is shown in the inset of Figure 4.4. This reaction is applied to modify the axle **4.4b** of rotaxane **4.4c**. The axle **4.4b** has two pyridines for coordination to the metal ion. The ring **4.4a** has a dipicolinc amide functional unit linked to a polyether part. These two components form rotaxane binds to palladium ion to form a rotaxane palladium (II) complex **4.4c**. The axle of the rotaxane has two sites with propargyl carbamate units which undergo a cyclisation recation in the presence of ethoxide. The palladium complex **4.4c**, in conjunction with magnesium ethoxide, catalyses the specific cyclo-isomerization reaction of the axle. The acetylinic functional units of propargyl functionalities of the axle are transformed to double bonds. This occurs due to intramolecular cyclisation. The amide groups act as nucleophiles to attack at the propargylic sites and form cyclic components on the axle. The cyclisation reactions occur at two different locations of the axle, and the catalytically active palladium site is exclusively present within the

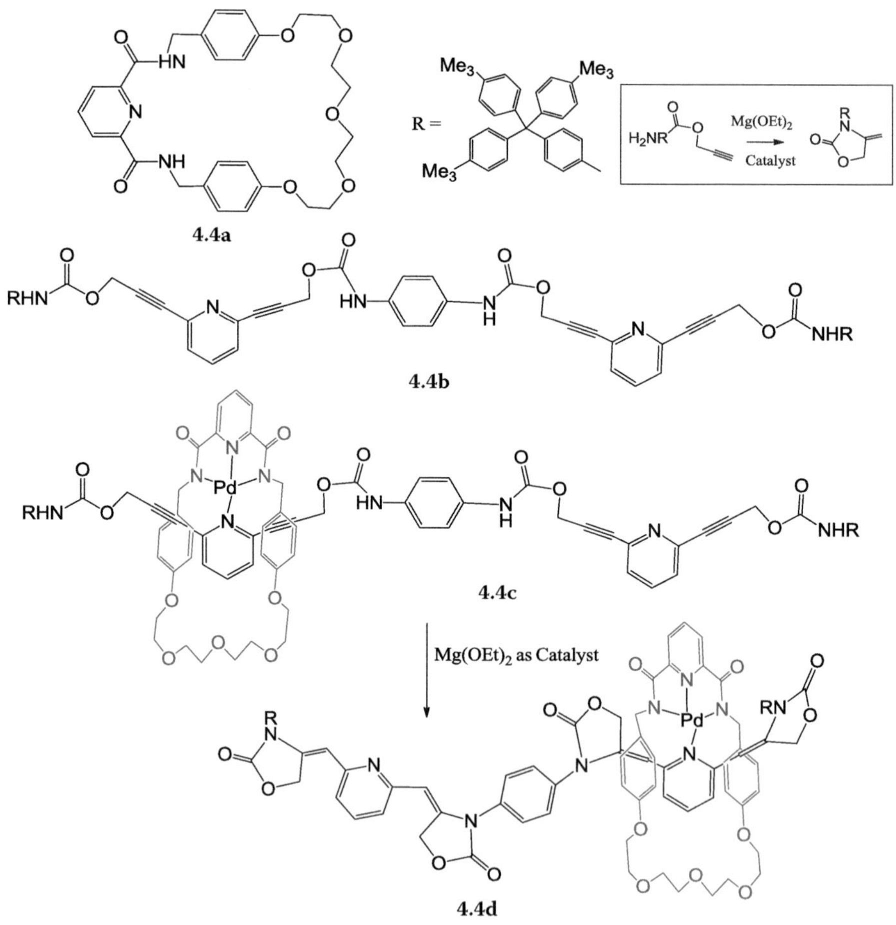

FIGURE 4.4 Palladium-rotaxane complex as catalyst. (From N. Miyagawa et al. *Chem. Commun.* 2010, 46, 1920–1922.[5])

circumference of the macrocycle. Due to the change of the structure of the axle, the orientation of the new rotaxane is all together different from the original rotaxane. The axle has two bulky groups at two ends; which acts as stoppers so that the movement of the macrocycle is restricted in a limited space between these stoppers. Hence, no isomerization of the product takes place; neither are multiple side products formed.

In supramolecular chemistry, the local environment around a catalytic site is important to modulate its catalytic activity. This is due to the directional behaviour of weak interactions. When the lone pair of electrons of an atom participates in catalysis, the direction of the orbitals and local electronic influences caused by molecules in the vicinity of the lone pair influence its ability to interact with substrates. For example, in the compound **4.5a**, the *cis*-geometry of the azo group provides an orientation so that the lone pair on the nitrogen atom is more exposed than the *trans*-form of the compound (Figure 4.5). Hence, the reactivity of the *cis*-compound in the condensation reaction between a nitro olefin and 4-nitrobenzaldehyde is slower than the *trans*-form of the compound. The *cis*- and *trans*-forms are interconvertible by irradiation or thermal process. Due to this, the reactivity of the condensation reaction is controlled.

Another approach to control the reactivity of a substrate in supramolecular catalysis by interlocked system is by partially or completely concealing a reactant by forming an inclusion complex. The catalytic hydrolysis reactions of esters are modulated by such an approach. Cyclodextrin molecules form an inclusion complex with nitro aromatic esters. In this inclusion process, the nitro-aromatic part projects inwards and goes beneath the ring, whereas the ester group remains encapsulated on the hydrophilic side of the ring. The ester group undergoes catalytic hydrolysis when the ester has the appropriate geometrical arrangement of hydroxyl functional groups

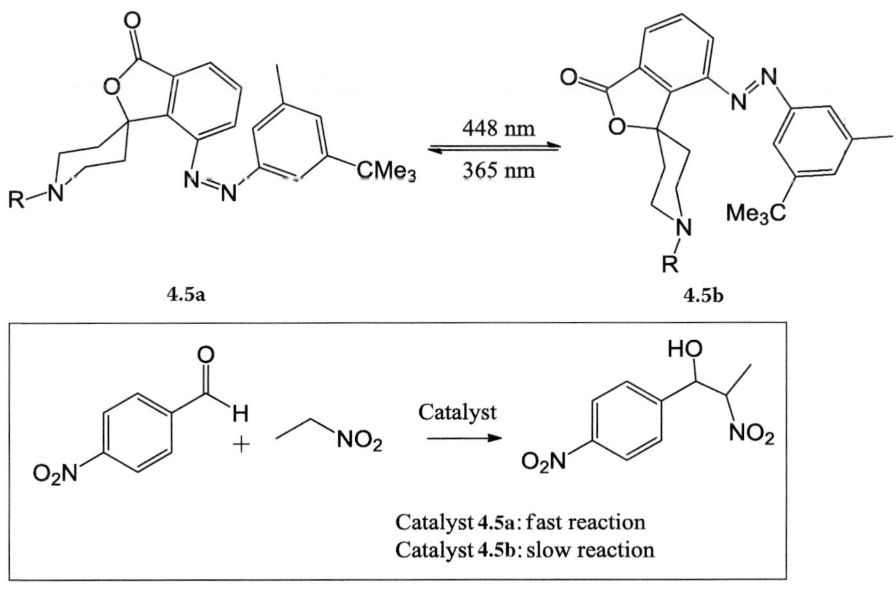

FIGURE 4.5 Geometrical isomers of a catalyst with different reactivity. (From M. V. Peters et al. *Angew. Chem. Int. Ed.* 2008, 47, 5968–5972.[6])

of cyclodextrin located at the outer periphery. Thus, the proximity of the hydroxyl group relative to the ester group is a key factor for such an activation. If the ester compound resides deep inside the cavity, as in **4.6b**, or if the ester group is not within the reach of the hydroxyl group of the cyclodextrin, then hydrolysis is difficult. This aspect is revealed in the reactivity study by designing cyclodextrin in different ways. For example, the guest binding ability of cyclodextrin with a *cis*-azo group **4.6a** or *trans*-azo group **4.6b** is different. This is due to the encapsulation of the *trans*-form prominently, which is not the case in the *cis* azo-group containing cyclodextrin. Such an effect pushes the nitro-aromatic ester partially towards the hydrophilic side and makes the geometry conducive **4.6b** to the interaction of the hydroxyl group with the ester (Figure 4.6). This facilitates catalytic reaction.

Analogous switching on reactivity is caused by competitive inclusion of a functional part of the cyclodextrin with a reactant. In order to cause hydrolysis of a nitro-aromatic ester catalysed by cyclodextrin, the substrate is to be included in the cavity of the catalyst. When the cavity is filled with a functional unit, inclusion of the reactant is not possible. Thus, competitive guest binding causes inhibition of the catalytic reaction. In the example given in Figure 4.7, the cyclodextrin has an attached part with azo functionality. The functional unit in the *trans*-form of the azo-group has a structure suitable to form an intramolecular inclusion complex **4.7a**. It is not suitable for hydrolysis of aromatic esters. The *cis*-derivative of the cyclodextrin derivative **4.7b** has the azo-group located at a position relatively away from the cyclodextrin. The cyclodextrin ring in this form is empty, and a catalytic reaction is possible. The 4-nitrophenyl acetate is inserted into the cavity and undergoes fast catalytic hydrolysis, whereas the form of the cyclodextrin derivative with the *trans*-azo group is an interlocked compound **4.7a**. This prevents inclusion of 4-nitrophenyl acetate in the cyclodextrin cavity; hence, it cannot perform a catalytic hydrolysis reaction. The geometrical changes across the azo-bond affect the reactivity. The *cis*-azo compound transforms to a *trans*-azo compound upon heating, and the reverse is possible by irradiation. Thus, the hydrolysis reaction can be switched on and off accordingly by suitably irradiating or heating the reaction mixture containing the particular isomer of cyclodextrin derivative.

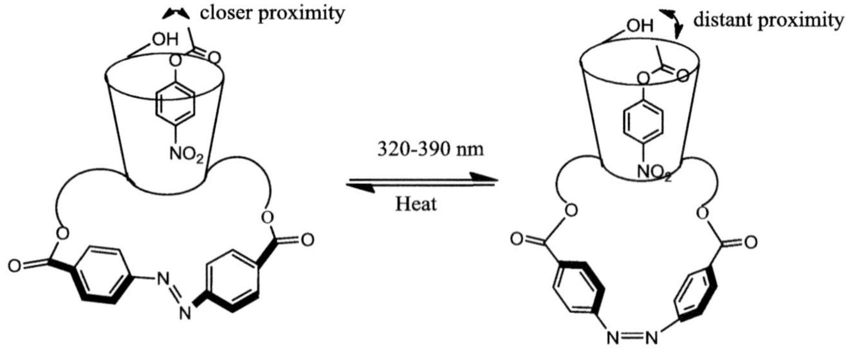

4.6a **4.6b**

FIGURE 4.6 Geometrical change outside periphery of cyclodextrin modulating catalytic reactivity. (From A. Ueno et al. *J. Chem. Soc. Chem. Commun.* 1981, 3, 94–96.[7])

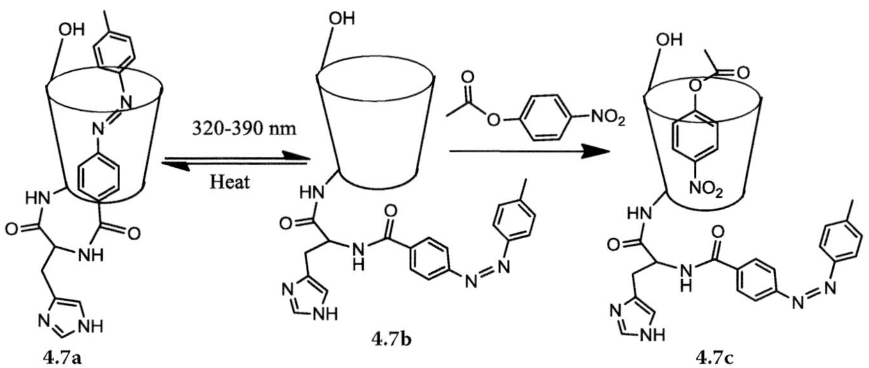

FIGURE 4.7 Geometrical changes across azo-group in a unit attached to cyclodextrin to modulate guest occupancy. (From O. Bistri, O. Reinau, *Org. Biomol. Chem.* 2015, 13, 2849–2865.[8])

4.1.1 Dynamic Aspects of Concealing and Revealing Catalytic Sites

The translational motion of the ring component of a rotaxane along the axis of the axle is utilised in different signalling processes and is also useful in modulating catalytic activities. In catalysis, the movement of ring over an axle causes changes in the environment, and circular movement of the ring could also change an environment. Thus, the multicomponents within the catalyst system provide different possibilities, based on which catalytic activity can be changed. The functional groups at the remote sites of the ring or axle modulate the incoming reactant to the active site and also cause stereochemical consequences. As a result of such movements, there are two extreme cases; in one situation, a ring occupies a position blocking the activity of the active site. Alternatively, the ring remains far away as a spectator component, leaving the catalytic site exposed. The ring in the second situation guides the reactivity to certain extent, providing the scope for region- and stereo-selective reactions. These aspects are realised from the catalytic reactions of the rotaxanes **4.8a** and **4.8b** (Figure 4.8).

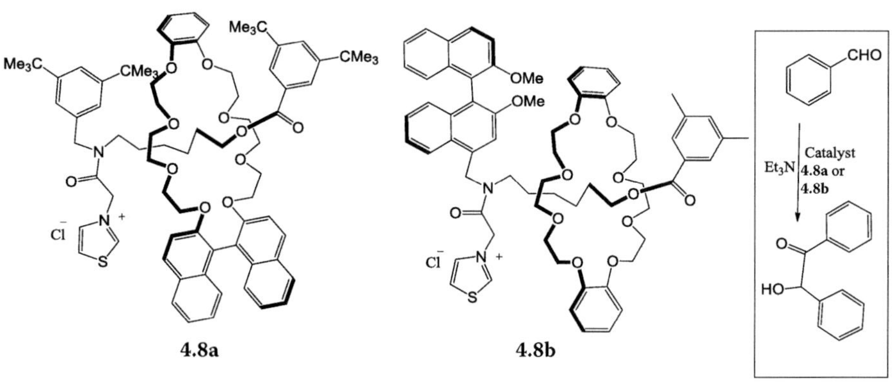

FIGURE 4.8 Two rotaxane-based catalysts performing a benzoin condensation reaction. (From D. A. Leigh et al. *ACS Catal.* 2014, 4, 4490–4497.[1])

Both rotaxanes are catalysts for the benzoin condensation reaction. In such a reaction, two aromatic aldehydes couple together to form a C—C bonded hydroxy-keto compound. The rotaxanes **4.8a** and **4.8b** each have a binol group which provides puckering to the ring component to provide chirality to the systems. In these catalytic reactions, the rotaxane **4.8a** gives a yield up to 90% and 32% enantiomeric excess of the R isomer, whereas the rotaxane **4.8b** gives a yield up to 82% enantiomer and an excess of 25% of the S isomer. The major difference between the two isomers is in the ring and axle having or not having a binol unit. In the case of **4.8a** the stereo-chemical consequence is result of an effect of binol on the ring component, whereas in the case of the rotaxane **4.8b**, the binol part is anchored to the axle component.

The axle of a rotaxane composed of multiple active sites provides scope to utilise those sites in a unitary or combinatory manner. The movable ring component stationed at an active site conceals the active site, exposing others. Hence, depending on the binding affinity of the ring to a particular site or by putting constraints on the movement of the ring over a limited space, active sites are exposed or concealed. The rotaxane shown in Figure 4.9 has three active sites: an amine, a triazole and a thiourea. Depending on the pH, the locations of the ring change. In a weakly acid medium containing acetic acid, the amine is protonated. The ammonium cation recognises

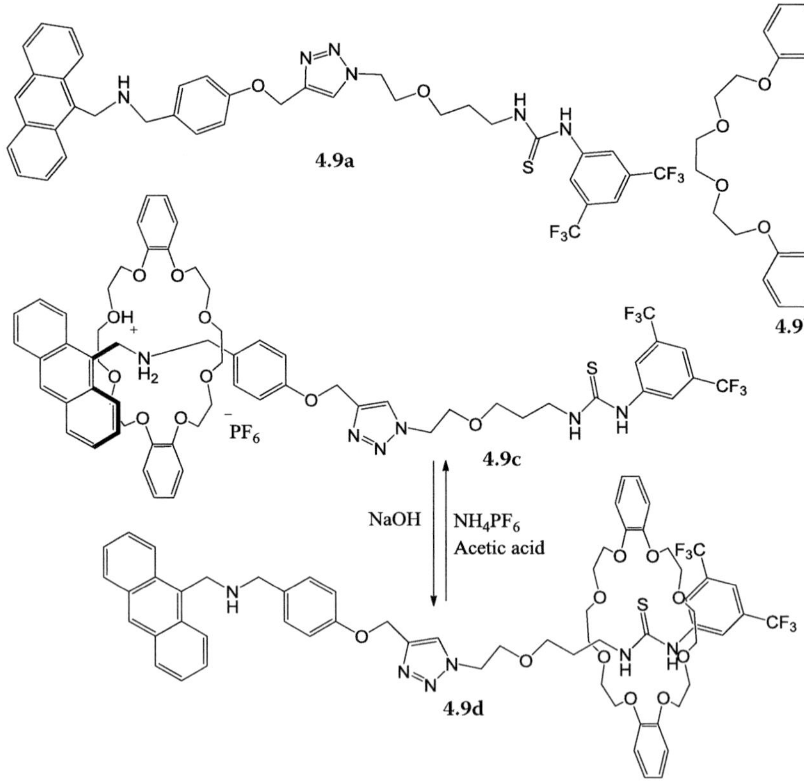

FIGURE 4.9 Concealing and revealing active sites for dual mode. (From C.-S. Kwan et al. *Org. Lett.* 2016, 18, 976–979.[10])

the crown ether. The crown ether takes up a position over the ammonium ion of the rotaxane **4.9c**, whereas upon the addition of sodium hydroxide, the ammonium group changes to amine and the ring moves to the thiourea site **4.9d**. Thus, the acidification conceals one site, whereas the site is exposed in basic conditions. This property of the rotaxane is useful in utilising the reactivity of the catalytically active amine and urea sites of the rotaxane.

Movement of a ring to conceal or expose catalytic sites is also done by reacting with a reactant to control catalytic activity. The rotaxane derived from the axle **4.10a** and ring component **4.10b** has interesting reactivity towards triphenylphosphine (Figure 4.10). Under neutral conditions, the ring is located at the end containing the azo-amide group, as shown in state **4.10c**. The azo-group reacts with triphenylphosphine to form ylide and pushes the ring to other end of the axle to generate the state **4.10d**. The ylide is stabilised by 4-nitro-phenyl carboxylate anion, as shown in state **4.10e**. The ylide reacts with benzylic alcohol to form the corresponding ester of 4-nitro-benzoic acid. The rotaxane reverts back to the original state after the coupling reaction; as a consequence, it gets ready for next cycle of the catalytic reaction. Thus, an originally concealed reactive site is exposed by reacting with a reactant, in this example with

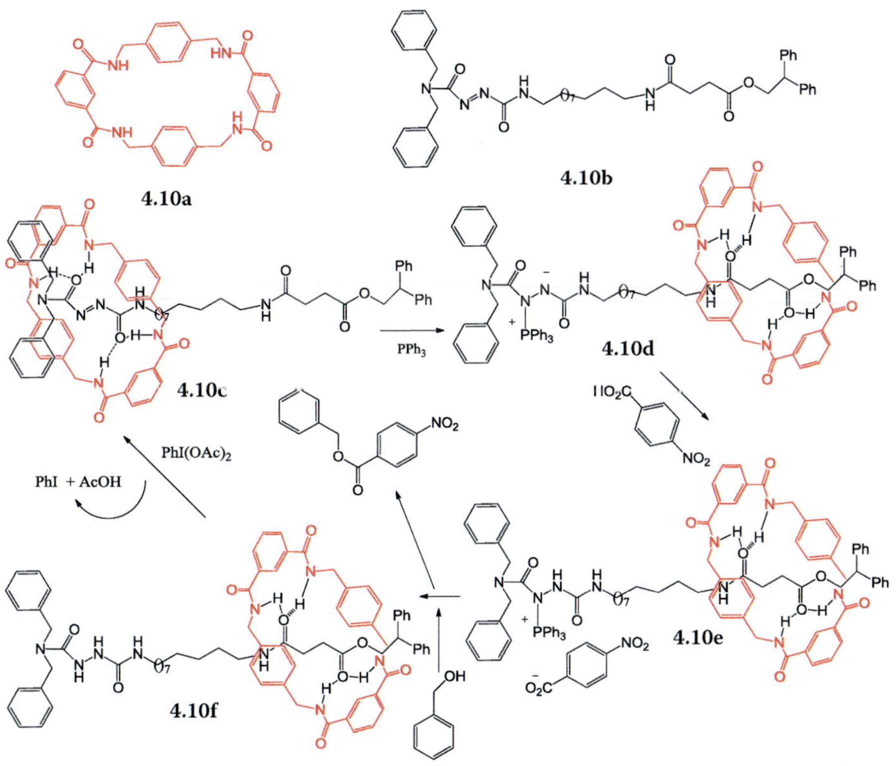

FIGURE 4.10 Modulation of catalytic activity of a rotaxane-based catalyst possessing two catalytic sites by exposing one of the catalytic site through a chemical reaction. (From J. Berna et al. *J. Am. Chem. Soc.* 2010, 132, 10741–10747.[11])

triphenylphosphine. The reaction of phosphine results in the generation of a modified active site for catalytic reaction to take place.

4.1.2 Utility of Interlocked Systems in Supramolecular Catalysis

The above principles of changing positions of a ring over an axle are useful in carrying out two different types of reactions at two different conditions by using the same catalyst. The rotaxane made up of an axle **4.11a** and ring **4.11b** is used for carrying out a C—C bond formation reaction between α,β-unsaturated carbonyl compounds or nitro-olefin with β-diketones like dibenzoylacetone. In acidic conditions, the ring of the rotaxane is located over the protonated amine group. This exposes the squaramide group to cause catalysis of C—C bond formation between nitro-olefins with β-diketones. The addition of a base generates neutral amine, and the ring moves to hydrogen-bond with the squaramide unit and conceals this part. This form of the rotaxane catalyses the Michael addition reaction between the α,β-unsaturated carbonyl compound with β-diketones (Figure 4.11).

Sequential reactions leading to the movements of rings towards different positions on an axle of rotaxane coupled with a group transfer reaction are utilised in polypeptide synthesis in a selective manner. The interlocked DNA molecules forming a catenane-type structure act as catalysts for switchable cyclic catalytic reactions. The active and passive states of the DNA-catenane cause a catalytic reaction of luminol with hydrogen peroxide by modification of the catenanelike assembly of the two DNA. The catalyst passivation is done by coordination to a magnesium ion by one of the DNA rings, and activation is done by releasing a magnesium ion from the DNA by using blocker/displacer strands.

A catenane-based phosphoric acid—derived catalyst **4.12b** is used in asymmetric catalysis of hydrogenation reactions of quinolone derivatives. It provides highly stereoselective (up to 95%) products, which is not achieved by using a catalytic amount of a noninterlocked system **4.12a** (Figure 4.12). The interlocked system with flexible ether linkages provides an adequate conformation to provide higher amounts of phosphoric acid groups in a local environment. The overall geometrical features of the chiral interlocked system are responsible for higher enantioselectivity. The higher local concentration of the acid is responsible for the performance of the catalyst at very low concentrations (2.5 mol %).

4.1.3 Metallacycles in Catalysis

Metallacycles are observed as reaction intermediates in many organometallic reactions. Ligand modification allows synthesis of metallacycles of different nuclearities. Modifications of ligands on a metallacycle recognise substrate. Metallacycles are able to adopt porous structures which provide advantages over conventional catalysts. For example, the metallacycle **4.13a** composed of a diazapyrenium ligand coordinated to platinum(II) is a reusable catalyst for the nucleophilic substitution reaction of 3,5-dinitro-bromobenzene with sodium azide (Figure 4.13). The substitution reaction is 1,000 times faster than an uncatalysed reaction. The nitrate salt of the ligand has poor catalytic activity. The addition of the nonreactive aromatic compound pyrene to

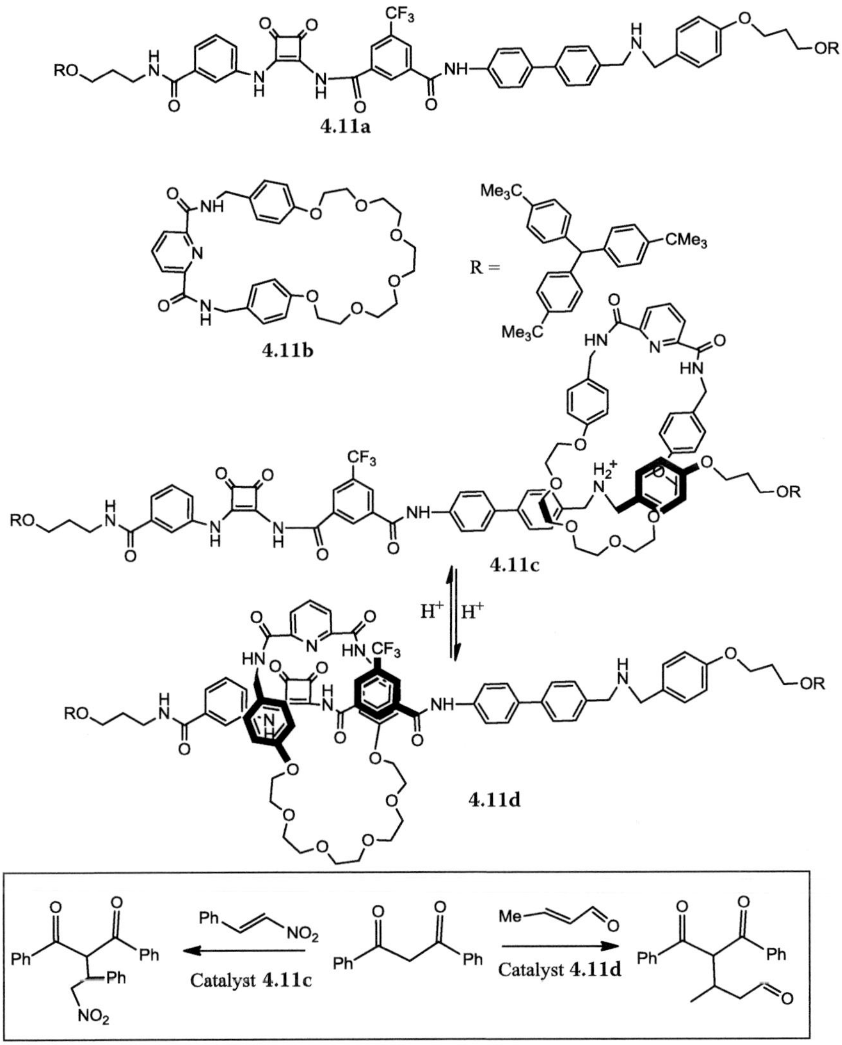

FIGURE 4.11 Dual catalysis by changing reaction conditions. (From J. Beswick et al. *Chem. Sci.* 2015, 6, 140–143.[12])

the reaction mixture of 3,5-dinitro-bromobenzene with sodium azide in the presence of catalytic amounts (10 mol %) of the platinum complex drastically decreases the reaction rate. Beyond the porous nature of the complex, the interior portion of the pores being filled with π-deficient aromatic molecules makes ways for pyrene, thereby decreasing reactivity. The catalytic ability of the complex increases by association of the azide to the diazapyrenium cationic subunit **4.13b**.

The complex **4.14b** is an extraordinary catalyst for selective acyl-transfer reaction. This Cp*Rh-based functional metalla-rectangle is composed of two coordinatively unsaturated copper centres at two opposite sides. The copper ions provide a

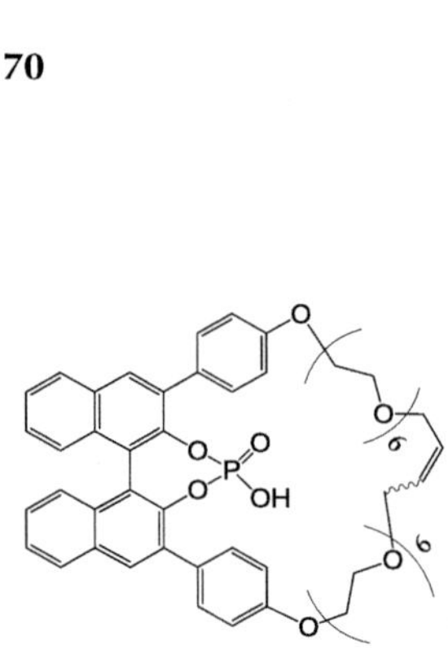

4.12a

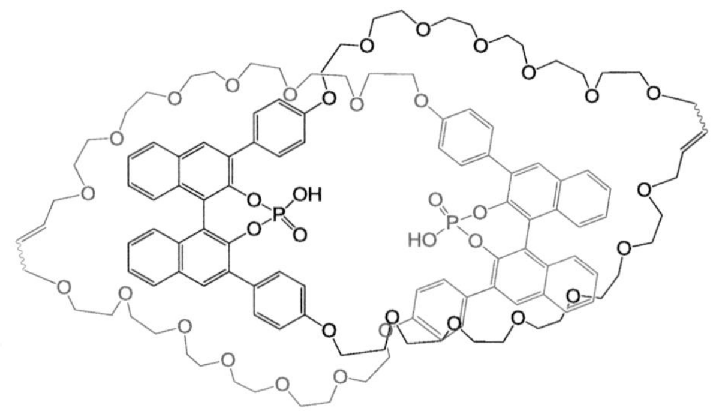

4.12b

FIGURE 4.12 The monomeric and catenated forms of a cyclic 1,11′-binaphthyl-phosphoric acid derivative. (From R. Mitra et al. *Angew. Chem. Int. Ed.* 2017, 56, 11456–11459.[13])

coordination site to (pyridin-3-yl)methanol and 1-(1H-imidazol-1-yl)ethanone (Figure 4.14). The alcohol group of the former reacts with the carbonyl functional group to stabilise the intermediate **4.14b**. The intermediate has a conducible distance to stabilise the condensed intermediate within the void. It releases imidazole and (pyridin-3-yl)methyl acetate after acetylation.

The cage-like complex [[(tmeda)Pd]$_3$(L)$_2$](BF$_4$)$_6$ **4.15b** has a difference in structure when it is in hydrate or anhydrous form (tmeda = tetramethylethylenediamine). The metallacycle has tridentate ligands L = 1,3,5-tris(isonicotinoyloxyethyl)cyanurate, and each ligand has a central core of cyanurate linking three pyridine units through flexible tethers as **4.15a**, shown in Figure 4.15. The cage effectively causes catalytic Suzuki-Miyaura C—C cross-coupling reactions. The efficiencies of the catalytic activity of the hydrate and anhydrous forms are different. The complex with the

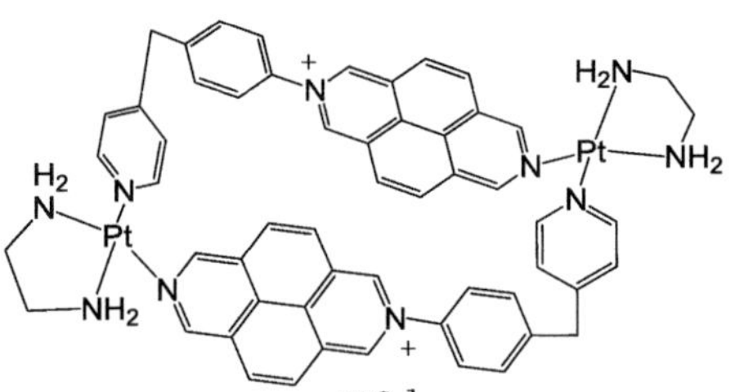

[NO3]6
4.13a

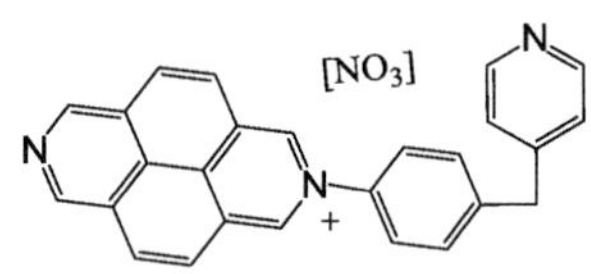

[NO3]
4.13b

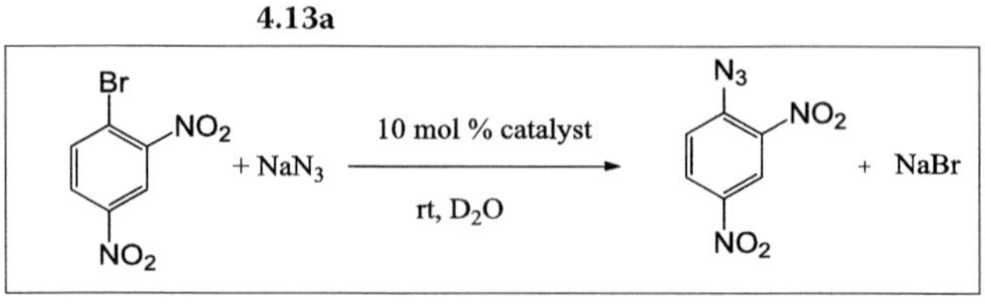

FIGURE 4.13 The nucleophilic substitution reaction of 2,4-dinitro-bromobenzene by azide catalysed by platinum metallacycle 4.13a derived from the ligand 4.13b. (From E. M. López-Vidal et al. *J. Org. Chem.* 2014, 79, 1265–1270.[14])

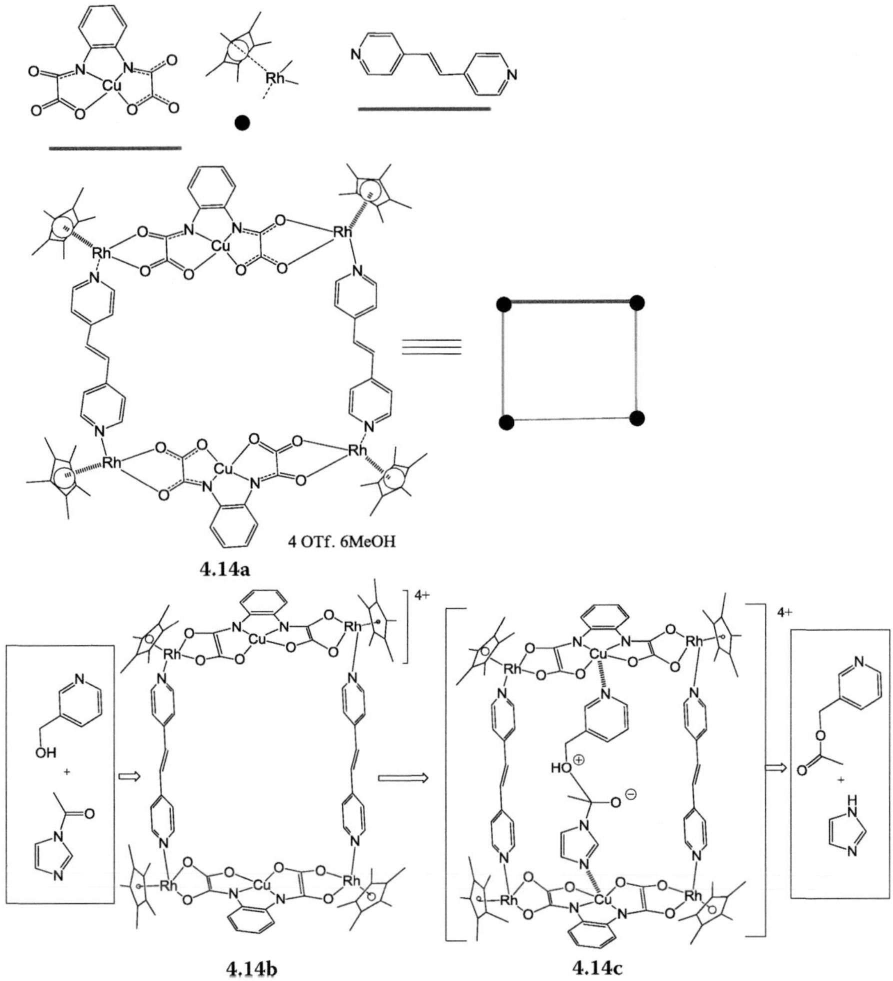

FIGURE 4.14 A catalytic acyl-transfer reaction by a rhodium-based rectangular-shaped metallacycle 4.14a and the intermediate steps involved in the catalytic activation. (From S.-L. Huang et al. *J. Am. Chem. Soc.* 2013, 135, 8125–8128.[15])

nitrate anion is reversibly dehydrated or hydrated on heating and cooling. The C—C bond-forming reaction by the hydrated catalysts is higher at low as well as high temperatures. The anhydrous form in each case gives a lower yield.

4.2 CONTROL OF CATALYTIC ACTIVITY BY SWITCHING

The control of reactivity by modifying the structural or electronic environment at localised sites is as important as the switching of reactivity by concealing and revealing reactive sites in interlocked molecules. Localised environmental changes, particularly by change of oxidation site or electronic properties of a delocalised

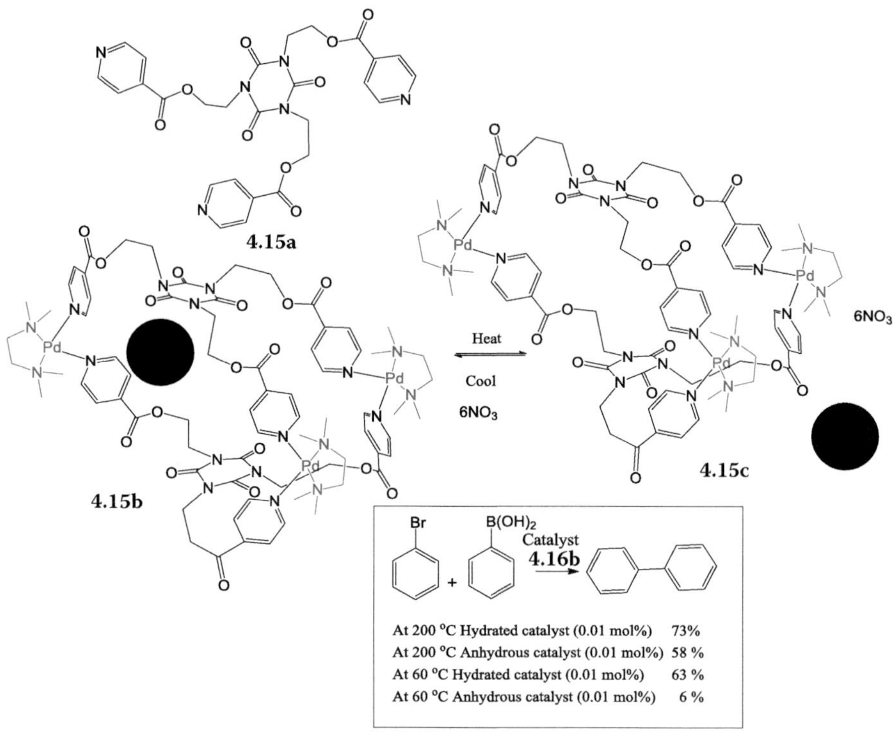

FIGURE 4.15 A palladacycle with different catalytic activity in hydrated form. (From T. H. Noh et al. *J. Am. Chem. Soc.* 2011, 133, 1236–1239.[16])

system, influence the overall reactivity of a catalyst. The acetonitrile ligands of the ruthenium complex **4.16a** are easily replaced by a ligand composed of imidazole tethered benzylic carbonyl compound. This complex activates the reactant by complexation but also serves as an electron sink to transfer electrons when the intermediate complex is irradiated. The transfer of electrons facilitates the generation of a CCl_3 radical from CCl_3Br. The CCl_3 radical reacts with the coordinated ligand, and the activation process is illustrated through an intermediate **4.16d**, which results in the product **4.16c** (Figure 4.16). The reaction continues with the coordination of another set of imidazole-carbonyl compounds and is photo-activated to perform the next cycle of the catalytic reaction.

Geometrical changes of metal complexes are often used in conventional catalytic systems to differentiate reactivity. Supramolecular features are greatly influenced in metal complexes by changes of structural features through photo-isomerization. A 3-(2-(2,5-dihydrothiophen-3-yl)-3,3,4,4,5,5-hexafluorocyclopent-1-en-1-yl)-2,3-dihydrothiophene derivative with two 4-isopropyl-4,5-dihydrooxazole adopts different structures under different conditions. One complex has a scissorlike geometry. It has two 4-isopropyl-4,5-dihydrooxazole units coordinating through the nitrogen atoms of the rings to the copper (I) ion. This complex **4.17a** changes geometry upon irradiation at 313 nm to form another copper complex **4.17b** in which the

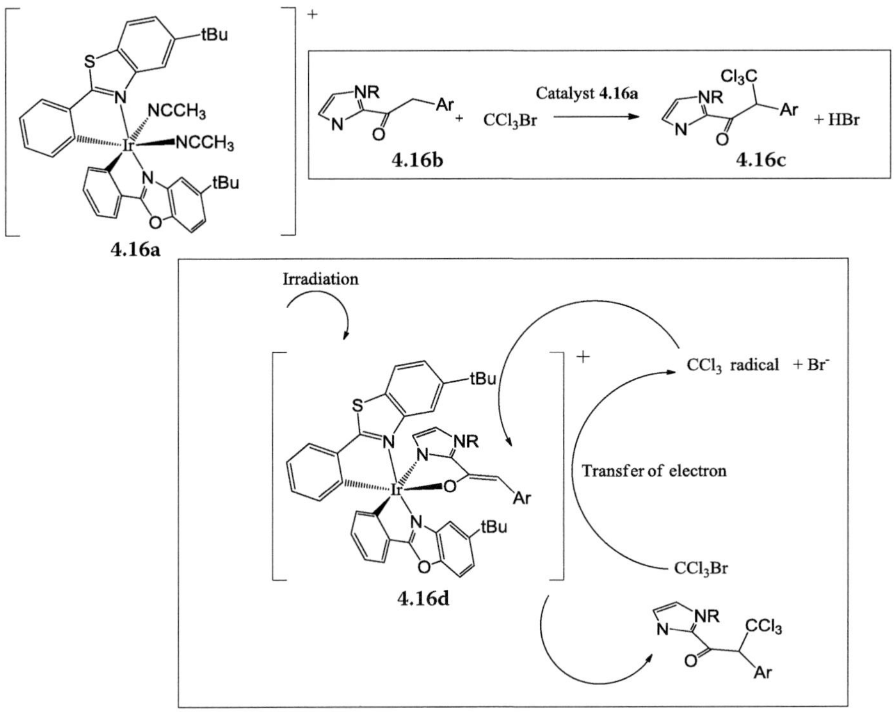

FIGURE 4.16 Localised switching of a radical reaction by photo-irradiation (top left: an iridium-based catalyst 4.16a, top right: a catalytic photo-halogenation reaction and below: the localised switching process). (From E. Meggers, *Angew. Chem. Int. Ed.* 2017, 56, 5668–5675.[17])

4-isopropyl-4,5-dihydrooxazole groups are at *trans*-disposition. The complex **4.17a** is a catalyst for the cyclopropanation reaction between styrene and ethyl 2-diazoacetate. The catalytic reaction yields the stereoisomer of ethyl 2-phenylcyclopropanecarboxylate. The catalytic reactions of both complexes yield higher amounts of (1S,2S)-ethyl 2-phenylcyclopropanecarboxylate over the other stereoisomer, but a higher selectivity is observed with the complex **4.17a** as catalyst (Figure 4.17).

As mentioned in an earlier section, the *cis* and *trans* isomers of complexes or substrates have different catalytic activity. A series of such catalysts is listed in Figure 4.18; these catalysts change geometry by application of stimuli: one isomer has higher catalytic activity.

Also, a series of reactions where the catalytic activity is controlled by *cis-trans* isomerization of catalysts is listed in Figure 4.19. The hydrolysis of amide derivatives is catalysed by the *cis*-form **4.18a** efficiently over the *trans*-form **4.18b**; the reaction between 3-phenylpropanal with cyclopenten 3-one is catalysed by the *cis*-form **4.18c**; hydrolysis of the 4-nitrophenyl ether of sugar derivative is efficiently catalysed by the *cis*-form of the diphenylcarboxylate derivative **4.18e**; and, finally, the coupling between the amine **4.19h** and acid derivative **4.19g** is favourably catalysed by the *cis*-form **4.18g**. In all these examples, the *cis*-form is transformed to the *trans*-form at thermal conditions; hence, in solution, the reactions are switched on by irradiation.

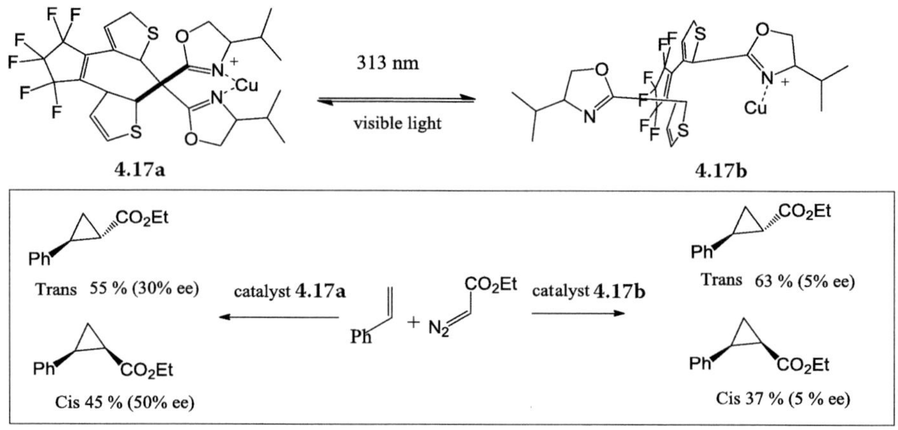

FIGURE 4.17 Geometrical transformation of catalyst to control stereospecificity. (From D. Sud et al. *Angew. Chem. Int. Ed.* 2005, 44, 2019–2021.[18])

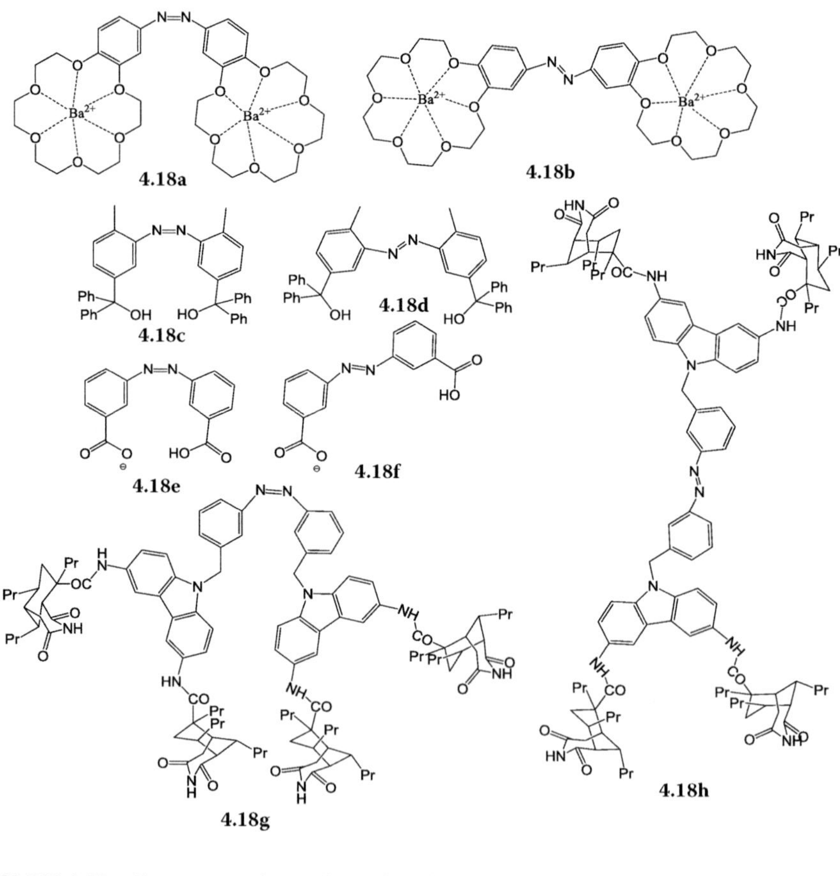

FIGURE 4.18 Some example catalysts that change catalytic reactivity upon irradiation or thermal treatment. (From V. Blanco et al. *Chem. Soc. Rev.* 2015, 44, 5341–5370.[19])

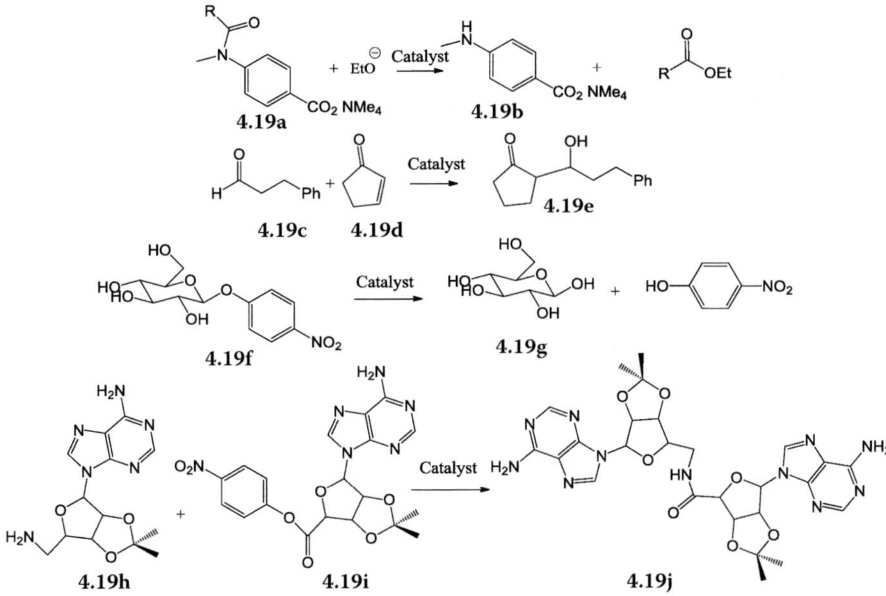

FIGURE 4.19 Specific examples of catalytic reactions in which control of reactivity is achieved by photo- and thermochemical means.

4.2.1 Chemical Switching

By adding a competing ligand or a complex, a catalytically inactive complex reorganises to another complex that is suitable to act as a catalyst. The copper complex **4.20a** is not a catalyst for C—C bond formation between diethylmalonate and 4-nitrobenzaldehyde. So is the complex **4.20b**. The complex **4.20b** has a zinc ion coordinated strongly to a porphyrin ring and pyridine ligand. The two complexes **4.20a** and **4.20b** together are also not catalysts for the reaction (Figure 4.20), whereas the addition of morpholine to the reaction mixture provides an active catalyst for the C—C bond. In this reaction, the morpholine occupies one of the axial positions of the zinc porphyrin complex **4.20c**, which releases the terpyridine end of the ligand that becomes free to bind a copper ion. This unit, together with the bipyridine unit of another arm of the ligand, abstracts the copper ion from the complex **4.20a** to form the complex **4.20d**. The ligand **4.20c** remains free via these exercises. The complex **4.20d** is an active catalyst for the coupling reaction between diethylmalonate and 4-nitrobenzaldehyde to form a C—C bond. The reaction sequence to generate the catalyst is like a chemical triggering of the porphyrin complex to generate a new catalyst.

Another example of chemical switching of chemical reactivity is shown in Figure 4.21. In this example, the catalyst **4.21a** has a catalytically active squaramide part concealed due to the geometrical constraints of the complex. The platinum complex **4.21a** reorganises the structure by accepting a chloride ligand to form a *trans*-form **4.21b**. The coordination of the chloride ion to the platinum site releases

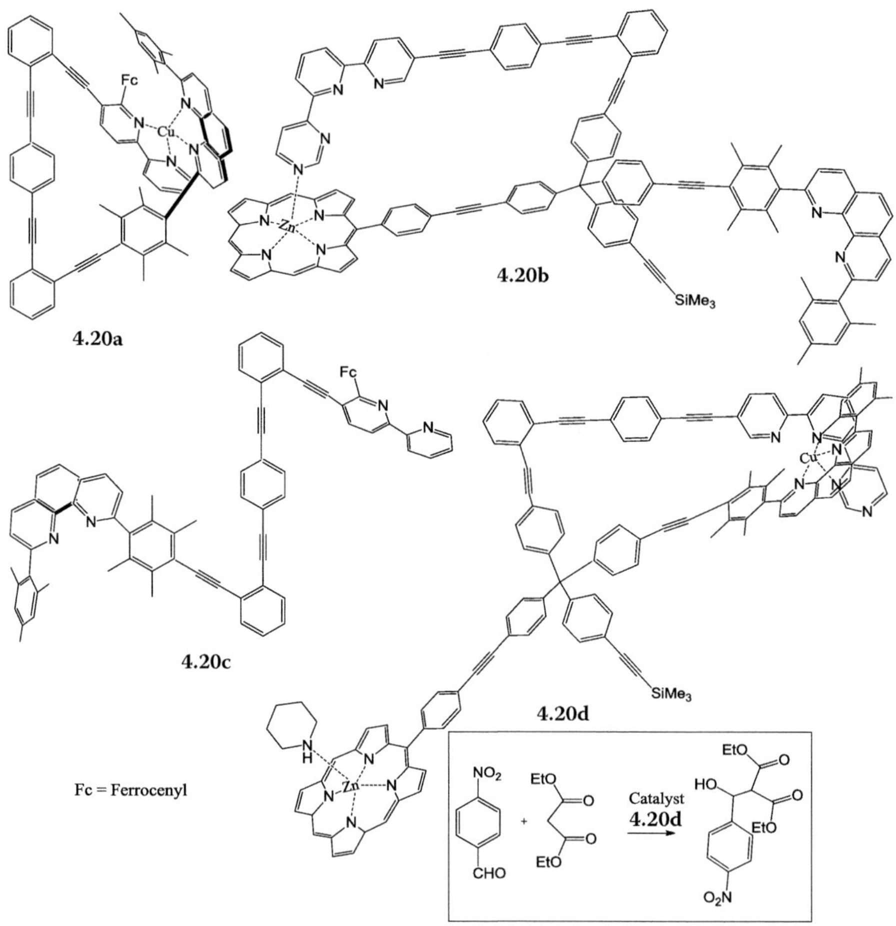

FIGURE 4.20 Modulation of catalytic activity of a complex by ligand exchange. (From S. Gaikwad et al. *Dalton Trans.* 2018, 47, 1786–1790.[20])

one binding site at each platinum centre to accommodate the chloride in the tetra coordination environment of platinum. One of the PS-ligands acts as a monodentate P-ligand on each platinum ion in **4.21b**. It happens as a consequence of the release of sulphur from the coordination site by the chloride ion. This exposes the squaramide part to activate the substrate for catalytic reaction.

4.2.2 Electrochemical and Photochemical Switching

The reactivity of large numbers of supramolecular catalysts is modulated by changing redox active sites. For example, the copper complex **4.22a** changes its coordination environment due to the change of the oxidation state, differentiable binding affinity of sulphur and oxygen atoms to copper (II) and copper (I) ions and also due to different coordination numbers of copper ions in different oxidation states (Figure 4.22). The complex **4.22a** has a copper (II) ion in an N3OX (X = Cl, Br) five-coordination

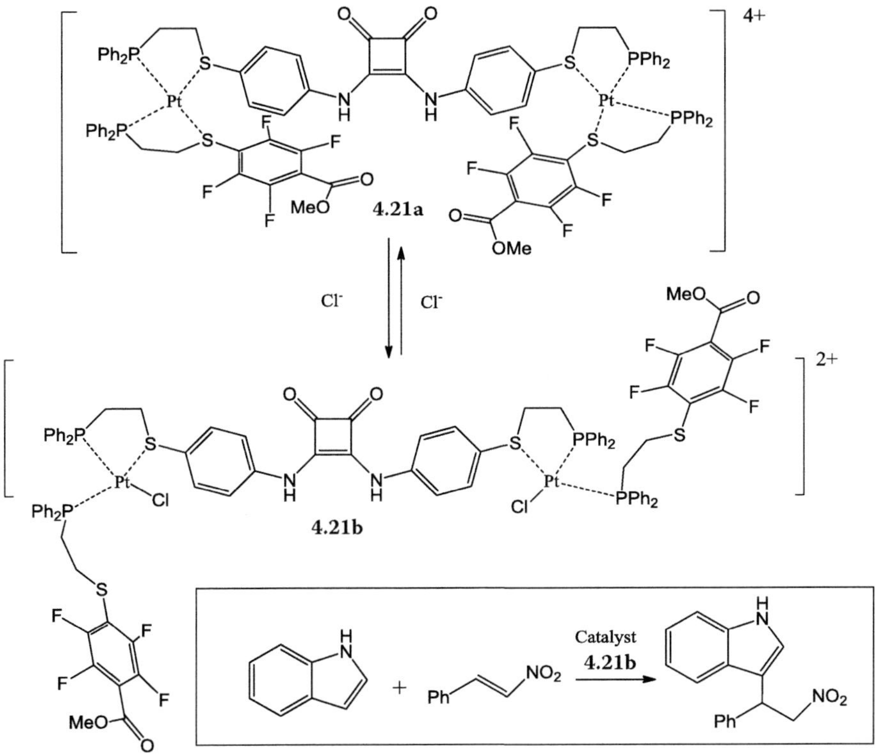

FIGURE 4.21 An example of a chloride ion-driven switching of catalyst. (From M. Zirngast et al. *Chem. Commun.* 2011, 47, 2261–2263.[21])

environment, whereas the copper (I) complex **4.22b** of the same ligand is in an N3S four-coordination environment. Both complexes catalyse the coupling reaction between (E)-(2-nitrovinyl) benzene and β-diketoesters to form specific isomers of dialkyl 2-(2-nitro-1-phenylethyl)malonate. In the case of copper (II), the complex provides the R-isomer in 70% enantiomeric excess, while the copper (I) complex results in the S-stereoisomer in 72% enantiomeric excess. Stereoisomer (3S,6R)-3,6-dimethyl-1,4-dioxane-2,5-dione undergoes a catalytic polymerization reaction by the titanium Schiff—base catalyst with two ferrocene units (Figure 4.23). This complex, when electrochemically oxidised, is not suitable for catalyzing the ring-opening reaction.

4.2.3 Stimulus-Guided Switching

A molecular motor composed of a dimethylaminopyridine (DMAP) and a thiourea, catalyse the Michael addition reaction of an aromatic thiol to a 2-cyclohexen-1-one. Depending on the light and temperature as stimuli, the motor adopts different orientations. The different orientations influence the stereochemistry of the products. The compound has a centrally located double bond connected to extend the two arms. The double bond is connected through two functionalised

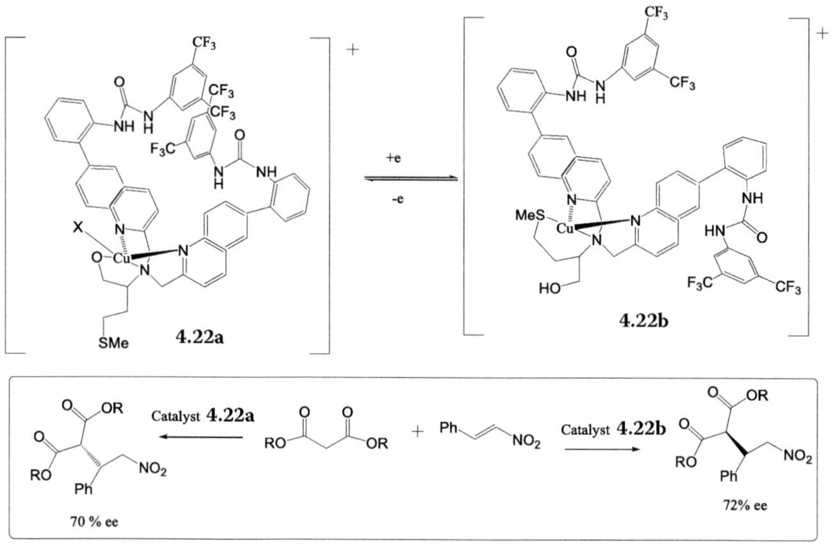

FIGURE 4.22 Changes in the reactivity of a copper complex by electrochemical means. (From S. Mortezaei et al. *J. Am. Chem. Soc.* 2012, 134, 8054–8057.[22])

2,4,7-trimethyl-2,3-dihydro-1H-indene units. The methyl group on each five-member ring makes the molecule to adopt different geometries other than the two obvious *cis*- and *trans*-orientations. The *cis*- and *trans*-orientations have two more options to project one portion of the tethers, namely, up or down orientations across the double bond. These arrangements are referred to here as stages I-IV, as illustrated

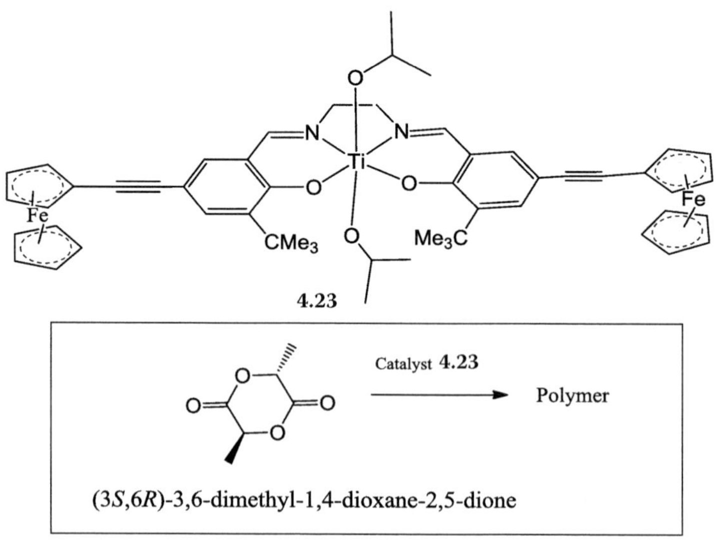

FIGURE 4.23 A switchable titanium catalyst in ring opening polymerization. (From C. K. A. Gregson et al. *J. Am. Chem. Soc.* 2006, 128, 7410–7411.[23])

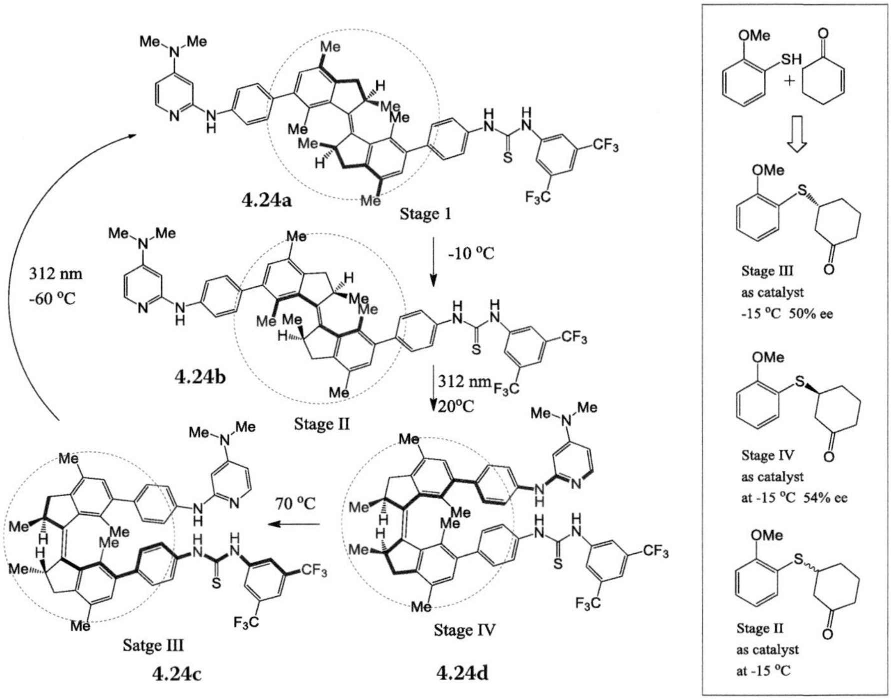

FIGURE 4.24 A molecular motor catalyzing enantioselective C—S bond formation reaction. (From J. Wang, B. L. Feringa. *Science* 2011, 331, 1429–1432.[24])

in Figure 4.24. The portion where the projection across the double bond is above the plane of the paper is shown in blue, and the geometry of the double bond is circled in red. These stages change under different reaction conditions. For example, stage I converts to stage II at −10°C and stage II to III at 20°C under irradiation with a 310 nm wavelength; stage III converts to stage IV at 70°C, and, finally, stage I is recovered from stage IV by irradiation with 312 nm at −60°C. The compound is a catalyst for the coupling reaction of 2-methoxyphenylthiol with α,β-cyclyohexeneone. The 0.3 mol % of catalyst stage I yields a product with no stereoselectivity and very poor yield. Upon irradiation with 312 nm UV at −15°C, stage II is generated, and this stage is catalytically active: 0.3 mol % catalyst affords 50% yield and 50% enantiomeric excess. The compound at stage III acts as a catalyst with higher activity, up to 83% yield under similar conditions. In the case of stage III as a catalyst, opposite enantiomers of the one that is formed by the stage II catalyst are observed.

The availability of a particular ligating site on the *cis*- and *trans*-isomers of a ligand depends on the steric factors of substituents. The (Z)-2-(3,5-dimethylstyryl) pyridine derivative **4.25a** is suitable ligand to coordinate to the porphyrin complex of aluminium through the nitrogen atom of the pyridine ring from one face. On the other face of the complex, there is a methoxy group. The complex catalyses the reaction between oxiranes with carbon dioxide to form 1,3-dioxolan-2-one (Figure 4.25). (Z)-2-(3,5-dimethylstyryl)pyridine transforms to the corresponding

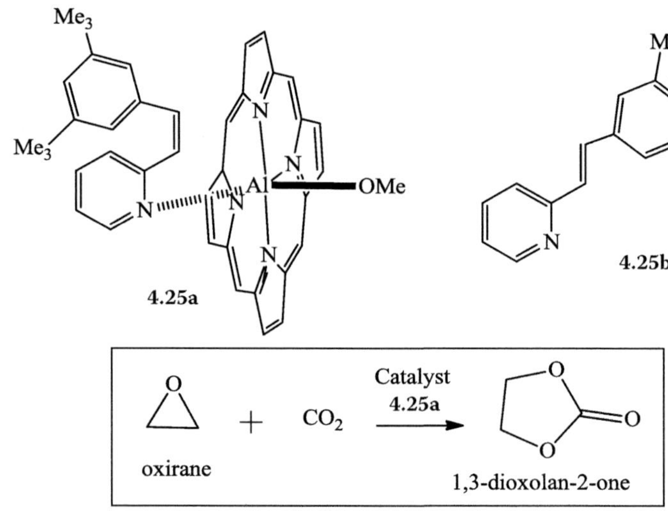

FIGURE 4.25 Steric effect of *cis-trans* isomers differentiating coordination behaviour to influence catalysis. (From H. Sgimoto et al. *J. Am. Chem. Soc.* 1999, 121, 2325–2326.[9])

trans-isomer. This isomer is not suitable to coordinate to the aluminium porphyrin unit. The reaction does not take place with a catalytic amount of the *trans*-isomer of the ligand with the aluminium porphyrin complex, but the *trans*-isomer can be converted to the *cis*-isomer upon irradiation. Thus, upon irradiation, the catalytic activity of the complex is regained.

REFERENCES

1. D. A. Leigh, V. Marcos, M. R. Wilson, Rotaxane catalysts. *ACS Catal.* 2014, 4, 4490–4497.
2. B. M. Neilson, C. W. Bielawski, Illuminating photoswitchable catalysis. *ACS Catal.* 2013, 3, 1874–1885.
3. S. Hoekman, M. O. Kitching, D. A. Leigh, M. Papmeyer, D. Roke, Goldberg active template synthesis of a [2]rotaxane ligand for asymmetric transition-metal catalysis. *J. Am. Chem. Soc.* 2015, 137, 7656–7659.
4. G. Hattori, T. Hori, Y. Miyake, Y. Nishibayashi, Design and preparation of a chiral ligand based on a pseudorotaxane skeleton: Application to rhodium-catalyzed enantioselective hydrogenation of enamides. *J. Am. Chem. Soc.* 2007, 129, 12930–12931.
5. N. Miyagawa, M. Watanabe, T. Matsuyama, Y. Koyama, T. Moriuchi, T. Hirao, Y. Furusho, T. Takata, Successive catalytic reactions specific to Pd-based rotaxane complexes as a result of wheel translation along the axle. *Chem. Commun.* 2010, 46, 1920–1922.
6. M. V. Peters, R. S. Stoll, A. Kühn, S. Hecht, Photoswitching of basicity. *Angew. Chem. Int. Ed.* 2008, 47, 5968–5972.
7. A. Ueno, K. Takahashi, T. Osa, Photocontrol of catalytic activity of capped cyclodextrin. *J. Chem. Soc. Chem. Commun.* 1981, 3, 94–96.
8. O. Bistri, O. Reinau, Supramolecular control of transition metal complexes in water by a hydrophobic cavity: A bio-inspired strategy. *Org. Biomol. Chem.* 2015, 13, 2849–2865.

9. H. Sgimoto, T. Kimura, S. Inoue, Photoresponsive molecular switch to control chemical fixation of CO_2. *J. Am. Chem. Soc.* 1999, 121, 2325–2326.
10. C.-S. Kwan, A. S. C. Chan, K. C.-F. Leung, A fluorescent and switchable rotaxane dual organocatalyst. *Org. Lett.* 2016, 18, 976–979.
11. J. Berna, M. Alajarín, R.-A. Orenes, Azodicarboxamides as template binding motifs for the building of hydrogen-bonded molecular shuttles. *J. Am. Chem. Soc.* 2010, 132, 10741–10747.
12. J. Beswick, V. Blanco, G. D. Bo, D. A. Leigh, U. Lewandowska, B. Lewandowski, K. Mishiro, Selecting reactions and reactants using a switchable rotaxane organocatalyst with two different active sites. *Chem. Sci.* 2015, 6, 140–143.
13. R. Mitra, H. Zhu, S. Grimme, J. Niemeyer, Functional mechanically interlocked molecules: Asymmetric organocatalysis with a catenated bifunctional Brønsted acid. *Angew. Chem. Int. Ed.* 2017, 56, 11456–11459.
14. E. M. López-Vidal, A. Fernández-Mato, M. D. García, M. Pérez-Lorenzo, C. Peinador, J. M. Quintela, Metallacycle-catalyzed S_NAr reaction in water: Supramolecular inhibition by means of host—guest complexation. *J. Org. Chem.* 2014, 79, 1265–1270.
15. S.-L. Huang, Y.-J. Lin, T. S. A. Hor, G.-X. Jin, Cp*Rh-based heterometallic metalla rectangles: Size-dependent Borromean link structures and catalytic acyl transfer. *J. Am. Chem. Soc.* 2013, 135, 8125–8128.
16. T. H. Noh, E. Heo, K. H. Park, O.-S. Jung, Motion of an isolated water molecule within a flexible coordination cage: Structural properties and catalytic effects of ionic palladium(II) complexes. *J. Am. Chem. Soc.* 2011, 133, 1236–1239.
17. E. Meggers, Exploiting octahedral stereocenters: From enzyme inhibition to asymmetric photoredox catalysis. *Angew. Chem. Int. Ed.* 2017, 56, 5668–5675.
18. D. Sud, T. B. Norsten, N. R. Branda, Photoswitching of stereoselectivity in catalysis using a copper dithienylethene complex. *Angew. Chem. Int. Ed.* 2005, 44, 2019–2021.
19. V. Blanco, D. A. Leigh, V. Marcos, Artificial switchable catalysts. *Chem. Soc. Rev.* 2015, 44, 5341–5370.
20. S. Gaikwad, S. Pramanik, S. De, M. Schmittel, A high-speed network of nanoswitches for on/off control of catalysis. *Dalton Trans.* 2018, 47, 1786–1790.
21. M. Zirngast, E. Pump, A. Leitgeb, J. H. Albering, C. Slugovc, Pyridine as trigger for chloride isomerisation in chelated ruthenium benzylidene complexes: Implications for olefin metathesis. *Chem. Commun.* 2011, 47, 2261–2263.
22. S. Mortezaei, N. R. Catarineu, J. W. Canary, A redox-reconfigurable, ambidextrous asymmetric catalyst. *J. Am. Chem. Soc.* 2012, 134, 8054–8057.
23. C. K. A. Gregson, V. C. Gibson, N. J. Long, E. L. Marshall, P. J. Oxford, A. J. P. White, Redox control within single-site polymerization catalysts. *J. Am. Chem. Soc.* 2006, 128, 7410–7411.
24. J. Wang, B. L. Feringa. Dynamic control of chiral space in a catalytic asymmetric reaction using a molecular motor. *Science* 2011, 331, 1429–1432.

5 Dendrimers in Supramolecular Catalysis

5.1 PRINCIPLES AND DEFINITIONS

Hyperbranched macromolecules with branches similar to trees are called dendrimers. The branches are grown step by step to build up as generation after generation and, depending on their having a skeleton of generation, they are classified as first generation, second generation and so on. As the dendrimer grows to higher dimensions, the types of pores increase and the sizes of pores at the interior sites are different as it progresses to the outer periphery.[1] Thus, the structures of dendrimers are very important. It is a challenge to prepare uniformly grown robust dendrimers. Certain dendrimers accommodate catalytic species at the interior, and the included guest substrates in such dendrimers react with them. Such catalysts are encountered in dendrimers with catalytic cores or dendrimers that are composed of dendritic wedges. The terminal functional groups of dendrimers are modified through functionalisation with supramolecular entities, complexation with metal ions or anchoring exterior parts to supports. Each modified material has its own characteristic features, leading to different catalytic entities. Metallic nanoparticles are used as cores of dendrimers or encapsulated in the voids. Catalytic reactions often take advantage of the varied internal structures of dendrimers. When a catalytic unit is linked to a dendrimer, the catalytic activity and selectivity of the catalyst are changed from the original catalyst. Such an effect on the changing reactivity of the catalyst by association with the dendrimer is called a dendritic effect. A major focus of interest in dendrimers lies in their use as heterogeneous catalysts with recyclability. A good number of colloidal solutions of nanoparticles encapsulated or linked to dendrimer systems are microheterogeneous. The majority of them are based on noble metal nanoparticles. They are useful catalysts for conventional reactions that are otherwise also catalysed by such metals with lower or equal efficiencies but with a different selectivity. Reactions such as hydrogenation, carbon—carbon and carbon—heteroatom (nitrogen, oxygen, sulphur, etc.) coupling reactions, oxidation of alcohols and cyclisation reactions for synthesis of heterocyclic compounds are common reactions catalysed by nanoparticle-dendrimer—based catalyst systems.[2]

5.2 CONVERGENT AND DIVERGENT APPROACHES FOR DENDRIMERS

The focal point of a dendrimerlike structure is the pivot point on which the branches are built. Thus, there are two approaches; one is called the convergent approach, in which small divergent building blocks of a repetitive nature are built independently and connected to a pivot or a template. The term *dendron* usually refers to a single

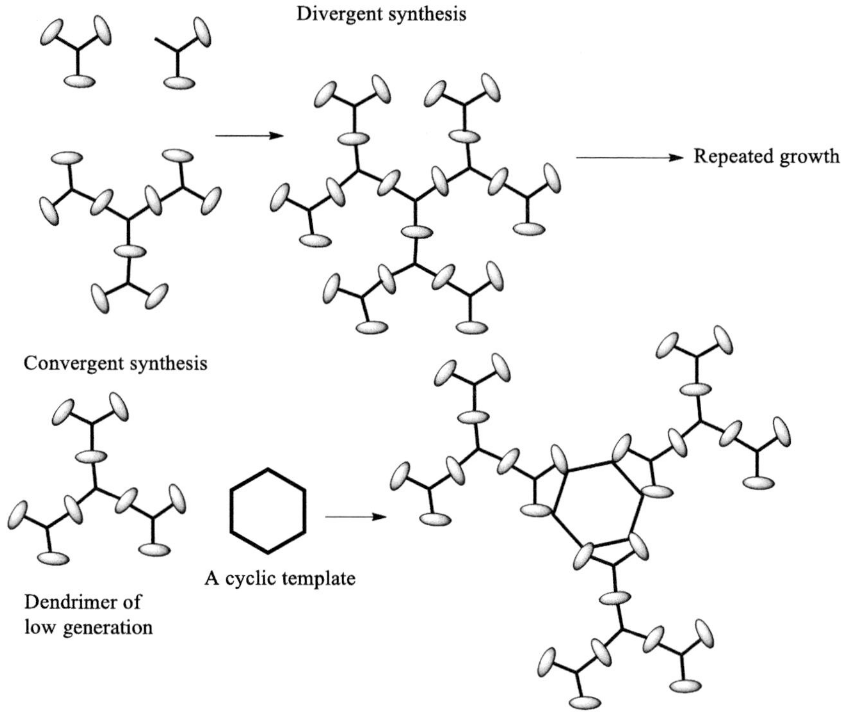

FIGURE 5.1 Schematic representations of convergent and divergent synthesis of dendrimers. (From S. M. Grayson, J. M. J. Frechet, *Chem. Rev.* 2001, 101, 3819–3868;[1] M. Bernechea et al. *Inorg. Chem.* 2009, 48, 4491–4496.[3])

chemically addressable group that is grown on a core or identified end. For example, in Figure 5.1, a hexagonal structure is taken as a template or pivot and smaller building units (dendrons) are attached to it. In the divergent approach, a pivot is constructed or taken from readily available multifunctional compounds, or a functional group is taken as a pivot to build the branches one by one in a repetitive manner to grow branches from the pivot. Thus, the first uniform polyfunctionalisation across the pivot provides a dendrimer called a first-generation dendrimer. Uniform functionalisation over all the branching sites of the first-generation dendrimer results in the second generation, sequential repeat of which gives third-, fourth-, fifth-, sixth- or higher-generation dendrimers.

5.3 MODIFICATION OF INTERIOR AND EXTERIOR OF DENDRIMERS

The conventional way to modify a dendrimer is through attachment of ligating sites at appropriate places to form metal complexes. Dendrimers grow symmetrically through branching hence make avenues to functionalise on each branch to create multiple binding sites to anchor to metal ions. In a sense, the entire outer periphery of a dendrimer, once decorated with metal complexes, provides multiple catalytic sites

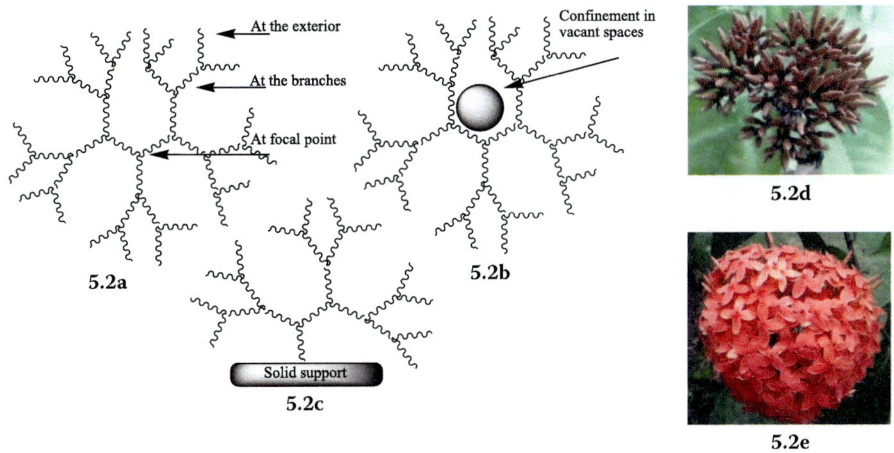

FIGURE 5.2 Schematic representations of different ways to functionalise dendrimers and comparison with floral arrangements.

to act collectively or independently as catalysts. The functionalisation of dendrimers is done at different sites and in different ways, as shown in Figure 5.2. The functional groups present at the exterior, interior or vacant coordination sites of metal ions at the interior are some ways to modify a dendrimer, as marked in **5.2a**. The introduction of guest molecules, including metallic nanoparticles in the voids, represented **5.2b** in Figure 5.2 is another way to modify a dendrimer. The immobilisation of dendrimers is done by anchoring them on solid metallic, oligomer, polymer, oxide or cluster surfaces through covalent or noncovalent interactions, as in **5.2c**. Dense branches are generated as the generation of the dendrimer increases. Depending on the dense nature of the dendrimer, its ability and mechanism to include guest molecules differ. As the dendrimer grows, the outer periphery is organised as a thick layer and is like blooming floral stages shown as two stages **5.2d** and **5.2e**.

Dendrimers are used as ligands to prepare metal complexes of different nuclearities. Conventionally, mononuclear complexes are formed when the pivot part acts as a binding site to occupy the coordination site of metal ions. Such complexes are used to provide hydrophobic confinement to metal complexes of prepared heterogeneous metal complexes. The nickel complex **5.3b** shown in Figure 5.3 is formed through a pincer-type ligand which is anchored through an amide bond to a dendrimer polyether. The complex is a useful catalyst of C—C bond formation between ethylene with styrene to form but-3-en-2-ylbenzene as a major product. The dendrimer **5.3a** has nickel complexes bound at its exterior. Preparing these complexes with uniform distributions of metal complexes is difficult due to steric and electronic factors. Carbosilane dendrimers functionalised with pincer-nickel(II) groups **5.3b** are created through the binding of NCN sites to the nickel(II) ion. Such complexes are used in regioselective catalysts in the reaction of alkenes with halogenated compounds under photochemical conditions to form polyhalogenated alkanes. In this catalysis, passivation of the catalyst occurs due to the irreversible formation of inactive nickel (III) species. Dendrimers with denser peripheries in metal sites are deactivated easily.

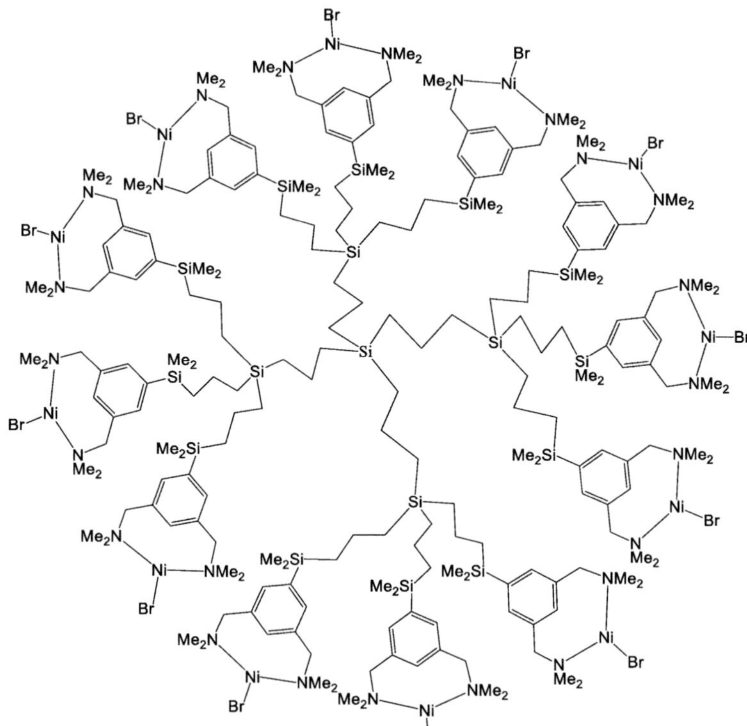

5.3a

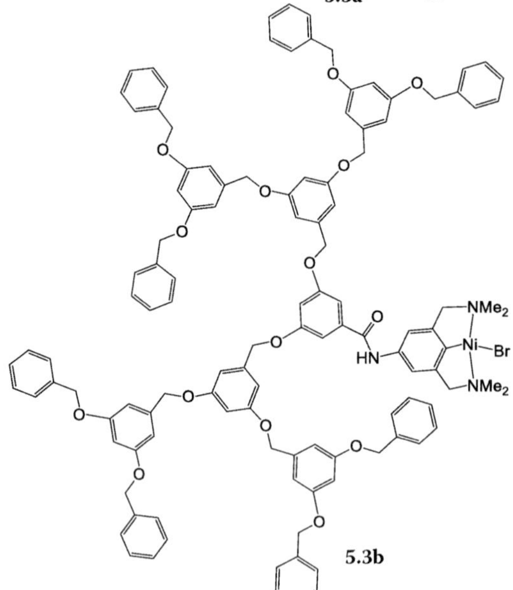

5.3b

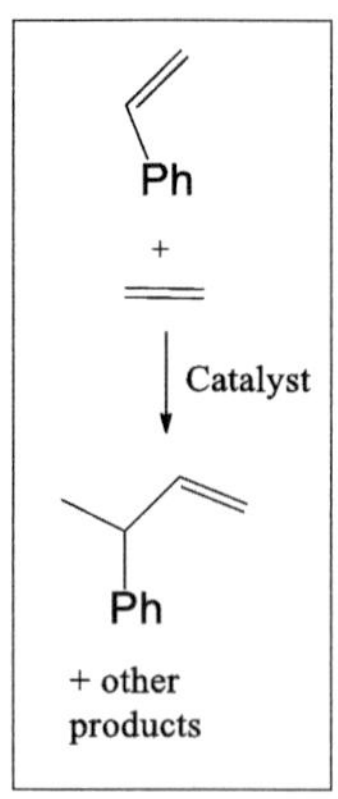

FIGURE 5.3 Pincer-like nickel complexes anchored at outer periphery of dendrimer; a nickel dendrimer complex and catalytic C—C bond formation reaction. (From R. van Heerbeek et al. *Chem. Rev.* 2002, 102, 3717–3756.[4])

Such an effect is called the *proximity effect* occurring between nickel centres. The addition of carbon tetrachloride to methyl methacrylate is catalysed by the dendrimer by these uniformly decorated nickel complexes at the outer periphery. The catalytic activity of the mononuclear NCN-pincer nickel complex is similar to the dendritic complex. The reaction passes through homogeneous catalytic paths, as revealed by a study on the reactivity of a third-generation nickel dendrimer by compartmentalising the reactant and catalyst via membrane. The dendrimer also catalyses hydro-vinylation of styrene and ethylene to form 3-phenyl-2-butene as a major product. Depending on the catalyst, concentration and reaction conditions, the dendrimer catalyses the racemization reaction of the chiral 2-phenyl-2-butene. Hence, there is a necessity to optimise catalyst performance to achieve stereo-selective products.

Carbosilane dendrimers with a [bis(diphenylphosphanyl)ferrocene] core are used as ligands with [crotyl-PdCl]$_2$ complex for high catalytic activity of allylic alkylation reactions (Figure 5.4). In the presence of the dendrimer ligand, the reaction of cinnamyl acetate with 2-methyl-diethylmalonate yields the compound (E)-diethyl 2-methyl-2-styrylmalonate (70% product). In the absence of the ligand, the catalytic reaction results in only 8% product. The dendrimer's external peripheries are also decorated by palladium complexes. Such complexes are also catalysts for C—C bond formation reactions (Figure 5.5).

Dendrimers or dendrons provide support to combine metallic nanoparticles on different supports. The functional groups in the dendrimers provide orientation and spaces for nanoparticles, and subtle changes brought about in the reaction condition change the courses of the catalytic reactions. Optimization of the reaction conditions to obtain the best results provides control of the catalytic centre environment so as to design new combinations that serve as superior catalysts. In general, the external

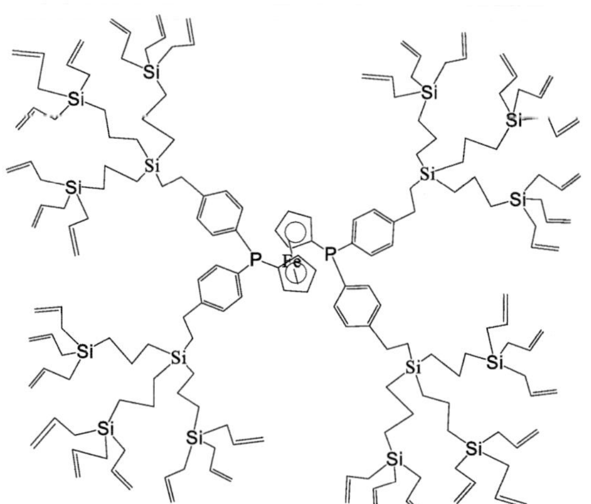

5.4a

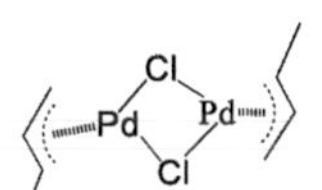

5.4b

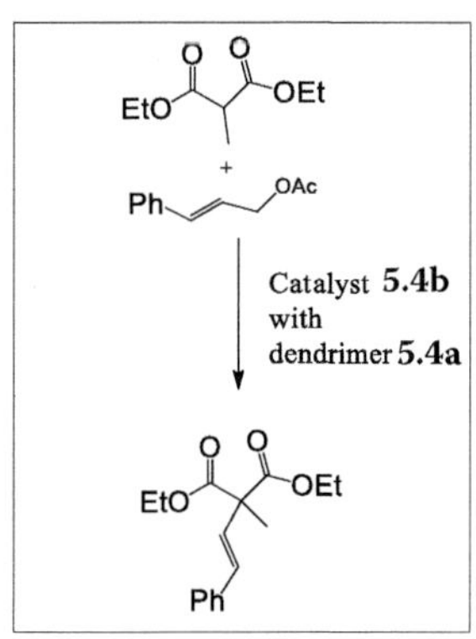

FIGURE 5.4 A dendrimeric diphosphine ligand and palladium-catalysed allylic alkylation reaction. (From G. E. Oosterom et al. *Chem. Commun.* 1999, 1119–1120.[5])

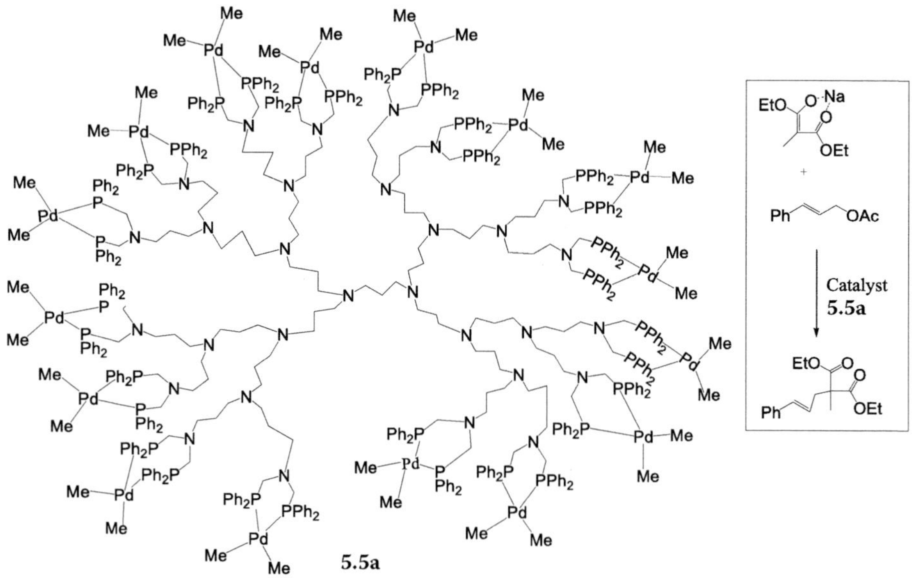

FIGURE 5.5 A dendrimeric palladium complex and catalysis in C—C bond formation. (From R. Andres et al. *New J. Chem.* 2007, 31, 1161–1191.[6])

peripheries of dendrimers do not prevent access of reagents to the metal nanoparticles that are held beneath the dendrimer. Rather, the outer peripheral part of dendrimers controls the process of the intake of substrates and expulsion of products assisting catalytic processes. Dendrimer-supported nanoparticles also provide a means to perform reactions at the interface or function as a combination of homogeneous and heterogeneous catalytic reactions. Heterogenisation of such nanoparticles is also done by polymeric or inorganic support or a combination of the two to generate new heterogeneous catalysts. Dendrons holding metal nanoparticles inhibit agglomeration of metal atoms, expose the maximum surface for catalytic reactions and allow nanoparticles to cause catalytic reactions.

The palladium-nanoparticle—cored third-generation dendrimer **5.6a** accommodates about 300 palladium atoms in the metallic core. The dendrimer has a spherelike arrangement of branches of the organic substrates composed of 14 third-generation dendrimers held to each nanoparticle to provide an average diameter of 2.0 nm for the nanoparticles. In this self-assembly, about 90% of the surfaces of the metal nanoparticles are exposed to cause catalytic reactions.

Palladium nanoparticles held as cores of dendrimers act as catalysts for olefin hydrogenation. In such hydrogenation reactions (Figure 5.6), higher-generation dendrimers show low performance with lower turnover frequency. The reactive palladium sites are being located in the interiors of the dendrimer, it is difficult for olefin substrates to get encapsulated to come into the proximity of the catalytic site. In the case of a higher-generation dendrimer with palladium nanoparticles (**5.6a**) as the core, it also becomes difficult for a hydrogenated product to get released from the reactive site. This is also the case with a substrate to reach the encapsulated reactive

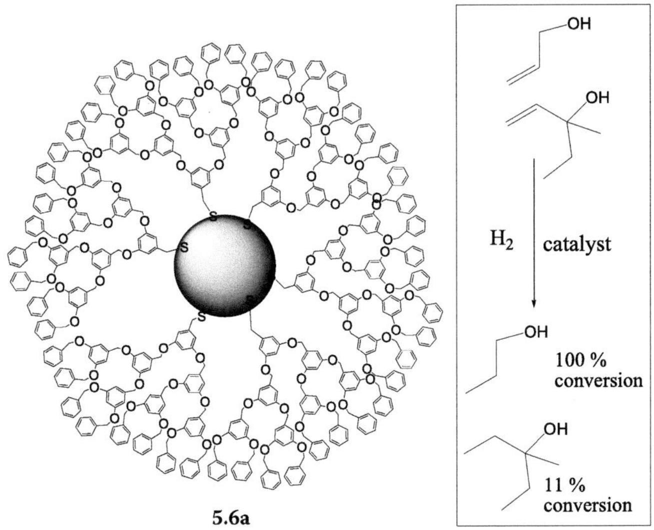

5.6a

FIGURE 5.6 Palladium nanoparticle decorated by dendrons and size-selective catalysis. (From K. R. Gopidas et al. *Nano Lett.* 2003, 3, 1757–1760.[7])

site and leave the host after hydrogenation. The turnover frequency of these catalytic reactions is dependent on the generation, which is also the case observed in the hydrogenation reaction of but-3-en-2-ol. The turnover frequencies for the fourth, sixth and eighth generations of the dendrimer shown in Figure 5.6 are 450, 380 and 93, respectively. The advantage of using a dendrimer containing palladium nanoparticles as a catalyst is the ability to perform size-selective hydrogenation reaction. When an equimolar amount of allyl alcohol and 3-methyl-1-penten-3-ol is hydrogenated together with a fourth-generation dendrimer with palladium nanoparticles with an average of 40 palladium atoms, there is a clear difference in the relative hydrogenation of the two substrates. In a hydrogenation reaction of a mixture of allyl alcohol and 3-methyl-1-penten-3-ol, a preference for the hydrogenation of allyl alcohol is observed. In the mixture, all the allyl alcohol is hydrogenated after 5 h, but 11% of the 3-methyl-1-penten-3-ol is hydrogenated within this time. Complete hydrogenation of 3-methyl-1-penten-3-ol requires 42 hours under similar reaction conditions. The palladium metal-catalysed reaction without a support is unselective and has much lower activity than palladium nanoparticles containing dendrimers.[10]

The dendrimer obtained from the sequential reactions of the Michael addition of methyl acrylate to ethylenediamine followed by subsequent amidation with ethylenediamine is known as a poly(amido)amine dendrimer. It has complexation sites at the interior from the tertiary amine and amide groups. These groups act as ligating sites to various noble-metal ions. The metal complexes formed at the interiors undergo reduction by reducing agents, allowing encapsulation of metallic particles at nanodimensions. Different palladium-based catalytic species such as palladium (II) acetate and palladium (0) nanoparticles encapsulated into the poly(amidoamine) dendrimer **5.7a** act as catalysts in the coupling reactions between trichloro(phenyl)

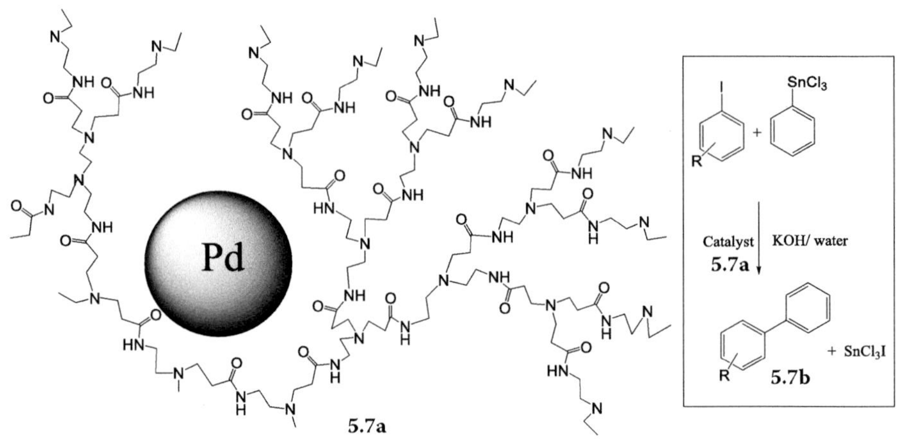

FIGURE 5.7 Poly(amidoamine) dendrimer-supported palladium nanoparticle catalyst for C—C bond formation. (From L. M. Bronstein, Z. B. Shifrina, *Chem. Rev.* 2011, 111, 5301–5344;[8] C. Deraedt et al. *J. Am. Chem. Soc.* 2017, 139, 18084–18092.[9])

stannane and iodoarenes in water (Figure 5.7). The palladium nanoparticles encapsulated in the dendrimer are inferior (or closely comparable) in catalytic activity to palladium (II) acetate, for a coupling reaction to form a compound **5.7b**. However, dendrimer-palladium nanoparticles have shown higher selectivity to form hetero-coupled products. Homo-coupling between the stannane is suppressed in such cases. It is also possible to reuse the catalyst. The catalytic action of dendrimer-palladium nanoparticles involves palladium species present in solution as a result of leaching, but despite leaching out in solution, some portion remain coordinated to the dendritic part.

Fourth-generation poly(amidoamine)dendrimers **5.8a** with encapsulated palladium, platinum or rhodium nanoparticles are catalysts in the dehydrogenation of tetrahydroquinoline/indoline derivatives (Figure 5.8). Generally, a dehydrogenation reaction is carried out by a hydrogen acceptor compound to abstract hydrogen affecting aromatisation or to form conjugated double bonds of the hydrogen-donating compound or unit. The dehydrogenation of tetrahydroquinoline derivative **5.8b** by dendrimer nanoparticles takes place at 130°C to form the product **5.8c**. Importantly, the dehydrogenation reaction caused by an organometallic catalyst requires rigorous anhydrous conditions, and reactions are carried out in an inert atmosphere. In the case of catalysis by dendrimer nanoparticles, the catalysts are air stable with higher reactivity. The catalysts are recyclable. Nanoparticle-based catalysts also perform the hydrogenation reaction of quinoline derivatives. The catalytic hydrogenation of quinoline **5.8c** also occurs with these catalysts at 60°C by hydrogen gas at 1 atmosphere pressure. The hot filtration test confirms the process as heterogeneous catalysis, and the recovered catalysts are recyclable. This reversible processes makes dendrimer nanoparticles a means for hydrogen storage.

Dendrimer-encapsulated palladium nanoparticles of a dendrimer **5.8a** are excellent catalysts for the coupling reaction between hypervalent iodine compound Ph_2IBF_4 (**5.9b**) with indole (Figure 5.9). The reaction results in arylation at the C-2 position of

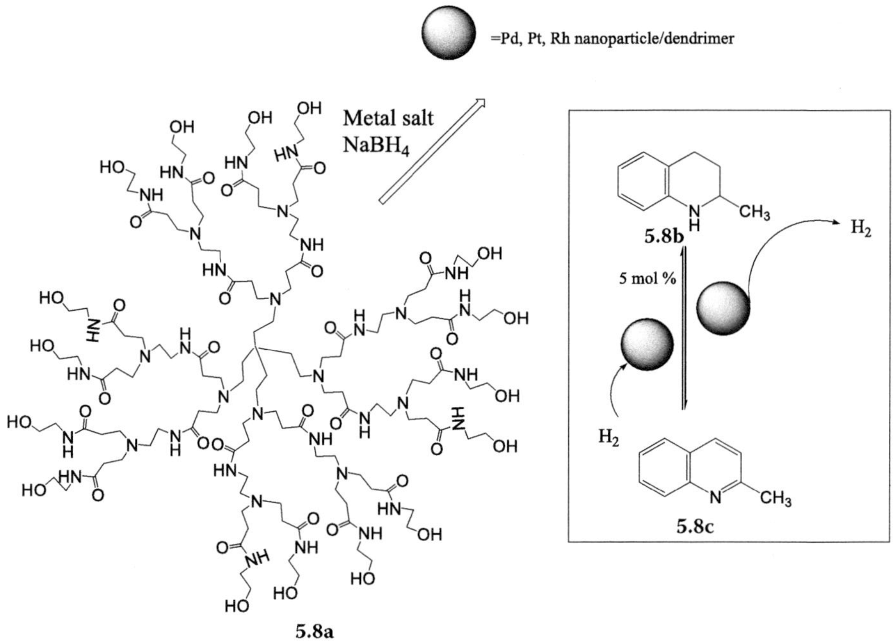

FIGURE 5.8 Dendrimer-supported nanoparticles as hydrogen storage catalyst. (From J. A. Johnson et al. *J. Phys. Chem. C.* 2013, 117, 22644–22651.[11])

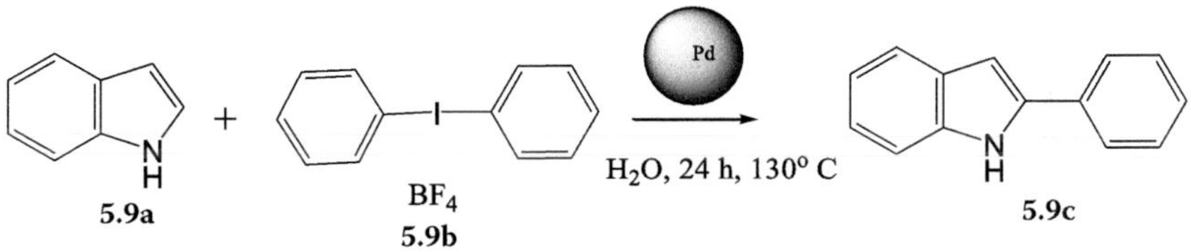

FIGURE 5.9 A C—C coupling reaction of indole in water by using dendrimer-palladium nanoparticle catalyst. (From J. A. Johnson et al. *J. Phys. Chem. C.* 2013, 117, 22644–22651.[11])

the indole in water, forming the product **5.9c**. This cross-coupling reaction involves C—H activation selectively at a carbon atom next to the nitrogen atom of the indole. Palladium (II) acetate without dendrimers also catalyses a similar reaction under homogeneous conditions. But the dehydrogenation of the respective indoline and arylation reactions is achieved through the reaction of Ph_2IBF_4 with the dendrimer-supported nanoparticles of palladium at a low catalyst concentration (5 mol%) and mild heterogeneous conditions. Overall, the activation process leading to the product involves three steps: dehydrogenation, C—H activation and a C—C cross-coupling reaction.

Poly(amido)amine dendrimers encapsulate monodisperse palladium nanoparticles consisting of 10–200 atoms. The dendrimer with palladium nanoparticles **5.8a** is a catalyst for reduction of 4-nitrophenol to 2-aminophenol. Comparisons of the reactivity of smaller clusters of nanoparticles with large-sized clusters have revealed that palladium nanoparticles composed of assemblies of 10 to 50 numbers of atoms are the most suitable catalysts. The experimentally observed rate constant of those catalytic reactions show linear dependence of the ratio of the concentration of the palladium (II) ions per dendrimer. This linear dependence in the rate equation proves the contribution of most of the atoms on the surface of the nanoparticles to the catalytic activity. The normalised rate constant calculated on a per-unit basis of palladium clusters of 10 and 50 atoms remains unchanged, as does that of the individual clusters. This supports all the atoms located on the surface of the nanoparticles contributing to catalysis. For nanoparticles with more palladium, numbering between 50 and 200 palladium atoms, the per-atom activity decreases as the catalyst has more particles. Hence, in these clusters, some nanoparticles located beneath the core are not available for catalysis.

Steric crowding at the periphery significantly affects catalytic activity. As the generation of dendrimer increases, steric crowding on the dendrimer periphery also increases. This makes a dendrimer to act like an adjustable mesh for nanofiltration of substrates. Accordingly, the reach of a substrate to a catalytic nanoparticle at the interior is controlled by the packing density on the dendrimer periphery. The monodispersed palladium nanoparticles (0.2 nm) located at the interiors of different generations of hydroxyl-terminated poly(amidoamine) dendrimers perform identical catalytic reactions differently on a substrate. For example, in general, in the catalytic hydrogenation of allyl alcohol, the use of the higher-generation catalysts or high substrate concentrations results in lower turnover frequencies (Figure 5.10). During

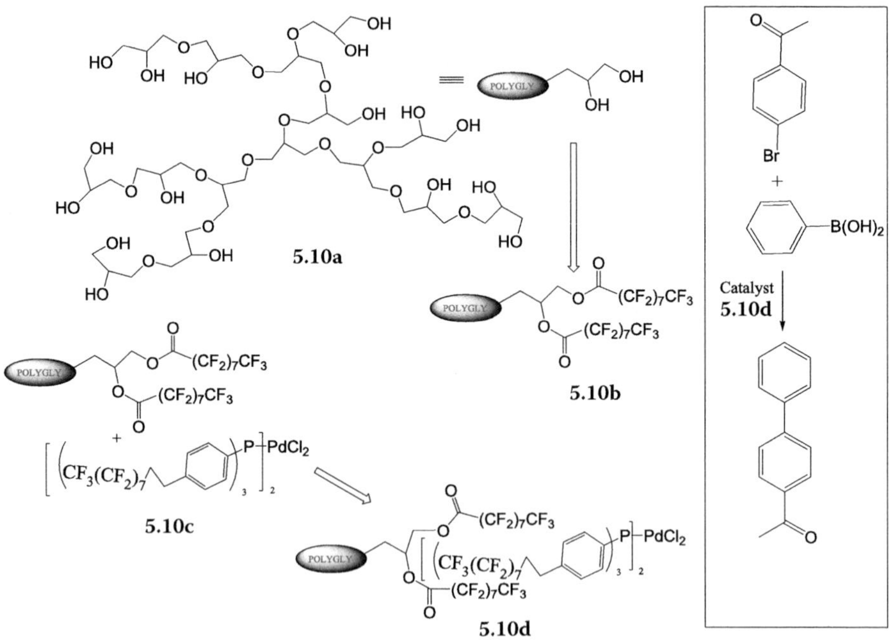

FIGURE 5.10 Polymer-supported palladium complex trapped in dendrimer and a catalytic reaction. (From A. Garcia et al. *Adv. Synth. Catal.* 2005, 347, 1389–1394.[12])

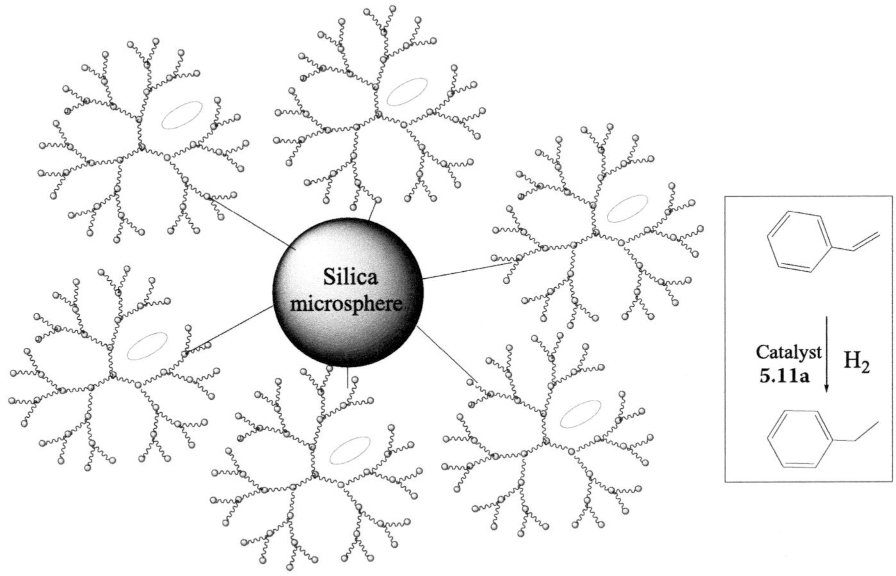

SiO_2@Pd-dendrimer **5.11a**

FIGURE 5.11 Core-shell modified silica with dendrimers and its catalytic hydrogenation. (From V. Biradar et al. *Langmuir* 2011, 27, 14408–14418.[13])

such hydrogenation reactions, the olefin group of allylic alcohols are hydrogenated, but ketones are formed as a side product when substituted allyl alcohols are used.

Poly(amido)amine dendrimers encapsulating palladium nanoparticles anchored to a core-shell SiO_2 microsphere SiO_2@Pd-poly(amido)amine dendrimer **5.11a** causes catalytic hydrogenation of styrene (Figure 5.11). The core shell SiO_2 microspheres are functionalised through vinyl groups by grafting the surfaces to vinyl triethoxysilane to link the dendrimer. Dendrimer-linked silica core shells of 100–400 nm are catalysts for the epoxidation of olefins. Functionalised silica coupled with amine-terminated fourth-generation poly(amidoamine) (PAMAM) dendrimers yield silica-supported dendrimers as SiO_2@PAMAM core shell microspheres. Treatment of the core shell with palladium (II) salt, followed by reaction with sodium borohydride, provides the nanoparticle incorporated in the core shell. The average size of the palladium nanoparticles is 1.56 nm. The core shell microsphere is a catalyst for hydrogenation of various olefins, alkynes, keto and nitro groups containing compounds. They hydrogenate with high atom economy. The catalysts are efficiently recycled up to five times without losing activity.

Water-soluble dendrimers with covalently linked small peptide units on the exterior periphery encapsulate palladium nanoparticles. The palladium nanoparticles@peptide-bearing dendrimers act as catalysts for various organic transformations. The nanoscale morphologies of the peptide surfaces differ depending on the density of atoms on the dendrimer. The design principle followed for this purpose involves conversion of a dendrimer **5.12a** to a peptide derivative **5.12b**, which forms a self-assembly

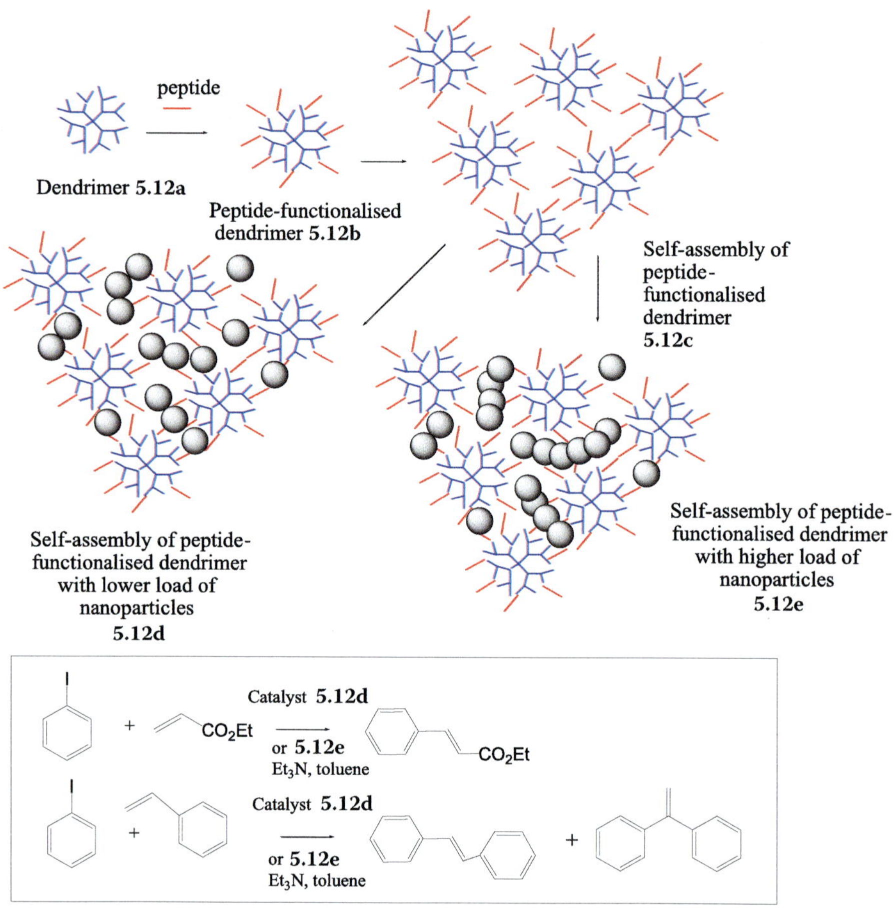

FIGURE 5.12 Self-assembly of palladium nanoparticles in peptide-modified dendrimer and catalysis. (From N. M. Bedford et al. *Chem. Mater.* 2014, 26, 4082–4091.[14])

5.12c. Different quantities of palladium nanoparticles are incorporated to prepare the nanoparticle-containing peptide-functionalised dendrimers (**5.12d** or **5.12e**). In this case, the palladium nanoparticles held in the fifth-generation dendrimer are relatively less aggregated in solution than the other higher-generation oligomers. Due to the change in the self-assemblies of peptides, the morphologies of the dendrimer assemblies prepared from different solutions change with solvents. Dissolution in solvents also alters the agglomeration of nanoparticles, which affects the morphology of palladium nanoparticles. Such dendrimers with palladium nanoparticles are catalysts for C—C bond formation by vinylic activation (Figure 5.12). The less agglomerated palladium nanoparticles have higher reactive palladium surface area for catalysis. Thus, the catalytic reactions of these dendrimer-supported nanoparticles are greatly influenced by the size of nanoparticles encapsulated in these dendrimers.

Appropriate functional units on the outer periphery of dendrimers participate in the formation of noncovalent self-assembly with appropriate functional units of ligand of

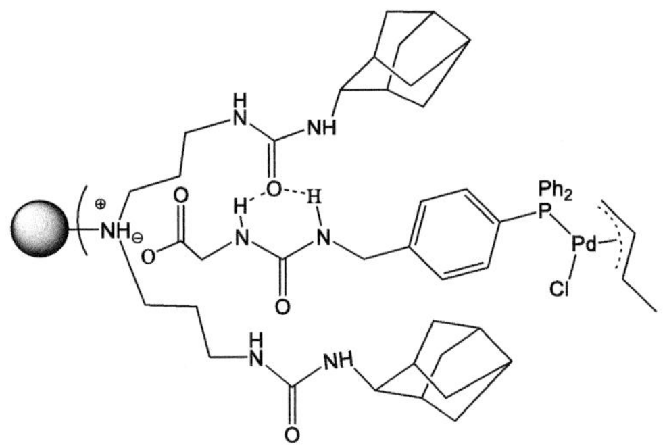

5.13a

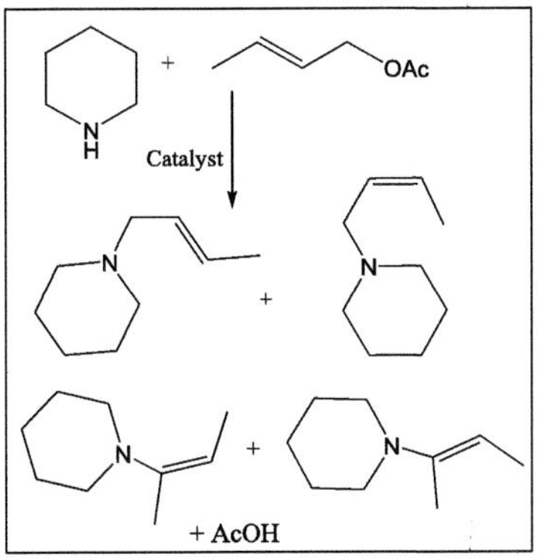

FIGURE 5.13 A dendrimer noncovalently linked to palladium catalytic sites and catalysis. (From R. van de Coevering et al. *Prog. Polym. Sci.* 2005, 30, 474–490.[15])

metal complexes. Such self-assembly immobilises the metal complexes. For example, in the example in Figure 5.13, urea tethered to an adamentane unit and a nitrogen atom located at the exterior part of a dendrimer are used to immobilise a palladium complex through noncovalent interactions. The terminal part of such an assembly is shown in **5.13a**, where the carboxylate functional group on a ligand of the allyl-palladium complex forms a salt through exchange of protons. The urea part tethered to the phosphine ligand of the complex acts as a template to form a hydrogen bond with the carbonyl of the urea group on the dendrimer. This palladium complex catalyses the amination reaction of crotyl acetate with piperidine. The dendrimer—ligand binding having an ionic character is able to retain near-quantitative integrity of the assembly depending on the amount of functionalised metal complexes loaded into the dendrimer. The catalytic reaction maintains a constant conversion rate for about 10 hours, after which the catalyst deteriorates and yields of the coupling products decrease.

5.4 MICELLES IN SUPRAMOLECULAR CATALYSIS

Reactions occurring in nature under aqueous conditions, irrespective of the reacting substrates, are soluble or insoluble. Micelles formed by surfactants in the form of nanosized nonpolar aggregates in water can accommodate catalysts and reagents to cause catalytic reactions. The hydrophobic effect and ion pairs operate in concentrated solutions. Surfactants are amphiphilic molecules; when they are dissolved in mixed solvent with water as the major component, the hydrophobic effect of the surfactant enforces formation of spontaneous micellar aggregates at a particular concentration called *critical micelle concentration*, abbreviated as *cmc*. The assemblies formed by aggregation are composed of 50–100 monomers of nanometre dimensions. They remain in thermodynamic equilibrium with rapidly exchanging monomers with the aggregates at a lifetime of exchange on the order of 10^{-3}–10^{-2} s. The hydrophobic cores of the micelle pores are shielded from contact with water. The hydrophilic surface has polar head groups that are exposed to water. Micelles change the behaviour of water molecules surrounding

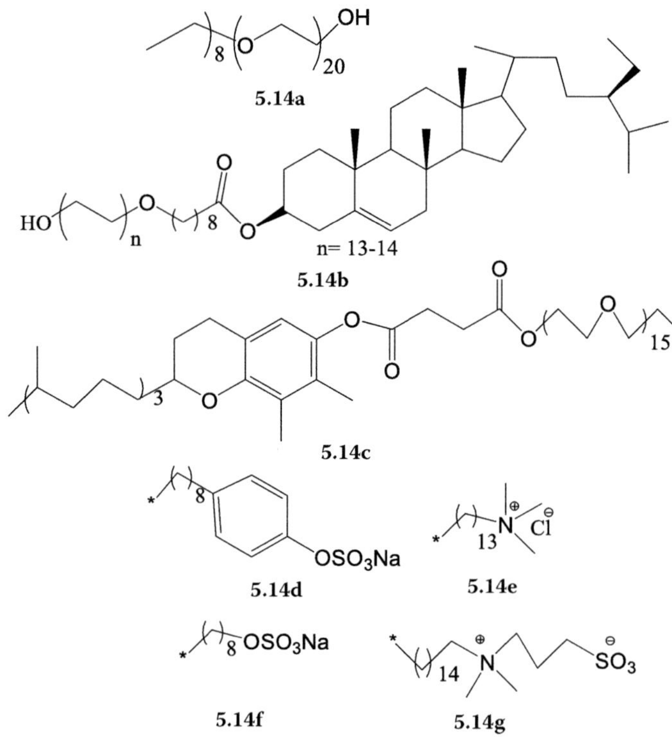

FIGURE 5.14 Some examples of surfactant molecules. (From G. La Sorella et al. *Green Chem.* 2015, 17, 644–683.[16] A. Sorrenti et al. *Chem. Soc. Rev.* 2013, 42, 8200–8219.[17])

the polar external surface of the aggregates. Parameters such as the nature of the amphiphile; charge on the hydrophilic part; length of the hydrocarbon chain and geometry of the surfactant; and experimental conditions such as temperature, ionic strength and pH control the dimension and shape of the aggregates in micelles. Surfactants are commercially available low-cost materials, and some common surfactants are listed in Figure 5.14. The reactions carried out under micellar catalysis in water are of commercial interest. One prominent industrial catalytic reaction, Ruhrchemie—Rhône-Poulenc rhodium catalysts with sulphonate-containing phosphine ligands, is used in hydroformylation of propene and 1-butene into the corresponding aldehydes.

Micelles are useful catalysts in many organic reactions. Esterification of carboxylic acid is a common reaction facilitated by micelles (Figure 5.15). The reaction passes smoothly in water media, and another advantage of the use of micelle catalysts is the low activation energy in micelle conditions. For such reactions, the azeotrope conditions needed in conventional acid-catalysed esterification reactions are not required.

Multicomponent reactions are also facilitated by micelles to form selective products. For example, the reaction between dimethyl-dimedone, malononitrile and indoline-2,3-dione carried out in sodium stearate micelle in water selectively

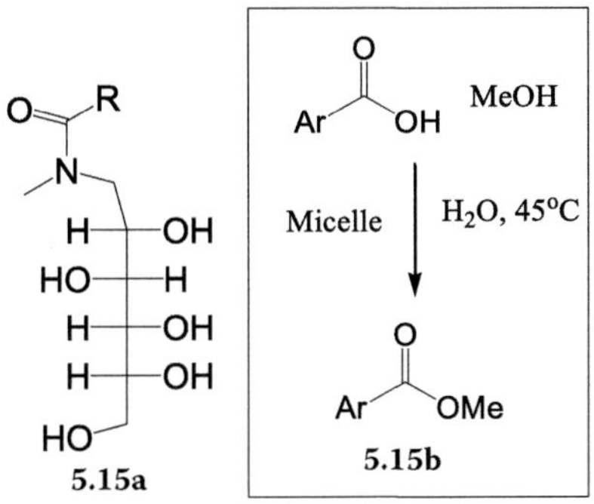

FIGURE 5.15 Ester hydrolysis mediated by a micelle. (From F. Rajabi, R. Luque. *RSC Adv.* 2014, 4, 5152–5155.[18])

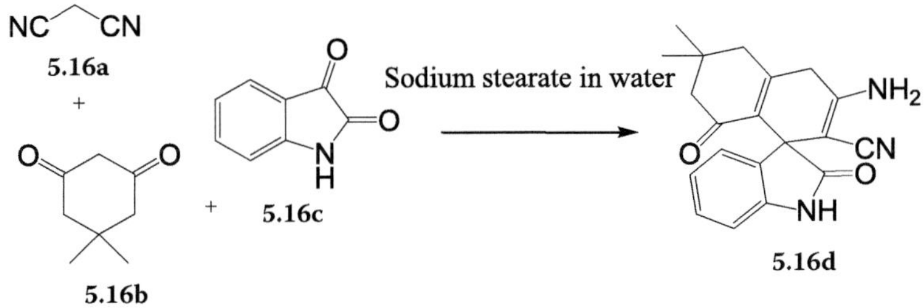

FIGURE 5.16 A multicomponent reaction mediated by a micelle. (From L.-M. Wang et al. *Tetrahedron* 2010, 66, 339–343.[19])

yields (Figure 5.16) the spiro-compound 3′-amino-6′,6′-dimethyl-2,8′-dioxo-5′,6′,7′,8′-tetrahydro-4′H-spiro[indoline-3,1′-naphthalene]-2′-carbonitrile (**5.16d**).

Size selectivity of reactant is an important aspect in micelle-facilitated reactions. The aggregation of the micelles provides hydrophobic voids of different sizes. Hence, a particular micelle can be very selective to a set of compounds that has matching sizes for chemical reactions. N,N-diallyl-N-(4-(dodecyloxy)benzyl)prop-2-en-1-aminium bromide forms micelles in water and facilitates the azideation reaction between benzylic bromides with sodium azide in the presence of 1,4-dimercaptobutane-2,3-diol under photochemical conditions (Figure 5.17). Such reactions are highly size selective. When a equimolar amount of (((5-(bromomethyl)-1,3-phenylene)bis(oxy)) bis(methylene))dibenzene and ((5-(bromomethyl)-1,3-phenylene)bis(oxy)) bis(tert-butyldiphenylsilane) reacts with sodium azide in a solution with a micelle **5.17b**, the [??] **5.17c** is selectively transformed to the corresponding azide (((5-(azidomethyl)-1,3-phenylene)bis(oxy))bis(methylene)) dibenzene (**5.17e**) with a sevenfold higher amount than the formation of ((5-(azidomethyl)-1,3-phenylene)bis(oxy))bis(tert-butyldiphenylsilane) (**5.17g**) from the compound **5.17d**. This is due to the size selectivity of the reactants in micelle conditions to undergo the nucleophilic substitution reaction.

Autocatalysis by micelle to transform a component is observed in many micelles. Micelles of octanoate anions cause autocatalysis to form additional octanoates.

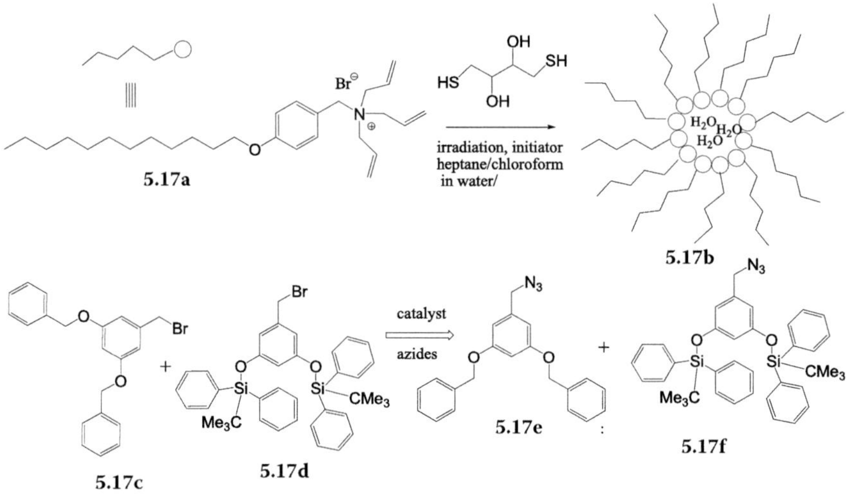

FIGURE 5.17 A micelle and its size-selective azidation reaction. (From L.-C. Lee, Y. Zhao. *Org. Lett.*, 2012, 14, 784–787.[20])

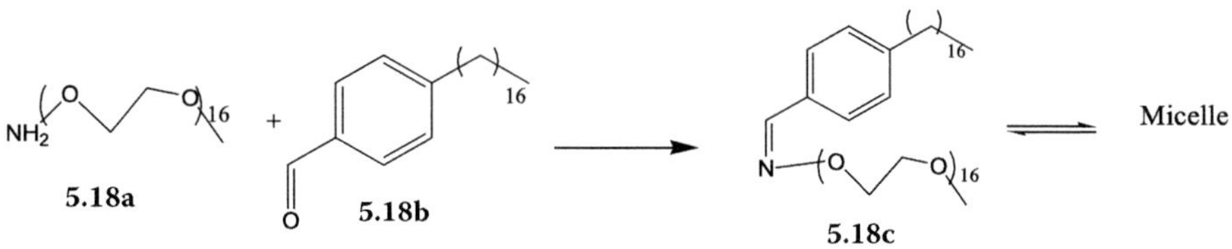

FIGURE 5.18 Autocatalysis by micelles to form imines. (From R. Nguyen et al. *Angew. Chem. Int. Ed.* 2009, 48, 1093–1096.[21])

The micelles formed by imines through the reaction between amine and aldehydes further catalyses imine formation reactions, and such observations are found in many systems. One example is listed in Figure 5.18.

The autocatalytic reactions of micelles formed by ester hydrolysis are the results of mixed micelles composed of product and ester formed during the reaction. Such effects solubilise the ester in the aqueous phase to act as autocatalysts. Catalytic reactions do not occur in the micelle by bulk solution where micelles are phase-transfer catalysts. Ethanol is released as a product in hydrolysis of an ethyl ester, which acts as cosolvent to increase the solubility of the ester (Figure 5.19). Support for such a process comes from the biphasic nature of the ester hydrolysis reaction proved by control experiments. In such systems, the product is hydrotropic and facilitates solubilisation of hydrophobic molecules in the aqueous phase.

5.5 GELS AND IONIC LIQUIDS AS CATALYSTS

A material which shows behaviour between that of a solid and a liquid is generally termed a *gel*. According to the definition given by International Union of Pure and Applied

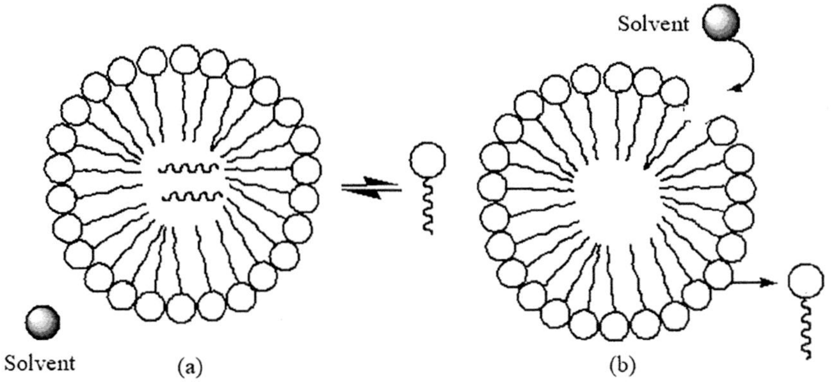

FIGURE 5.19 (a) A micelle in equilibrium with individual surfactant; (b) cleavage and reformation of micelle to autocatalyse formation of surfactant. (From P. Stano, P. L. Luisi. *Chem. Commun.* 2010, 46, 3639–3653.[22])

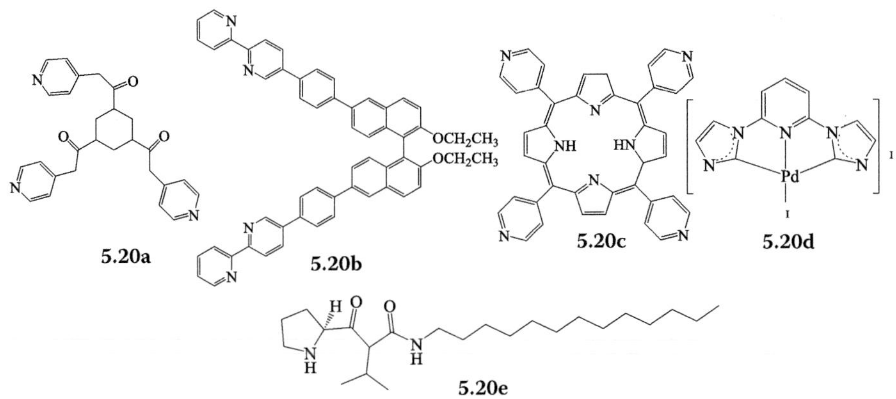

FIGURE 5.20 Different compounds that form gel catalysts. (From W. Fang et al. *Chem. Asian J.* 2018, 13, 712–729.[23])

Chemistry (IUPAC), a gel is a nonfluid colloidal or polymer network that is expanded throughout its whole volume by a fluid. Gels with water as swelling are hydrogels; gels after removal of swelling agents are called xerogels. Gels are heterogeneous materials with a two or more components which may be individually solid or liquid, but the solid component is invisible to the naked eye. Gels have 3-D networks and sustain shear stress.

Gels have found extensive application in supramolecular catalysis.[23–25] The compounds listed in Figure 5.20 are examples of some that form gels under specific conditions. These gels act as catalysts of different types of reactions. The compounds in **5.20a** and **5.20b,** in the presence of cadmium sulphate and copper (I) salt, respectively, form gel in water; the compound **5.20c** forms a gel in the presence of palladium acetate in dimethylsulphoxide. The gel of **5.20a**, with cadmium sulphate, is a catalyst for carbon—carbon bond formation between active hydrogens containing nitrile with aldehydes. The combination of the copper (I) salt and gel of the compound **5.20b** is useful in the click reaction between acetylic compounds with azide to form five-membered heterocyclic compounds. The compound **5.20c** with palladium acetate in dimethylsulpoxide is utilised in the catalytic oxidation reaction of benzyl alcohol to benzaldehyde. The palladium complex **5.20d** itself forms a gel and catalyses the Michael addition reaction between ethylcyanoacetate and methyl-vinyl ketone. This catalytic reaction proceeds at room temperature in dichloromethane in the presence of a catalytic amount of the gel but requires isopropyl amine as the base in the reaction. This gel-catalysed reaction is superior to a reaction where the soluble palladium complex of a similar skeleton is used. The compound **5.20d** forms a gel upon heating followed by cooling; this gel is a catalyst for aldol-type reactions under sonicating conditions.

Ionic liquids are salts in liquid form, conventionally with a melting point below 100°C. The gel of the ionic liquid comprises of two components, an ionic liquid and a continuous solid phase. Ionic liquid gels easily trap nanoparticles and enhance homogeneous catalyst performance; they have Brønsted acidity that is free from contamination.[26]

Ionogels are composed of an ionic liquid within a porous oxide. Catalytically active molecules trapped within ionic liquid provide new heterogeneous catalysts. Silica gel—based ionic liquids are used in sol-gel methods for the preparation of solid catalysts. The methodology involves trapping an ionic liquid within an inorganic oxide matrix. This type of catalysis provides catalysts with hierarchical pores. The ionic liquid is miscible with common organic compounds. The ionic liquid gel triethylammonium propane sulfone bistriflamide is a good catalyst for the dehydration of 1-phenylethanol to form styrene, (oxybis(ethane-1,1-diyl))dibenzene and (E)-but-2-ene-1,3-diyldibenzene. The ratios of the three products depend on the ratio of ionic liquid to the 1-phenylethanol used in the catalytic reactions. The salt **5.21a** deposited on silica shows better catalysis. The ionic liquid is prepared by reacting triethylammonium propanesulphonic acid with bistriflamidic acid (Figure 5.21). The triethylammonium propanesulphonic acid is prepared by the ring-opening reaction of 1,2-oxathiolane 2,2-dioxide.

Adsorption of the acidic ionic liquid 1,3-bis(3-sulfopropyl)-1H-imidazol-3-ium trifluoromethanesulphonate into silica gel provides a solid acid which is used in catalytic reactions of bromoacetic acid with benzyl alcohol. The reaction is a

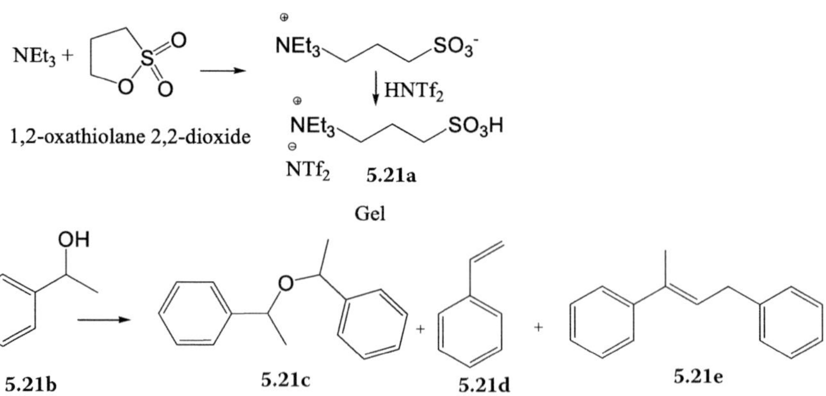

FIGURE 5.21 An ionic gel and its catalytic reactions. (From Y.-M. Wang et al. *ACS Sustainable Chem. Eng.* 2015, 3, 792–796.[27])

heterogeneous reaction which occurs due to enhanced acidity of the composite. The resulting silica-supported ionic liquid has 10 times higher reactivity than the unsupported ionic liquid. The catalyst can be reused for several batches of reactions.

Pyridine-bridged benzimidazolium salts with long hydrocarbon chains attached to quaternary ammonium sites form gels. These gels are used as phase-transfer catalysts. The reaction between 2,6-bis(bromomethyl)pyridine and benzimidazole in the presence of a gel (5 wt%) as catalyst results in an N-arylated product in 3 hrs at room temperature to give 60% yield (Figure 5.22), but the same reaction without a catalyst requires 50 h to give a similar yield. The long chains of hydrocarbons are like pincers; in a reaction medium, they bring substrates to the organic solvent/water interface for reaction. During the reaction, the gel fibres reorganise to increase the specific surface area of the catalytic centres. As a result of self-assembly, multiple catalytic centres in the fibre aggregates are formed to improvise catalytic activity.

Proline-based gels generate interest from asymmetric catalysis. Such gels have higher basicity than the corresponding solutions. Upon sonication, the l-proline—based compound (2R)-N-((S)-2-methyl-1-(tridecylamino)propyl)pyrrolidine-2-carboxamide forms a hydrogel.[29] This gel is a reusable heterogeneous catalyst for the aldol reaction of cyclohexanone and 4-nitro benzaldehyde (Figure 5.23). Structurally similar gels perform enantioselective catalysis of the aldol reaction in solution,[30] but in the presence of a base, the gels catalyse reactions to form racemic products. Gels of this kind provide scope for variations of reactivity in aggregated states than the free state of the catalyst.

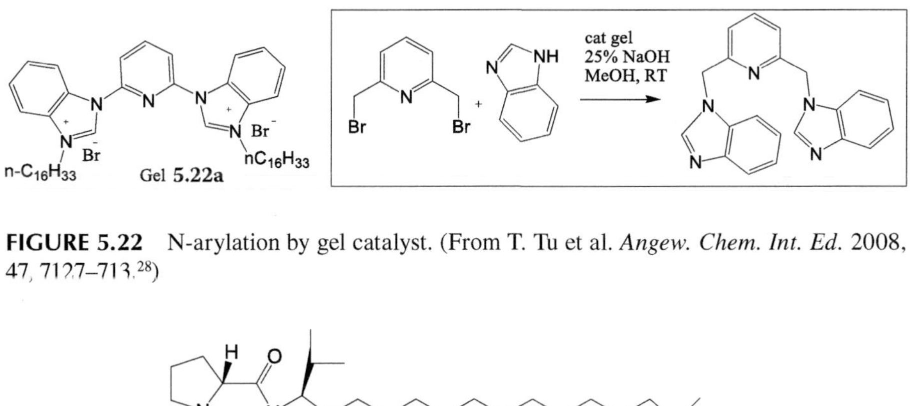

FIGURE 5.22 N-arylation by gel catalyst. (From T. Tu et al. *Angew. Chem. Int. Ed.* 2008, 47, 7127–713.[28])

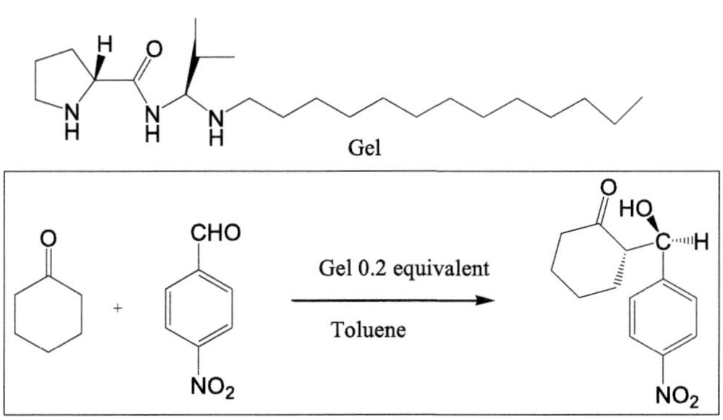

FIGURE 5.23 Proline-based gel catalyzing aldol condensation reaction. (From F. Rodríguez-Llansola et al. *Chem. Commun.* 2009, 7303–7305.[30])

Subtle structural changes caused by variation of solvent compositions influence chemical reactions and can reach an extent to switch the catalysis process on or off. The difference in reactivity in such a case occurs due to differences in substrate diffusion within the swollen and shrunken gel affected by solvents. These properties are observed in polymer gel consisting of two species of monomers. The role of one component is to reversibly swell and shrink the gel in response to change stimuli. The expansion and contraction result in changing the local microenvironments. Expansion also changes the local density of the hydrophobic moiety. This process influences the substrate binding as the affinity of a site changes. The other minor component acts as a catalyst and is copolymerised into the gel network formed by the major component. In the event of a substrate reversibly binding to the polymer cavity, the reactivity is switched on and off. The reaction of 4-nitrophenolate to form 4-aminophenolate catalysed by a gel made up of *N*-isopropylacrylamide and vinylimidazole is such an example (Figure 5.24). A gel consisting of *N*-isopropylacrylamide and vinylimidazole acts as catalyst whose reactivity is modulated by the swelling and shrinking of the gel. The gel with *N*-isopropylacrylamide as a major component swells in response to the amount of ethanol present in water. The minor component, vinylimidazole, catalyses the hydrolysis reaction and during the process it is copolymerised and penetrates into the interior of the gel network. The swollen gel shows lower catalytic activity for hydrolysis of *p*-nitrophenyl caprylate. The shrunken form increases the reaction rate fivefolds. The mechanism is like enzyme—substrate interactions. A kinetics study revealed that 4-nitrophenyl caprylate is absorbed in the hydrophobic environment of *N*-isopropylacrylamide polymer and the adsorption is not a factor affecting the reaction rate in the case of swollen gel.[18]

Cinchona alkaloid-based quaternary ammonium salt anchored different polymeric supports to cause asymmetric phase-transfer catalysis.[32–35] Silica-gel is used as support in such reactions are also useful and one such silica gel-support

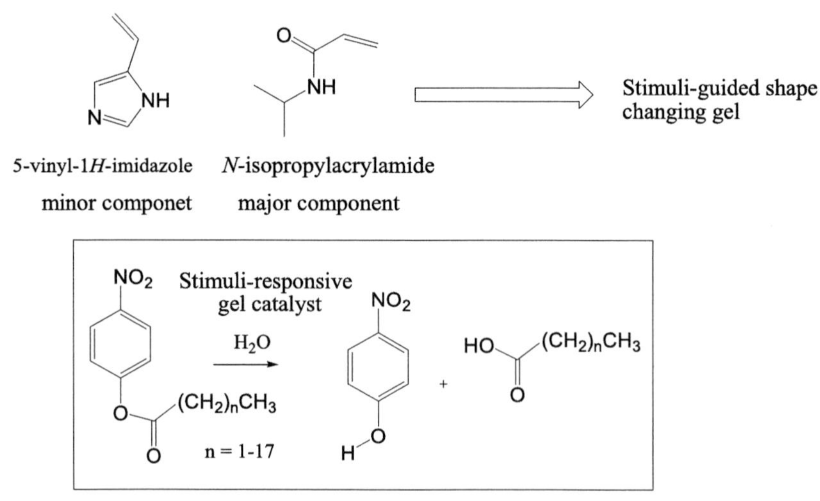

FIGURE 5.24 Modulation of ester hydrolysis by swelling and shrinking of gel. (From G. Wang et al. *Proc. Nat. Acad. Sci. (USA)* 2000, 97, 9861–9864.[31])

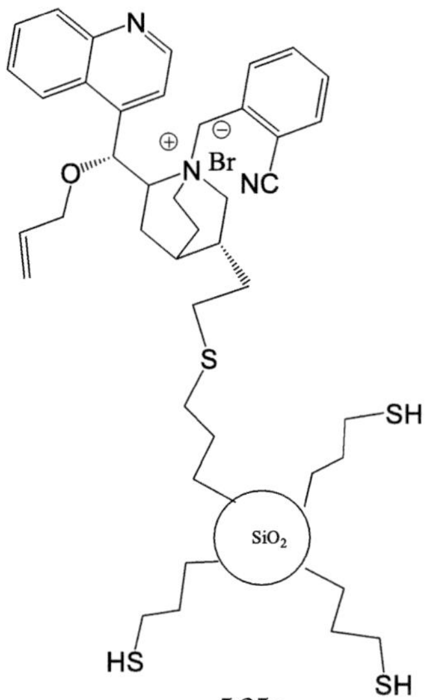

5.25a

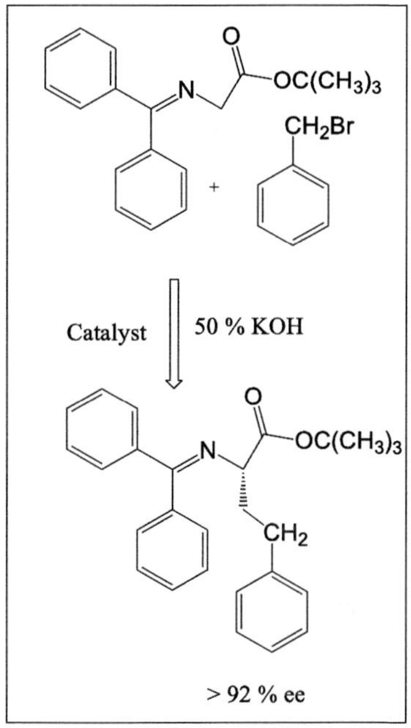

FIGURE 5.25 Silica-supported catalyst for asymmetric C—C bond formation reaction. (From D. Feng et al. *Catal. Sci. Technol.* 2015, 5, 2141–2148.[34])

5.25a catalyses α-alkylation of imines. The catalyst facilitates the reaction of tert-butyl 2-((diphenylmethylene)amino) acetate with benzyl bromide to form (S)-tert-butyl 2-((diphenylmethylene)amino)-4-phenylbutanoate (Figure 5.25). The catalyst is reusable five times and is recovered completely after each cycle. The reaction of 3-mercaptopropyl trimethoxysilane with *N*-(2-cyanobenzyl)-*O*(9)-allyl-cinchonidinium bromide is followed by hydrolysis of trimethoxysilane to yield the catalyst **5.25a**.

REFERENCES

1. S. M. Grayson, J. M. J. Frechet, Convergent dendrons and dendrimers: From synthesis to applications. *Chem. Rev.* 2001, 101, 3819–3868.
2. D. Feng, J. Xu, J. Wan, B. Xie, X. Ma, Facile one-pot fabrication of a silica gel-supported chiral phase-transfer catalyst-*N*-(2-cyanobenzyl)-*O*(9)-allyl-cinchonidinium salt. *Catal. Sci. Technol.* 2015, 5, 2141–2148.
3. M. Bernechea, E. de Jesus, C. Lopez-Mardomingo, P. Terreros, Dendrimer-encapsulated Pd nanoparticles versus palladium acetate as catalytic precursors in the Stille reaction in water. *Inorg. Chem.* 2009, 48, 4491–4496.
4. R. van Heerbeek, P. C. J. Kamer, P. W. N. M. van Leeuwen, J. N. H. Reek, Dendrimers as support for recoverable catalysts and reagents. *Chem. Rev.* 2002, 102, 3717–3756.
5. G. E. Oosterom, R. J. van Haaren, J. N. H. Reek, P. C. J. Kamer, P. W. N. M. van Leeuwen, Catalysis in the core of a carbosilane dendrimer. *Chem. Commun.* 1999, 1119–1120.
6. R. Andres, E. de Jesus, J. C. Flores, Catalysts based on palladium dendrimers. *New J. Chem.* 2007, 31, 1161–1191.

7. K. R. Gopidas, J. K. Whitesell, M. A. Fox, Synthesis, characterization, and catalytic applications of a palladium-nanoparticle-cored dendrimer. *Nano Lett.* 2003, 3, 1757–1760.
8. L. M. Bronstein, Z. B. Shifrina, Dendrimers as encapsulating, stabilizing, or directing agents for inorganic nanoparticles. *Chem. Rev.* 2011, 111, 5301–5344.
9. C. Deraedt, R. Ye, W. T. Ralston, F. D. Toste, G. A. Somorjai, Dendrimer-stabilized metal nanoparticles as efficient catalysts for reversible dehydrogenation/hydrogenation of N-heterocycles. *J. Am. Chem. Soc.* 2017, 139, 18084–18092.
10. Y. Niu, L. K. Yeung, R. M. Crooks, Size-selective hydrogenation of olefins by dendrimer-encapsulated palladium nanoparticles. *J. Am. Chem. Soc.* 2001, 123, 6840–6846.
11. J. A. Johnson, J. J. Makis, K. A. Marvin, S. E. Rodenbusch, K. J. Stevenson, Size-dependent hydrogenation of p-nitrophenol with Pd nanoparticles synthesized with poly(amido)amine dendrimer templates. *J. Phys. Chem. C* 2013, 117, 22644–22651.
12. A. Garcia, C. C. Tzschucke, W. Bannwarth, R. Haag, Supramolecular immobilization of a perfluoro-tagged Pd-catalyst with dendritic architectures and application in Suzuki reactions. *Adv. Synth. Catal.* 2005, 347, 1389–1394.
13. V. Biradar, A. Biradar, T. Asefa, Silica-dendrimer core-shell microspheres with encapsulated ultrasmall palladium nanoparticles: Efficient and easily recyclable heterogeneous nanocatalysts. *Langmuir* 2011, 27, 14408–14418.
14. N. M. Bedford, R. Bhandari, J. M. Slocik, S. Seifert, R. R. Naik, M. R. Knecht, Peptide-modified dendrimers as templates for the production of highly reactive catalytic nanomaterials. *Chem. Mater.* 2014, 26, 4082–4091.
15. R. van de Coevering, R. J. M. Klein Gebbink, G. van Koten, Soluble organic supports for the non-covalent immobilization of homogeneous catalysts; modular approaches towards sustainable catalysts. *Prog. Polym. Sci.* 2005, 30, 474–490.
16. G. La Sorella, G. Strukul, A. Scarso, Recent advances in catalysis in micellar media. *Green Chem.* 2015, 17, 644–683.
17. A. Sorrenti, O. Illa, R. M. Ortuno, Amphiphiles in aqueous solution: well beyond a soap bubble. *Chem. Soc. Rev.* 2013, 42, 8200–8219.
18. F. Rajabi, R. Luque, An efficient renewable-derived surfactant for aqueous esterification reactions. *RSC Adv.* 2014, 4, 5152–5155.
19. L.-M. Wang, N. Jiao, J. Qiu, J.-J. Yu, J.-Q. Liu, F.-L. Guo, Y. Liu, Sodium stearate-catalyzed multicomponent reactions for efficient synthesis of spirooxindoles in aqueous micellar media. *Tetrahedron* 2010, 66, 339–343.
20. L.-C. Lee, Y. Zhao, Size-selective phase-transfer catalysis with interfacially cross-linked reverse micelles. *Org. Lett.*, 2012, 14, 784–787.
21. R. Nguyen, L. Allouche, E. Buhler, N. Giuseppone, Dynamic combinatorial evolution within self-replicating supramolecular assemblies. *Angew. Chem. Int. Ed.* 2009, 48, 1093–1096.
22. P. Stano, P. L. Luisi, Achievements and open questions in the self-reproduction of vesicles and synthetic minimal cells. *Chem. Commun.* 2010, 46, 3639–3653.
23. W. Fang, Y. Zhang, J. Wu, C. Liu, H. Zhu, T. Tu, Recent advances in supramolecular gels and catalysis. *Chem. Asian J.* 2018, 13, 712–729.
24. T. Tu, W. Assenmacher, H. Peterlik, R. Weisbarth, M. Nieger, K. H. Dotz, An air-stable organometallic low-molecular-mass gelator: Synthesis, aggregation, and catalytic application of a palladium pincer complex. *Angew. Chem. Int. Ed.* 2007, 46, 6368–6371.
25. F. Rodriguez-Llansola, J. F. Miravet, B. Escuder, A supramolecular hydrogel as a reusable heterogeneous catalyst for the direct aldol reaction. *Chem. Commun.* 2009, 1, 7303–7305.
26. P. C. Marr, A. C. Marr, Ionic liquid gel materials: Applications in green and sustainable chemistry. *Green. Chem.* 2016, 1, 8105–8112.
27. Y.-M. Wang, V. Ulrich, G. F. Donnelly, F. Lorenzini, A. C. Marr, P. C. Marr, A recyclable acidic ionic liquid gel catalyst for dehydration: Comparison with an analogous SILP catalyst. *ACS Sustainable Chem. Eng.* 2015, 3, 792–796.

28. T. Tu, W. Assenmacher, H. Peterlik, G. Schnakenburg, K. H. Dotz, Pyridine-bridged benzimidazolium salts: Synthesis, aggregation, and application as phase-transfer catalysts. *Angew. Chem. Int. Ed.* 2008, 47, 7127–7131.
29. F. Rodrıguez-Llansola, B. Escuder, J. F. Miravet, Remarkable increase in basicity associated with supramolecular gelation. *Org. Biomol. Chem.* 2009, 7, 3091–3094.
30. F. Rodríguez-Llansola, J. F. Miravet, B. Escuder, A supramolecular hydrogel as a reusable heterogeneous catalyst for the direct aldol reaction. *Chem. Commun.* 2009, 7303–7305.
31. G. Wang, K. Kuroda, T. Enoki, A. Grosberg, S. Masamune, T. Oya, Y. Takeoka, T. Tanaka, Gel catalysts that switch on and off. *Proc. Nat. Acad. Sci. (USA)* 2000, 97, 9861–9864.
32. M. J. O'donnell, The enantioselective synthesis of α-amino acids by phase-transfer catalysis with achiral Schiff base esters. *Acc. Chem. Res.* 2004, 37, 506–517.
33. M. Ferre, R. Pleixats, M. W. C. Man, X. Cattoen, Recyclable organocatalysts based on hybrid silicas. *GreenChem.* 2016, 18, 881–892.
34. D. Feng, J. Xu, J. Wan, B. Xie, X. Ma, Facile one-pot fabrication of a silica gel-supported chiral phase-transfer catalyst-N-(2-cyanobenzyl)-O(9)-allyl-cinchonidinium salt. *Catal. Sci. Technol.* 2015, 5, 2141–2148.
35. R. Chinchilla, P. Mazon, C. Najera, Polystyrene-anchored cinchona ammonium salts: Easily recoverable phase-transfer catalysts for the asymmetric synthesis of α-amino acids. *Adv. Synth. Catal.* 2004, 346, 1186–1194.

6 Versatility in Supramolecular Catalysis

6.1 MATERIAL DESIGN AND PROCESSES

6.1.1 Biomaterials

There is interest in catalytic reactions in the synthesis of supramolecular hydrogelators.[1] Hydrogelators prepared by catalytic reactions control the formation, dynamics and fate of supramolecular structures. For example, hydrazone bonds containing gelators are prepared by catalytic reactions by controlling pH. In general, the condensation reactions of hydrazides with aldehydic compounds are slow at pH ranging from 7 to 10; whereas in the range of pH 3 to 6, such condensation reactions smoothly occur (Figure 6.1). The preparation of gel by condensing the oligoethylene glycol chain containing aldehyde with tris-hydrazide derivative cyclohexane 1,3,5-tricarboxylic acid takes advantage of this fact. A reaction product **6.1a** is prepared by an acid-catalysed reaction forms a gel. The phenyl groups present in the compound stabilise the fibrelike structures of the gel through π–π interactions. The fibrelike structures form a network of cross-linked assembly by interlinking with the solvent molecules. The rate of gel formation is controlled through catalysis of the condensation reaction with hydrazine. Such control on the catalytic reaction rate allows the gel to adopt reproducible self-assemblies under control conditions.

Supramolecular catalysis leading to dynamic assemblies is important in medicinal chemistry. For example, oligopeptide **6.2a**, upon dephosphorylation, acts as a hydrogelator; this reaction is catalysed in certain cancer cells by alkaline phosphatase (ALP). During the course of such a reaction and the self-assembling process, selective actions on the cancer cells over normal cells in a coculture occur (Figure 6.2).

Insights into the many catalytic chemical reactions occurring in nature are found by integrating catalytic enzymatic reactions with molecular self-assembly of synthetic macromolecules such as dendrimers, oligomers and nanomaterials. Catalytic nanoparticles are used for detection of DNA, for which the electro-catalytic properties of hydrazine for the catalytic reduction of hydrogen peroxide gold-nanoparticles is utilised. For such purposes, densely functionalised dendrimers with large numbers of amine groups are used to immobilise gold nanoparticles. The hydrazine label, when immobilised on a sensing layer of conducting polymer such as polythiophene through avidin-biotin interaction, is useful for electrochemical detection of DNA or proteins.

6.2 CONFINEMENT OF NANOPARTICLES AS CATALYSTS

The self-assembling process has been extensively used to prepare catalysts from the perspective of nanocatalysts. The layer-by-layer self-assembly on polymeric nanofibre

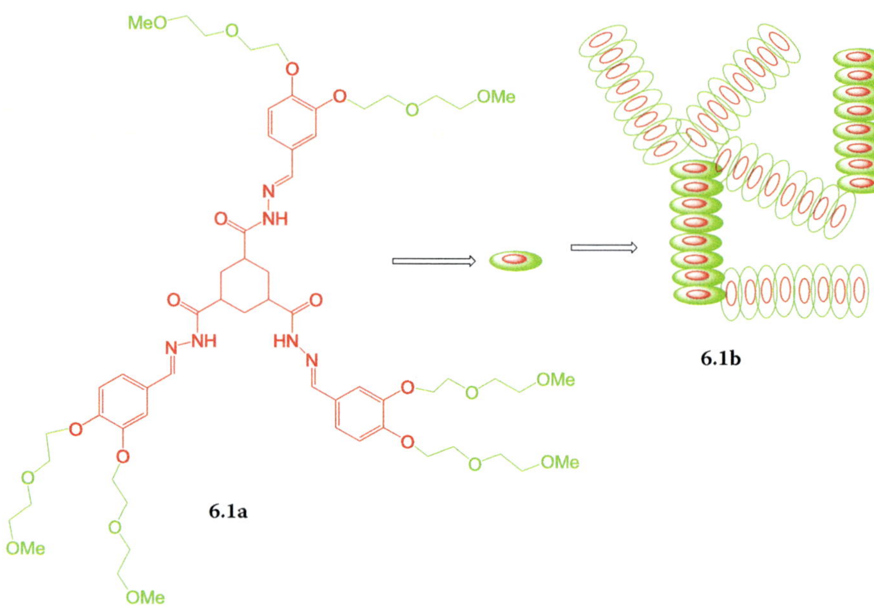

6.1a

6.1b

FIGURE 6.1 A substrate forming fibrelike structure assembling to a network-like structure gelling solvent molecules. (From F. Trausel et al. *Acc. Chem. Res.* 2016, 49, 1440–1447.[2])

of poly(methyl methacrylate) by WO_3 upon calcination provides a WO_3 nanotube which is a prospective catalyst. Nanocatalysts are also prepared by loading metallic nanoparticles of Pt, Pd, Au, Ag and Rh into such a nanotube.[4]

Surfactant-encapsulated polyoxometalates are useful catalysts for oxidation reactions. Surfactant-encapsulated polyoxometalates are prepared by linking tetraethyl orthosilicate with hydroxyl-terminated polyoxometalate complexes by a sol-gel process. The polyoxometalates of the assembly are replaced by quaternary di(11-hydroxyundecyl)methyl ammonium cations to form a reverse micelle. This reverse micelle is a supramolecular catalyst for oxidation reactions of organic functional groups. Due to the hydrophobic nanoenvironment in the immobilised

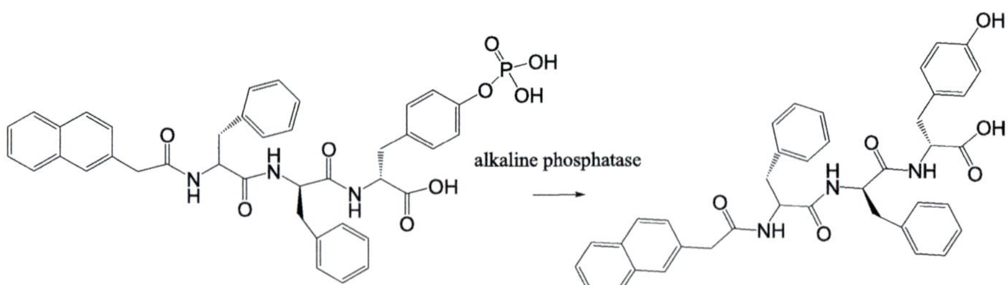

6.2a

6.2b

FIGURE 6.2 Catalytic dephosphorylation leading to assembly that has a selective effect on cancer cells over normal cells. (From M. J. A. Shiddiky et al. *Anal. Chem.* 2007, 79, 6886–6890.[3])

polyoxometalate, they catalyse selective oxidation of alkenes, alcohols and sulphides with high efficiency.[5]

Catalytically active metals and metallic nanoparticles inside a framework cavity such as in MOF-5, MOF-177, MIL-101(Cr), ZIF-8 and ZIF-90 are used for various catalytic reactions. Au@MOF and Pd@MOF are generally prepared by reducing the corresponding metal complexes loaded into MOFs to generate metallic particles. The metal organic chemical vapour deposition technique of the Au(CO)Cl complex is used to synthesise Au@ZIF-8 and Au@ZIF-90. In these cases, the loading of metal particles to the MOF is between 5% and 33% wt, respectively. Au@ZIF-8 is useful in catalytic oxidation of alcohols, but Au@ZIF-90 is inactive. Synthesis of nanostructures based on palladium metal ions with peptide templates produces nanomaterials ranging from nanospheres to nanoparticles. Exceptionally, in the case of platinum, such immobilisation provides particles with a spherical shape. The catalytic turnover for reduction of allyl alcohol to n-propanol by palladium and platinum nanoparticles supported by peptides is independent of morphology.[6]

Zeolite frameworks are used to encapsulate various catalytically active species. The encapsulations of metal nanoparticles control the size of the nanoparticles and limit aggregation, but in a highly rigid confined medium, it becomes increasingly difficult for a reactant to access the encapsulated nanoparticles. This difficulty is overcome by modifying zeolite with hierarchical structures or unusual morphologies with a smaller mean diffusion path. Poisoning or sintering of nanoparticles is less likely in core–shell materials. Core–shell catalysts act as nanoreactors, where the permeability of the shell is more important to allow substrate and product diffusing in and out. Metal-encapsulated zeolites act as functional catalysts due to the presence of nanoparticles in the vicinity of the original active sites of zeolite catalysts. Hierarchical zeolites are zeolites that contain two or more types of pores of different sizes. They encapsulate nanoparticles, and these nanoparticles@zeolites are useful to catalytically produce hydrocarbons in Fischer–Tropsch synthesis.[7]

6.3 SUPRAMOLECULAR CATALYST IMMOBILISATION

Surface modification provides a means to immobilise catalysts. These immobilised catalysts change the course of conventional reactions and provide the means to carry out catalytic reactions in heterogeneous conditions. Surface modifications are done on various metallic, polymeric and nonmetallic surfaces. Gold and silver nanoparticles immobilised on the surface of zinc oxide are catalysts for decomposition of hydrogen peroxide. This immobilisation has relevance as supramolecular catalysis to mimic catalaselike behaviour. Immobilisation of nickel complex on carbon nanotubes such as **6.3** is useful to mimic the hydrogenase enzyme (Figure 6.3). The carbon nanotube not only acts as a support but also acts as an electron source and participates in the electron transfer process in hydrogenation catalysis. In the immobilised complex **6.3**, a nickel-hydride complex in a P4H coordination environment (four phosphorus atoms and hydride) is anchored to a carbon nanotube through covalent linkage. The phthalimide part of the ligand is used for encapsulating the active site and also acts as an electroactive site assisting in the electron transfer process. This immobilised nickel complex is a mimic of the enzyme hydrogenase.

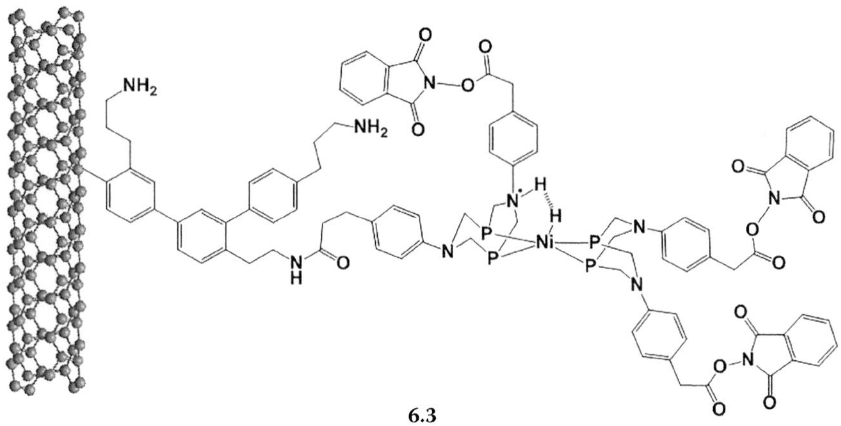

6.3

FIGURE 6.3 A nickel complex as hydrogenase model immobilised on carbon nanotube. (From S. Gentil et al. *Angew. Chem. Int. Ed.* 2017, 56, 1845–1849.[8])

Surface-active ionic liquids are useful templates to modify the environments around polyoxometalates. For example, phosphotungstic acid is used to immobilise the ionic liquid N-decyl-N′-carboxymethylimidazolium bromide (**6.4**) (Figure 6.4). This immobilised ionic liquid is a catalyst for the decomposition of methyl orange in aqueous solution. The special feature of this immobilised catalyst is that the electrostatic repulsions between polyoxometalate and the surface active ionic liquid cause pH-induced assembly and disassembly of the hybrid nanomaterial formed **6.4** with tungstate. This assembly and disassembly modulate the catalyst regeneration; they provide an example of a renewable nanocatalyst of ionic liquid (Figure 6.5).

Polymeric materials have special interest in supramolecular chemistry, as they not only provide heterogenity, stability and selectivity to a catalyst, but also in certain cases enhance catalytic performances by providing self-encapsulating pockets. These pockets retaining the catalytic sites are utilised as specific microenvironments for the catalytic centre for selective transformations. As an example, the polymeric aluminium-porphyrin complex **6.5a** has an acetate ligand located at the axial position of the aluminium coordination sphere. The complex is an active catalyst for the methanolysis of nerve-agent simulant **6.5b**. The polymer, upon supercritical CO_2 processing, increases the pore size and volume. The transformed form of the polymer has higher catalytic activity than the unprocessed polymer. The aluminium ion of **6.5a** activates the organophosphate substrate and delivers the methoxide nucleophile

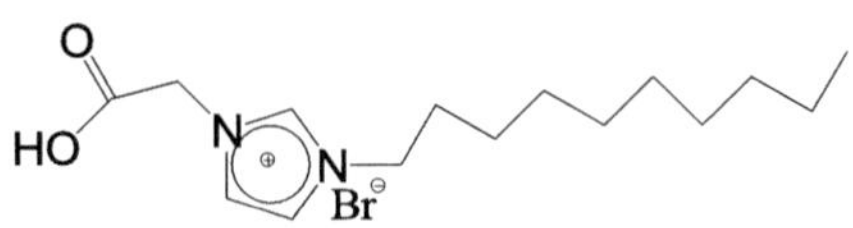

6.4

FIGURE 6.4 N-decyl-N′-carboxymethylimidazolium bromide. (From Y. Gong et al. *ACS Sustainable Chem.* 2017, 5, 3650–3658.[9])

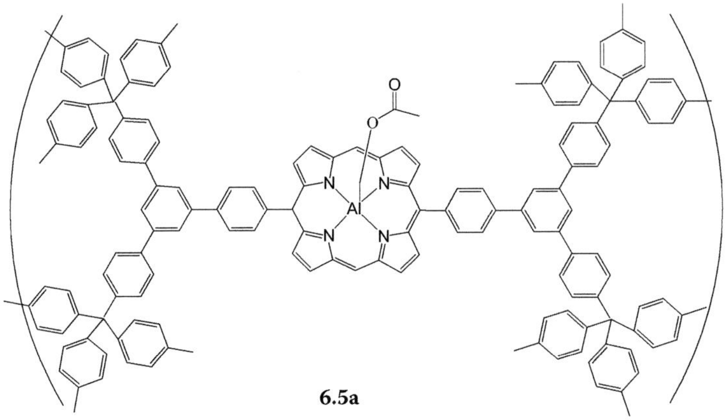

6.5a

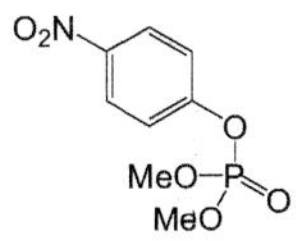

6.5b

FIGURE 6.5 Polymeric self-support of catalytically active site. (From R. K. Totten et al. *J. Am. Chem. Soc.* 2013, 135, 11720–11723.[10])

to the phosphorus centre in a cooperative fashion. The change in catalytic activity by supercritical CO_2 is due to tuning the microporous environment around the active site.

Nerve agents are used in chemical warfare; hence, finding selective means to destroy them is a challenge. Furthermore, there are large numbers of insecticides and pesticides that have phosphate ester groups, and their catalytic degradation is in general sought after. One of the means for their degradation is the utilisation of supramolecular catalysts with enzymelike activity. The catalysis shown by **6.5a** to decompose the nerve gas demonstrates a step towards such catalyst design.

6.4 SUPRAMOLECULAR AUTOCATALYSIS

Self-replication is an indispensable natural process in biology. There are large numbers of self-replicating biological processes in peptides, amyloid fibrils and DNA analogues where autocatalytic building takes place. Supramolecular autocatalysis concerns not only the products but also links replication activities observed in several processes in biological sciences. Autocatalysis in chemical science is a process analogous to the propagation of life through the concept of evolution.[11] Supramolecular autocatalysis concerns not only the products but also links replication activities observed in several processes in biological sciences. For a chemist, the acid-catalysed oxidation reaction of potassium permanganate is a well-known autocatalytic reaction. The same reaction may be carried out by using supramolecular features by adding crown ether. The crown ether binds to potassium ions and allows the reaction to occur efficiently in nonaqueous medium. In this reaction, the manganese (II) ions formed during the initial decomposition of potassium permanganate through the oxidation reaction catalyse further oxidation reactions.

In conventional practice, a product formed in a reaction catalyses further reaction between the reactants, an autocatalytic reaction occurs. However, there are many more ways autocatalysis occurs; for example, it may be through an intermediate complex that participates in catalysis. In a conventional catalytic reaction, one of the components A

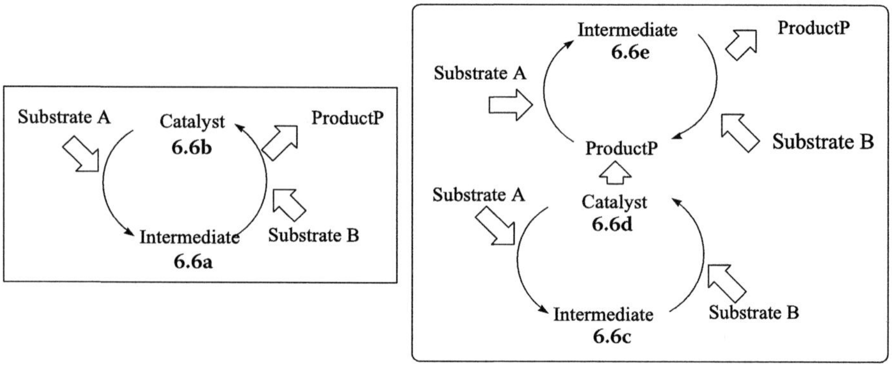

FIGURE 6.6 Paths followed by simple catalytic and autocatalytic reactions.

is activated by a catalyst gives product or it reacts with B to give the products. But in autocatalytic reaction, the activation process is different; the product participates in the catalytic cycle to make multiple paths. Hence, autocatalysis occurs through multiple paths; the prominent paths in an autocatalytic reaction are listed in Figure 6.6.

As illustrated in Figure 6.6 (right side), the autocatalytic reaction of the substrate A forms an intermediate **6.6c** with the catalyst that reacts with substrate B to generate a catalyst to form product P. Product P interacts with substrate A again to form an intermediate **6.6e**. The intermediate forms product P by reacting with substrate B. By following two cycles of activations, an autocatalytic scheme is set up to form the product.[12]

Self-generated polymers also act as catalysts during the course of catalytic reactions; two such possibilities are shown in Figure 6.7. In the first case, schematically shown as (a), certain metal complexes catalyse polymerisation reactions where polymer formed

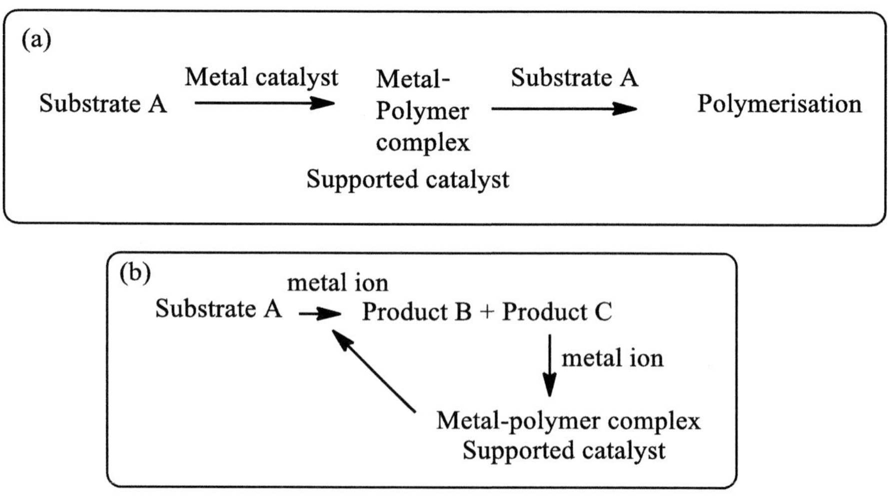

FIGURE 6.7 Autocatalysis through self-generated metal-polymer catalysed reactions.

in situ binds to the metal catalyst and acts as a supported catalyst. Hence, there is an initiation process in which the metal complex alone catalyses such reactions, and there is an increase in the rate with time due to the polymer-supported catalyst being formed *in situ*. Such catalysis occurs in palladium(0)-catalysed dehydrogenative reactions of polysilanes forming siloxane.[13] In another sequence of reaction, a metal complex causes a reaction and results in two products from a reactant. One of the undesired products polymerises *in situ* and participates as a catalyst, as shown in Figure 6.7b. Such a reaction occurs in the oxidation of phenylaldimine to benzoic acid by copper(II)-catalysed reaction in the presence of hydrogen peroxide. In this particular reaction, hydrolysis of imines produces aldehyde and aniline. The aldehyde is oxidised to benzoic acid, which is the final product. As the aniline is formed as a product, under the reaction condition *in situ*, it is oxidised to polyaniline. The polyaniline acts as support or directly activates the hydrogen peroxide to oxidise the benzaldehyde to benzoic acid. Thus, the reaction is catalysed not only by the copper(II) ion but also by the polyaniline generated in the reaction.[14]

In certain reactions, additives are added along with the catalyst to enhance the rate of reactions. In many such reactions, additives are polymerised to form *in-situ* polymer-bound catalysts to enhance catalytic activity. This concept is found in the copper-catalysed oxidation of benzylalcohol in the presence of hydrogen peroxide, where additives such as 1,6-naphthalenediol promote the reactions.[15] In such a case, poly-naphthaleneethers are formed *in situ*, which provides multiple cycles to perform catalytic oxidation.

There are many reactions in which the two reacting components form a product which serves as a template to bind the reactants. This process further provides templates to enhance the rate of the reaction between the reactants. Generally, in those reactions complementary hydrogen bond formation is the main factor, along with the geometrical arrangements of the products to have complementary hydrogen bonds with reactants. Though many products have complementary hydrogen bond sites to bind to reactants, such bindings are not necessarily a factor to cause autocatalysis. The product must provide the template so that the reactants come into close proximity to each other with the correct geometry for the catalytic reaction to happen. By bringing the reactants into close vicinity, the products must not block the active sites through hydrogen bonds (Figure 6.8).

The ester-containing compound phenyl 2-(3,4-dihydro-5-methyl-2,4-dioxopyrimidin-1(2H)-yl)acetate **6.8a** reacts with 9-(4-(aminomethyl)-tetrahydro-2,2-dimethylfuro[3,4-d][1,3]dioxol-6-yl)-5,9-dihydro-4H-purin-6-amine **6.8b** to form

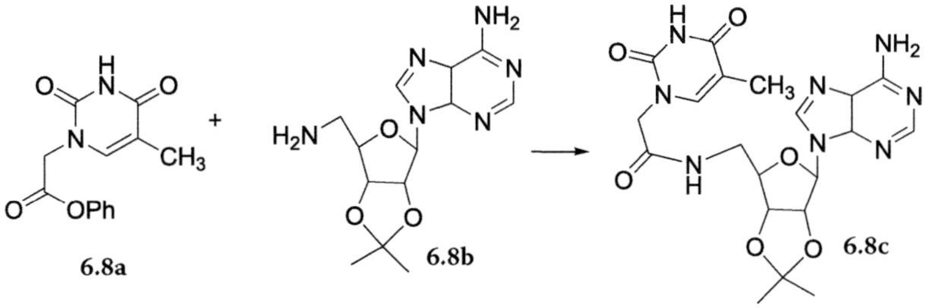

FIGURE 6.8 An autocatalytic amidation reaction.

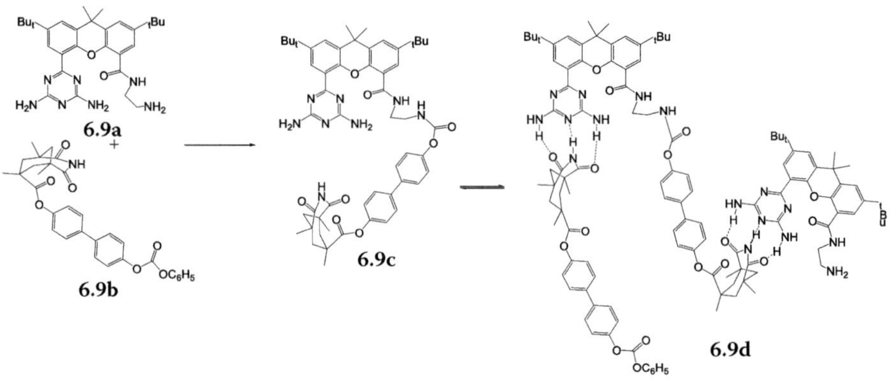

FIGURE 6.9 An autocatalytic reaction where the particular conformation of product provides anchoring sites to reactants for further reaction. (From Q. Feng et al. *Science* 1992, 256, 1179–1180.[17])

N-((4-(6-amino-4H-purin-9(5H)-yl)-tetrahydro-2,2-dimethylfuro[3,4-d][1,3]dioxol-6-yl)methyl)-2-(3,4-dihydro-5-methyl-2,4-dioxo-pyrimidin-1(2H)-yl)acetamide **6.8c**. The product **6.8c** acts as a template and forms the adduct. The hydrogen-bonded reactants of the adduct transform to products and it also assist the reactants for further amidation reactions (Figure 6.9).[16]

The free amine group of the ethylenediamine part of the compound **6.9a** reacts with the ester group of the compound **6.9b** to form a compound **6.9c**. The compound **6.9c** binds the two reactants **6.9a** and **6.9b** through complementing hydrogen bonds to form an adduct **6.9d**. The adduct **6.9d** facilitates the reaction between the bound reactant and thereby causes an autocatalytic reaction. The adduct has an S-like geometry to form hydrogen-bonds between the two reactants in which the reacting functional groups of the two reacting substrates are in the appropriate geometry to facilitate the reaction (Figure 6.10).

The coupling reaction between (E)-methyl 2-(4-methoxyphenylimino)acetate (**6.10a**) and acetone (**6.10b**) forms methyl 2-(4-methoxyphenylamino)-4-oxopentanoate

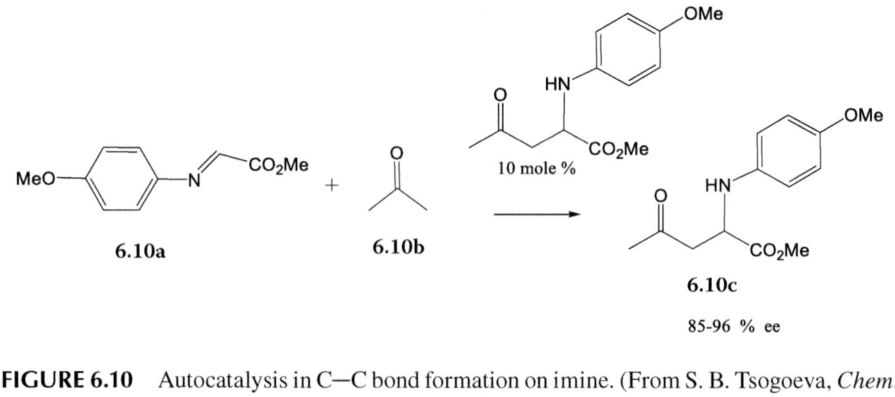

FIGURE 6.10 Autocatalysis in C—C bond formation on imine. (From S. B. Tsogoeva, *Chem. Commun.* 2010, 46, 7662–7669.[18])

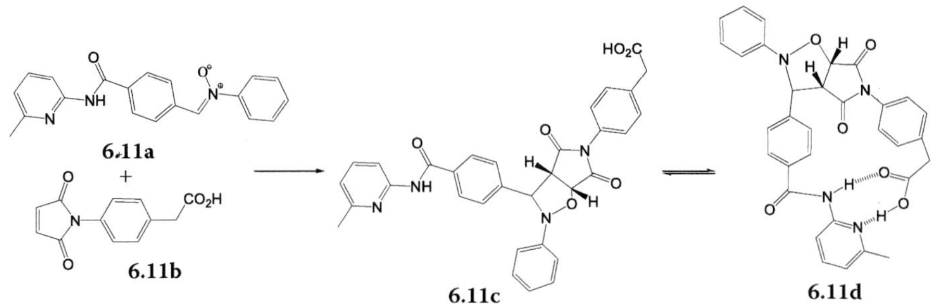

FIGURE 6.11 An autocatalytic reaction by one form of a product. (From E. Kassianidis, D. Philp, *Angew. Chem. Int. Ed.* 2006, 45, 6344–6348.[19])

(**6.10c**). This reaction is catalysed by 10 mole % of the product **6.10c**. The reaction is very slow in the absence of the added product to the reaction mixture, and it also occurs with poor enantioselectivity; when the product **6.10c** is added as a catalyst to the reaction mixture, a catalytic reaction takes place which yields 85%–90% enantiomeric excess (Figure 6.11).

The heterocyclic derivative **6.11c** is obtained from the reaction between the N-oxide derivative **6.11a** with 2-(4-(2,5-dioxo-2H-pyrrol-1(5H)-yl)phenyl)acetic acid **6.11b**. The product has two different geometrical forms, namely **6.11c** and **6.11d**. One form **6.11d** has intramolecular hydrogen bonds. This intramolecular hydrogen-bonded form does not participate in autocatalysis, whereas the other form performs autocatalysis of the cyclisation reaction. This form has an open structure which has hydrogen bonding sites available to bind the substrates. The pyridylamide part of **6.11a** forms hydrogen bonds with the carboxylic acid group of **6.11c**, and the carboxylic acid group **6.11b** forms hydrogen bonds with the pyridylamide part of **6.11c**. This assembling process facilitates the autocatalytic reaction.

The $[Co_8L_{12}](BF_4)_{16}$ (**6.12c**) cage performs the autocatalytic ring-opening reaction of benzisoxazole (Figure 6.12) to form 2-cyanophenol. In this reaction, the cage acts as a molecular flask. The reaction proceeds in basic conditions through the reaction steps shown in Figure 6.12. The benzisoxazole groups are bound to the interior side of the cavity of the cage. The cage has also surface-bound hydroxyl ions that are formed as a default of the requirement of reaction. They provide the active catalyst sites at the initial phase of the reaction. But with the lapse of time, the 2-cyanophenolate formed in the reaction catalyses the reaction further. The evidence comes from the study of kinetics but also from the fact that the addition of anions like chloride and fluoride inhibits the catalytic ring-opening reaction, but the addition of hydroxide ions does not. It is also a matter of fact that the 2-cyanophenolate anion has the highest affinity, followed by chloride, fluoride and hydroxide, to bind to the surface of the cage. Hydroxide is generated in the reaction by default of the reaction path, and hydroxide ions help the catalytic reaction. In this reaction, the 2-cyanophenolate produced is also a base. As the catalytic reaction by the cage proceeds, these anions accumulate on the surface of the cage, and these basic anions on the surface autocatalyse the reaction.

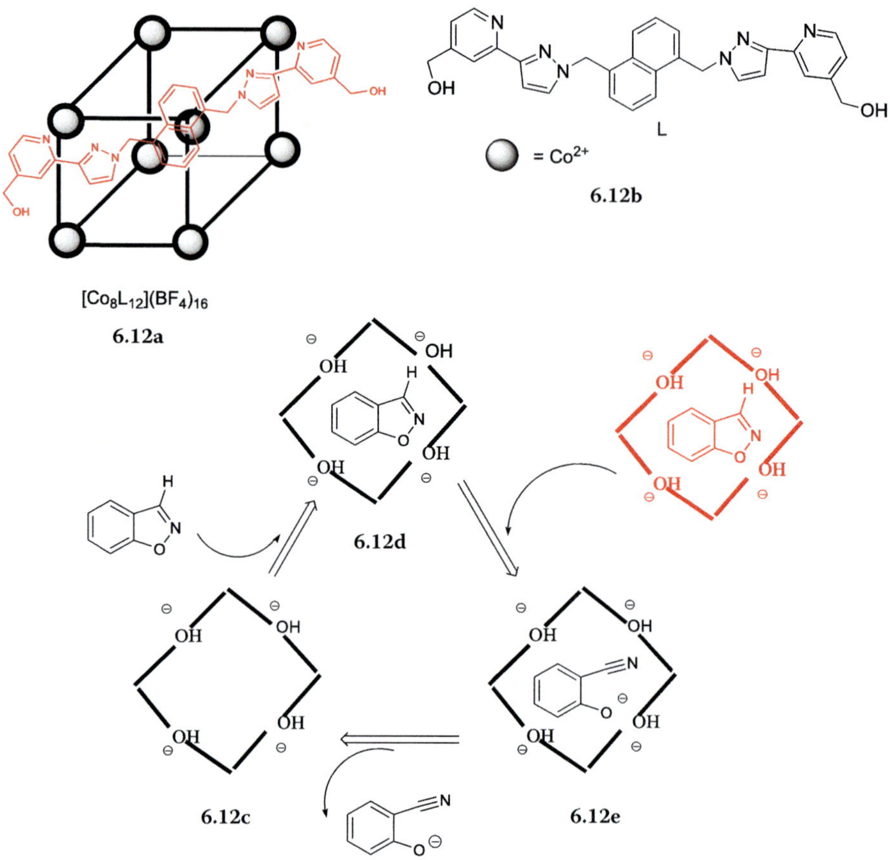

FIGURE 6.12 An autocatalytic reaction of an inorganic cage. (From W. Cullen et al. *J. Am. Chem. Soc.* 2018, 140, 2821–2828.[20])

6.5 SENSING THROUGH SUPRAMOLECULAR CATALYSIS

The changes in the UV-visible absorptions or fluorescence emissions of a product or reacting substrate in a catalytic reaction are useful to design sensors for real-time signal transductions. The mixed metal rhodium and zinc complex **6.13a** has four-coordinated rhodium sites chelated by the P and S atoms of the ligands. The complex reacts reversibly with chloride ions and carbon monoxide to form six-coordinated rhodium complex **6.13b**. The six-coordinated rhodium complex **6.13b** is a catalyst for the acetylation reaction of picolyl alcohol with acetic anhydride to picolyl acetate, whereas the complex **6.13a** is not a catalyst for such a reaction (Figure 6.13). This acetylation reaction generates acetic acid, making the reaction medium acidic. If the reaction is carried out in the presence of a pH-responsive fluorescent compound, the emission property of such a compound changes as the reaction proceeds. For example, 1-(anthracen-9-yl)-N,N-dimethylmethanamine is nonfluorescent due to the photo-electron transfer mechanism operative in it. As the lone pair of electrons

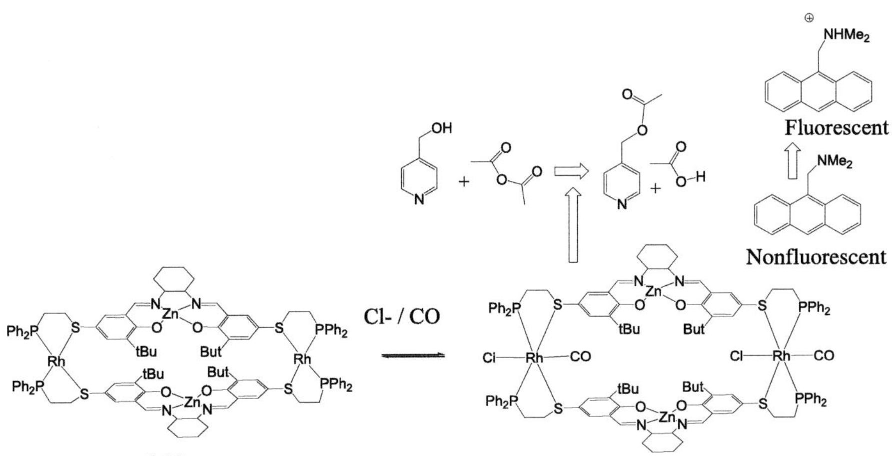

6.13a

6.13b

FIGURE 6.13 Carbon monoxide sensing through supramolecular catalysis. (From N. C. Gianneschi et al. *J. Am. Chem. Soc.* 2005, 127, 1644–1645.[21])

interferes in the emission process, the effect is nullified in the acidic medium where the NMe_2 group is protonated. With the progress of hydrolysis, the acidity of the medium increases and the solution becomes fluorescent. Hence, the consumption of CO and chloride can be sensed by following the reaction with fluorescence spectroscopy (Figure 6.14).

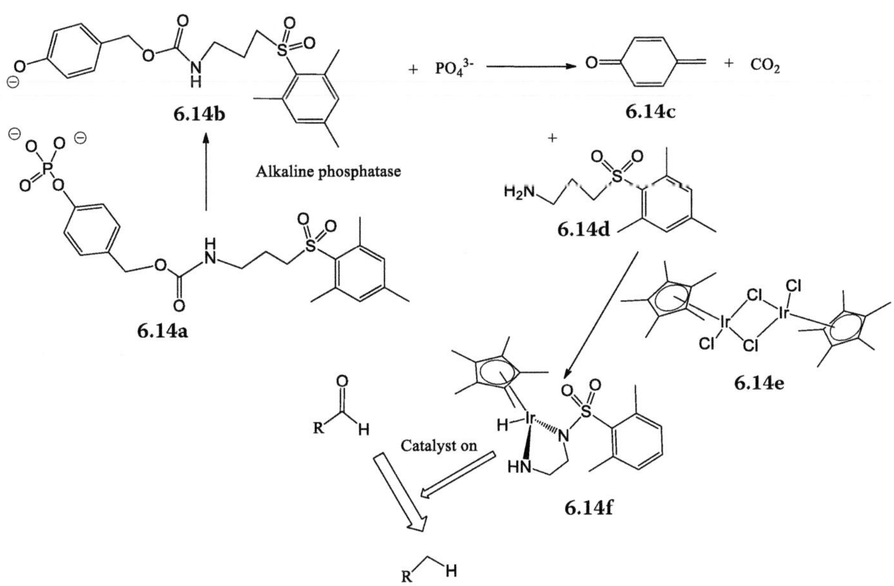

FIGURE 6.14 The enhancement of catalysis by enzyme-mediated degradation of ligand followed by complexation. (From S. Goggins et al. *Chem. Sci.* 2015, 6, 4978–4985.[22])

The enzyme-triggered signal amplification process is utilised in catalytic hydrogenation reactions. The catalytic activity of the dinuclear iridium complex is modified by generating phosphate hydrolysis of a proligand **6.14a**. The proligand undergoes hydrolysis of phosphate by phosphatase enzyme. The hydrolysed product **6.14b** liberates quinone-methide and releases carbon dioxide to yield a new ligand **6.14d**. This ligand forms an iridium-hydride complex **6.14f** upon reaction with the dinuclear complex **6.14e**. The hydride complex is an active catalyst for hydrogenation of carbonyl compounds to corresponding alcohols. The complex **6.14e** is not a good catalyst to perform the hydrogenation reaction. Changes in the binding of ligand results in the signal transduction observed during the catalytic reaction.

6.6 MODIFIED BIOMOLECULES AS CATALYSTS

Self-assembly through the biocatalytic process generates various enzyme-mediated assemblies. Enzymes modified by polydopamine and polyphenol release enzymes for bulk gelation. Catalytically active biological molecules such as cytochrome P450 monooxygenase are immobilised by such modifications to prepare water-insoluble gel. Protein gel catalyses monooxygenation in the presence of phosphite and nicotinamide adenine dinucleotide (NAD).[23]

Bovine serum albumin (BSA) and human serum albumin (HSA) are globular, water-soluble proteins with molecular weight 65,000 Dalton. They have binding abilities to different (depending on the structure and functional group on) substrates. Serum albumins are good catalysts for many catalytic reactions. These proteins lack catalytic sites, which makes them versatile catalysts in conjunction with another catalytic component for a broad range of chemical reactions. The basic hydrophobic pockets, in particular the binding site containing a lysine and Lys-222 in BSA and the homologous Lys-199 in HSA, contribute to the catalytic reactivity. Albumin causes asymmetric oxidation, reduction, addition, condensation and elimination reactions.

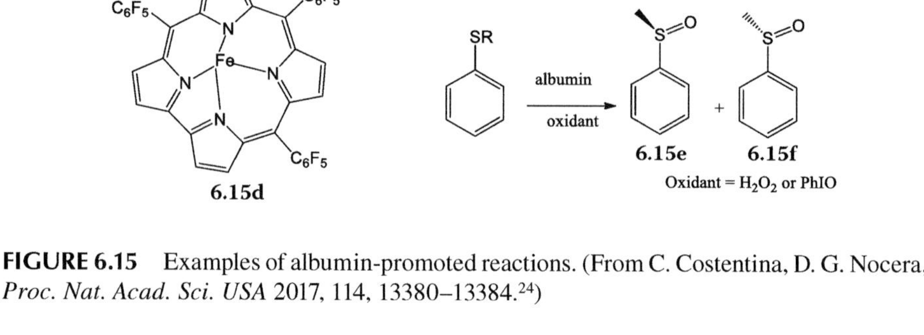

FIGURE 6.15 Examples of albumin-promoted reactions. (From C. Costentina, D. G. Nocera, *Proc. Nat. Acad. Sci. USA* 2017, 114, 13380–13384.[24])

BSA causes enantioselective reduction of prochiral ketones and enantioselective sulphoxidation in an aqueous medium (Figure 6.15).

BSA- and HSA-catalysed aldol reactions between 6-methoxynaphthalene-2-carbaldehyde (**6.15a**) with acetone form 4-hydroxy-4-(2-methoxynaphthalen-6-yl) butan-2-one, but the stereo selectivity of the reaction is dependent on the type of albumin; that is, BSA yields the (S) isomer **6.15c**, whereas HSA yields the R-isomer **6.15b**. Though the albumins do not have active sites of their own, in the presence of different metal complexes, they show specific enzyme-mimicking properties. The complex prepared by incorporating the iron porphyrin complex in **6.15d** to albumin causes oxidation of alkylthiol to form sulphoxide.

6.7 CONCLUSION AND OUTLOOK

The future of catalysis in different areas of the environment, energy and health care to mimic biological phenomena is immense.[25–29] As an example, water is a source for obtaining fuel in large-scale hydrogen as a fuel or transforming to liquid fuel by combining with carbon dioxide or in the form of renewable energy in stored materials and indirectly utilised as fertilisers. There is a necessity for catalyst design to ensure such energy-related process are robust, green and inexpensive, and the concept of self-healing of catalysts to achieve robustness has started to emerge. NOx decomposition, low-cost fuel cells and replacement of noble metals by inexpensive nontoxic catalysts will continue to emerge. Materials such as thermites are sources of the release of high amounts of heat. The use of thermites by using nanoaluminium and molybdenum exists, but alternatives that create cost-effective systems will continue to generate interest. Synthetic methodology for selective products with nanostructured solid catalysts to perform with high turnover and minimal waste is going to concern the catalyst community. One foresees the growth of research into the manipulation of nanostructured catalysts to inhibit undesirable paths to push reactions through low-energy catalysed paths for specific products in a predictive direction. Light harvesting through immobilised catalyst systems has emerged as challenging area. Research carried out on anchoring chromophoric linkers to a TiO_2 surface holding a dimer Mn^{2+} catalyst has shown promise. Large numbers of reactions are devoted to ligands stabilizing radicals in metal complexes. The modulation of those complexes with new additions will lead to the development of new synthetic protocols and catalytic processes.[30–31]

New catalyst development will continue to emphasise three aspects (Figure 6.16): immobilisation, the search for carriers and signalling units and, finally, multiple performance of control activity. The robust domain of activities of catalysis will be focused on getting universal solutions close to mimicking nature. The entire process can be compared to the Goddess Durga, who is believed by certain section of the society to represent the strength and power to deliver well for the universe. Supramolecular concepts in catalysis have started to disseminate to concerns on global warming to maintain a green environment to create balance in the ecosystem. With major amounts of biomass of the earth being solid, there are difficulties in treating solid catalysts for which the engineering aspects are emerging in a synergistic manner with catalysis. In addition to conventional tailor-made catalytic processes, there are issues related to bringing out the potential link between biological catalytic

FIGURE 6.16 Top is idol of Goddess Durga, also known as Shakti in the Hindu religion; the bottom is the three major aspects of catalyst development.

materials to heterogeneous catalysis such as by immobilisation of enzymes through dendrimers and polymers or inorganic support that are expected to give congenial solutions to many waste-management issues.

REFERENCES

1. J. M. Poolman, J. Boekhoven, A. Besselink, A. G. L. Olive, J. H. van Esch, R. Eelkema, Variable gelation time and stiffness of low molecular-weight hydrogels through catalytic control over self-assembly. *Nat. Protoc.* 2014, 9, 977–988.
2. F. Trausel, F. Versluis, C. Maity, J. M. Poolman, M. Lovrak, J. H. van Esch, R. Eelkema, Catalysis of supramolecular hydrogelation. *Acc. Chem. Res.* 2016, 49, 1440–1447.
3. M. J. A. Shiddiky, M. A. Rahman, Y. -B. Shim, Hydrazine-catalyzed ultrasensitive detection of DNA and proteins. *Anal. Chem.* 2007, 79, 6886–6890.

4. W.-T. Koo, S.-J. Choi, N.-H. Kim, J.-S. Jang, I.-D. Kim, Catalyst-decorated hollowWO_3 nanotubes using layer-by-layer self-assembly on polymeric nanofiber templates and their application in exhaled breath sensor. *Sensors and Actuators B* 2016, 223, 301–310.
5. W. Qi, Y. Wang, W. Li, L. Wu, Surfactant-encapsulated polyoxometalates as immobilized supramolecular catalysts for highly efficient and selective oxidation reactions. *Chem. Eur. J.* 2010, 16, 1068–1078.
6. R. Bhandari, D. B. Pacardo, N. M. Bedford, R. R. Naik, M. R. Knecht, Structural control and catalytic reactivity of peptide-templated Pd and Pt nanomaterials for olefin hydrogenation. *J. Phys. Chem. C.* 2013, 117, 18053–18062.
7. D. Farrusseng, A. Tuel, Perspectives on zeolite-encapsulated metal nanoparticles and their applications in catalysis. *New J. Chem.* 2016, 40, 3933–3949.
8. S. Gentil, N. Lalaoui, A. Dutta, Y. Nedellec, S. Cosnier, W. J. Shaw, V. Artero, A. Le Goff, Carbon-nanotube-supported bio-inspired nickel catalyst and its integration in hybrid hydrogen/air fuel cells. *Angew. Chem. Int. Ed.* 2017, 56, 1845–1849.
9. Y. Gong, Y. Guo, Q. Hu, C. Wang, L. Zang, L. Yu, pH-responsive polyoxometalate-based supramolecular hybrid nanomaterials and application as renewable catalyst for dyes. *ACS Sustainable Chem.* 2017, 5, 3650–3658.
10. R. K. Totten, Y.-S. Kim, M. H. Weston, O. K. Farha, J. T. Hupp, S. B. T. Nguyen, Enhanced catalytic activity through the tuning of micropore environment and supercritical CO_2 processing: Al(porphyrin)-based porous organic polymers for the degradation of a nerve agent simulant. *J. Am. Chem. Soc.* 2013, 135, 11720–11723.
11. A. J. Bissette, S. P. Fletcher, Mechanisms of autocatalysis. *Angew. Chem. Int. Ed.* 2013, 52, 12800–12826.
12. D. G. Blackmond, An examination of the role of autocatalytic cycles in the chemistry of proposed primordial reactions. *Angew. Chem. Int. Ed.* 2009, 48, 386–390.
13. A. Purkayastha, J. B. Baruah, Role of self-supported metallic particles in Si-H bond activation by coinage metal complexes. *Polymer Preprints (American Chemical Society, Division of Polymer Chemistry).* 2004, 45, 666–667.
14. J. G. Handique, L. Borah, R. J. Sarma, J. B. Baruah, Participation of self-generated polyaniline support in copper(II) catalysed oxidative reaction of phenyl aldimines. *J. Mol. Catal, A: Chem.* 2002, 188, 63–69.
15. J. G. Handique, J. B. Baruah, The oligomer of 1,6-naphthalenediol as a support in oxidative reactions of cis-bisglycinato copper(II) and hydrogen peroxide. *J. Mol Catal. A: Chem.* 2001, 172, 19–23.
16. T. K. Park, Q. Feng, J. Rebek Jr., Synthetic replicators and extrabiotic chemistry. *J. Am. Chem. Soc.* 1992, 114, 4529–4532.
17. Q. Feng, T. K. Park, J. Rebek, Crossover reactions between synthetic replicators yield active and inactive recombinants. *Science* 1992, 256, 1179–1180.
18. S. B. Tsogoeva, When chiral product and catalyst are the same: Discovery of asymmetric organoautocatalysis. *Chem. Commun.* 2010, 46, 7662–7669.
19. E. Kassianidis, D. Philp, Design and implementation of a highly selective minimal self-replicating system. *Angew. Chem. Int. Ed.* 2006, 45, 6344–6348.
20. W. Cullen, A. J. Metherell, A. B. Wragg, C. G. P. Taylor, N. H. Williams, M. D. Ward, Catalysis in a cationic coordination cage using a cavity-bound guest and surface-bound anions: Inhibition, activation, and autocatalysis. *J. Am. Chem. Soc.* 2018, 140, 2821–2828.
21. N. C. Gianneschi, S. T. Nguyen, C. A. Mirkin, Signal amplification and detection via a supramolecular allosteric catalyst. *J. Am. Chem. Soc.* 2005, 127, 1644–1645.
22. S. Goggins, B. J. Marsh, A. T. Lubbena, C. G. Frost, Signal transduction and amplification through enzyme-triggered ligand release and accelerated catalysis. *Chem. Sci.* 2015, 6, 4978–4985.
23. D. C. M. Albanesea, N. Gaggero, Albumin as a promiscuous biocatalyst in organic synthesis. *RSC Adv.* 2015, 5, 10588–10598.

24. C. Costentina, D. G. Nocera, Self-healing catalysis in water. *Proc. Nat. Acad. Sci. USA* 2017, 114, 13380–13384.
25. T. Ennaert, J. V. Aelst, J. Dijkmans, R. De Clercq, W. Schutyser, M. Dusselier, D. Verboekend, B. F. Sels, Potential and challenges of zeolite chemistry in the catalytic conversion of biomass. *Chem. Soc. Rev.* 2016, 45, 584–611.
26. Z. Mycroft, M. Gomis, P. Mines, P. Law, T. D. H. Bugg, Biocatalytic conversion of lignin to aromatic dicarboxylic acids in Rhodococcusjostii RHA1 by re-routing aromatic degradation pathways. *Green Chem.* 2015, 17, 4974–4979.
27. K. Ariga, H. Ito, J. P. Hill, H. Tsukube, Molecular recognition: From solution science to nano/materials technology. *Chem. Soc. Rev.* 2012, 41, 5800–5835.
28. M. P. Conte, K. H. A. Lau, R. V. Ulijn, Biocatalytic self-assembly using reversible and irreversible enzyme immobilization. *ACS Appl. Mater. Interfaces* 2017, 9, 3266–3271.
29. K. J. Young, B. J. Brennan, R. Tagore, G. W. Brudvig, Photosynthetic water oxidation: Insights from manganese model chemistry. *Acc. Chem. Res.* 2015, 48, 567–574.
30. A. Z. Haddad, S. P. Cronin, M. S. Mashuta, R. M. Buchanan, C. A. Grapperhaus, Metal-assisted ligand-centered electrocatalytic hydrogen evolution upon reduction of a bis(thiosemicarbazonato)Cu(II) complex. *Inorg. Chem.* 2017, 56, 11254–11265.
31. A. I. O. Suarez, V. Lyaskovskyy, J. N. H. Reek, J. I. van der Vlugt, B. de Bruin, Complexes with nitrogen-centered radical ligands: Classification, spectroscopic features, reactivity, and catalytic applications. *Angew. Chem. Int. Ed.* 2013, 52, 12510–12529.

Index

T

U

V

W

Y

Z